Advances in Physical Geochemistry

Volume 10

Editor-in-Chief

Surendra K. Saxena

Editorial Board

L. Barron   P.M. Bell   N.D. Chaterjee   R. Kretz   D.H. Lindsley
Y. Matsui   A. Navrotsky   R.C. Newton   G. Ottonello   L.L. Perchuk
R. Powell   R. Robie   A.B. Thompson   B.J. Wood

Advances in Physical Geochemistry

Series Editor: Surendra K. Saxena

Volume 1 R.C. Newton/A. Navrotsky/B.J. Wood (editors)
　　　　　Thermodynamics of Minerals and Melts
　　　　　　1981. xii, 304 pp. 66 illus.
　　　　　　ISBN 0-387-90530-8

Volume 2 S.K. Saxena (editor)
　　　　　Advances in Physical Geochemistry, Volume 2
　　　　　　1982. x, 353 pp. 113 illus.
　　　　　　ISBN 0-387-90644-4

Volume 3 S.K. Saxena (editor)
　　　　　Kinetics and Equilibrium in Mineral Reactions
　　　　　　1983. vi, 273 pp. 99 illus.
　　　　　　ISBN 0-387-90865-X

Volume 4 A.B. Thompson/D.C. Rubie (editors)
　　　　　Metamorphic Reactions: Kinetics, Textures, and Deformation
　　　　　　1985. xii, 291 pp. 81 illus.
　　　　　　ISBN 0-387-96077-5

Volume 5 J.V. Walther/B.J. Wood (editors)
　　　　　Fluid-Rock Interactions during Metamorphism
　　　　　　1986. x, 211 pp. 59 illus.
　　　　　　ISBN 0-387-96244-1

Volume 6 S.K. Saxena (editor)
　　　　　Chemistry and Physics of Terrestrial Planets
　　　　　　1986. x, 405 pp. 94 illus.
　　　　　　ISBN 0-387-96287-5

Volume 7 S. Ghose/J.M.D. Cohe/E. Salje (editors)
　　　　　Structural and Magnetic Phase Transitions in Minerals
　　　　　　1988. xiii, 272 pp. 117 illus.
　　　　　　ISBN 0-387-96710-9

Volume 8 J. Ganguly (editor)
　　　　　Diffusion, Atomic Ordering, and Mass Transport:
　　　　　Selected Topics in Geochemistry
　　　　　　1991. xiv, 584 pp. 170 illus.
　　　　　　ISBN 0-387-97287-0

Volume 9 L.L. Perchuk/I. Kushiro (editors)
　　　　　Physical Chemistry of Magmas
　　　　　　1991. 352 pp. 137 illus.
　　　　　　ISBN 0-387-97500-4

Volume 10 S.K. Saxena (editor)
　　　　　Thermodynamic Data:
　　　　　Systematics and Estimation
　　　　　　1992. xiv, 367 pp. 126 illus.
　　　　　　ISBN 0-387-97696-5

Surendra K. Saxena
Editor

Thermodynamic Data
Systematics and Estimation

With Contributions by
F. Baccarin A.B. Belonoshko M. Blander S.L. Chaplot N. Choudhury
S. Usha-Devi G. Fiquet D.G. Fraser S. Ghose P. Gillet A. Della Giusta
C.M. Gramaccioli G. Grimvall A.F. Guillermet K. Heinzinger E. Ito
A.G. Kalinichev Y.H. Kim M.H. Manghnani L.C. Ming A. Dal Negro
D.R. Neuville G. Ottonello T. Pilati K.R. Rao K. Refson P. Richet C.R. Stover
V.S. Urusov J.-A. Xu

With 126 Illustrations

Springer-Verlag
New York Berlin Heidelberg London Paris
Tokyo Hong Kong Barcelona Budapest

Surendra K. Saxena
Planetary Geochemistry Program
Institute of Geology
Uppsala University
Box 555
S-75122 Uppsala
Sweden

Library of Congress Cataloging-in-Publication Data
Thermodynamic data: systematics and estimation / Surendra K. Saxena.
   p. cm. — (Advances in physical geochemistry; v. 10)
   Includes bibliographical references and index.
   ISBN 0-387-97696-5
   1. Geochemistry. 2. Thermodynamics. I. Saxena, Surendra Kumar,
  1936– . II. Series.
  QE515.T433   1992
  551.9—dc20   91-31481

Printed on acid-free paper

© 1992 Springer-Verlag New York Inc.
All rights reserved. This work may not be translated or copied in whole or in part without the written permission of the publisher (Springer-Verlag New York Inc., 175 Fifth Avenue, New York, NY 10010, USA), except for brief excerpts in connection with reviews or scholarly analysis. Use in connection with any form of information storage and retrieval, electronic adaptation, computer software, or by similar or dissimilar methodology now known or hereafter developed is forbidden.
The use of general descriptive names, trade names, trademarks, etc., in this publication, even if the former are not especially identified, is not to be taken as a sign that such names, as understood by the Trade Marks and Merchandise Marks Act, may accordingly be used freely by anyone.

Production managed by Christin R. Ciresi; manufacturing supervised by Robert Paella.
Typeset by Asco Trade Typesetting Ltd., Hong Kong.
Printed and bound by Edwards Brothers, Inc., Ann Arbor, MI.
Printed in the United States of America.

9 8 7 6 5 4 3 2 1

ISBN 0-387-97696-5 Springer-Verlag New York Berlin Heidelberg
ISBN 3-540-97696-5 Springer-Verlag Berlin Heidelberg New York

Preface

With the rapid development of fast processors, the power of a mini-supercomputer now exists in a lap-top box. Quite sophisticated techniques are becoming accessible to geoscientists, thus making disciplinary boundaries fade. Chemists and physicists are no longer shying away from computational mineralogical and material science problems "too complicated to handle." Geoscientists are willing to delve into quantitative physico-chemical methods and open those "black boxes" they had shunned for several decades but with which had learned to live. I am proud to present yet another volume in this series which is designed to break the disciplinary boundaries and bring the geoscientists closer to their chemist and physicist colleagues in achieving a common goal.

This volume is the result of an international collaboration among many physical geochemists (chemists, physicists, and geologists) aiming to understand the nature of material. The book has one common theme: namely, how to determine quantitatively through theory the physico-chemical parameters of the state of a solid or fluid.

The book begins with a study of fluids in the first three chapters. Chapter 1 by Kalinichev and Heinzinger provides us with an extensive review of the current methods for simulating the pressure-volume-temperature (P-V-T) of fluids. After reviewing the various techniques, the authors focus on the Monte Carlo simulation method and provide useful data on aqueous fluids. Fluids are also the topic of discussion in the next two chapters. Both deal with the technique of molecular dynamics. In Chapter 2, Fraser and Refson use the technique to simulate the high P-V-T properties of fluids, and in Chapter 3, Belonoshko and Saxena do the same. The two chapters differ in using different types of intermolecular potentials, and it is interesting to compare the results. Such comparisons are instructive in providing us with insight into the operation of these new tools of theory.

In Chapter 4, Richet, Gillet, and Fiquet review the current information on linking the macroscopic and microscopic properties of minerals. The latter section prepares the reader for the study of the last five chapters of the book. The authors begin with the experimental methods of obtaining the heat capacity of solids and then follow through with a review of the empirical methods of estimation and calculations. The next chapter, also by Richet in coauthorship with Neuville, reviews the latest work on the thermodynamics of silicate melts. The authors have chosen the configurational heat capacity of the melts as their theme. This property of materials is of paramount importance in geochemical and industrial applications because it not only constrains the phase-equilibrium relations, but is also related to viscosity via the configurational entropy. Chapter 5 provides an excellent opportunity for an interested reader to start his or her study of this rapidly developing branch of physical chemistry.

V.S. Urusov presents in Chapter 6 the energetics of binary solid solutions. His analysis of the various energy contributions arising from the mixing of components with different volumes and crystal structures lays the foundation for a quantitative treatment of solid solution behavior. Such attempts are now beginning to meet with considerable success, of which Chapter 7 is an excellent example. Ottonello, Della Giusta, Dal Negro, and Baccarin show how a constrained ionic model simulation of crystal structures of pure components and of their solution in a multicomponent chemical system can be performed with a surprising degree of accuracy.

Chapters 8 through 11 are devoted to the estimation of the thermodynamic properties of a solid. In the first of these contributions, Gramaccioli and Pilati review the current approaches to calculating the energetics of crystals. Their review outlines the problems encountered in rigorous calculations and presents useful simplifications and computer techniques that lead to improved calculations of energy with minimum errors.

In Chapter 9, Blander and Stover outline an effective method for the prediction of high-temperature entropies of vapor molecules and for some classes of solids with a given stoichiometry. The method has great potential for further development to include other solids and fluids.

In Chapter 10, Grimvall and Fernandez-Guillermot continue with the theme of predicting entropy. They introduce an entropy Debye temperature that changes atomic masses and arrive at another important approach to predicting the vibrational entropy of solids at intermediate and high temperatures.

The next chapter by Ghose reviews the work on the phonon density of states and demonstrates the application of the lattice dynamicals approach to calculate thermodynamic functions for silicates.

The book ends with a final chapter by Ming, Manghnani, Kim, Usha-Devi, Xu and Ito who have joined their forces to obtain experimental data on thermal expansion on the spinels. Such data are as yet a rarity because of the experimental difficulties of measuring the volumes of solids in-situ at high pressure and temperatures.

I thank the editorial staff of Springer-Verlag (Victor Van Beuren, Lorraine

Walsh and Jessica Downey) for their invaluable support and the production department for the care with which it handled the book.

July 9, 1991 S.K. Saxena
Uppsala, Sweden

Contents

Preface — v

Contributors — xi

1. Computer Simulations of Aqueous Fluids at High Temperatures and Pressures — 1
   A.G. KALINICHEV and K. HEINZINGER

2. Estimating Thermodynamic Properties by Molecular Dynamics Simulations: The Properties of Fluids at High Pressures and Temperatures — 60
   D.G. FRASER and K. REFSON

3. Equations of State of Fluids at High Temperature and Pressure (Water, Carbon Dioxide, Methane, Carbon Monoxide, Oxygen, and Hydrogen) — 79
   A.B. BELONOSHKO and S.K. SAXENA

4. Thermodynamic Properties of Minerals: Macroscopic and Microscopic Approaches — 98
   P. RICHET, P. GILLET, and G. FIQUET

5. Thermodynamics of Silicate Melts: Configurational Properties — 132
   P. RICHET and D.R. NEUVILLE

6. Crystal Chemical and Energetic Characterization of Solid Solution — 162
   V.S. URUSOV

7. A Structure Energy Model for C2/c Pyroxenes in the System
   Na–Mg–Ca–Mn–Fe–Al–Cr–Ti–Si–O 194
   G. OTTONELLO, A. DELLA GIUSTA, A. DAL NEGRO, and F. BACCARIN

8. Practical Problems in Calculating Thermodynamic Functions for
   Crystalline Substances from Empirical Force Fields 239
   C.M. GRAMACCIOLI and T. PILATI

9. Predictions of the Entropies of Molecules and Condensed Matter 264
   M. BLANDER and C.R. STOVER

10. Systematics of Bonding Properties and Vibrational Entropy
    in Compounds 272
    G. GRIMVALL and A. FERNÁNDEZ GUILLERMET

11. Phonon Density of States and Thermodynamic Properties
    of Minerals 283
    S. GHOSE, N. CHOUDHURY, S.L. CHAPLOT, and K.R. RAO

12. Thermal Expansion Studies of $(Mg, Fe)_2SiO_4$-Spinels Using
    Synchrotron Radiation 315
    L.C. MING, M.H. MANGHNANI, Y.H. KIM, S. USHA-DEVI,
    J.-A. XU, and E. ITO

Appendix 335

Index 359

Contributors

F. BACCARIN	E.N.E.L., Unita Nazionale Geotermica, Via Andrea Pisano 120, 56100 Pisa, Italy
A.B. BELONOSHKO	Department of Mineralogy and Petrology, Uppsala University, Uppsala S75122, Sweden
M. BLANDER	Argonne National Laboratory, 9700 South Cass Avenue, Argonne, IL 60439, USA
S.L. CHAPLOT	Solid State Physics Division, Bhabha Atomic Research Centre, Trombay, Bombay 400085, India
N. CHOUDHURY	Solid State Physics Division, Bhabha Atomic Research Centre, Trombay, Bombay 400085, India
S. USHA-DEVI	National Aeronautical Laboratory, Bangalore 560017, India
G. FIQUET	Laboratorie de Minéralogie Physique, Universite de Rennes I, 35042 Rennes Cedex, France
D.G. FRASER	Department of Earth Sciences, University of Oxford, Parks Road, Oxford OX1 3PR, United Kingdom
S. GHOSE	Mineral Physics Group, Department of Geological Sciences, University of Washington, Seattle, WA 948195, USA
P. GILLET	Laboratoire de Minéralogie Physique, Universite de Rennes I, 35042 Rennes Cedex, France

A. Della Giusta	Dipartimento di Mineralogia e Petrologia, Universita di Padova, Corso Garibaldi 37, 35100 Padova, Italy
C.M. Gramaccioli	Department of Earth Sciences, University of Milan, Via Botticelli 23, I-20133 Milan, Italy
G. Grimvall	Department of Theoretical Physics, The Royal Institute of Technology, S-100 44 Stockholm, Sweden
A. Fernández Guillermet	Censejo Nacional de Investigaciones Cientificas y Tecnicas, Centro Atomico Bariloche, 8400 San Carlos de Bariloche, Rio Negro, Argentina
K. Heinzinger	Max-Planck-Institut fuer Chemie, (Otto-Hahn-Institut), D-6500 Mainz, Germany
E. Ito	Institute of Study of the Earth's Interior, Okayama University, Misasa, Tottorri-Ken 682-02, Japan
A.G. Kalinichev	Institute of Experimental Mineralogy, USSR Academy of Sciences, 142432 Chernogolovka, Moscow District, Russia
Y.H. Kim	Department of Geology, Gyeong Sang National University, Inju 660-701, Korea
M.H. Manghnani	Mineral Physics Group, Department of Geology and Geophysics, School of Ocean and Earth Science and Technology, University of Hawaii, Honolulu, HI 96822, USA
L.C. Ming	Mineral Physics Group, Department of Geology and Geophysics, School of Ocean and Earth Science and Technology, University of Hawaii, Honolulu, HI 96822, USA
A. Dal Negro	Dipartimento di Mineralogia e Petrologia, Universita di Padova, Corso Garibaldi 37, 35100 Padova, Italy
D.R. Neuville	Institut de Physique du Globe, 4, Place Jussieu, 75252 Paris Cedex 05, France

G. Ottonello	Dipartimento di Scienze della Terra, Universita di Cagliari, Via Trentino 51, 09100 Cagliari, Italy
T. Pilati	CNR, Centro per lo Studio delle Relazioni tra Struttura e Reattivita Chimica, Via Golgi 19, I-20133 Milan, Italy
K.R. Rao	Solid State Physics Division, Bhabha Atomic Research Centre, Trombay, Bombay 400085, India
K. Refson	Department of Earth Sciences, University of Oxford, Parks Road, Oxford OX1 3PR, United Kingdom
P. Richet	Institut de Physique du Globe, 4, Place Jussieu, 75252 Paris Cedex 05, France
C.R. Stover	Department of Mathematics, University of Chicago, Chicago, IL 60637, USA
V.S. Urusov	Department of Crystallography and Crystal Chemistry, Geological Faculty, Moscow State University, 119899 Moscow, Russia
J.-A. Xu	Mineral Physics Group, Department of Geology and Geophysics, School of Ocean and Earth Science and Technology, University of Hawaii, Honolulu, HI 96822, USA

Chapter 1
Computer Simulations of Aqueous Fluids at High Temperatures and Pressures

A.G. Kalinichev* and K. Heinzinger

Introduction

Water is a unique substance in many respects. It is the only chemical compound that occurs in all three physical states (solid, liquid, and vapor) under the thermodynamic conditions unique to the Earth's surface. It has played a principal role in major natural processes during the long geological and biological history of the planet. Its oustanding properties as a solvent and its general abundance almost everywhere on our planet's suface have made it also an integral part of many technological processes since the very beginning of human civilization.

Only the above-mentioned reasons could be enough to understand the great scientific interest devoted to studies of water properties. A comprehensive review of our knowledge of water and aqueous solutions, together with a discussion of experimental and theoretical techniques used for their investigation, can be found in the series of monographs edited by Franks (1972–1982). Computer simulations using the molecular dynamics (MD) and Monte Carlo (MC) methods have become more and more powerful tools in these investigations during the last two decades.

Being neither experiment nor theory, computer "experiments" can take over the task of both, studying the properties of complex many-body systems of interacting particles. The advantage of simulation techniques over conventional theoretical approaches lies in the limited number of approximations used, the crucial ones being intermolecular potential functions. Provided one has a reliable way to calculate intermolecular potentials, the simulations can lead to information on a wide variety of properties (thermodynamic, structural, trans-

---

Permanent address: Institute of Experimental Mineralogy, Russian Academy of Sciences, 142432 Chernogolovka, Moscow District, Russia.

port, spectroscopic, etc.) of the systems under study. In the case of simple liquids—say, liquid noble gases—where the interactions between particles are well known and where many-body non additive interactions do not play an important role, the results of computer simulations have a high degree of reliability and can be used as an "experimental" check against analytical theories (see, e.g., Hansen and McDonald, 1976). In the case of complicated systems—say, aqueous solutions in a wide range of temperatures, densities, and concentrations that cannot yet be treated on a molecular level analytically—the computer simulations can play the role of theory. They can predict the properties of fluids that cannot be or not directly be measured, and they can explain macroscopically measured properties on a molecular level.

Since the first MC and MD simulations of pure liquid water (Barker and Watts, 1969; Rahman and Stillinger, 1971), great progress has been made in the simulation studies of aqueous systems. One of the earliest significant results was the ruling out of "iceberg" formation in liquid water. Computer simulations—in spite of quite different intermolecular potentials employed—have unequivocally shown that liquid water consists of a macroscopically connected, random network of hydrogen bonds continuously undergoing topological reformations (Geiger et al., 1979b). The anomalous properties of water arise from the competition between nearly tetrahedrally coordinated local patterns characterized by strong hydrogen bonds and more compact arrangements characterized by more strained and broken bonds (Stillinger, 1980).

Apart from numerous simulations of pure liquid water under ambient conditions, much work has been done to study the thermodynamic, structural, transport, and spectroscopic properties of dilute aqueous solutions of ions (Mezei and Beveridge, 1981; Impey et al., 1983; Bounds, 1985; Mills et al., 1986; Guàrdia and Padró, 1990a) and hydrophobic solutes (Geiger et al., 1979a; Pangali et al., 1979; Okazaki et al., 1981; Rapaport and Scheraga, 1982; Swope and Andersen, 1984; Jorgensen et al., 1985). These studies have been reviewed already extensively (see, e.g., Wood, 1979; Beveridge et al., 1983; Levesque et al., 1984). Futhermore, in recent years the dielectric properties of several water models employed in computer simulations have been checked (Neumann, 1986; Alper and Levy, 1989; Ruff and Distler, 1990). Guissani et al. (1988) have simulated the ionic equilibrium of water and found the MD results in very good agreement with the experimental pH value.

The effects of temperature and pressure on the properties of water and aqueous solutions were also the subject of computer simulations. However, in most studies either high pressures (Stillinger and Rahman, 1974b; Impey et al., 1981; Jancsó et al., 1984, 1985; Pálinkás et al., 1984; Rami Reddy and Berkowitz, 1987; Madura et al., 1988) or high temperatures (Stillinger and Rahman, 1972, 1974a; Szász and Heinzinger, 1983a; Jorgensen and Madura, 1985; De Pablo and Prausnitz, 1989) were applied to the system. The term "high pressure" here indicates pressures up to about 20 kbar, which is for pure water equivalent to densities up to about 1.35 g/cm^3, whereas "high temperature" usually means the range from normal temperature to only about 500 K.

Except for LiCl solutions (Tamura et al., 1988), the solute concentrations studied by computer simulations never exceeded 2.2 molal (Heinzinger and Pálinkás, 1985; Heinzinger, 1990), and thus, the effects of a solute concentration on simulated properties have been scarcely studied so far.

On the other hand, in recent years the physical chemistry of aqueous solutions at high temperatures and pressures has become a subject of great scientific interest (Franck, 1970, 1981, 1987). While under normal conditions both ionic and nonionic solutes mostly have a rather low solubility in water, such solutions can become homogeneous in the whole concentration range at supercritical temperatures. The importance of understanding the behavior of hot, concentrated aqueous systems, known in geochemistry as hydrothermal solutions, on the molecular level has been stressed many times in the literature (Valyashko, 1977, 1990; Hamann, 1981; Tödheide, 1982; Holloway, 1987).

The near-critical region of the phase diagram, where the properties of water as well as those of aqueous solutions undergo drastic changes in a very narrow range of temperatures and pressures, seems to be of primary importance from various points of view. Because of the high compressibility of water under such conditions, the solutions have very large excess volumes (Franck, 1970, 1987; Zakirov and Kalinichev, 1980). This gives rise to significant changes in the activity coefficients of the components (Shmulovich et al., 1980a,b, 1982a), which have to be taken into account in any calculations of mineral-fluid equilibria (see, e.g., Ferry and Baumgartner, 1987). Recent measurements (Tivey et al., 1990) show that temperatures of seafloor hydrothermal vents can reach near-critical values of 350 to 400°C, which is of great importance for their complex chemistry (Von Damm, 1990). The possible significance of submarine near-critical springs for the origin of life has also been discussed recently (Corliss, 1990).

Except for purely geological interest, there is an increasing demand for data and models for aqueous solutions at high temperatures and pressures from various groups of engineering and environmental* scientists (Tödheide, 1982; Levelt Sengers, 1990). However, up to now fundamental work has lagged far behind that devoted to solutions in ordinary liquid water.

Although computer simulation techniques have already proved to be one of the most powerful tools to study usual aqueous solutions (Heinzinger and Pálinkás, 1985; Heinzinger, 1990) and permit transition from the "McMillan–Mayer" to "Born–Oppenheimer" level of solution models (Friedman, 1981), these methods still have not been applied systematically over a sufficiently wide range of thermodynamic conditions required in geochemistry. Even the thermodynamic and structural properties of the solvent itself—pure water—have not yet been studied extensively by these methods at high temperatures and pressures, although several important contributions should be mentioned in this context.

The first MC simulation of supercritical steam (Beshinske and Lietzke, 1969)

---

*Information about the use of supercritical water reactors to destroy hazardous wastes has already appeared in the literature [*Environ. Sci. Technol.* **24**, 1277 (1990)].

was published almost simultaneously with the first simulations of liquid water (Barker and Watts, 1969). Only pressures and internal energies for very dilute water vapor with a maximum density as low as 0.038 g/cm^3 were calculated in this early paper. O'Shea and Tremaine (1980) have studied supercritical water at higher densities. Kataoka (1987, 1989) has made the most extensive simulation study of water over a very wide range of the phase diagram. However, in both cases structural results were not reported. On the other hand, in the MD simulations of Mountain (1989), only the radial distribution functions of expanded water at high temperatures were analyzed. De Pablo et al. (1989, 1990) have calculated from MC simulations the densities of water along the liquid–vapor coexistence curve from 25°C to the critical point, while some properties of water directly at the critical point were simulated in the work of Evans et al. (1988). In this way, the thermodynamic properties and structure of dense supercritical water have also been studied recently (Kalinichev, 1985b; 1986; 1991b; Kalinichev and Heinzinger, 1992).

Unlike silicate melts (Kubicki and Lasaga, 1991), computer simulation methods have not yet been employed for the investigation of complex hydrothermal solutions. The aim of this chapter is to introduce this powerful technique into studies of aqueous systems of geochemical interest. Taking into account the current progress in computer technology, we feel that these methods could in a short time contribute a lot to the fundamental understanding of correlations between the thermodynamic, structural, spectroscopic, and transport properties of such systems on a molecular level.

The simulation methodology will be described briefly first, followed by a discussion of the intermolecular potentials used in high-temperature and high-pressure simulations. Then recent results on the thermodynamics of supercritical water are presented. In the final two sections a review of the effects of temperature, pressure, and solute ions on the structure and dynamics of water and aqueous fluids will be given.

The Simulation Techniques

Two methods of computer simulation have been developed: Monte Carlo (MC) and molecular dynamics (MD). In both cases, the simulations are performed on a small number of particles (atoms, ions, or molecules) $10 \leq N \leq 10^4$ confined in a (usually) cubic box. The interparticle interactions are described by pair potentials, and it is generally assumed that the total potential energy of the system can be developed as a sum of these pair potentials. Large numbers of particle configurations are generated in both methods, and from this microscopic information macroscopic properties (pressure, temperature, internal energy, heat capacity, etc.) can be calculated with the help of statistical mechanics. It is not the aim of this chapter to discuss the theoretical fundamentals of statistical physics and computer simulations, since many excellent sources are currently available (see, e.g., McQuarrie, 1976; Hansen and McDonald, 1976; Binder and Stauffer, 1984; Allen and Tildesley, 1987; Catlow et al., 1990). However, some basic concepts and relationships should be mentioned here for completeness.

Molecular Dynamics Simulations

In the MD simulations the classical equations of motion are integrated numerically for the particles in the box. The length of the time step for integration depends on a number of factors: temperature and density, the masses of the particles, and the nature of the interparticle potential. In the simulations of aqueous systems, it is typically on the order of 10^{-15} to 10^{-16} sec and the trajectories are followed (after preequilibration) for any number between 10^3 and 10^5 steps, depending on the properties that are under investigation.

The resulting knowledge of the trajectories for each of the particles (i.e., particle positions, velocities, as well as orientations and angular velocities if molecules are involved) means a complete description of the system in a classical sense. The thermodynamic properties of the system can then be calculated from the corresponding time averages. For example, the temperature is related to the average value of the kinetic energy

$$T = \frac{2}{3Nk_B} \left\langle \sum_{i=1}^{N} \frac{m_i v_i^2}{2} \right\rangle, \tag{1}$$

where k_B is the Boltzmann constant, m_i and v_i are the masses and the velocities of the particles in the box, respectively, and angular brackets denote the time average of the system.

The pressure can be derived from the virial theorem

$$P = \frac{Nk_B T}{V} - \left(\frac{1}{3V}\right) \left\langle \sum_{i=1}^{N} \mathbf{r}_i \cdot \mathbf{F}_i \right\rangle, \tag{2}$$

where V is the volume of the box and $(\mathbf{r}_i \cdot \mathbf{F}_i)$ means the dot product of the position and the force vectors of particle i.

The constant volume heat capacity C_V can be obtained from the temperature fluctuation by

$$C_V = R\left(\frac{2}{3} - N\frac{\langle T^2 \rangle - \langle T \rangle^2}{\langle T \rangle^2}\right)^{-1}, \tag{3}$$

where R is the gas constant.

Molecular dynamics simulations may be performed under a variety of conditions and constraints, corresponding to different *ensembles* in statistical mechanics. Most commonly the *microcanonical* (NVE) ensemble is used, i.e., the number of particles, volume, and total energy remain constant during the simulation. The above relationships are valid for this case. There are modifications of the method, allowing one to carry out the simulations in the *canonical* (NVT)

or *isothermal-isobaric* (*NPT*) ensembles. Relationships similar to Eq. (1–3) and many others can be derived for these ensembles (Allen and Tildesley, 1987). However, up to now these latter ensembles have rarely seen used in the simulations of aqueous systems (Ruff and Diestler, 1990).

The application of various statistical ensembles in MD simulations, as well as the numerical algorithms commonly used to integrate the equations of motion of the particles, have been discussed recently in the geochemical literature in connection with simulations of silicate melts (Kubicki and Lasaga, 1991).

Monte Carlo Simulations

In the MC simulations a large number of particle configurations are generated by the following scheme. Starting from a given configuration, a trial move of a randomly (or cyclically) chosen particle to a new position—as well as to a new orientation if molecules are involved—is made. The energy change ΔU connected with this move is then calculated, and if the energy is decreased, the move is unconditionally accepted. If ΔU is positive, the Boltzmann factor $\exp(-\Delta U/k_B T)$ is calculated and compared with a randomly chosen number between 0 and 1. The move is accepted if the Boltzmann factor is larger than this number and rejected otherwise. In other words, the trial configuration is accepted with probability

$$p = \begin{cases} 1, & \Delta U \leq 0; \\ \exp(-\Delta U/k_B T), & \Delta U > 0. \end{cases} \qquad (4)$$

The reiteration of such a procedure gives a Markov chain of molecular configurations distributed in the phase space of the system, with the probability density proportional to the Boltzmann weight factor corresponding to the *NVT* statistical ensemble. Typically, about 10^6 configurations are generated after some preequilibration stage of about the same length. The thermodynamic properties of the system can then be calculated as the averages over the ensemble of configurations*.

The ranges of maximum molecular displacement and rotation are usually adjusted at the equilibration stage for each run to yield an acceptance ratio of approximately 0.5. If these ranges are too small or too large, the acceptance ratio becomes closer to 1 or 0, respectively, and the phase space of the system is explored less effectively.

It is an advantage of the MC method that it can be readily adapted to the calculation of averages in any statistical ensemble (Allen and Tildesley, 1987). For example, to perform simulations in the *NPT* ensemble, one can introduce volume-changing trial moves. All intermolecular distances are scaled to the new box size during this move. The acceptance criterion is then also changed accord-

---

*The equivalence of ensemble and time averages, the so-called *ergodic hypothesis*, constitutes the basis of statistical mechanics (e.g., McQuarrie, 1976).

ingly. Instead of the energy difference ΔU in Eq. (4), one should use the instant enthalpy difference

$$\Delta H = \Delta U + P\Delta V - k_B T \ln(1 + \Delta V/V)^N, \tag{5}$$

where P is the pressure (which is kept constant in this case) and V the volume of the system.

In this ensemble, besides the trivial averages for configurational (i.e., due to intermolecular interactions) enthalpy,

$$H = \langle U \rangle + P\langle V \rangle \tag{6}$$

and molar volume,

$$V_m = \langle V \rangle N_A/N, \tag{7}$$

such thermodynamic properties as the isobaric heat capacity C_P, isothermal compressibility k, and thermal expansivity α can be easily calculated from the corresponding fluctuation relations

$$C_P = \left(\frac{\langle H^2 \rangle - \langle H \rangle^2}{N k_B T^2} \right), \tag{8}$$

$$k \equiv -\frac{1}{V}\left(\frac{\partial V}{\partial P}\right)_T = \left(\frac{\langle V^2 \rangle - \langle V \rangle^2}{N k_B T \langle V \rangle} \right), \tag{9}$$

$$\alpha \equiv \frac{1}{V}\left(\frac{\partial V}{\partial T}\right)_P = \left(\frac{\langle HV \rangle - \langle H \rangle\langle V \rangle}{N k_B T^2 \langle V \rangle} \right). \tag{10}$$

The *grand canonical* (μVT) ensemble, in which the chemical potential of the particles is fixed and the number of particles may fluctuate, seems to be very attractive for simulations of geochemical fluids. Up to now it has not been used even for pure water simulations. However, another technique, called the *Gibbs ensemble* Monte Carlo simulation (Panagiotopoulos, 1987), which permits the direct calculation of the phase coexistence properties of pure components and mixtures from a single simulation, was recently successfully used for calculations of the vapor–liquid coexistence properties of water (De Pablo and Prausnitz, 1989; De Pablo et al., 1990).

MD simulations have the great advantage of allowing the study of time-dependent phenomena. However, if thermodynamic and structural properties alone are required, the MC method might be more useful, primarily because of the ease with which various statistical ensembles can be introduced (Hansen and McDonald, 1976; Allen and Tildesley, 1987; Catlow et al., 1990).

Boundary Conditions and Long-Range Corrections

One of the most obvious difficulties arises in both simulation methods from the small system size, a consequence of the limited availability of computer time. So-called *periodic boundary conditions* are therefore used in order to minimize surface effects and to simulate more closely the properties of a bulk macroscopic system. This means that the basic cubic box is assumed to be surrounded by identical boxes in all directions infinitely. If a particle leaves the box through one side, it enters simultaneously through the opposite side because of the identity of the boxes. In this way, the problem of surfaces is circumvented at the expense of the introduction of periodicity.

Whether the properties of a small infinitely periodic system and the macroscopic system which the model is designed to represent are the same depends on the range of the intermolecular potential and property under investigation. For short-range interactions either spherical or minimum image cutoff criteria are commonly used (Allen and Tildesley, 1987). The latter means that each molecule interacts only with the closest image of every other molecule in the basic box or in its periodic replica. However, any realistic potential for water (not to mention electrolyte solutions) contains long-range Coulomb interactions, which should be properly taken into account. The treatment of these interactions is still under discussion, and several methods are commonly used (see, e.g., Allen and Tildesley, 1987), of which the Ewald summation is usually considered the most satisfactory.

All existing schemes for handling long-range interactions are rather computer time-consuming. Therefore, it makes sense to consider carefully the question of whether they have to be included for the system under investigation before extensive simulations are started. In the case of ionic interactions, where unscreened Coulomb potential is used, there is no choice but to implement the Ewald method for accurate calculations of thermodynamic and structural properties (Heinzinger and Pálinkás, 1985; Heinzinger, 1990; Kubicki and Lasaga, 1991). On the other hand, as numerous studies with several intermolecular potentials truncated at various distances and with different system sizes from 64 to 512 molecules have shown (Pangali et al., 1980; Andrea et al., 1983; Jorgensen and Madura, 1985), thermodynamic properties and atom–atom distribution functions of liquid water are very insensitive to the range of the potential beyond ≈ 6 Å. It means that even $N = 64$ already gives a good approximation to bulk water as far as thermodynamic and structural properties are concerned. This is, however, not the case for the dynamical properties of the system, orientational correlations, and connected to them, dielectric properties, where long-range interactions must be carefully accounted for (Andrea et al., 1983; Neumann, 1986; Anderson et al., 1987; Alper and Levy, 1989; Ruff and Diestler, 1990).

Contrary to the results for water under normal conditions, the analyses of MC simulations for supercritical water have shown a slight, but noticeable, system-size dependence of thermodynamic and structural properties at densities significantly lower than that of normal liquid water (Kalinichev, 1992). This finding is in accordance with recent results on simple model fluids (Schaink and Hoheisel, 1990), where the governing role of long-range attractive interactions has been established for the correct description of the structural and dynamical behavior of fluids.

The thermodynamic values calculated from classical simulations have to be corrected for quantum effects. There are several methods to compute the quantum corrections to the thermodynamic properties of liquid water (Owicki and Scheraga, 1977; Beveridge et al., 1983; Berens et al., 1983). However, in the high-temperature case, the relative importance of these corrections decreases significantly, and they will be neglected therefore in this chapter.

Statistical Errors of Simulations and the Convergence Analysis

Similar to every experimental method, the results of computer simulations may be subject to systematic and statistical errors. Some sources of systematic errors are briefly discussed above. These should, of course, be estimated and eliminated when possible. However, as simulation averages are taken over runs of finite length, it is essential to estimate the statistical significance of the results.

The statistical uncertainties of simulated properties are usually estimated by the method of block averages (Allen and Tildesley, 1987). The chain of molecular configurations is subdivided into several nonoverlapping blocks of equal length, and the averages of every property are computed for each block. If $\langle A \rangle_i$ is the mean value of the property A computed over the block i, then the statistical error δA of the mean value $\langle A \rangle$ over the whole chain of configurations can be estimated as follows:

$$(\delta A)^2 = \frac{1}{M(M-1)} \sum_{i=1}^{M} [\langle A^2 \rangle_i - \langle A \rangle_i^2], \tag{11}$$

where M is the number of blocks.

Strictly speaking, Eq. (11) is valid only if all $\langle A \rangle_i$ are statistically independent and show a normal Gaussian distribution. Thus in computer simulations of insufficient length, these error bounds are to be taken with caution (Beveridge et al., 1983), especially for the properties calculated from fluctuations, such as heat capacity. Therefore, the analysis of convergence profiles giving the running averages of the properties during a simulation might be quite useful. Then one can also estimate roughly the limits of error as maximal variations in the average values at the final stage of the simulation.

Figure 1 shows convergence profiles of thermodynamic properties during a NPT-ensemble MC simulation of water at 773 K and 30 kbar (Kalinichev, 1985a). It is clearly seen from the plots that at this high temperature and pressure a satisfactory convergence of all average values (including such "badly converging" properties as C_P, κ, and α) is attainable after $\approx 1000 \cdot N$ configurations (N is the number of molecules), in contrast to the $8000 \cdot N$ usually recommended for simulations of liquid water under normal conditions (Jorgensen, 1982b). The convergence of the average value for the pressure, according to Eq. (2), during an MD simulation of water at 680 K and 0.9718 g/cm^3 (Kalinichev and Heinzinger, 1992) is shown in Fig. 2. Under the conditions studied, about 0.2 ps are already sufficient to attain the average value of pressure with a rather good accuracy. It should be stressed that in both cases the diagrams demonstrate the convergence from preequilibrated starting configurations.

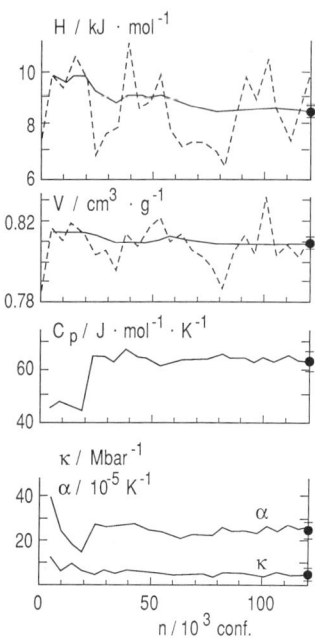

Fig. 1. Convergence of thermodynamic properties during a MC simulation of water at 773 K and 30 kbar. Full lines, running averages; dashed lines, block averages; $N = 64$ (Kalinichev, 1985a).

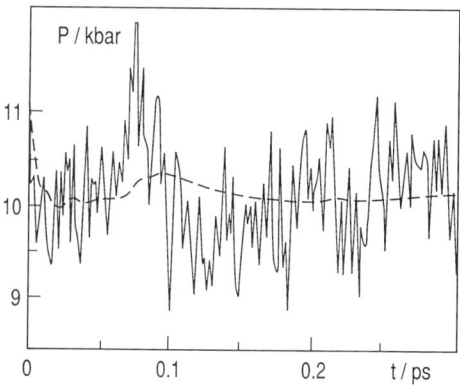

Fig. 2. Convergence of pressure during a MD simulation of water at 680 K and 0.9718 g/cm^3. Full line, instant pressure; dashed line, running average.

It has been noted (Kalinichev, 1985a) that contrary to normal conditions, where enthalpy and volume fluctuate during an NPT-ensemble MC water simulation fairly independently (Jorgensen, 1982b), the fluctuations of both properties are significantly correlated under high-temperature conditions. Though this correlation is not very much pronounced at high densities, it becomes very strong at low densities (Kalinichev, 1992). Preliminary estimates show that at least in the case of low-density simulations, the main reason for the correlation between the fluctuations of H and V is not the increasing contribution of the PV term to the configurational enthalpy [Eq. (5)], but the strong dependence of the configurational internal energy U on V. Therefore, more frequent than is usually recommended (Jorgensen, 1982b; Allen and Tildesley, 1987) attempts to change the volume of the system may help to speed up significantly the convergence of high-temperature NPT simulations.

Volume-changing moves in the NPT-ensemble MC simulations of molecular liquids are much more computer-time-consuming than the usual MC moves, because of the nesessity to rescale the coordinates of all interaction sites in the system and to recalculate correspondingly all pair interaction energies. However, if a vectorization is possible, a tenfold increase of the frequency of volume-changing moves slows down the average speed of the calculations only by $\approx 25\%$ (Kalinichev, 1992), which may be considered an acceptable price for the improvement of the convergence.

It is well known (see, e.g., Jorgensen, 1982b) that structural properties converge very rapidly, even in the case of liquid water simulations under normal conditions. The situation can only improve with temperature. Therefore, the convergence of atom–atom distribution functions deserves no special discussion.

Intermolecular Potentials for Water and Electrolyte Solutions

Decisive for reliability of the properties calculated from the simulations are the potential functions employed. The intermolecular interactions between water molecules are far more complicated than those between particles of simple liquids. The most important difference is the ability to form hydrogen bonds, which makes water an associated liquid. An additional difficulty in the description of water interactions is the presence of substantial nonadditive three- and higher-body terms, studied in detail by several authors (Hankins et al., 1970; Gellatly et al., 1983; Clementi, 1985), which may raise doubts on the applicability of the pair-additivity approximation ordinarily used in computer simulations.

On the other hand, the analysis of experimental shockwave data of water has shown (Ree, 1982) that at high temperature ($\gtrsim 1300$ K) and high pressure ($\gtrsim 80$ kbar), the intermolecular interactions of water become simpler. In this case, it is possible to use a spherically symmetric model potential for the calculations of the water properties, either from computer simulations (Belonoshko and Saxena, 1991) or from thermodynamic perturbation theory in a way similar to that applied to nonpolar fluids (Shmulovich et al., 1982b; Kalinichev, 1991a). However, such simplifications exclude the possibility of understanding the complex phenomena in aqueous fluids on a true molecular level, which is, actually, the strongest advantage and the main objective of computer simulations.

The pair potential functions for the description of intermolecular interactions used in computer simulations of aqueous systems can be grouped into two broad classes as far as their origin is concerned: empirical and quantum mechanical potentials. In the first case, all parameters of a model are adjusted to fit experimental data for water from different sources, and thus necessarily incorporate effects of many-body interactions in some average way. The second class of potentials, obtained from *ab initio* quantum mechanical calculations, represent purely the pair energy of the water dimer and they do not take into account any many-body effects. However, such potentials can be regarded as the first term in a systematic many-body expansion of the total quantum mechanical potential (Clementi, 1985).

In the last two decades, both types of potentials have been used extensively in computer simulations of aqueous systems. Several reviews comparing the abilities of different potentials for reproducing a wide range of gas, liquid, and solid-state properties of water are currently available (Wood, 1979; Reimers et al., 1982; Morse and Rice, 1982; Beveridge et al., 1983; Jorgensen et al., 1983; Clementi, 1985). These comparisons have shown that none of the models gives a satisfactory account of all three phases simultaneously. They revealed, on the other hand, that many properties of liquid water and aqueous solutions can be qualitatively and even quantitatively reproduced in computer simulations, irrespective of the intermolecular potential used, verifying in this way the reliability of the models.

The first simulations of water under high pressure were performed with the empirical ST2 model (Stillinger and Rahman, 1974b). It is a four-point charge rigid model, with the charges arranged tetrahedrally around the oxygen atom [Fig. 3(a)]. The positive charges are located at the hydrogen atom positions at a distance of 1 Å from the oxygen atom, nearly the real distance in the water molecule. The negative charges are located at the other two vertices of the tetrahedron but at a distance of only 0.8 Å from the oxygen atom. The charges were chosen to be 0.23 elementary charges leading to roughly the correct dipole moment of the water molecule. The tetrahedrally arranged point charges render possible the formation of hydrogen bonds in the right directions. The ST2 model

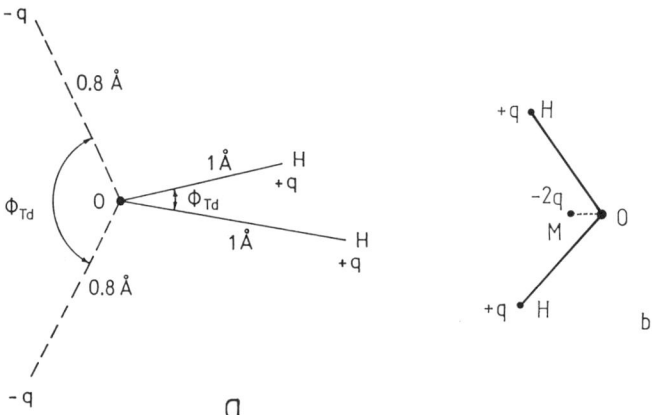

Fig. 3. Schematic reperesentation of a water molecule in the (a) ST2 and (b) TIP4P and MCY models.

is completed by adding a (12-6) Lennard–Jones (LJ) potential, the center of which is located at the oxygen atom, with $\sigma = 3.10$ Å and $\varepsilon = 0.317$ kJ/mol.

In the simulations of aqueous alkali halide solutions where the ST2 model is employed, the ions are modeled as LJ spheres with a point charge at the center (Heinzinger and Vogel, 1976). With these models for the two kinds of particles—water and ions—it is easy to formulate the effective pair potentials for the six different kinds of interactions: cation–cation, anion–anion, cation–anion, cation–water, anion–water, and water–water. An six pair potentials consist of an LJ term

$$U^{LJ}_{ij}(r) = 4\varepsilon_{ij}\left[\left(\frac{\sigma_{ij}}{r}\right)^{12} - \left(\frac{\sigma_{ij}}{r}\right)^{6}\right], \tag{12}$$

where i and j refer either to ions or water molecules, and a Coulomb term, different for water–water, ion–water, and ion–ion interactions, given by

$$U^{C}_{ww}(r, d_{11}, d_{12}, \ldots) = S_{ww}(r) \cdot q^2 \sum_{\alpha,\beta=1}^{4} \frac{(-1)^{\alpha+\beta}}{d_{\alpha\beta}}, \tag{13a}$$

$$U^{C}_{\substack{+w \\ (-w)}}(d_{\substack{+1 \\ (-1)}}, d_{\substack{+2 \\ (-2)}}, \ldots) = -\sum_{\alpha=1}^{4}(-1)^{\alpha}\frac{q \cdot e}{d_{\substack{+\alpha \\ (-\alpha)}}}_{(+)}, \tag{13b}$$

$$U^{C}_{\substack{\pm\pm \\ (+-)}}(r) = \genfrac{}{}{0pt}{}{+}{(-)}\frac{e^2}{r}. \tag{13c}$$

The switching function $S_{ww}(r)$, in the Coulomb term of the water pair potential has been introduced by Rahman and Stillinger in order to reduce unrealistic Coulomb forces between very close water molecules. d and r denote distances between point charges and LJ centers, respectively, q being the charge in the ST2 model. The sign of the Coulomb term is correct if α and β are chosen as odd for positive and even for negative charges.

The LJ parameters for the cations are taken from the isoelectronic noble gases (Hogervorst, 1973). If we compare e.g., Pauling radii, it is obvious that halide ions have a larger ionic radius than the isoelectronic alkali ions. In order to describe all interactions consistently, new LJ parameters had to be determined for the halide ions on the basis of the Pauling radii (Pálinkás et al., 1977). With a knowledge of the parameters for cation–cation and anion–anion interactions, the parameters for cation–water and anion–water interactions have been determined by applying Kong's combination rules (Kong, 1973). The results of this procedure are given in Table 1.

Evans (1986) has proposed recently a modified version of the ST2 potential that includes atom–atom LJ terms centered both on the oxygen and hydrogen atoms, thus eliminating the need to use the switching function in Eq. (13a). This model has been employed in MD simulations of water at temperatutes up to 1273 K and at constant densities of 1.0 and 0.47 g/cm^3 (Evans et al., 1988) and

Table 1. Lennard–Jones parameters in the pair potentials for cation–cation, anion–anion, cation–water, and anion–water interactions (Pálinkás et al., 1977).

Ion	Pauling radius (Å)	σ_{II} (Å)	ε_{II} (kJ mol^{-1})	σ_{IW} (Å)	ε_{IW} (kJ·mol^{-1})
Li$^+$	0.60	2.37	0.149	2.77	0.224
Na$^+$	0.95	2.73	0.358	2.92	0.330
F$^-$	1.36	4.00	0.050	3.53	0.123
K$^+$	1.33	3.36	1.120	3.25	0.568
Cl$^-$	1.81	4.86	0.168	4.02	0.185
Rb$^+$	1.48	3.57	1.602	3.39	0.641
Br$^-$	1.95	5.04	0.270	4.16	0.215
Cs$^+$	1.69	3.92	2.132	3.61	0.662
I$^-$	2.16	5.40	0.408	4.41	0.228

has shown, within statistical uncertainty, the satisfactory reproducibility of the experimental pressure in this range and at the critical point of water.

Another empirical water model often used in simulations at high temperature and pressure is the TIP4P model (Jorgensen et al., 1983). It differs from the ST2 model in several respects. The rigid geometry employed is that of the gas-phase monomer, with an O–H distance of 0.9572 Å and an H–O–H angle of 104.52°. The two negative charges are reduced to a single one at a point M positioned on the bisector of the H–O–H angle at a distance of 0.15 Å in the direction of the H atoms [Fig. 3(b)], which bear a charge of +0.52 elementary charges. This simplification of the charge distribution even seems to be an improvement as there are indications that the negative charges in the ST2 model in the tetrahedral vertices exaggerate the directionality of the lone-pair orbitals of the water molecule. There is again a (12-6) Lennard–Jones potential centered at the oxygen atom, with the parameters $\sigma = 3.1536$ Å and $\varepsilon = 0.649$ kJ/mol. This larger value for ε compared with the ST2 model compensates for the reduced Coulomb energy because of the fact that here the opposite charges cannot approach as near as in the ST2 model. This is also the reason why in the TIP4P model a switching function is not necessary. As there are only three charged sites in the TIP4P model, the sum in Eq. (13a) reduces to only nine terms.

The TIP4P water model has already proven its reliability in numerous simulations of a wide range of water properties. It reproduces the density maximum of liquid water around 0°C (Jorgensen and Madura, 1985), just as the ST2 model does (Stillinger and Rahman, 1974a). De Pablo and Prausnitz (1989) have studied the vapor–liquid equilibrium properties of TIP4P water, while Motakabbir and Berkowitz (1991) and Karim and Haymet (1988) have simulated vapor–liquid and ice–liquid interfaces, respectively. Dielectric properties for this water model have been simulated by Neumann (1986) and Alper and Levy (1989). Mountain (1989) has used the TIP4P model in his investigation of hydrogen bonding in expanded water at high temperatures. The thermodynamic and structural properties of TIP4P water at normal temperature and pressures up to 10 kbar (Rami Reddy and Berkowitz, 1987; Madura et al., 1988) as well as at normal density and temperatures up to 2300 K (Brodholt and Wood, 1990) have also been studied. Sprik and Klein (1988) proposed a polarizable version of this potential using variable charge sites.

Recently, the empirical simple point-charge (SPC) model (Berendsen et al., 1981) has been used successfully in MC simulations of water along the liquid–vapor coexistence curve from 25°C to the critical point (De Pablo et al., 1990). This model has a rigid geometry and LJ parameters quite similar to those of the TIP4P model. The main difference is in regard to the position of the negative charge, which is now located not on the point M [see Fig. 3(b)], but on the oxygen atom itself. The number of interaction sites is thus reduced to three, which speeds up the calculation of the intermolecular interaction energy of the system during the simulation. Flexible versions of the SPC model have also been proposed (Toukan and Rahman, 1985; Dang and Pettitt, 1987) that proved to give better results for the dielectric constant of water up to 350 K than the TIP4P model (Anderson et al., 1987). Ahlström et al. (1989) developed a polarizable version of the SPC model in their attempts to take into account the many-body interactions in liquid water. Guissani et al. (1988) made the first attempt to calculate the pH value of water from MD simulations and, after all polarization effects were included, achieved rather good agreement with experiments up to 593 K.

The role of molecular flexibility on the thermodynamic, structural, and dynamical properties of simulated water is still under discussion (Teleman et al., 1987; Guàrdia and Padró, 1990b; Barrat and McDonald, 1990). However, flexible water models have the great advantage of permitting the investigation of the effect of temperature, pressure, and ions on the *intramolecular* properties of a water molecule like geometry and vibrations. The first model of this kind, the so-called central force (CF) model, was proposed by Lemberg and Stillinger as early as 1975 and modified later by Stillinger and Rahman (1978). This model consists of oxygen and hydrogen atoms, bearing partial charges, where the water molecule geometry is solely preserved by an appropriate set of oxygen–hydrogen and hydrogen–hydrogen pair potentials. Thus, the CF model has the advantage that the influence of temperature, pressure, and ions on the intramolecular properties of water can be studied. Unfortunately, it is unable to describe correctly the gas–liquid frequency shifts of the intramolecular vibrations.

In order to improve CF potential, the BJH water model has been developed (Bopp et al., 1983), in which the total potential is separated into an intermolecular and an intramolecular part. The intermolecular pair potential is an only slightly modified version of the CF model and is given by

$$U_{OO}(r) = \frac{604.6}{r} + \frac{111889}{r^{8.86}} - 1.045 \{\exp[-4(r-3.4)^2] - \exp[-1.5(r-4.5)^2]\},$$

(14a)

$$U_{OH}(r) = -\frac{302.2}{r} + \frac{26.07}{r^{9.2}} - \left\{\frac{41.79}{1 + \exp[40(r - 1.05)]}\right\}$$
$$- \left\{\frac{16.74}{1 + \exp[5.493(r - 2.2)]}\right\}, \tag{14b}$$

$$U_{HH}(r) = \frac{151.1}{r} + \left\{\frac{418.33}{1 + \exp[29.9(r - 1.968)]}\right\}. \tag{14c}$$

The intramolecular part is based on the water potential in the formulation of Carney et al. (1976)

$$U_{intra} = \sum L_{ij}\rho_i\rho_j + \sum L_{ijk}\rho_i\rho_j\rho_k + \sum L_{ijkl}\rho_i\rho_j\rho_k\rho_l, \tag{15}$$

with $\rho_1 = (r_1 - r_e)/r_1$, $\rho_2 = (r_2 - r_e)/r_2$, and $\rho_3 = \alpha - \alpha_e = \Delta\alpha$, where r_1, r_2, and α are the instantaneous O–H bond lengths and H–O–H angle; the quantities r_e and α_e are the corresponding equilibrium values ($r_e = 0.9572$ Å, $\alpha_e = 104.52°$). The finally adopted parameter set is given in Table 2.

In the simulations of the electrolyte solutions with the BJH model for water, the ion–oxygen and ion–hydrogen pair potentials (as well as the ion–ion ones that are not very important in the case of the dilute solution) are derived from *ab initio* calculations. They are presented in Table 3.

Table 2. Potential constants used for the intramolecular part of the BJH model of water in units of kJ/mol (Bopp et al., 1983). The notations are according to Eq. (15).

$\rho_1\rho_2(\rho_1 + \rho_2)$	−55.7272	$(\rho_1^2 + \rho_2^2)$	2332.27
$(\rho_1^2 + \rho_2^2)\Delta\alpha$	237.696	$\rho_1\rho_2$	−55.7272
$(\rho_1^4 + \rho_2^4)$	5383.67	$(\rho_1 + \rho_2)\Delta\alpha$	126.242
$\rho_1\rho_2(\rho_1^2 + \rho_2^2)$	−55.7272	$(\Delta\alpha)^2$	209.860
$(\rho_1^3 + \rho_2^3)\Delta\alpha$	349.151	$(\rho_1^3 + \rho_2^3)$	−4522.52

Table 3. Ion–oxygen and ion–hydrogen pair potentials employed in the simulations of electrolyte solutions with the BJH model. Energies are given in kJ/mol and distances in Å

$U_{NaO}(r) = -916.5/r - 153.6/r^2 + 48.93 \cdot 10^4 \exp(-4.526r)$
$U_{NaH}(r) = 458.2/r + 31.31/r^2 + 41.68 \cdot 10^4 \exp(-7.07r)$
$U_{MgO}(r) = -1832/r - 890.7/r^2 + 26.95 \cdot 10^4 \exp(-4.08r)$
$U_{MgH}(r) = 916.5/r + 82.02/r^2 + 73.83 \exp(-0.349r)$
$U_{CaO}(r) = -1832/r - 1572/r^2 + 25.97 \cdot 10^4 \exp(-3.49r)$
$U_{CaH}(r) = 916.5/r + 626/r^2 + 12.02 \cdot 10^4 \exp(-6.79r)$
$U_{SrO}(r) = -1832/r - 1192/r^2 + 17.64 \cdot 10^4 \exp(-3.11r)$
$U_{SrH}(r) = 916.5/r + 352.3/r^2 + 14.45 \exp(-0.158r)$
$U_{ClO}(r) = 916.5/r - 111.3/r^2 + 37.96 \cdot 10^4 \exp(-3.21r)$
$U_{ClH}(r) = -458.2/r + 18.90 \cdot 25 \exp(-34r)$

From the family of quantum mechanical water potentials, the MCY model (Matsuoka et al., 1976) should be mentioned in the context of high-pressure and high-temperature simulations. This model has the same geometry as the TIP4P potential [Fig. 3(b)], but a much more complicated functional form with parameters derived from *ab initio* calculations. A flexible version of this model (MCYL) has also been proposed (Lie and Clementi, 1986). The MCY model was used by Impey et al. (1981) in their MD studies of the structure of water at elevated temperatures and high density, and by O'Shea and Tremaine (1980) in the MC simulations of the thermodynamic properties of liquid and supercritical water. It is well known, however, that this potential reproduces poorly the pressure at a given density (or the density at a given pressure) (Beveridge et al., 1983; Owicki and Sheraga, 1977). Even the addition of quantum mechanical three- and four-body terms to the potential, though extremely demanding in terms of computer time, did not improve the situation significantly (Clementi, 1985). A slightly different *ab initio* CC potential (Carravetta and Clementi, 1984) has been used by Kataoka (1987, 1989) in extensive MD simulations of the thermodynamic and transport properties of fluid water in a wide range of conditions. A qualitative reproduction of the anomalous behavior of these properties has been achieved.

Thus, despite the great importance of quantum mechanical potentials from a purely theoretical point of view, simple effective two-body potential functions for water seem to be preferable for the extensive simulations of complex aqueous systems of geochemical interest in the near future.

Thermodynamic Properties of Supercritical Water

The thermodynamics of water in a wide range of temperatures ($630 \leq T \leq 780$ K) and pressures ($0.3 \leq P \leq 30$ kbar) corresponding to densities $0.166 \leq \varrho \leq 1.284$ g/cm^3 have recently been studied by both computer simulation methods (Kalinichev, 1991b; Kalinichev and Heinzinger, 1992). The rigid TIP4P potential (Jorgensen et al., 1983) has been used in connection with NPT-ensemble MC simulations, whereas conventional NVE-ensemble MD simulations have been performed with the flexible BJH water model (Bopp et al., 1983).

In the MC simulations, two different system sizes were used ($N = 64$ and 216) to assure the reliability of the minimum image cutoff principle applied here to calculate the potential energy of the system. No long-range corrections were made, as it was shown (Andrea et al., 1983) that this method leads to only small errors in the resulting thermodynamic and structural properties of liquid water, even with $N = 64$.

The systems studied in the MD simulations consisted of 200 water molecules in a cubic box with the side length adjusted to give the required density. The time step of the integration was 1.5×10^{-16} s and the simulations extended over 10,000 to 15,000 time steps. Ewald summation in tabulated form was used for the

Table 4. Thermodynamic results of Monte Carlo simulations with the TIP4P potential.

Run	T (K)	P (kbar)	N	n (10^6 conf.)	V_m (cm^3/mol)	H_{conf} (kJ/mol)	C_P (J/(mol·K))	κ (1/Mbar)	$\alpha \cdot 1000$ (1/K)
MC1	773	30.0	216	1.10	14.03 ± 0.02 (13.52)^a	12.35 ± 0.10 (15.17)	61.1 ± 2.2 (61.4)	6.7 ± 0.5 (8.0)	0.25 ± 0.03 (0.12)
MC2	773	10.0	64	0.37	18.20 ± 0.05 (17.51)	−8.91 ± 0.18 (−9.55)	69.0 ± 5.2 (63.6)	32.7 ± 3.5 (23.1)	0.66 ± 0.08 (0.41)
MC3	773	1.0	64	1.60	43.13 ± 0.67 (34.11)	−11.53 ± 0.22 (−14.70)	97.6 ± 4.6 (100.1)	780 ± 70 (543)	3.40 ± 0.3 (3.28)
MC4	773	0.5	64	1.10	83.34 ± 1.70 (70.11)	−6.07 ± 0.22 (−7.35)	91.3 ± 5.3 (130.4)	2520 ± 340 (3263)	4.20 ± 0.4 (7.06)
MC5	673	10.0	64	1.10	17.12 ± 0.04 (16.82)	−13.14 ± 0.13 (−12.98)	67.9 ± 2.3 (63.9)	19.7 ± 1.3 (19.8)	0.50 ± 0.04 (0.39)
MC6	673	1.0	64	1.10	31.23 ± 0.53 (26.01)	−18.39 ± 0.29 (−21.22)	99.0 ± 7.3 (88.5)	420 ± 50 (211)	3.10 ± 0.4 (2.14)
MC7	673	0.5	64	1.60	47.20 ± 0.99 (31.17)	−14.60 ± 0.25 (−19.72)	126.3 ± 7.7 (122.3)	1680 ± 160 (713)	6.20 ± 0.6 (4.89)
MC8	673	0.3	216	1.60	89.70 ± 1.50 (50.31)	−9.04 ± 0.18 (−14.74)	120.2 ± 10.8 (451.8)	4500 ± 600 (11712)	7.25 ± 1.0 (36.75)

^{a}Values calculated from the HGK equation of state (Haar et al., 1984) are given in parentheses for comparison as "experimental."

Table 5. Thermodynamic results of MD simulations with the BJH potential.

	Run				
	MD1	MD2	MD3	MD4	MD5
$\varrho/\text{g}\cdot\text{cm}^{-1}$	1.2840	0.5282	0.9718	0.1666	0.6934
$V_m/\text{cm}^3\cdot\text{mol}^{-1}$	14.03	34.11	18.54	108.12	25.98
T/K	771[a]	772	680	673	630
P/kbar	24.1	3.3	10.1	0.66	4.6
	(25.7)[b]	(1.00)	(6.16)	(0.25)	(0.57)
$U_{\text{conf}}/\text{kJ}\cdot\text{mol}^{-1}$	−29.1	−16.9	−28.3	−7.8	−24.2
	(−26.35)	(−18.13)	(−28.74)	(−9.74)	(−24.81)
$C_V/\text{J}\cdot\text{mol}^{-1}\cdot\text{K}^{-1}$	53	49	54	51	49
	(58.70)	(47.74)	(52.97)	(58.93)	(52.05)

[a] Statistical errors of the simulated properties are estimated to be of the order of the last digit given.
[b] Values calculated from the HGK equation of state (Haar et al., 1984) are given in parentheses for comparison as "experimental."

coulombic part of the interactions, and the "shifted-force" method (see, e.g., Allen and Tildesley, 1987) was used for the other parts of the BJH potential. With this procedure, the total energy change $\Delta E/E$ during the simulation was smaller than $5\cdot 10^{-5}$ in all cases and the average temperature remained constant without rescaling, which is very important for the reliability of the dynamical properties calculated from velocity autocorrelation functions.

The calculated values of the thermodynamic properties for all thermodynamic states studied are reported in Table 4 and 5. Thermodynamic properties calculated via the HGK equation of state for water (Haar et al., 1984), approved by the International Association for the Properties of Steam (IAPS) as an international standard up to 1273 K and 10 kbar, are also given there for comparison. This equation reproduces virtually all available thermodynamic measurements within the limits of their experimental accuracy and is used in the present study as a reliable source of self-consistent "experimental" data.

The nonconfigurational contributions to the internal energy, enthalpy, and heat capacities C_V and C_P were assumed to be identical with those for real gas in its standard state (Woolley, 1980). The experimental configurational properties were estimated from the equation of state by

$$U_{\text{conf}} = U(T,V) - U_{\text{ig}}(T) \qquad (16a)$$

and

$$H_{\text{conf}} = H(T,P) - H_{\text{ig}}(T) + RT, \qquad (16b)$$

with the ideal-gas reference state at the given temperature. No other corrections were added to any property reported.

Macroscopic Properties

P-V-T Relationships

A comparison of simulated and experimental molar volumes and pressures is presented in Tables 4 and 5, and in Fig. 4 the positions of state points on the thermodynamic surface of water, where the simulations have been performed, are shown. The good agreement of calculated and experimental volumes for runs MC1, MC2, and MC5 cannot be considered surprising, because the TIP4P potential has already been carefully tested at liquidlike and higher densities up to 373 K (Jorgensen and Madura, 1985) and up to 10 kbar at 298 K (Madura et al., 1988). In very recent MD simulations (Brodholt and Wood, 1990), the TIP4P potential was successfully used even at a density almost twice as high as the normal one and at a temperature of about 2000 K.

At lower pressures the volumes obtained in MC simulations become systematically higher than the experimental values. The reason for the agreement with experiment in the first case, and for the discrepancy in the second, most probably lies in the "effective" nature of the TIP4P potential, which was specially developed to reproduce correctly thermodynamic and structural properties of liquid water under normal conditions and implicitly includes many-body effects of intermolecular interactions. The relative influence of these effects should obviously be density-dependent, which is not readily taken into account by the present potential model and may lead to increasing disagreement with experimental values as the density decreases. The same conclusion can be drawn from the results of De Pablo and Prausnitz (1989) who made Gibbs-ensemble MC

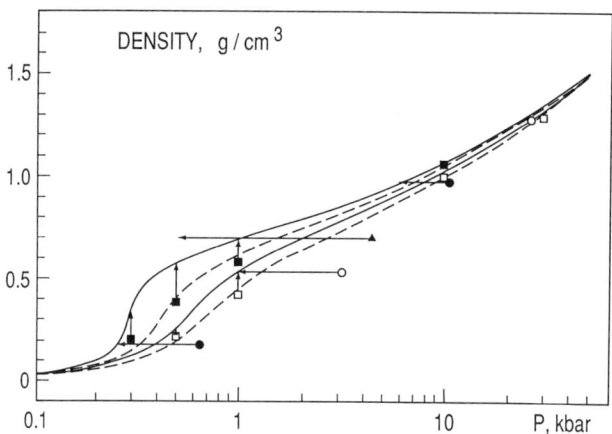

Fig. 4. Simulated (symbols) and HGK (Haar et al., 1984) (curves) isotherms of water density. Full lines, 673 and 773 K; dashed lines, 723 and 823 K. Squares: MC simulations at 673 K (filled) and 773 K (open). Circles: MD simulations at ≈ 673 K (filled) and ≈ 773 K (open). Triangle: MD simulation at 630 K. Arrows show the discrepancy in density (MC) and pressure (MD) between simulations and experimental data.

simulations of TIP4P water along the vapor–liquid coexistence curve. They noted that vapor densities are predicted less accurately than liquid ones and that agreement with experiment deteriorates rapidly as the temperature rises to around 500 K.

At 673 K and 0.3 kbar (run MC8), the simulated volume is almost twice as large as the experimental one. In this thermodynamic state, the water density changes drastically in a very narrow interval of pressure around 0.3 kbar (Fig. 4) because of the extremely high supercritical compressibility (see below). Therefore, the observed discrepancy between simulated and experimental densities may not only be due to the reasons mentioned above. It may also be an indication of the lower critical temperature of TIP4P water, compared with the experimental value of 647 K. The tendency toward such behavior is clearly seen from the simulated $\varrho - P$ isotherms in Fig. 4, which do not show the characteristic inflection with the rapid change of density around it. The shape of the simulated isotherms corresponds more closely to higher-temperature isotherms of water (such curves for 723 and 823 K are shown in Fig. 4 as dashed lines), suggesting that the critical point of TIP4P water may lie about 50 to 70 degrees lower than the experimental one. The same conclusion can be drawn from the recently published data on the coexistence curve of TIP4P water (De Pablo and Prausnitz, 1989). Estimates for the SPC potential also give lower values for critical density and temperature (De Pablo et al., 1990).

The failure to reproduce critical parameters correctly seems to be a general feature of all empirical potentials, which are parameterized to describe the properties of water at liquidlike densities. The reason for this lies in the above-mentioned effective nature of such potentials, which include only implicitly many-body effects. Indirect confirmation of this statement can be found in the work of Kataoka (1987), who estimated the critical temperature to be about 50° lower than the experimental one (as well as having a somewhat lower critical density) for the *ab initio* Carravetta–Clementi pair potential of water, which inherently does not include many-body effects either explicitly or implicitly. A similar temperature shift has been found by Impey et al. (1981) in high-temperature MD simulations of water structure with the MCY *ab initio* potential.

It is interesting to note that in their simulations of the ice–water interface, Karim and Haymet (1988) also found the freezing point of TIP4P water to be about 35° lower than the experimental value. On the other hand, it is known (Hankins et al., 1970) that three- and higher-body interactions act to keep ice solid above the melting point for "hypothetical ice," in which only pair interactions were operative.

So far, the phase diagram for any of the widely used potential models of water is still not known over a sufficient range of temperatures and pressures. The very recent study of De Pablo et al. (1990) is only a first step in this direction.

The results of the present MD simulations (Table 5 and Fig. 4) show that the BJH water model, while reproducing pressure with a reasonable accuracy at liquid like densities, is even less sensitive to the nearness of the critical point than the TIP4P model. We can only guess that the critical point of BJH water is much lower in temperature than the experimental one. In the recent NPT-ensemble MD simulations of BJH water, Ruff and Diestler (1990) have slightly changed the parameters of the non–Coulomb O–O interaction [Eq. (14a)] for better agreement with experimental values of liquid water density. One can expect, however, that more profound changes in the intermolecular part of the BJH potential are required to achieve a better description of the P-V-T relations of fluid water in a wide range of temperatures and densities (pressures). Nevertheless, it will be shown below that energetical and dynamical properties of water are reproduced well with this model in the whole range of thermodynamic states studied, which justifies the usage of the BJH potential under high-temperature and high-pressure conditions even in its present form.

Enthalpy and Internal Energy

As in the case of the molar volume, the simulated values of configurational enthalpy [Eq. (6)] show very good agreement with experiment at near-liquid water densities (see Table 4 and Fig. 5). Here again, agreement rapidly deteriorates as pressure decreases along an isotherm. However, if the calculated densities are taken instead of pressures as the basis for comparison with experiment, the agreement still remains good at lower densities (Kalinichev, l991b). That is, in constant-volume MC or MD simulations the TIP4P potential can reproduce correct values of enthalpy (or internal energy) over the whole density-temperature range studied at the price of an error in the pressure. The same conclusion can be drawn from the MD simulations with the BJH potential (Table 5), where quite good agreement with experimental values of internal energy is obtained.

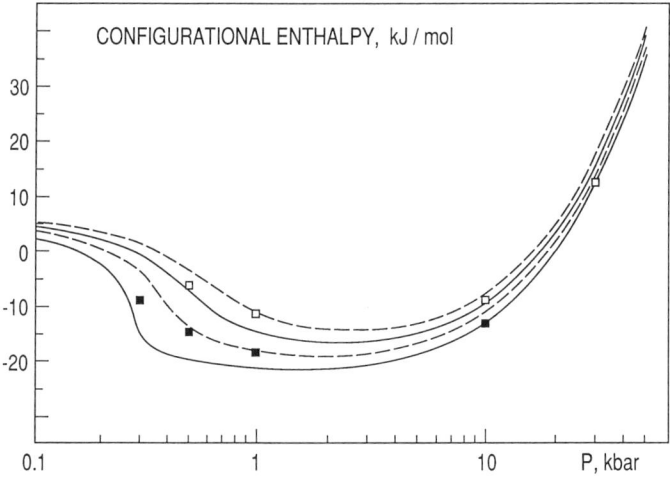

Fig. 5. Simulated (squares) and HGK (Haar et al., 1984) (curves) isotherms of configurational enthalpy (Kalinichev, 1991b). Filled symbols, 673 K; open symbols, 773 K. Full lines, 673 and 773 K; dashed lines, 723 and 823 K.

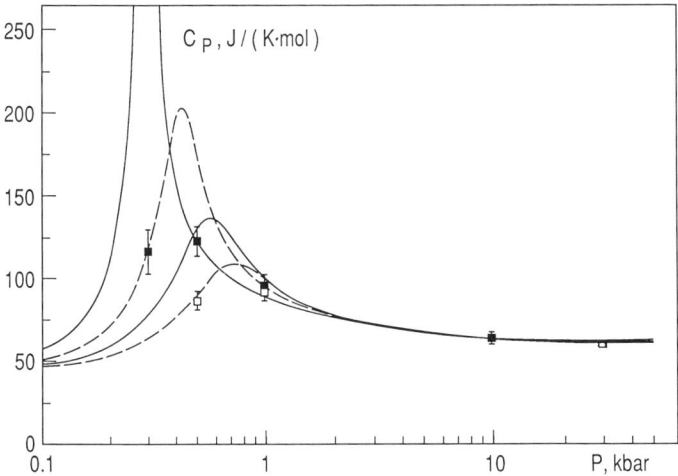

Fig. 6. Pressure dependence of isobaric heat capacity. Denotations as in Fig. 5.

Thus, the two water models used repoduce with reasonable accuracy both temperature and density dependencies of enthalpy and internal energy over a very wide range of thermodynamic parameters.

Heat Capacity

The agreement with experiment of the C_P values calculated from the MC simulations [Eq. (8)] is again excellent at liquidlike densities and becomes significantly worse only at the supercritical maxima of the heat capacity for both isotherms (Table 4 and Fig. 6). As in the case of densities and enthalpies, the experimental 723 and 823 K isotherms (dashed lines in Fig. 6) suggest that the thermodynamic properties of TIP4P water correspond to those of real water at a temperature about 50° higher. The absence of the sharp maximum of C_P in the simulated 673 K isotherm clearly indicates that the critical point for this water model lies at least 50° lower than the real one.

The values of C_V for the BJH potential (Table 5) are also in reasonable agreement with experiment, although this property does not change much under the conditions studied, and the error bounds for the calculated values are difficult to estimate.

Isothermal Compressibility and Thermal Expansion Coefficients

Confirmation of the conclusion above that the critical point of the TIP4P water model lies about 50° lower than the real one can be found from the behavior of κ and α, as calculated from Eqs. (9) and (10). The agreement with experiment is quite good, if we consider that these properties change by several orders of magnitude between the different thermodynamic states investigated (see Table 4

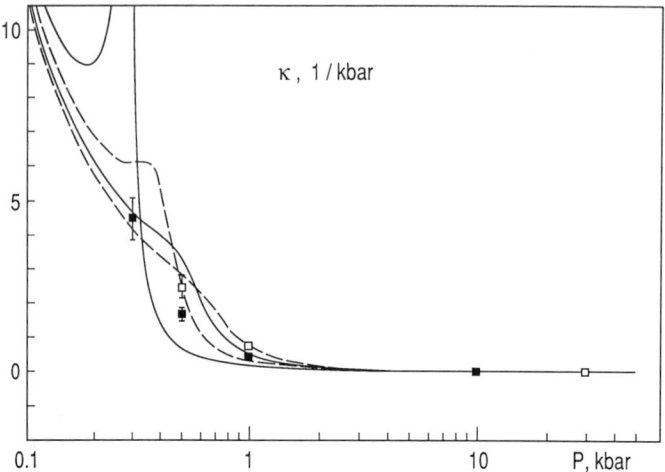

Fig. 7. Pressure dependence of isothermal compressibility. Denotations as in Fig. 5.

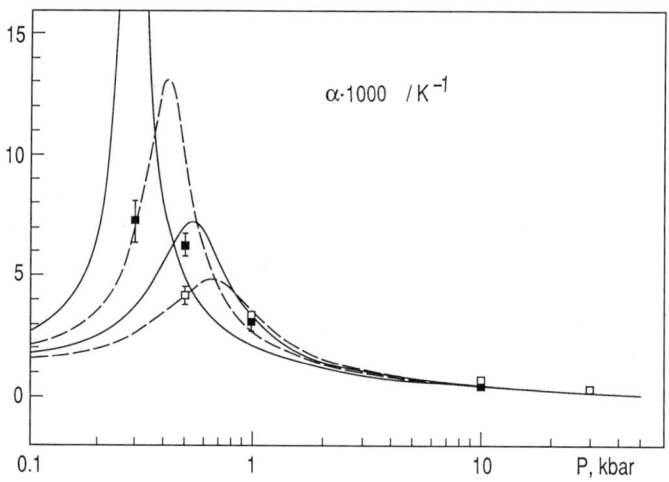

Fig. 8. Pressure dependence of thermal expansivity. Denotations as in Fig. 5.

and Figs. 7 and 8). Once again, the simulated isotherms at 673 and 773 K are much closer to the experimental 723 and 823 K isotherms, respectively (dashed lines in Figs. 7 and 8).

In accordance with earlier simulations with the TIPS2 water model (Jorgensen, 1982a), the simulated thermal expansion coefficient at 30 kbar has been reported to be approximately twice as large as the one calculated from the HGK equation of state (Kalinichev, 1986). It was suggested that the HGK equation may be incorrect in the region above ≈ 20 kbar, because none of the existent shock wave data at high pressures and temperatures were taken into account during its parametrization. This assumption has been confirmed recently by the work of Saul and Wagner (1989), who have shown that the HGK equation can even lead to negative values of the thermal expansion coefficient at higher pressures. Therefore, the values of α given in Table 4 as "experimental" may be underestimated at 30 kbar.

Microthermodynamic Properties

Computer simulations provide a unique opportunity to obtain detailed thermodynamic information on properties that are not readily measured in any experiment. One such property is the bonding energy

$$E_{b,i} = \sum_{j \neq i}^{N} U_{ij}, \quad (17)$$

which represents the energetic environment experienced by a single water molecule.

Bonding Energy Distribution

These distributions are shown in Figs. 9(a) and (b) for all thermodynamic states studied in the present simulations. The thermodynamic conditions of the runs MD1, MD2, and MD3 closely correspond to those of the runs MC1, MC3, and MC5, respectively (see Tables 4 and 5). Despite the difference in the potential models used in the MC and MD simulations, they resulted in the same bonding energy distributions, and therefore, only the results of the MC simulations are shown in Fig. 9. However, the distribution obtained from the run MD4, which has been calculated at the lowest density ever simulated, is shown as the dotted line in Fig. 9(b). The effect of pressure on the distribution along the 673 and 773 K isotherms is clearly seen from these curves. As in the case of normal

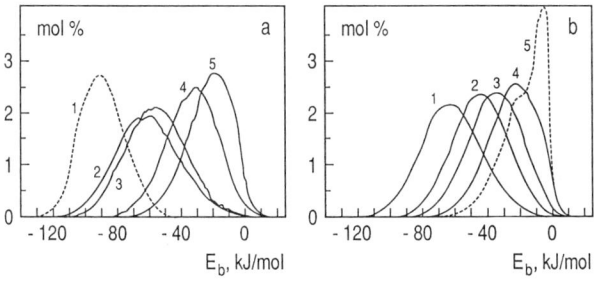

Fig. 9. Normalized distributions of total intermolecular bonding energies for a water molecule in supercritical water. (**a**) 773 K; 2, 30 kbar; 3, 10 kbar; 4, 1 kbar; 5, 0.5 kbar. Dotted line (1) gives the distribution for normal liquid water. (**b**) 673 K; 1, 10 kbar; 2, 1 kbar; 3, 0.5 kbar; 4, 0.3 kbar. Dotted line (5) results from the run MD4 (see Table 5). Units for the ordinate are mol % per *kJ/mol*.

temperature (Madura et al., 1988), the maxima of the distributions shift to lower energies and their widths increase with increasing pressure.

It is interesting to note that under high-temperature conditions a certain number of molecules even have a positive bonding energy. As the fraction of such molecules obviously cannot increase significantly with decreasing pressure, asymmetry of the distribution results at lower densities. At the very low density of the run MD4 [dotted line in Fig. 9(b)], the distribution becomes even bimodal, indicating that a water molecule can be found in two energetically different environments with bonding energies distributed around ≈ -20 and ≈ -5 kJ/mol. While the latter part of the distribution is obviously due to nonbonded, free water molecules, the former one might be considered an indication of the presence of weakly hydrogen-bonded molecules in significant amounts even under these high-temperature, low-density conditions.

The bonding energy distribution for the water molecules in normal liquid water is shown as a dotted line in Fig. 9(a). At 773 K and 10 kbar, the density of supercritical water is virtually the same (≈ 1 g/cm^3) as that of liquid water at 298 K and 1 bar. Therefore, the difference between these two distributions [curves 1 and 3 in Fig. 9[a]) can be considered an effect of temperature alone. At the high temperature, the maximum is shifted by about 40 kJ/mol to higher (less negative) energies and the distribution becomes significantly wider. The comparison of both distributions clearly shows that a water molecule experiences completely different energetic environments in these two thermodynamic states, despite the fact that densities (and hence average intermolecular distances) are virtually the same in both cases.

Pair Energy Distributions

These are another type of microthermodynamic information easily obtainable from computer simulations. Such functions, which represent the distribution of energies between every pair of water molecules (dimer energies), are shown as full lines in Fig. 10 for the thermodynamic states studied in the MC simulations. The distributions calculated from the MD simulations at two different densities are shown in Fig. 10(b) for comparison as the dotted and dashed lines. The difference between the TIP4P and BJH water potentials used in the MC and MD simulations, respectively, seems to be more pronounced here than in the bonding energy distributions discussed above.

Under the supercritical conditions considered here, the temperature is too high to obtain a bimodal distribution known for normal water (e.g., Stillinger and Rahman, 1972) and shown as a dotted line in Fig. 10(a). The maximum at low energies, which reflects the hydrogen-bonded neighbors, has already completely disappeared, but a distinct shoulder is clearly seen at the same range of energies and indicates that hydrogen bonding persists to some extent under the conditions studied. As one would expect, this shoulder is more pronounced at the lower temperature of 673 K [Fig. 10(b)]. With the pressure (density) decrease along each isotherm, the distributions become narrower with higher maxima

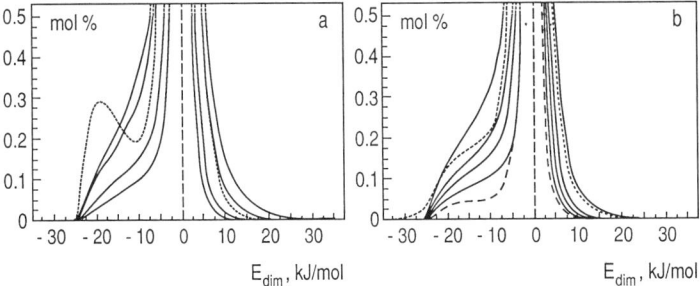

Fig. 10. Normalized distributions of pair interaction energies (dimerization energies) in supercritical water. (**a**) 773 K; full lines (from the top to the bottom), 30, 10, 1, and 0.5 kbar. Dotted line gives the distribution for normal liquid water. (**b**) 673 K; full lines (from the top to the bottom), 10, 1, 0.5, and 0.3 kbar. Dotted and dashed lines result from the runs MD3 and MD4, respectively (see Table 5). Units for the ordinate are mol % per *kJ/mol*.

(beyond the scale in Fig. 10), because relatively more pairs of molecules are found at large intermolecular distances, having a near-zero interaction energy.

A comparison of the distributions at 773 K and 10 kbar and at 298 K and 1 bar [the second curve from the top and the dotted curve in Fig. 10(a), respectively] indicates once again the opportunity to see the pure effect of a significant temperature increase on the shape of the distribution along an isochore, as densities at both thermodynamic states are virtually equal. The width and height of the main maximum remain unchanged. However, in the attractive branch of the distribution (negative energies) a significant amount of molecular pairs are redistributed from the "hydrogen-bonding" range of energies ($\approx -25 - -15$ kJ/mol) to the "non bonding" range ($\approx -15 - -6$ kJ/mol). In the repulsive (positive) branch, the probability for a molecular pair to have a rather high interaction energy between 10 and 20 kJ/mol noticeably increases, confirming that repulsive interactions become a more important contribution to the thermodynamics of water at high temperatures.

Average Potential Energy

Curves representing the distance dependence of the average energy of pair interaction $\langle U(r) \rangle$ are also a source of helpful information characterizing the energetic environment of a single water molecule (Heinzinger and Vogel, 1976). These functions along with the contributions arising from Coulombic and non-Coulombic interactions (full, dotted, and dashed lines, respectively) are shown in Fig. 11 for MD simulations of BJH water at densities of 0.9718 g/cm^3 (a) and 1.346 g/cm^3 (b) and a temperature of about 350 K (Jancsó et al., 1984). The same functions from the present high-temperature MD simulations are shown as dash-dotted lines in Figs. 11 (a) and (b) (runs MD3 and MD1, respectively; see Table 5). It is seen that the value of the average potential energy at the minimum becomes less negative with increasing density or temperature.

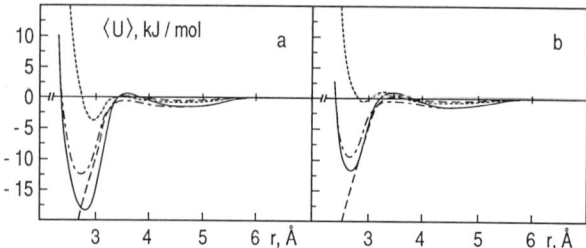

Fig. 11. Average potential energy of a water molecule for normal- (left) and high- (right) density water as a function of the O–O distance: total (solid lines), Coulombic (dashed lines), and non–Coulombic (dotted lines) components (Jancsó, et al., 1984). Dash-dotted lines in (**a**) and (**b**) represent high-temperature simulations MD3 and MD1, respectively (see Table 5).

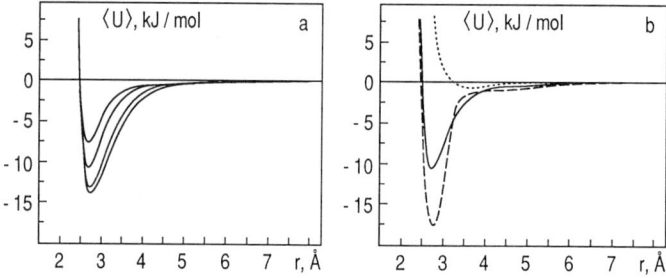

Fig. 12. Distance dependence of the average potential energy between two water molecules for supercritical water. (**a**) 773 K and 30, 10, 1, and 0.5 kbar (from the top to the bottom). (**b**) Full line, 773 K and 10 kbar; dashed line, 298 K and 1 bar. The water density is virtually the same in both thermodynamic states. Dotted line, non–Coulombic contribution to $\langle U(r) \rangle$ (Kalinichev, 1991b).

Average potential energy functions obtained from the MC simulations at 773 K and several pressures are shown in Fig. 12(a). As the TIP4P potential has been used, the only non–Coulombic contribution to the water–water interaction is the oxygen–oxygen Lennard–Jones term [Eq. (12)] and, hence, its contribution to $\langle U(r) \rangle$ is both temperature- and density-independent [shown in Fig. 12(b) as a dotted line]. Therefore, in contrast to the BJE intermolecular potential, where changes with pressure in both Coulombic and non-Coulombic components of the average potential energy occur (Fig. 11), in the case of the TIP4P potential, only the change of the Coulombic part determines the differences between the $\langle U(r) \rangle$ for different thermodynamic states. The general trend is the same: As the pressure (density) decreases, the Coulombic attraction becomes stronger (more negative), as it becomes easier for two water molecules to arrange in an energetically more favorable relative orientation. One can expect, therefore, that in order to describe correctly the thermodynamic properties of supercritical water over a wide enough range of temperatures and pressures with an effective spherically symmetric potential (e.g., Belonoshko and Saxena, 1991), density-dependent parameters of the potential might be necessary.

Together with bonding energy distributions (Fig. 9), the average potential experienced by a single water molecule under various thermodynamic conditions. The full and dashed lines in Fig. 12(b) represent, respectively, the average tions. The full and dashed lines in Fig. 11(b) represent, respectively, the average potential energy for supercritical water at 773 K and 10 kbar, and for water under normal conditions with the TIP4P model employed in both cases. It can be concluded from the comparison of the two curves that the average energetic environment of a water molecule changes drastically as the temperature increases, despite the fact that the average intermolecular distances remain unchanged because of the constant density. This is a consequence of a change in the average orientation because of thermal motion and a redistribution of the intermolecular distances (see the discussion of structural properties below). The shallow minimum around 4.5 Å in the coulombic part of $\langle U(r) \rangle$, which results from the tetrahedral ordering of the molecules in liquid water and remains unchanged even for high-density water at a relativety low temperature of 350 K (Fig. 11), completely disappears under supercritical conditions. The position of the potential minimum remains virtually unchanged while its depth decreases significantly with temperature, indicating that, on the average, water molecules are bonded to each other almost half as strong under supercritical conditions compared with normal liquid water, despite only minor differences in density between both thermodynamic states.

The analysis of the water-water pair energy distributions and average water-water potential energies obtained from the simulations of aqueous electrolyte solutions (Heinzinger and Vogel, 1976; Szász et al., 1981; Szász and Heinzinger, 1983) shows that the effect of ions on these functions is very similar to the effects of temperature and pressure. Therefore, one can expect even more profound changes in these properties in the case of high-temperature, high-pressure electrolyte solutions. The same conclusion can be drawn from the structural properties of water and solutions, which are discussed in the next section.

Structure of Aqueous Fluids at High Temperatures and Pressures

The first properties derived from a computer simulation as far as the structure of an aqueous system is concerned are the various radial distribution functions (RDF), $g_{ij}(r)$. These functions give the probability of finding a pair of atoms i and j a distance r apart, relative to the probability expected for a completely random distribution at the same density. The description of the structure in terms of radial distribution functions is an essential constituent of any statistical-mechanical treatment of a fluid (e.g., McQuarrie, 1976; Hansen and McDonald, 1976; Allen and Tildesley, 1987).

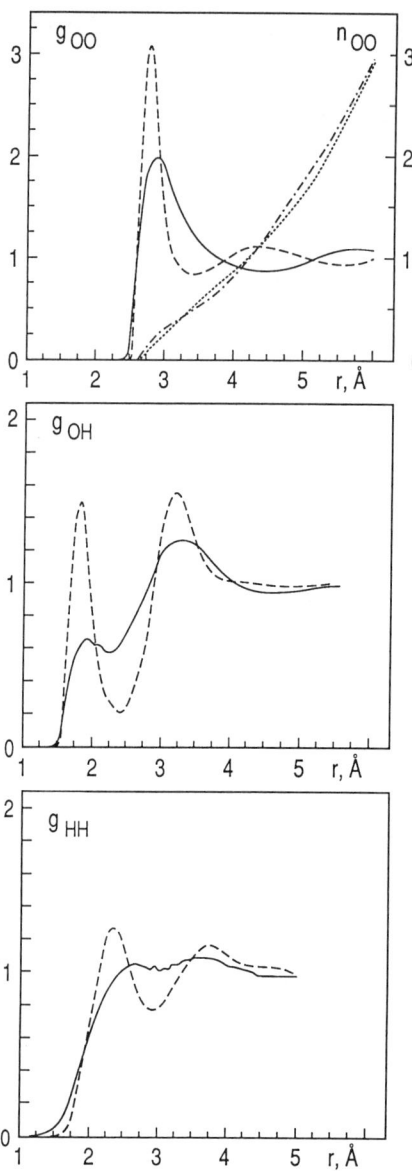

Fig. 13. Temperature dependence of atom–atom radial distribution functions. Full lines, 773 K and 10 kbar; dashed lines, 298 K and 1 bar. The water density is virtually the same in both thermodynamic states. Dotted and dash-dotted lines represent running coordination numbers for the high-temperature and low-temperature states, respectively (Kalinichev, 1991b).

Oxygen–oxygen, oxygen–hydrogen, and hydrogen–hydrogen RDFs for 773 K and 10 kbar (Kalinichev, 1991b), where supercritical water has the same density as liquid water under normal conditions, are shown in Fig. 13 (full lines). Dashed lines represent the MC results for 298 K and 1 bar. Running coordination numbers, calculated as

$$n_{ij}(r) = 4\pi\varrho_0 \int_0^r r^2 g_{ij}(r)\,dr, \tag{18}$$

where ϱ_0 is the number density of particles j, are also shown in Fig. 13 for O–O distributions as dotted and dash-dotted lines corresponding to the high- and low-temperature conditions.

The effect of temperature on the water structure at a constant density can be clearly seen from these curves. The characteristic second maximum of the oxygen–oxygen RDF at ≈ 4.5 Å reflecting the tetrahedral ordering of water molecules due to hydrogen bonding is completely absent under supercritical conditions. Moreover, a pronounced minimum of the distribution appears in its place. The comparison of $g_{OO}(r)$ and $n_{OO}(r)$ at low and high temperatures shows that a significant redistribution of first and second neighbors from the "hydrogen-bonding" regions of ≈ 2.6 to 2.9 Å and ≈ 3.8 to 5.0 Å to the intermediate "nonbonding" region of ≈ 3.1 to 3.8 Å takes place. Tanaka and Ohmine (1987) have shown recently that the water molecular pairs at intermediate distances (≈ 3.1–3.4 Å) are primarily responsible for repulsive interactions in liquid water. Therefore, additional molecular density at these distances gives rise not only to the increase of weakly bonded molecular pairs, but also to the increase of repulsive interactions as well (see Fig. 10 and the discussion of pair interaction energy distributions in the previous section).

The sharp first intermolecular peak of $g_{OH}(r)$ in normal liquid water becomes significantly lower and much less pronounced under supercritical conditions, although its presence alone clearly indicates that hydrogen bonding persists up to 773 K. This peak disappears, with only a shoulder remaining, at a slightly higher temperature (Mountain, 1989).

In Fig. 14 oxygen–oxygen, oxygen–hydrogen, and hydrogen–hydrogen RDFs for the thermodynamic points studied in the supercritical MC simulations (Kalinichev, 1991b, see also Table 4) are presented. The supercritical MD simulations with the BJH potential (Table 5) resulted in RDFs very similar to those obtained from the MC simulations with the TIP4P potential; therefore, they are not presented here. Available experimental data on the X-ray diffraction of water under supercritical conditions at 1 kbar pressure (Gorbaty and Demianets, 1983; 1985) are shown as dashed lines in Figs. 14 (a) and (d) for comparison. Though the agreement is sufficiently satisfactory as far as the heights of the first maximum and the general shape of the curves are concerned, the disagreement in the positions of the main maximum and in the different steepnesses of the repulsive branches of the curves is rather disappointing. The comparison of simulated RDFs at normal conditions with the X-ray data of Gorbaty and Demianets shows a similar discrepancy in the steepness at small distances. A recent determination of the liquid water structure from neutron diffraction measurements (Soper and Phillips, 1986) also gives a less steep repulsion branch of the RDF, as compared with computer simulations even using different water models (e.g., Ruff and Diestler, 1990).

The general distance, temperature, and density dependence of $g_{OO}(r)$ [Figs. 14 (a) and (d)] under supercritical conditions closely resemble that of a simple liquid (e.g., argon), with the second maxima at distances approximately twice as large as the first ones. Little can be said about the behavior of $g_{HH}(r)$ [Figs. 14

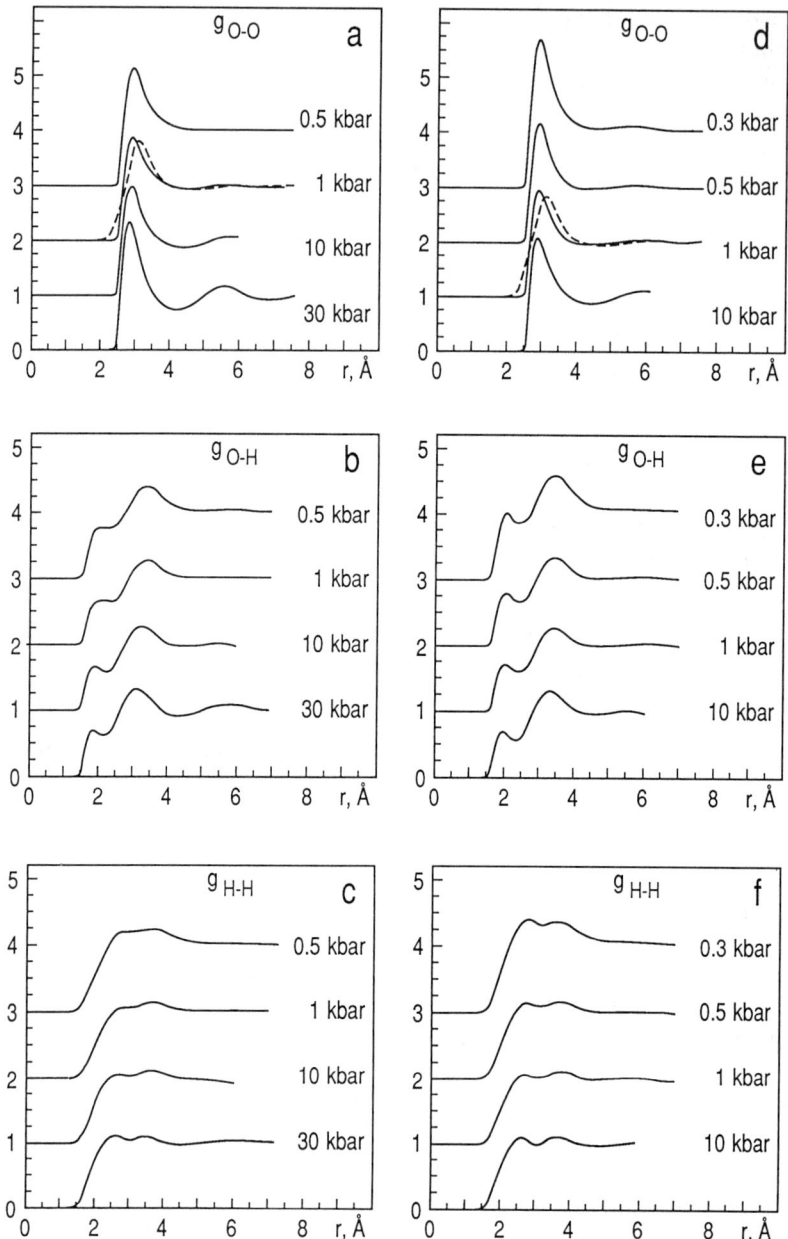

Fig. 14. Simulated atom-atom radial distribution functions (full lines) for supercritical water at 773 K (a, b, c) and 673 K (d, e, f) (Kalinichev, 1991b). Dashed lines represent X-ray diffraction measurements (Gorbaty and Demianets, 1983; 1985; and private communication) under the same conditions. Each curve is shifted by 1 relative to the previous one.

(c) and (f)], except a general notion that these functions show almost no H–H correlations under supercritical conditions. The shape of $g_{OH}(r)$ [Figs. 14 (b) and (e)] remains virtually the same along the isotherms of 673 and 773 K over a very wide range of pressures from 0.3 to 30 kbar. The small first peak at ≈ 1.8 Å indicates that hydrogen bonding still persists under these conditions.

If one adopts the simple geometrical convention that a hydrogen bond exists if the O–H pair is separated by less than 2.4 Å [the position of the first minimum of $g_{OH}(r)$ in normal liquid water; see Fig. 13], the average number of hydrogen bonds N_B experienced by an oxygen atom in a molecule can be easily estimated. Using this definition, Mountain (1989) has found that the number of hydrogen bonds per molecule scales as a single function of temperature in the density range $1.0 \geq \varrho \geq 0.6$ g/cm^3 and the temperature range $270 \leq T \leq 1130$ K. The scaling does not work for densities less than ≈ 0.45 g/cm^3 and breaks down for densities greater than ≈ 1.0 g/cm^3 when N_B saturates at a value of 2, while the density can increase with growing pressure.

The pressure increase has a strong effect on the hydrogen bond structure of water even at moderate temperatures. The projection densities for the eight nearest-neighbor water molecules around a central one and separately for the first four and the second four onto the xy-plane of a coordinate system as defined in the insertion are shown in Fig. 15, calculated from MD simulations of pure BJH water at 1 bar and 22 kbar (Pálinkás et al., 1984). Eight is about the number of nearest neighbors in the high-pressure (HP) case. It can be clearly observed that there is a decrease in the preference for the occupation of tetrahedral positions in the HP water. The neighbors five to eight in HP water are seen to be closer to the central molecule than the second neighbors in the normal-pressure (NP) water, but they do not show any preference for occupying tetrahedral positions. Accordingly, it seems reasonable to conclude that the first coordination sphere of HP water consists of neighbors of two different types. Similar conclusions have been drawn from the results of other MD and MC simulations, in which an interpenetration of the subsections of the random hydrogen bond network has been proposed (Stillinger and Rahman, 1974b; Impey et al., 1981; Madura et al., 1988).

A long standing-question was concerned with the effect of pressure on the hydration sphere of ions. Conflicting conclusions—from a complete breakdown to a strong enhancement—were drawn from various experimental investigations (Heinzinger, 1986). In order to clarify this point, two MD simulations of a 2.2 molal NaCl solution have been performed with the BJH model of water and the Na$^+$-water and Cl$^-$-water pair potentials as given in Table 3. In each case, the basic periodic cube contained 200 water molecules, 8 anions, and 8 cations. The densities of the simulations were 1.0792 and 1.3067 g/cm^3 that are at 298 K equivalent to pressures of 1 bar and 10 kbar, respectively. The simulations extended over about 5 ps. The average temperatures of the simulations were 299 and 303 K for the normal-pressure and high-pressure runs, respectively (Jancsó et al., 1985).

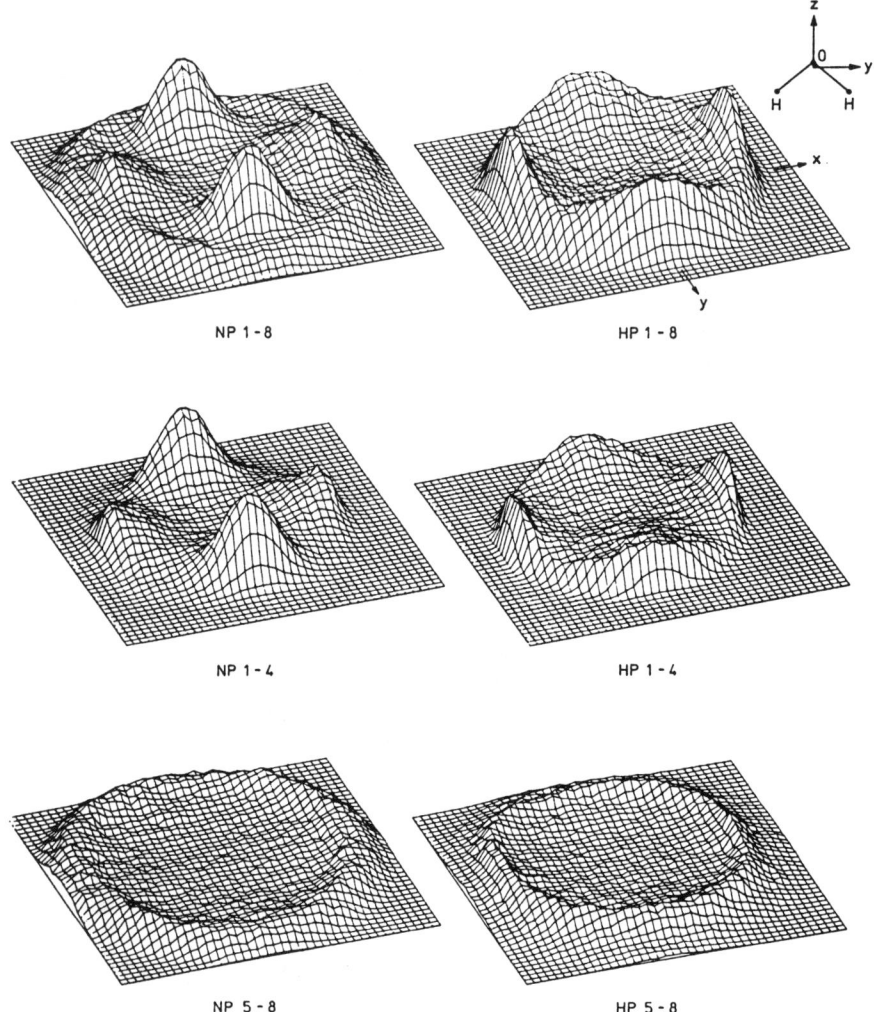

Fig. 15. Densities of the projections of the oxygen atom positions of the eight, and separately for the first and second four, nearest neighbor-water molecules around a central one onto the xy-plane of a coordinate system as defined in the insertion, calculated from MD simulations of pure BJH water at 1 bar (NP, left) and 22 kbar (HP, right) (Pálinkás et al., 1984).

The ion–oxygen and ion–hydrogen RDFs, together with the corresponding running coordination numbers as defined by Eq. (18), are shown in Fig. 16 for the normal-pressure (NP) and high-pressure (HP) simulations. The characteristic values of the RDFs are listed in Table 6. The position and width of the first peak in $g_{NaO}(r)$ remain unchanged, while its height decreases slightly with increasing pressure. The coordination number increases by about 0.5. This small

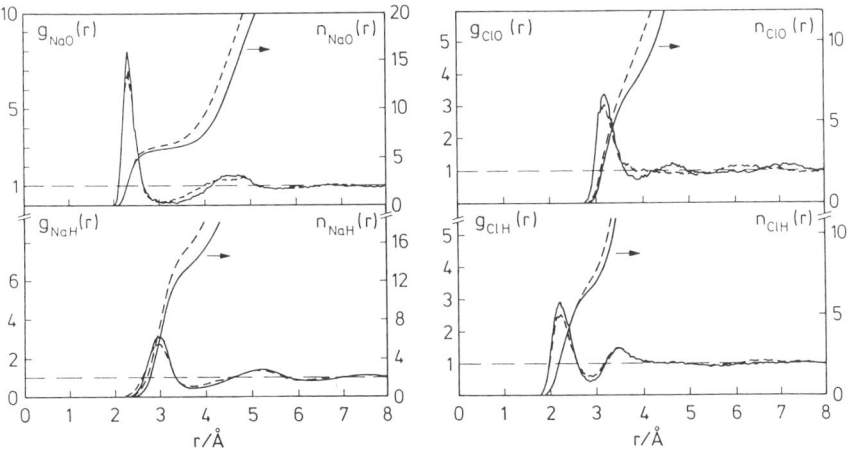

Fig. 16. Ion–oxygen and ion–hydrogen radial distribution functions and running coordination numbers for high-pressure (dashed) and normal-pressure (full) 2.2 molal NaCl solutions (Jancsó et al., 1985).

increase in the coordination number (the density is 21% higher), as well as the slight decrease in peak height, seem to indicate a certain loss of structure of the Na-hydration sphere with increasing pressure. The chloride hydration seems to undergo greater changes with a pressure increase, suggesting that the hydration layer of this ion is even less developed under pressure than it is normally, though the preference for forming linear hydrogen bonds between Cl^- and water is not to be affected significantly (Fig. 17). Unlike Na^+, the hydration number of Cl^- increases by almost two.

The orientation of the water molecules in the hydration shell of an ion can be characterized by the cosine of the angle between the dipole moment direction of the water molecule and the vector pointing from the oxygen toward the center of the ion. In Fig. 17 distributions of $(\cos \Theta)$ are shown for the hydration shells of sodium and chloride ions in the NP and HP solutions. The increase in pressure broadens the $(\cos \Theta)$ distributions. This means that with increasing pressure, a decrease in the preference for the trigonal orientation in the case of Na^- and for linear hydrogen bond formation in the case of Cl^- can be observed.

A second computer simulation in which the effect of pressure on the hydration shells of ions has been investigated is reported by Bounds (1985). He performed MD simulations where a Ca^{2+} and Cl^- were surrounded by 64 TIP4P water molecules at room temperature and also pressures of 1 bar and about 10 kbar. As far as the Ca^{2+} is concerned, the results are very similar to those found for Na^+: no change in the position of the first maximum (r_{M1}) in $g_{CaO}(r)$, a small decrease (hardly outside the limits of uncertainty) for r_{M1} in $g_{CaH}(r)$ and in the average value of Θ, as well as a slightly larger hydration number $(+0.3)$ in the high-pressure case.

The effect of elevated temperature and pressure on the ion–oxygen RDFs has

Table 6. Comparison of characteristic values of the radial distribution functions (intermolecular part) for normal-pressure (1 bar; denoted by NP) and high-pressure (10 kbar; HP) 2.2 molal NaCl solutions (Janscó et al., 1985), as well as for pure water at normal pressure (0.9718 g/cm³; 336 K) and high pressure (1.346 g/cm³; 350 K) (Janscó et al., 1984). R_k, r_{Mk}, and r_{mk} give the distances where $g_{ij}(r) = 1$ has its kth maximum and minimum, respectively. The distances are given in Å with an uncertainty of at least ±0.02 Å. $n_{ij}(r_{m1})$ is the running coordination number at the first minimum of $g_{ij}(r)$.

i	j		R_1	r_{M1}	$g_{ij}(r_{M1})$	R_2	r_{m1}	$g_{ij}(r_{m1})$	r_{M2}	$g_{ij}(r_{M2})$	$n_{ij}(r_{m1})$
Na	O	NP	2.10	2.30	8.00	2.66	3.10	0.11	4.47	1.58	5.8
		HP	2.10	2.30	7.02	2.63	3.15	0.18	4.8	1.46	6.3
Na	H	NP	2.64	2.95	3.10	3.34	3.70	0.45	5.18	1.42	13.9
		HP	2.59	2.93	2.73	3.32	3.68	0.49	5.05	1.35	16.2
Cl	O	NP	2.95	3.18	3.37	3.60	3.87	0.71	4.62	1.24	7.7
		HP	2.91	3.16	3.10	3.8	—	—	—	—	(9.6)[a]
Cl	H	NP	1.98	2.20	2.93	2.60	2.88	0.45	3.50	1.50	6.7
		HP	1.95	2.15	2.45	2.54	2.85	0.57	3.40	1.50	7.2
Pure water											
O	O	NP	2.64	2.85	2.93	3.13	3.28	0.71	4.63	1.23	4.2
		HP	2.60	2.80	2.95	3.49	4.05	0.71	5.48	1.14	12.2 (7.0)[a]
O	H	NP	1.80	1.92	1.40	2.05	2.53	0.19	3.23	1.51	4.0
		HP	1.95	1.98	1.01	2.02	2.40	0.51	3.15	1.56	4.6
H	H	NP	2.17	2.32	1.56	2.73	3.07	0.73	3.72	1.16	6.8
		HP	2.14	2.25	1.92	2.93	3.18	0.90	3.65	1.07	11.5

[a] These numbers result if $n_{ij}(r_{m1})$ is taken at r_{m1} of these NP cases.

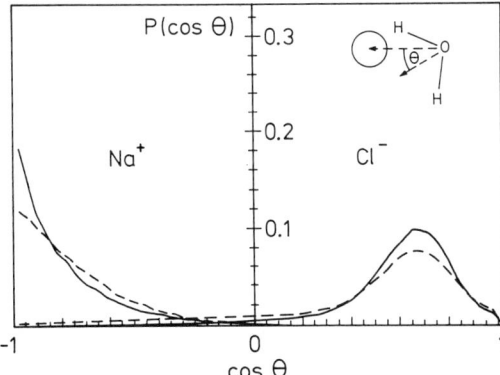

Fig. 17. Distribution of (cos Θ) in the first hydration shells of the ions from MD simulations of 2.2 molal NaCl solutions at normal pressure (solid line) and high pressure (dashed line). Θ is defined in the insertion. The distributions are normalized and given in arbitrary units (Heinzinger, 1986).

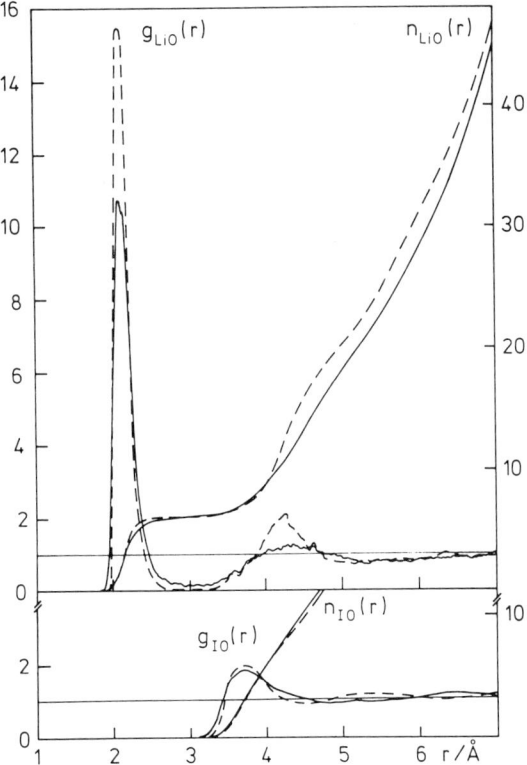

Fig. 18. Ion-oxygen radial distribution functions and running coordination numbers for a 0.55 molal LiI solution at 508 K (full) and 308 K (dashed) with the same density of 1.05 g/cm^3. At the high temperature, the pressure is about 3 kbar (Szász and Heinzinger, 1983a).

been investigated for a 0.55 molal LiI solution by MD simulations at 308 and 508 K and constant density, where the higher temperature corresponds to a pressure of about 3 kbar (Szász and Heinzinger, 1983a). It can be seen from Fig. 18 that the increase in temperature (and pressure) reduces significantly the height of the first peak in $g_{\text{LiO}}(r)$ and broadens it (full line). At the same time, the gap between the first and second hydration shell begins to get filled up. In this way, the number of water molecules (six) in the first hydration shell of Li^+ remains constant, as can be seen from $n_{\text{LiO}}(r = 3\ \text{Å})$. The second hydration shell of Li^+ almost disappears at high temperature and pressure. Accordingly, there is a significant difference between the two curves for $n_{\text{LiO}}(r)$ at $r = 4.5\ \text{Å}$ that slowly disappears with increasing distance. In the case of the iodide ion, the first hydration shell is, even at normal temperature and pressure, not well pronounced (lower part of Fig. 18). Consequently, not much change can be observed. Only a slight smearing out of the first peak results from the temperature (and pressure) increase.

Single Molecule Dynamics in Aqueous Fluids at High Temperatures and Pressures

Unlike the MC method, MD simulations can reveal the particle-averaged short-time behavior of various types of molecular relaxation phenomena. Diffusion coefficients, shear and bulk viscosities, and thermal conductivity can be calculated from the corresponding time correlation functions (see, e.g., Allen and Tildesley, 1987). Different types of molecular motions, such as molecular translations and librations (hindered rotations), as well as intramolecular vibrations if flexible potential models are employed, can also be analyzed in detail (see, e.g., Bopp, 1987; Heinzinger, 1990).

Self-Diffusion Coefficients

The self-diffusion coefficients can be determined from a MD simulation either by the mean square displacement

$$D = \lim_{t \to \infty} \left\langle \frac{[\mathbf{r}(t) - \mathbf{r}(0)]^2}{6t} \right\rangle \tag{19}$$

or, as done here, through the velocity autocorrelation functions (VACF) with the help of the Green–Kubo relation

$$D = \lim_{t \to \infty} \frac{1}{3} \int_0^t \langle \mathbf{v}(0) \cdot \mathbf{v}(t') \rangle\, dt'. \tag{20}$$

The averages have been calculated according to

$$\langle \mathbf{v}(0) \cdot \mathbf{v}(t) \rangle = \frac{1}{N_T N} \sum_{i=1}^{N_T} \sum_{j=1}^{N} \mathbf{v}_j(t_i) \cdot \mathbf{v}_j(t_i + t), \tag{21}$$

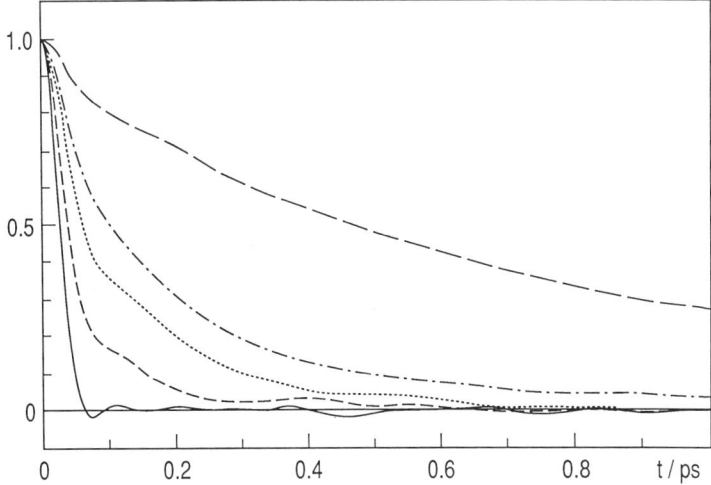

Fig. 19. Normalized velocity autocorrelation functions for water molecules at high temperatures and pressures. MD1 (full), MD3 (dashed), MD5 (dotted), MD2 (dash-dotted), and MD4 (long-dashed). See Table 5 for the thermodynamic conditions of the MD runs.

where N denotes the number of particles, N_T the number of time averages, and $\mathbf{v}_j(t)$ the velocity of particle j at time t.

The velocity autocorrelation functions for water molecules at different temperatures and densities calculated from MD simulations are shown in Fig. 19. The effect of temperature on the VACFs is in close agreement with the results of Stillinger and Rahman (1972) who simulated water at a temperature of about 600 K and a density of 1 g/cm^3 with an earlier version of the ST2 potential, known as BNS. The density dependence of these functions is very similar to that for water at normal temperatures (Jancsó et al., 1984), in agreement with the similarity in the pressure-induced changes of the structural properties as mentioned above. It can be seen from Fig. 19 that the VACFs decay faster at the higher density.

The corresponding self-diffusion coefficients of water, calculated from these simulations, are given in Table 7. In the case of the lowest density studied (run MD4), only a rough estimate of D can be made because of the very slow decay of the velocity autocorrelation function under such conditions (the uppermost curve in Fig. 19). The calculated values of D for the BJH water model employed in the simulations are in good agreement with the simulations of Stillinger and Rahman (1972) for the BNS potential and those of Brodholt and Wood (1990) for the TIP4P potential, as well as with the results of Kataoka (1989) who has also calculated the viscosities and thermal conductivities of water in a wide range of thermodynamic conditions from the MD simulations with the *ab initio* Carravetta–Clementi water potential.

The VACFs for the sodium ions, chloride ions, and solvent water at normal

Table 7. Self-diffusion coefficients of supercritical water.

	Run				
	MD1	MD2	MD3	MD4	MD5
$\varrho/\text{g}\cdot\text{cm}^{-1}$	1.2840	0.5282	0.9718	0.1666	0.6934
T/K	771	772	680	673	630
$D/10^{-5}\cdot\text{cm}^2\cdot\text{s}^{-1}$	11	76	23	>150	37
	—	$(74)^a$	(32)	(137)	(47)

[a]There are no experimental data for D under the thermodynamic conditions of the MD simulations. The values given in parentheses are estimated from the viscosity data via the Stokes–Einstein relation $D = k_B T/(6\pi\eta a)$, where $a = 1.4$ Å is taken as the molecular radius and η is substituted by $\eta' = \zeta\eta$, as described by Hausser et al. (1966). The value of $\zeta = 0.8$ is found to give the best fit of available measurements of viscosity (Dudziak and Franck, 1966) and self-diffusion (Hausser et al., 1966; Krynicki et al., 1978) up to the critical point of water.

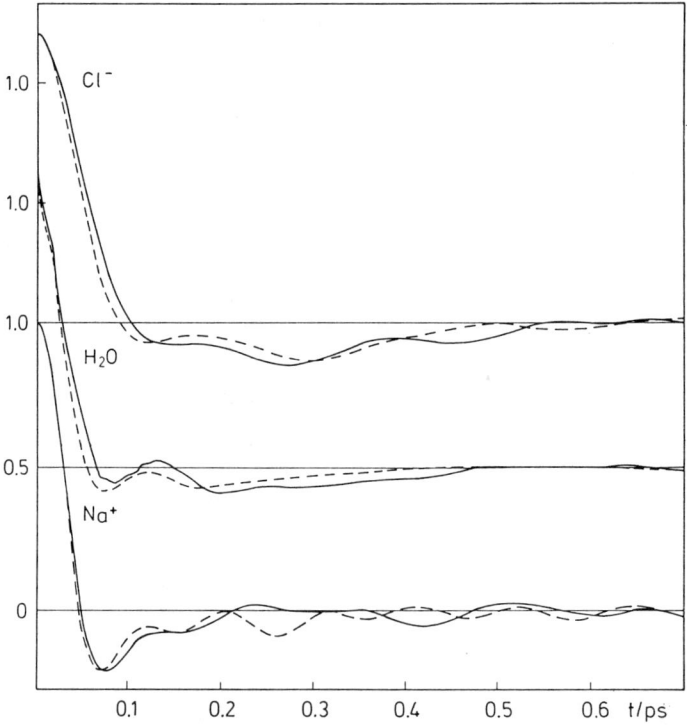

Fig. 20. Normalized velocity autocorrelation functions for Na^+, Cl^-, and oxygen from MD simulations of a normal-pressure (solid line) and high-pressure (dashed line) 2.2 molal NaCl solution (Jancsó et al., 1985).

Table 8. Self-diffusion coefficients of the ions (D_{Na+}, D_{Cl-}) and water (D_w) from MD simulations at normal pressure (NP, 1 bar) and high pressure (HP, 10 kbar) and from experiments at room temperature in units of 10^{-5} cm^2/s (Jancsó et al., 1985).

		D_w	D_{Na+}	D_{Cl-}
MD	NP (299 K)	1.60 (1.83[a])	0.82	1.08
	HP (303 K)	1.51	0.55	1.25
Exp.	NP (298 K)	1.92[b]	1.12[c]	1.60[c]
			1.30[d]	1.85[d]

[a] Extrapolated to 303 K according to Weingärtner (1982).
[b] Tanaka (1975).
[c] Robinson and Stokes (1955).
[d] Hawlicka, private communication.

pressure (1 bar) and high pressure (10 kbar) calculated from MD simulations of 2.2 molal NaCl solutions at normal temperature are shown in Fig. 20 (Jancsó et al., 1985). The pressure dependence found for solvent water is very similar to the one for pure water. In all three cases, the velocity autocorrelation function decays faster at the higher pressure. The pressure dependence of the self-diffusion coefficients of the ions and of solvent water has been calculated from the two simulations of the 2.2 molal NaCl solution with the help of the VACFs according to Eq. (20). The results are compared with the available experimental data in Table 8. Measurements of the pressure dependence of the self-diffusion coefficients of the ions in aqueous solutions are not known.

The self-diffusion coefficient of the solvent water has been corrected for the different temperatures of the NP- and HP-MD simulations by using the temperature dependence of D of pure water (Weingärtner, 1982). The comparison of the self-diffusion coefficients obtained from the MD simulations with the experimental results shows that the simulation leads to smaller values in all cases. By considering the uncertainties in the experimental results for D_{Na+} and D_{Cl-} as indicated by the values reported from different laboratories, the differences between simulation and experiments seem to be small enough to justify the conclusion that the changes in the self-diffusion coefficients for water and the ions with increasing pressure, as calculated from the simulations, are reliable, at least qualitatively.

Table 8 shows that at normal temperature the 21% density increase is accompanied by about a 20% decrease in the self-diffusion coefficient of solvent water D_w. (Separate calculations lead to similar changes for all three water subsystems: bulk water, hydration water of Na$^+$ and of Cl$^-$). In pure water at 350 K, a 38% density increase yielded about a 35% decrease in the self-diffusion coefficient (Jancsó et al., 1984), which suggests that at high pressures the increased steric hindrances between the molecules play an important role in controlling the diffusional motion of water molecules, both in pure water and the solution. In the NP solution, D_{Na+} is found to be significantly smaller than D_{Cl-}, in good agreement with experimental results (Robinson and Stokes, 1955; Hawlicka, private communication). This can be rationalized as in the case of the LiI solution (Szász and Heinzinger, 1983b) by assuming that the Na^+ is moving through the solution with its hydration shell attached, while the Cl^- diffuses essentially without its hydration shell. It is also remarkable that D_{Na+} decreases while D_{Cl-} increases with increasing pressure. It is interesting to note that in a recent study of the effect of pressure on the conductance of KCl in heavy water (Nakahara et al., 1985), it was found that the residual friction coefficient for K^+ slightly increases with pressure, while that for Cl^- decreases in qualitative agreement with the simulation results for the self-diffusion coefficients.

Impey et al. (1983) have estimated the residence times of water molecules in the hydration shells of several alkali and halide ions. As expected, a strong negative correlation with ion size has been found, the residence times decreasing from 33 to 5 ps in the order $Li^+ \to Na^+ \to K^+$ for the cations and as $F^- \to Cl^-$ for the anions studied. However, the simulations at elevated temperature have shown that even for Li^+, the coordination shell becomes significantly less stable with temperature; a more than fivefold decrease of the residence time has been obtained for a temperature increase only up to 368 K.

Hindered Translations

The pronounced effects of temperature and pressure (density) are also reflected in the spectral densities of the hindered translational motions for the water molecules that have been calculated by Fourier transformation

$$f(v) = \sqrt{2\pi c} \int_0^\infty \frac{\langle \mathbf{v}(0) \cdot \mathbf{v}(t) \rangle}{\langle \mathbf{v}(0)^2 \rangle} \cos(2\pi c v t) \, dt \tag{22}$$

from the normalized velocity autocorrelation functions. Such spectra for normal- and high-density simulations at low temperature (Jancsó et al., 1984) are shown in Fig. 21 as the solid and dashed lines, respectively. Similar spectra from high-temperature simulations (Kalinichev and Heinzinger, 1992) are shown in Fig. 21 as dotted and dash-dotted lines, respectively, for the normal- and high-density cases. In normal liquid water, the distinct peak at ≈ 50 cm^{-1} is usually assigned to the hydrogen bond O–O–O flexing motion and the broad peak around ≈ 200 cm^{-1} to O–O stretching motions (see, e.g., Sceats and Rice, 1980). The first peak strongly decreases with isothermally rising density, and both of the peaks have completely disappeared at high temperature, indicating a significant breakdown of the hydrogen bond structure of water due to increased density and temperature. The same temperature dependence has been observed already by Stillinger and Rahman (1972, 1974a) in MD simulations of water at elevated temperatures using BNS and ST2 potentials. Both the low- and high-

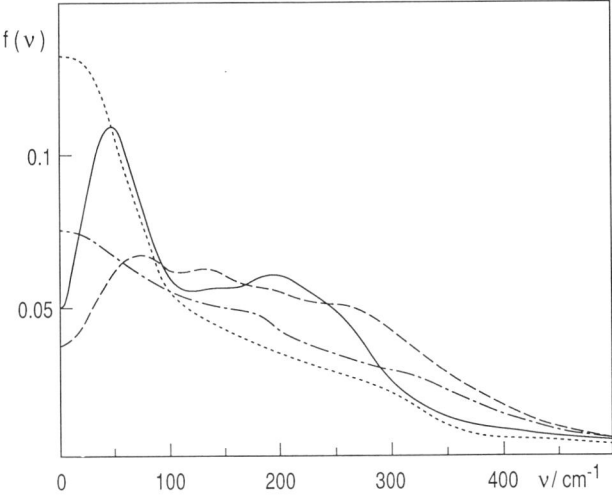

Fig. 21. Fourier transform of the VACFs, in arbitrary units, for normal- (full) and high- (dashed) density water at low temperatures (Jancsó et al., 1984). Similar high-temperature spectra for normal- (dotted) and high- (dash-dotted) density water are calculated from the MD3 and MD1 runs, respectively.

temperature simulations show in the region between 300 and 400 cm^{-1} at high density about twice the intensity than at normal density.

It is interesting to note that the effects of pressure and temperature on the spectral densities in the frequency region corresponding to the hydrogen bond bending and stretching motions are very similar to that of Li$^+$ as obtained from an MD simulation of a 2.2 molal LiI solution with the ST2 water model (Szász and Heinzinger, 1983b). The changes with pressure in the spectral densities of Na$^+$ and Cl$^-$ calculated by Fourier transformation [Eq. (22)] of the corresponding velocity autocorrelation functions (Fig. 20) are rather small. They do not give any hint for understanding the different pressure dependencies of the self-diffusion coefficients of the two ions discussed above (Jancsó et al., 1985).

Librational Motions and Intramolecular Vibrations

As the BJH water model used in the MD simulations at high temperatures and pressures allows distortions of the individual water molecules, changes in the average geometry of the molecules and in the modes of their intramolecular motions brought about by the changes of the thermodynamic conditions can be calculated. In order to separate the various modes of molecular librations (hindered rotations) and vibrations, the following scheme has been employed (Bopp, 1986; Spohr et al., 1988). The instantaneous velocities of the two hydrogen atoms in the center-of-mass system are projected onto the instantaneous unit vectors: 1) in the direction of the corresponding O–H bond ($\mathbf{u}_1$ and $\mathbf{u}_2$), 2) perpendicular to the O–H bonds in the plane of the molecule (v_1 and v_2), and 3) perpendicular to the plane of the molecule (p_1 and p_2).

By using capital letters to denote the projections of the hydrogen velocities onto the corresponding unit vectors, the following quantities are defined:

$$\begin{aligned} Q_1 &= U_1 + U_2, & R_x &= V_1 - V_2, \\ Q_2 &= V_1 + V_2, & R_y &= P_1 + P_2, \\ Q_3 &= U_1 - U_2, & R_z &= P_1 - P_2, \end{aligned} \qquad (23)$$

where Q_1, Q_2, and Q_3 describe approximately the three normal mode vibrations

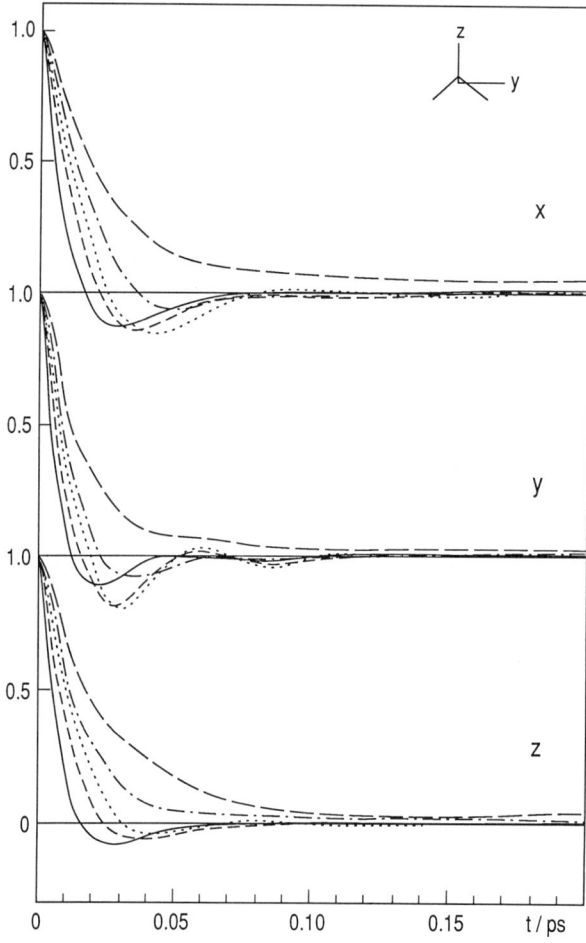

Fig. 22. Velocity autocorrelation functions for the three librations of water molecules at high temperatures and different densities. MD1 (full), MD3 (dashed), MD5 (dotted), MD2 (dash-dotted), and MD4 (long-dashed). See Table 5 for the thermodynamic conditions of the MD runs. The librations are calculated separately for the three components in a molecule-fixed coordinate system as defined in the insertion.

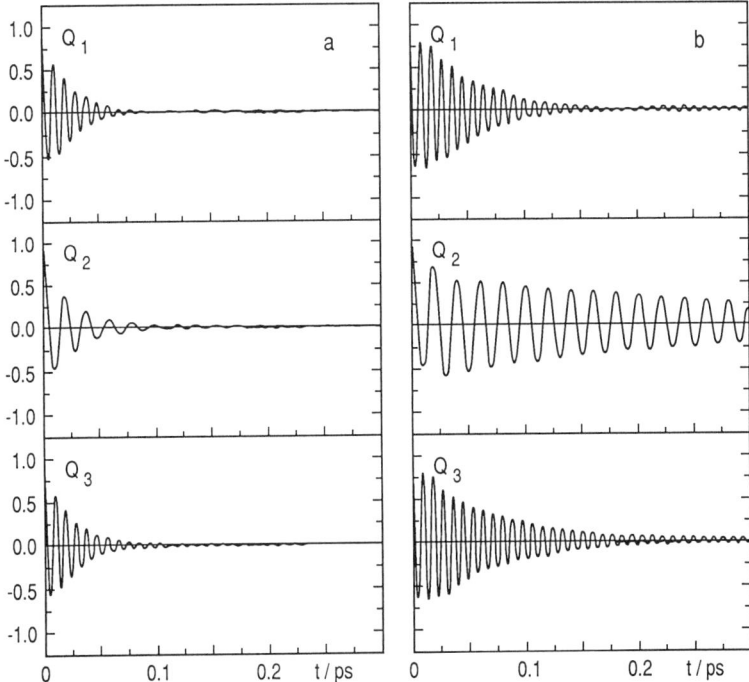

Fig. 23. Velocity autocorrelation functions for the three intramolecular vibrations of water molecules at high temperatures and highest (**a**) and lowest (**b**) densities studied (runs MD1 and MD4, respectively). Q_1, Q_2, and Q_3 denote the symmetric stretching, bending, and asymmetric stretching modes, respectively.

usually referred to as symmetric stretch, bend, and asymmetric stretch, respectively. R_x, R_y, and R_z approximate the instantaneous rotations around the three principal axes of the water molecule, as defined in the insertion of Fig. 22. The normalized velocity autocorrelation functions calculated in this way for water molecules under the supercritical thermodynamic conditions studied in the MD simulations (see Table 5) are shown in Figs. 22 and 23. In the case of vibrational modes, only the results for the highest and lowest densities studied are presented. The differences between x, y, and z reflect the different moments of inertia around the three axes.

The Fourier transforms of these velocity autocorrelation functions result in the spectral densities of the corresponding modes. They are shown in Figs. 24 and 25 for the librational and vibrational modes, respectively. The frequencies of the vibrational peak maxima are given in Table 9. In the case of the symmetric stretching modes (v_1), the widths at half-height, $\Delta v_{1/2}$, are given additionally and compared with available IR and Raman spectroscopic data. Because of the rather short simulation times of the high-temperature MD runs, the error bounds for the frequencies given in Table 9 are estimated to be about $\pm 30 \, \text{cm}^{-1}$.

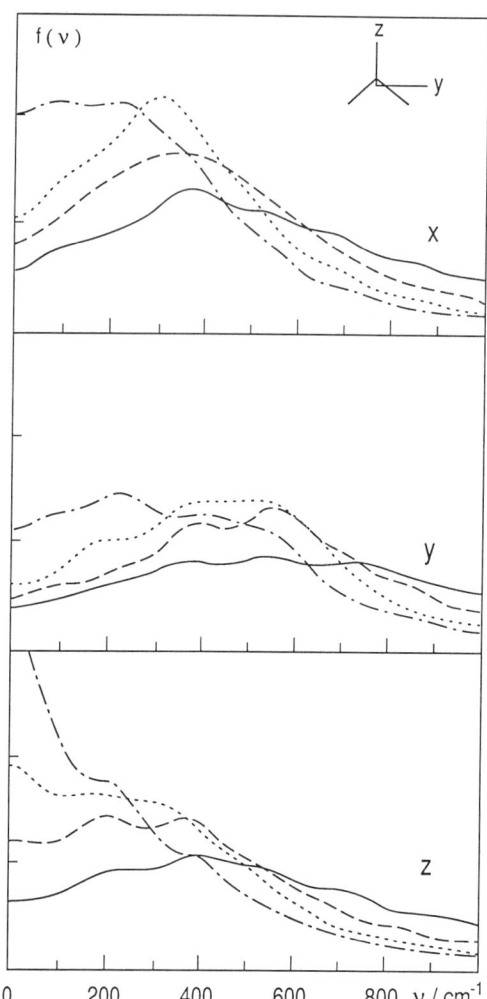

Fig. 24. Spectral densities (in arbitrary units) of molecular librations in high-temperature water at different densities. See Fig. 22 for the notation.

At high temperatures, a density decrease causes a significant redshift of the librational frequencies around the x- and y-axes (Fig. 24), and the maxima completely disappear at the lowest density studied (not shown in Fig. 24). The same trend is seen for the librations around the dipole moment axis (z), where the maximum disappears already at much higher densities, which means that the water molecule is able to rotate freely around this axis, even when its rotation around the other two axes is still hindered by the neighboring molecules.

The average intramolecular geometrical parameters of the water molecules at high temperatures and pressures as given in Table 9 can be compared with the corresponding values for an isolated BJH water molecule and for the molecules in normal liquid water, which are 0.9572 and 0.9755 Å for the intramolecular O–H distance, 104.52 and 100.78° for the intramolecular H–O–H angle, and

Fig. 25. Spectral densities (in arbitrary units) of the intramolecular vibrations of water molecules from high-temperature MD simulations. v_1, v_2, and v_3 denote the symmetric stretching, bending, and asymmetric stretching modes, respectively. See Fig. 22 for the notation.

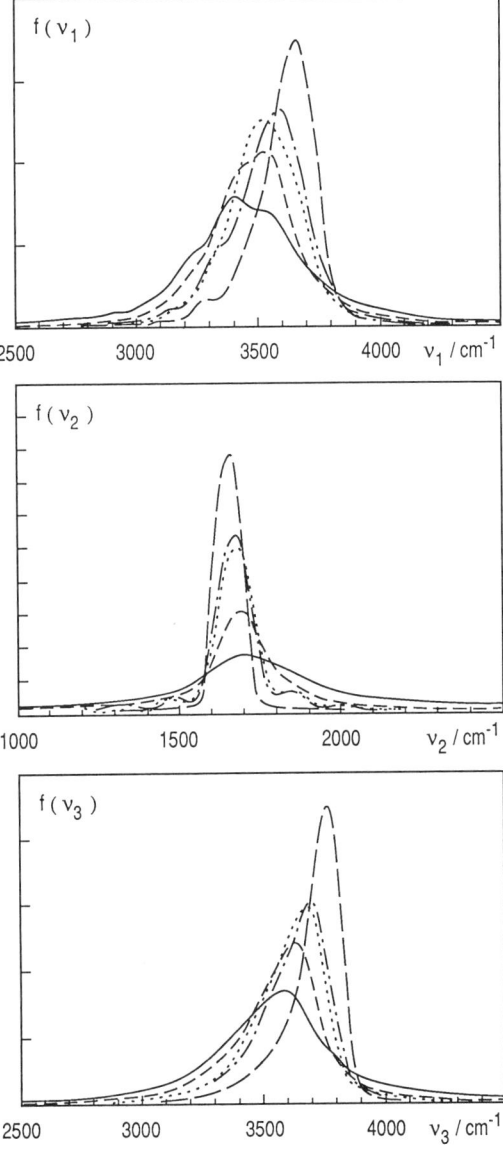

1.86 and 1.97 D for the dipole moment, respectively (Jancsó et al., 1984). Both the increases of temperature and of density lead to an increase in the intramolecular O–H distance and a decrease in the H–O–H angle. These changes in geometry, together with the partial charges on the O and H atoms, lead to the temperature and density dependence of the dipole moment for the water molecules. From an empirical relationship for the rate of decrease of the O–H stretching frequency with the intramolecular O–H distance (≈ 20.000 cm^{-1}/Å)

Table 9. Average intramolecular geometry, dipole moment, and frequencies of vibrational modes in supercritical water derived from MD simulation.[a]

	Run				
	MD1	MD2	MD3	MD4	MD5
$\varrho/\text{g}\cdot\text{cm}^{-1}$	1.2840	0.5282	0.9718	0.1666	0.6934
T/K	771	772	680	673	630
$\langle R_{OH}\rangle/\text{Å}$	0.9811	0.9750	0.9781	0.9705	0.9755
$\langle \angle HOH\rangle/°$	98.26	100.82	99.51	102.00	100.28
$\langle \mu\rangle/\text{D}^{b}$	2.07	2.02	2.05	1.99	2.03
v_1^{max}/cm^{-1}	3415	3580	3530	3640	3500
	—	—	(3550[c])	(3600[d])	(3569[d])
$\Delta v_{1/2}/\text{cm}^{-1}$	500	290	350	230	350
	(200[e])	(70[e])	(180[e])	(50[c]; 170[d])	(200[d])
		(170[f])	(170[f])	(150[f])	(160[f])
v_2^{max}/cm^{-1}	1730	1670	1700	1660	1690
v_3^{max}/cm^{-1}	3570	3690	3630	3760	3675

[a] In parentheses available experimental data are given for comparison.
[b] 1 Debye (D) $\equiv 10^{-18}$ esu cm = 3.33564×10^{-30} C m.
[c] Raman spectroscopic measurements for H_2O and HDO (Lindner, 1970).
[d] IR-data (Gorbaty, 1979, and private communication).
[e] Extrapolation of the Raman data of Lindner (1970).
[f] IR-data for HDO (Franck and Roth, 1967; Roth, 1969).

(La Placa et al., 1973), a redshift of about 55 cm^{-1} relative to liquid water at normal temperature is expected for the vibrational frequencies of the run MD3 at a temperature of 680 K, but at a density close to that of normal water. However, a blueshift of the same magnitude is estimated, based on the symmetric stretching frequency for normal liquid BJH water ($v_1 = 3475$ cm^{-1}) (Bopp, 1987) in reasonable agreement with the experimental data of Lindner (1970), from which a blueshift of about 120 cm^{-1} might be estimated. These results seem to indicate that the simple empirical relationship is not valid any more when both temperature and pressure (density) can influence independently the average intramolecular geometry.

The temperature and density dependence of the position of the maximum in the spectral densities of the symmetric stretching frequency v_1 (Fig. 25 and Table 9) is in a good agreement with available experimental data, whereas all simulated spectra are approximately twice as wide at half-height as the corresponding experimental ones. The very weak temperature and density dependence of the bending frequency v_2 is in agreement with the Raman measurements of Ratcliffe and Irish (1982) who found that this frequency (1638 cm^{-1}) does not change within experimental error up to 573 K. No experimental data exist for the asymmetric stretching frequencies v_3 for H_2O under the thermodynamic conditions of the MD simulations reported. However, the temperature and density dependence of the calculated spectra is very similar to those obtained in IR measurements of v_3 for HDO at high pressures and temperatures (Bondarenko and Gorbaty, 1973).

As in the case of the water structure, discussed above, the observed effects of temperature and pressure on the librational and vibrational motions of water molecules are very similar to those of ions (Heinzinger, 1990; Tamura et al., 1988). For example, Mills et al. (1986) have found that the vibrational spectrum simulated for a 0.87 molal KCl solution at 298 K is similar to that calculated for pure water at 350 K (Reimers and Watts, 1984).

Conclusions

In this chapter, the concept and application of computer simulation methods are presented with particular attention to aqueous systems at high temperatures and pressures. The statistical mechanical formulas for thermodynamic, structural, and kinetic properties are presented, including the treatment of time correlation functions for the analysis of the spectroscopic properties of water. The current status of available intermolecular potentials for aqueous electrolyte solutions is briefly summarized. Particular emphasis is placed on the properties of supercritical water in a wide range of densities.

It is obvious that computer simulation studies of water and aqueous solutions at high temperatures and pressures are just at their earliest stage as far as geochemical relevance is concerned. However, the results presented here seem to demonstrate that simulation methods can contribute significantly to a better understanding of the properties of hydrothermal and metamorphic fluids.

There is undoubtedly strong need for the improvement of the potentials employed in the simulations, since all empirical intermolecular potentials used so far in computer simulations have been parameterized to reproduce correctly a set of liquid water properties under normal conditions. This leads to some quantitative disagreements between the simulated and measured properties of aqueous systems at high temperatures and pressures. The reason of these disagreements lies, most probably, in the "effective" nature of the potentials used, which takes into account the many-body effects of intermolecular interactions only implicitly. Many-body effects can be accounted for by the polarization models of water, which can also describe the dissociation of water molecules (Stillinger and David, 1978). The effect of dissociation, which can be neglected under the thermodynamic conditions presently studied (Holzapfel, 1969; Tanger and Pitzer, 1989), has obviously to be taken into account at higher temperatures that are of geochemical interest (Helgeson, 1981; Sverjensky, 1987). So far, however, only one simulation of this kind has been performed (Demetros and David, 1982).

The direct successive inclusion of three- and higher-body terms from *ab initio* calculations (Clementi, 1985), though very impressive, can hardly be considered cost-effective in terms of computer time for most large-scale simulations, and such possibilities like the reparametrization of empirical potentials for various ranges of temperature and density seem to be preferable. However, even in their present form, available intermolecular potentials can provide invaluable qualitative information on the correlations between the thermodynamic, structural, transport, and spectroscopic properties of hydrothermal fluids on a fundamental molecular level. The ability to predict such correlations, which cannot be easily studied in the laboratory, seems to be a much more important feature of simulation methods than the ability to reproduce available experimental data within experimental error bounds.

There are several obvious areas of research where computer simulations of the sort described in this chapter can be used most effectively in the near future. First of all, they could easily be extended to study the thermodynamic, structural, transport, and spectroscopic properties of complex hydrothermal solutions of electrolytes and nonelectrolytes. The effects of ion association, which are very important for the understanding of mineral solubility and speciation in supercritical fluids (Sverjensky, 1987; Eugster and Baumgartner, 1987), can be analyzed on a molecular level by these methods (e.g., Berkowitz et al., 1984; Dang et al., 1990). Such simulations could develop to the "Born–Oppenheimer" level of solution models (Friedman, 1981) that might be considered a valuable supplement to the "McMillan–Mayer" type of models presently used in the geochemical literature (e.g., Helgeson, 1981; Pitzer, 1987).

Another area of future applications is the computer simulation of water/mineral interfaces (see e.g., Lasaga, 1990). The structural and dynamical properties of interfacial water differ from those of bulk water (Spohr and Heinzinger, 1988). This effect might be important for the analysis of the composition and behavior of aqueous fluids in small inclusions (e.g., Roedder, 1981) or in cracks. As the investigations showed that the effect of the wall on the properties of water does not extend beyond more than three molecular layers, a significant influence on the fluid will only be recognized when the cracks are narrow and the volume of the inclusions is very small. The size dependence of several properties of water in quartz micropores in the range 8 to 80 Å at high temperatures has been investigated recently through MD simulations by Belonoshko and Shumlovich (1987). Of geochemical relevance might also be the MD studies of water on β-Tridymite (Grivtsov et al., 1988), on a model mineral surface corresponding to that of mica (Kjellander and Marčelja, 1985), and on an insoluble NaCl crystal (Anastasiou et al., 1983).

Water and aqueous solutions can also react with surfaces. Computer simulations dealing with problems of this kind have been scarce up to now. The reaction of water with a silica surface was studied by Garofalini (1990). Fukushima et al. (1991) followed the kinetics of the dissolution process of several alkali halide crystals by MD. On the other hand, the nucleation of crystals from a supersaturated NaCl solution has been studied by Schwendinger and Rode (1989).

So far, the computer simulation studies of silicate melts (Kubicki and Lasaga,

1991) and aqueous solutions have developed in physical geochemistry quite independently. It is well known, however, that dissolved volatile components, water being the most abundant of them, have a very strong effect on the properties of melts (e.g., Persikov et al., 1990; Lange and Carmichael, 1990). Therefore, the computer simulations of water in silicate melts and glasses seem to be yet another attractive area of future applications, where molecular mechanisms of the effects of volatiles on the thermodynamic, structural, transport, and spectroscopic properties of magmas can be revealed.

Being the source of information intermediate between theory and experiment, computer simulation methods could obviously greatly stimulate the development of both theoretical and experimental studies of geochemical fluids in the near future. Thus, the prospects of computer simulations in physical geochemistry look very promising, if we take into account the rapid progress in computer technology, which makes powerful computers less expensive and accessible for extensive simulation studies by almost any geochemical laboratory.

Acknowledgments

This study was carried out during the research stay of one of the authors (A.K.) at the Max-Planck-Institut für Chemie (Otto-Hahn-Institut) within the scope of the Alexander von Humboldt Fellowship Program. Financial support by the Alexander von Humboldt-Foundation and Deutsche Forschungsgemeinschaft is gratefully acknowledged.

References

Ahlström, P., Wallqvist, A., Engström, S., and Jönsson, B. (1989). A molecular dynamics study of polarizable water. *Mol. Phys.* **68**, 563–581.

Allen, M.P. and Tildesley, D.J. (1987). *Computer Simulation of Liquids*. Oxford University Press, New York.

Alper, H.E. and Levy, R.M. (1989). Computer simulations of the dielectric properties of water: Studies of the simple point charge and transferrable intermolecular potential models. *J. Chem. Phys.* **91**, 1242–1251.

Anastasiou, N., Fincham, D., and Singer, K. (1983). Computer simulation of water in contact with a rigid-ion crystal surface. *J. Chem. Soc., Faraday Trans. II* **79**, 1639–1651.

Anderson, J., Ullo, J.J., and Yip, S. (1987). Molecular dynamics simulation of dielectric properties of water. *J. Chem. Phys.* **87**, 1726–1732.

Andrea, T.A., Swope, W.S., and Andersen, H.C. (1983). The role of long ranged forces in determining the structure and properties of liquid water. *J. Chem. Phys.* **79**, 4576–4584.

Barker, J.A. and Watts, R.O. (1969). Structure of water; a Monte Carlo calculation. *Chem. Phys. Lett.* **3**, 144–145.

Barrat, J.-L. and McDonald, I.R. (1990). The role of molecular flexibility in simulations of water. *Mol. Phys.* **70**, 535–539.

Belonoshko, A. and Shmulovich, K.I. (1987). A molecular dynamics study of a dense fluid in micropores. *Geochem. Internat.* **24**(6), 1–12.

Belonoshko, A. and Saxena, S.K. (1991). A molecular dynamics study of the P-V-T properties of supercritical fluids: I. H_2O. *Geochim. Cosmochim. Acta* **55**, 381–387.

Berendsen, J.C., Postma, J.P.M., Van Gunsteren, W.F., and Hermans, J. (1981). Interaction models for water in relation to protein hydration, in *Intermolecular Forces*, B. Pullman, ed., Riedel, Dordrecht, pp. 331–342.

Berens, P.H., Mackay, D.H.J., White, G.M., and Wilson, K.R. (1983). Thermodynamics and quantum corrections from molecular dynamics for liquid water. *J. Chem. Phys.* **79**, 2375–2389.

Berkowitz, M., Karim, O.A., McCammon, J.A., and Rossky, P.J. (1984). Sodium chloride ion pair interaction in water: Computer simulation. *Chem. Phys. Lett.* **105**, 577–580.

Beshinske, R.J. and Lietzke, M.H. (1969). Monte Carlo calculation of some thermodynamic properties of steam using a dipole-quadrupole potential. *J. Chem. Phys.* **51**, 2278–2279.

Beveridge, D.L., Mezei, M., Mehrotra, P.K., Marchese, F.T., Ravi-Shanker, G., Vasu, T., and Swaminathan, S. (1983). Monte Carlo computer simulation studies of the equilibrium properties and structure of liquid water, in *Molecular-Based Study of Fluids*, J.M. Haile and G.A. Mansoori, eds., *Advances in Chemistry Series*, Vol. 204, American Chemical Society, Washington, D.C., pp. 297–351.

Binder, K. and Stauffer, D. (1984). A simple introduction to Monte Carlo simulation and some specialized topics, in *Applications of the Monte Carlo Method in Statistical Physics*, K. Binder, ed., *Topics in Current Physics*, Vol. 36, Springer-Verlag, Berlin, pp. 1–36.

Bondarenko, G.V. and Gorbaty, Yu. E. (1973). Infrared spectra of v_3 HDO at high pressures and temperatures. *Dokl. Akad. Nauk SSSR* **210**, 132–135 (in Russian).

Bopp, P. (1986). A study of the vibrational motions of water in an aqueous $CaCl_2$ solution. *Chem. Phys.* **106**, 205–212.

Bopp, P. (1987). Molecular dynamics computer simulations of solvation in hydrogen bonded systems. *Pure & Appl. Chem.* **59**, 1071–1082.

Bopp, P., Jancsó, G., and Heinzinger, K. (1983). An improved potential for non-rigid water molecules in the liquid phase. *Chem. Phys. Lett.* **98**, 129–133.

Bounds, D.G. (1985). A molecular dynamics study of the structure of water around the ions Li^+, Na^+, K^+, Ca^{++}, Ni^{++} and Cl^-. *Mol. Phys.* **54**, 1335–1355.

Brodholt, J. and Wood, B. (1990). Molecular dynamics of water at high temperatures and pressures. *Geochim. Cosmochim. Acta* **54**, 2611–2616.

Carney, G.D., Curtiss, L.A., and Langhoff, S.R. (1976). Improved potential functions for bent AB_2 molecules: water and ozone. *J. Mol. Spectr.* **61**, 371–379.

Carravetta, V. and Clementi, E. (1984). Water-water interaction potential: An approximation of electron correlation contribution by a functional of the SCF density matrix. *J. Chem. Phys.* **81**, 2646–2651.

Catlow, C.R.A., Parker, S.C., and Allen, M.P., eds. (1990). *Computer Modelling of Fluids, Polymers and Solids*. Kluwer Academic Publishers, Dordrecht.

Clementi, E. (1985). *Ab initio* computational chemistry. *J. Phys. Chem.*, **89**, 4426–4436.

Corliss, J.B. (1990). Hot springs and the origin of life. *Nature* **347**, 624.

Dang, L.X. and Pettitt, B.M. (1987). Simple intermolecular model potentials for water. *J. Phys. Chem.* **91**, 3349–3354.

Dang, L.X., Rice, J.E., and Kollman, P.A. (1990). The effect of water models on the interaction of the sodium-chloride ion pair in water: Molecular dynamics simulations. *J. Chem. Phys.* **93**, 7528–7529.

Demetros, G. and David, C.W. (1982). Polarization model for high temperature and high density water droplets. *J. Chem. Phys.* **77**, 6340–6341.
De Pablo, J.J. and Prausnitz, J.M. (1989). Phase equilibria for fluid mixtures from Monte Carlo simulation. *Fluid Phase Equil.* **53**, 177–189.
De Pablo, J.J., Prausnitz, J.M., Strauch, H.J., and Cummings, P.T. (1990). Molecular simulation of water along the liquid–vapor coexistence curve from 25°C to the critical point. *J. Chem. Phys.* **93**, 7355–7359.
Dudziak, K.H. and Franck, E.U. (1966). Messungen der Viskosität des Wassers bis 560°C und 3500 bar. *Ber. Bunsenges. Phys. Chem.* **70**, 1120–1128.
Eugster, H.P. and Baumgartner, L. (1987). Mineral solubilities and speciation in supercritical metamorphic fluids, in *Thermodynamic Modeling of Geological Materials: Minerals, Fluids and Melts*, I.S.E. Carmichael and H.P. Eugster, eds., *Reviews in Mineralogy* Vol. 17, Mineralogical Society of America, Washington, D.C., pp. 367–403.
Evans, M.W. (1986). Molecular dynamical simulation of new auto and cross correlations in liquid water. *J. Mol. Liq.* **32**, 173–181.
Evans, M.W., Lie, G.C., and Clementi, E. (1988). Molecular dynamics simulation of water from 10 to 1273 K. *J. Chem. Phys.* **88**, 5157–5165.
Ferry, J.M. and Baumgartner, L. (1987). Thermodynamic models of molecular fluids at the elevated pressures and temperatures of crustal metamorphism, in *Thermodynamic Modeling of Geological Materials: Minerals, Fluids and Melts*, I.S.E. Carmichael and H.P. Eugster, eds., *Reviews in Mineralogy* Vol. 17, Mineralogical Society of America, Washington, D.C, pp. 323–365.
Franck, E.U. (1970). Water and aqueous solutions at high pressures and temperatures. *Pure & Appl. Chem.* **24**, 13–30.
Franck, E.U. (1981). Survey of selected non-thermodynamic properties and chemical phenomena of fluids and fluid mixtures, in *Chemistry and Geochemistry of Solutions at High Temperatures and Pressures*, D.T. Rickard and F.E. Wickman, eds., *Physics and Chemistry of the Earth*, Vol. 13/14, Pergamon Press, Oxford, pp. 65–88.
Franck, E.U. (1987). Fluids at high pressures and temperatures. *Pure & Appl. Chem.* **59**, 25–34.
Franck, E.U. and Roth, K. (1967). Infrared absorption of HDO in water at high pressures and temperatures. *Disc. Farad. Soc.* **43**, 108–114.
Franks, F., ed. (1972–1982). *Water. A Comprehensive Treatise*, Vols. 1–7. Plenum, New York.
Friedman, H.L. (1981). Electrolyte solutions at equilibrium. *Ann. Rev. Phys. Chem.* **32**, 179–204.
Fukushima, N., Tamura, Y., and Ohtaki, H. (1991). Dissolution of alkali fluoride and chloride crystals in water studied by molecular dynamics simulations. *Z. Naturforsch.* **46a**, 193–202.
Garofalini, S.H. (1990). Molecular dynamics computer simulations of silica surface structure and adsorption of water molecules. *J. Non-Cryst. Solids* **120**, 1–12.
Geiger, A., Rahman, A., and Stillinger, F.H. (1979a). Molecular dynamics study of the hydration of Lennard–Jones solutes. *J. Chem. Phys.* **70**, 263–276.
Geiger, A., Stillinger, F.H., and Rahman, A. (1979b). Aspects of the percolation process for hydrogen-bond networks in water. *J. Chem. Phys.* **70**, 4185–4193.
Gellatly, B.J., Quinn, J.E., Barnes, P., and Finney, J.L. (1983). Two, three, and four body interactions in model water interactions. *Mol. Phys.* **59**, 949–970.
Gorbaty, Yu. E. (1979). Some new data on the structure of liquid and supercritical

water, in *Problems of Physical-Chemical Petrology*, V.A. Zharikov, ed., Vol. II, Nauka, Moscow, pp. 15–24 (in Russian).

Gorbaty, Yu. E. and Demianets, Yu. N. (1983). The pair-correlation functions of water at a pressure of 1000 bar in the temperature range 25–500°C. *Chem. Phys. Lett.* **100**, 450–454.

Gorbaty, Yu. E. and Demianets, Yu. N. (1985). An X-ray study of the effect of pressure on the structure of liquid water. *Mol. Phys.* **55**, 571–588.

Grivtsov, A.G., Zhuravlev, L.T., Gerasimova, G.A., and Khazin, L.G. (1988). Molecular dynamics of water: Adsorption of water on β-Tridymite. *J. Colloid and Interface Sci.* **126**, 397–407.

Guàrdia, E. and Padró, J.A. (1990a). Molecular dynamics simulation of ferrous and ferric ions in water. *Chem. Phys.* **144**, 353–362.

Guàrdia, E. and Padró, J.A. (1990b). Molecular dynamics simulation of single ions in aqueous solutions: Effects of the flexibility of the water molecules. *J. Phys. Chem.* **914**, 6049–6055.

Guissani, Y., Guillot, B., and Bratos, S. (1988). The statistical mechanics of the ionic equilibrium in water: A computer simulation study. *J. Chem. Phys.* **88**, 5850–5856.

Haar, L., Gallagher, J.S., and Kell, G.S. (1984). *NBS-NRC Steam Tables. Thermodynamic and Transport Properties and Computer Programs for Vapor and Liquid States of Water in SI Units*. Hemisphere, Washington, D.C.

Hamann, S.D. (1981). Properties of electrolyte solutions at high pressures and temperatures, in *Chemistry and Geochemistry of Solutions at High Temperatures and Pressures*, D.T. Rickard and F.E. Wickman, eds., *Physics and Chemistry of the Earth*, Vols. 13/14, Pergamon Press, Oxford, pp. 89–111.

Hankins, D., Moskowitz, J.W., and Stillinger, F.H. (1970). Water molecule interactions. *J. Chem. Phys.* **53**, 4544–4554.

Hansen, J.P. and McDonald, I.R. (1976). *Theory of Simple Liquids*. Academic Press, London.

Hausser, R., Maier, G., and Noack, F. (1966). Kernmagnetische Messungen von Selbstdiffusions-Koeffizienten in Wasser und Benzol bis zum kritischen Punkt. *Z. Naturforsch.* **21a**, 1410–1415.

Heinzinger, K. (1986). MD simulations of the effect of pressure on the structural and dynamical properties of water and aqueous electrolyte solutions, in *Supercomputer Simulations in Chemistry*, M. Dupuis, ed., *Lecture Notes in Chemistry*, Vol. 44, Springer-Verlag, Berlin, pp. 261–279.

Heinzinger, K. (1990). Molecular dynamics simulation of aqueous systems, in *Computer Modelling of Fluids, Polymers and Solids*, C.R.A. Catlow et al., eds., Kluwer Academic Publishers, Dordrecht, pp. 357–394.

Heinzinger, K. and Pálinkás, G. (1985). Computer simulation of ion-solvent systems, in *The Chemical Physics of Solvation*, R.G. Dogonadze et al., eds., Part A, Elsevier, Amsterdam, pp. 313–353.

Heinzinger, K. and Vogel, P.C. (1976). A molecular dynamics study of aqueous solutions. III. A comparison of selected alkali halides. *Z. Naturforsch.* **31a**, 463–475.

Helgeson, H.C. (1981). Prediction of the thermodynamic properties of electrolytes at high pressures and temperatures, in *Chemistry and Geochemistry of Solutions at High Temperatures and Pressures*, D.T. Rickard and F.E. Wickman, eds., *Physics and Chemistry of the Earth*, Vols. 13/14, Pergamon Press, Oxford, pp. 133–177.

Hogervorst, W. (1973). Diffusion coefficients of noble-gas mixtures between 300 and 1400 K. *Physics* **51**, 59–89.

Holloway, J.R. (1987). Igneous fluids, in *Thermodynamic Modeling of Geological Materials: Minerals, Fluids and Melts*, I.S.E. Carmichael and H.P. Eugster, eds., *Reviews in Mineral* Vol. 17, Mineralogical Society of America, Washington, D.C., pp. 211–233.

Holzapfel, W.B. (1969). Effect of pressure and temperature on the conductivity and ionic dissociation of water up to 100 kbar and 1000°C. *J. Chem. Phys.* **50**, 4424–4428.

Impey, R.W., Klein, M.L., and McDonald, I.R. (1981). Molecular dynamics study of the structure of water at high temperatures and density. *J. Chem. Phys.* **74**, 647–652.

Impey, R.W., Madden, P.A., and McDonald, I.R. (1983). Hydration and mobility of ions in solution. *J. Phys. Chem.* **87**, 5071–5083.

Jancsó, G., Bopp, P., and Heinzinger, K. (1984). Molecular dynamics study of high-density liquid water using a modified central-force potential. *Chem. Phys.* **85**, 377–387.

Jancsó, G., Heinzinger, K., and Bopp, P. (1985). Molecular dynamics study of the effect of pressure on an aqueous NaCl solution. *Z. Naturforsch.* **40a**, 1235–1247.

Jorgensen, W.L. (1982a). Revised TIPS for simulations of liquid water and aqueous solutions. *J. Chem. Phys.* **77**, 4156–4163.

Jorgensen, W.L. (1982b). Convergence of Monte Carlo simulations of liquid water in the NPT ensemble. *Chem. Phys. Lett.* **92**, 405–410.

Jorgensen, W.L., Chandrasekhar, J., Madura, J.D., Impey, R.W., and Klein, M.L. (1983). Comparison of simple potential functions for simulating liquid water. *J. Chem. Phys.* **79**, 926–935.

Jorgensen, W.L. and Madura, J.D. (1985) Temperature and size dependence for Monte Carlo simulations of TIP4P water. *Mol. Phys.* **56**, 1381–1392.

Jorgensen, W.L., Gao, J., and Ravimohan, C. (1985). Monte Carlo simulations of alkanes in water: Hydration numbers and the hydrophobic effect. *J. Phys. Chem.* **89**, 3470–3473.

Kalinichev, A.G. (1985a). Convergence acceleration in the Monte Carlo studies of aqueous systems at high pressures and temperatures, in *Application of Mathematical Methods to Description and Study of Physical-Chemical Equilibria*, Vol. III Novosibirsk, SO AN SSSR, pp. 89–93 (in Russian).

Kalinichev, A.G. (1985b). A Monte Carlo study of dense supercritical water. *High Temp.* **23**, 544–548.

Kalinichev, A.G. (1986). Monte Carlo study of the thermodynamics and structure of dense supercritical water. *Internat. J. Thermophys.* **7**, 887–900.

Kalinichev, A.G. (1991a). Theoretical modeling of geochemical fluids under high-pressure, high-temperature conditions. *High Press. Res.* **7**, 378–380.

Kalinichev, A.G. (1991b). Monte Carlo simulations of water under supercritical conditions. I. Thermodynamic and structural properties. *Z. Naturforsch.* **46a** 433–444.

Kalinichev, A.G. (1992). Monte Carlo simulation of water under supercritical conditions. II. Convergence characteristics and the system size effects (in preparation).

Kalinichev, A.G. and Heinzinger, K. (1992). Molecular dynamics of supercritical water with the flexible BJH potential (in preparation).

Karim, O.A. and Haymet, A.D. (1988). The ice/water interface: A molecular dynamics simulation study. *J. Chem. Phys.* **89**, 6889–6896.

Kataoka, Y. (1987). Studies of liquid water by computer simulations. V. Equation of state of fluid water with Carravetta–Clementi potential. *J. Chem. Phys.* **87**, 589–598.

Kataoka, Y. (1989). Studies of liquid water by computer simulations. VI. Transport properties of Carravetta-Clementi water. *Bull. Chem. Soc. Jpn.* **62**, 1421–1431.

Kjellander, R. and Marčelja, S. (1985). Perturbation of hydrogen bonding in water near polar surfaces. *Chem. Phys. Lett.* **120**, 393–396.

Kong, C.L. (1973). Combining rules for intermolecular potential parameters. II. Rules for the Lennard–Jones (12-6) potential and the Morse potential. *J. Chem. Phys.* **59**, 2464–2467.

Krynicki, K., Green, C.D., and Sawyer, D.W. (1978). Pressure and temperature dependence of self-diffusion in water. *Faraday Disc. Chem. Soc.* **66**, 199–208.

Kubicki, J.D. and Lasaga, A.C. (1991). Molecular Dynamics and Diffusion in Silicate Melts, in *Diffusion, Atomic Ordering, and Mass Transport*, J. Ganguly, ed., *Advances in Physical Geochemistry*, Vol. 8, Springer-Verlag, New-York, pp. 1–50.

Lange, R.L. and Carmichael, I.S.E. (1990). Thermodynamic properties of silicate liquids with emphasis on density, thermal expansion and compressibility, in *Modern Methods of Igneous Petrology: Understanding Magmatic Processes*, J. Nicholls and J.K. Russell, eds., Review in Mineralogy, Vol. 24, Mineralogical Society of America, Washington, D.C., pp. 25–64.

La Placa, S.J., Hamilton, W.C., Kamb, B., and Prakash, A. (1973). On a nearly proton-ordered structure for ice IX. *J. Chem. Phys.* **58**, 567–580.

Lasaga, A.C. (1990). Atomic treatment of mineral-water surface reactions, in Mineral-Water Interface Geochemistry, M.F. Hochella, Jr. and A.F. White, eds., Reviews in Mineralogy, Vol. 23, Mineralogical Society of America, Washington, D.C., pp. 17–85.

Levelt Sengers, J.M.H. (1990). Thermodynamic properties of aqueous solutions at high temperatures: Needs, methods, and challenges. *Internat. J. Thermophys.* **11**, 399–415.

Levesque, D., Weis, J.J., and Hansen, J.P. (1984). Recent developments in the simulation of classical fluids, in *Applications of the Monte Carlo Method in Statistical Physics*, K. Binder, ed., *Topics in Current Physics*, Vol. 36, Springer-Verlag, Berlin, pp. 37–91.

Lie, G.C. and Clementi, E. (1986). Molecular-dynamics simulation of liquid water with an *ab initio* flexible water–water interaction potential. *Phys. Rev.* **A33**, 2679–2693.

Lindner, H.A. (1970). Ramanspektroskopische Untersuchungen an HDO, gelöst in H_2O, an HDO in wässrigen Kaliumjodidlösungen und an reinem H_2O bis 400°C und 5000 bar. Ph.D. Thesis, Karlsruhe.

Madura, J.D., Pettitt, B.M., and Calef, D.F. (1988). Water under high pressure. *Mol. Phys.* **64**, 325–336.

Matsuoka, O., Clementi, E., and Yoshimine, M. (1976). CI study of the water dimer potential surface. *J. Chem. Phys.* **64**, 1351–1361.

McQuarrie, D.A. (1976) *Statistical Mechanics*. Harper & Row, New York.

Mezei, M. and Beveridge, D.L. (1981). Monte Carlo studies of dilute aqueous solutions of Li^+, Na^+, K^+, F^-, and Cl^-. *J. Chem. Phys.* **74**, 6902–6910.

Mills M.F., Reimers, J.R., and Watts, R.O. (1986). Monte Carlo simulation of the OH stretching spectrum of solutions of KCl, KF, LiCl and LiF in water. *Mol. Phys.* **57**, 777–791.

Morse, M.D. and Rice, S.A. (1982). Tests of effective pair potentials for water: Predicted ice structures. *J. Chem. Phys.* **76**, 650–660.

Motakabbir, K.A. and Berkowitz, M.L. (1991). Liquid–vapor interface of TIP4P water: Comparison between a polarizable and a nonpolarizable model. *Chem. Phys. Lett.* **176**, 61–66.

Mountain, R.D. (1989). Molecular dynamics investigation of expanded water at elevated temperatures. *J. Chem. Phys.* **90**, 1866–1870.

Nakahara, M, Zenke, M., Ueno, M., and Shimizu, K. (1985). Solvent isotope effect on ion mobility in water at high pressure. Conductance and transference number of potassium chloride in compressed heavy water. *J. Chem. Phys.* **83**, 280–287.

Neumann, M. (1986). Dielectric relaxation in water. Computer simulations with the TIP4P potential. *J. Chem. Phys.* **85**, 1567–1580.

Okazaki, S., Nakanishi, K., Touhara, H., Watanabe, N., and Adachi, Y. (1981). A Monte Carlo study on the size dependence in hydrophobic hydration. *J. Chem. Phys.* **74**, 5863–5871.

O'Shea, S.F. and Tremaine, P.R. (1980). Thermodynamics of liquid and supercritical water to 900°C by a Monte Carlo method. *J. Phys. Chem.* **84**, 3304–3306.

Owicki, J.C. and Scheraga, H.A. (1977). Monte Carlo calculations in the isothermal-isobaric ensemble. 1. Liquid water. *J. Amer. Chem. Soc.* **99**, 7403–7412.

Pálinkás, G., Riede, W.O., and Heinzinger, K. (1977). A molecular dynamics study of aqueous solutions. VII. Improved simulation and comparison with X-ray investigations of a NaCl solution. *Z. Naturforsch.* **32a**, 1137–1145.

Pálinkás, G., Bopp, P., Jancsó, G., and Heinzinger, K. (1984). The effect of pressure on the hydrogen bond structure of liquid water. *Z. Naturforsch.* **39a**, 179–185.

Panagiotopoulos, A.Z. (1987). Direct determination of phase coexistence properties of fluids by Monte Carlo simulation in a new ensemble. *Mol. Phys.* **61**, 813–826.

Pangali, C., Rao, M., and Berne, B.J. (1979). A Monte Carlo simulation of the hydrophobic interaction. *J. Chem. Phys.* **71**, 2975–2981.

Pangali, C., Rao, M., and Berne, B.J. (1980). A Monte Carlo study of structural and thermodynamic properties of water: Dependence on the system size and on the boundary conditions. *Mol. Phys.* **40**, 661–680.

Persikov, E.S., Zharikov, V.A., Bukhtiyarov, P.G., and Pol'skoy S.F. (1990). The effect of volatiles on the properties of magmatic melts. *Eur. J. Mineral.* **2**, 621–642.

Pitzer, K.S. (1987). A thermodynamic model for aqueous solutions of liquid-like density, in *Thermodynamic Modeling of Geological Materials: Minerals, Fluids and Melts*, I.S.E. Carmichael and H.P. Eugster, eds., *Review Mineral*, Vol. 17, Mineralogical Society of America, Washington, D.C., pp. 97–142.

Rahman, A. and Stillinger, F.H. (1971). Molecular dynamics study of liquid water. *J. Chem. Phys.* **55**, 3336–3359.

Rami Reddy, M. and Berkowitz, M. (1987). Structure and dynamics of high-pressure TIP4P water. *J. Chem. Phys.* **87**, 6682–6686.

Rapaport, D.C. and Scheraga, H.A. (1982). Hydration of inert solutes. A molecular dynamics study. *J. Phys. Chem.* **86**, 873–880.

Ratcliffe, C.I. and Irish, D.E. (1982). Vibrational spectral studies of solutions at elevated temperatures and pressures. 5. Raman studies of liquid water up to 300°C. *J. Phys. Chem.* **86**, 4897–4905.

Ree, F.H. (1982). Molecular interaction of dense water at high temperature. *J. Chem. Phys.* **76**, 6287–6302.

Reimers, J.R., Watts, R.O., and Klein, M.L. (1982). Intermolecular potential functions and the properties of water. *Chem. Phys.* **64**, 95–114.

Reimers, J.R. and Watts, R.O. (1984). The structure, thermodynamic properties and infrared spectra of liquid water and ice. *Chem. Phys.* **91**, 201–223.

Robinson, R.A. and Stokes, R.H. (1955). *Electrolyte Solutions*. Butterworths, London.

Roedder, E. (1981). Natural occurence and significance of fluids indicating high pressure and temperature, in *Chemistry and Geochemistry of Solutions at High Temperatures and Pressures*, D.T. Rickard and F.E. Wickman, eds., *Physics and Chemistry of the Earth*, Vols. 13/14, Pergamon Press, Oxford, pp. 9–39.

Roth, K.H. (1969). Die Infrarotabsorption von HDO in H_2O und der Zustand des Wassers bis 500°C und 4000 bar. Ph.D. Thesis, Karlsruhe.

Ruff, I. and Diestler, D.J. (1990). Isothermal-isobaric molecular dynamics simulation of liquid water. *J. Chem. Phys.* **93**, 2032–2042.

Saul, A. and Wagner, W. (1989) A fundamental equation for water covering the range from the melting line to 1273 K at pressures up to 25,000 MPa. *J. Phys. Chem. Ref. Data* **18**, 1537–1564.

Sceats, M.G. and Rice, S.A. (1980). The water–water pair potential near the hydrogen bonded equilibrium configuration. *J. Chem. Phys.* **72**, 3236–3247.

Schaink, H. and Hoheisel, C. (1990). Structural and dynamical behaviour of model fluids at high and low densities. *J. Chem. Phys.* **93**, 2754–2761.

Schwendinger, M.G. and Rode, B.M. (1989). A Monte Carlo simulation of a supersaturated sodium chloride solution. *Chem. Phys. Lett.* **155**, 527–532.

Shmulovich, K.I., Mazur, V.A., Kalinichev, A.G., and Khodorevskaya, L.I. (1980a). P-V-T and component activity-concentration relations for systems of H_2O-nonpolar gas type. *Geochem. Internat.* **17**(6),18–31.

Shmulovich, K.I., Shmonov, V.M., Mazur, V.A., and Kalinichev, A.G. (1980b). P-V-T and activity-concentration relations in the H_2O-CO_2 system. *Geochem. Internat.* **17**(6), 123–139.

Shmulovich, K.I., Shmonov, V.M., and Zharikov, V.A. (1982a). The thermodynamics of supercritical fluid systems, in *Advances in Physical Geochemisitry*, S.K. Saxena, ed., Vol. 2, Springer-Verlag, New York, pp. 173–190.

Shmulovich, K.I., Tereshchenko, E.N., and Kalinichev, A.G. (1982b). The equation of state and isochores for nonpolar gases up to 2000 K and 10 GPa. *Geochem. Internat.* **19**(6), 49–64.

Soper, A.K. and Phillips, M.G. (1986). A new determination of the structure of water at 25°C. *Chem. Phys.* **107**, 47–60.

Spohr, E. and Heinzinger, K. (1988). Computer simulations of water and aqueous solutions at interfaces. *Electrochimica Acta* **33**, 1211–1222.

Spohr, E. Pálinkás, G., Heinzinger, K., Bopp, P., and Probst, M.M. (1988). Molecular dynamics study of an aqueous $SrCl_2$ solution. *J. Phys. Chem.* **92**, 6754–6761.

Sprik, M. and Klein, M.L. (1988). A polarizable model for water using distributed charge sites. *J. Chem. Phys.* **89**, 7556–7560.

Stillinger, F.H. (1980). Water revisited. *Science* **209**, 451–457.

Stillinger, F.H. and David, C.W. (1978). Polarization model for water and its ionic dissociation products. *J. Chem. Phys.* **69**, 1473–1484.

Stillinger, F.H. and Rahman, A. (1972). Molecular dynamics study of temperature effects on water structure and kinetics. *J. Chem. Phys.* **57**, 1281–1292.

Stillinger, F.H. and Rahman, A. (1974a). Improved simulation of liquid water by molecular dynamics. *J. Chem. Phys.* **60**, 1545–1557.

Stillinger, F.H. and Rahman, A. (1974b). Molecular dynamics study of liquid water under high compression. *J. Chem. Phys.* **61**, 4973–4980.

Stillinger, F.H. and Rahman, A. (1978). Revised central force potentials for water. *J. Chem. Phys.* **68**, 666–670.

Sverjensky, D.A. (1987). Calculation of the thermodynamic properties of aqueous species and the solubilities of minerals in supercritical electrolyte solutions, in *Thermodynamic Modeling of Geological Materials: Minerals, Fluids and Melts*, I.S.E. Carmichael and H.P. Eugster, eds., *Reviews in Mineral* Vol. 17, Mineralogical Society of America, Washington, D.C., pp. 177–209.

Swope, W.C. and Andersen, H.C. (1984). A molecular dynamics method for calculating the solubility of gases in liquids and the hydrophobic hydration of inert-gas atoms in aqueous solution. *J. Phys. Chem.* **88**, 6548–6556.

Szász, G.I., Heinzinger, K., and Riede, W.O. (1981). Structural properties of an aqueous LiI solution derived from a molecular dynamics simulation. *Z. Naturforsch.* **36a**, 1067–1075.

Szász, G.I. and Heinzinger, K. (1983a). Hydration shell structures in LiI solution at elevated temperature and pressure: A molecular dynamics study. *Earth Planet. Sci. Lett.* **64**, 163–167.

Szász, G.I. and Heinzinger, K. (1983b). A molecular dynamics study of the translational and rotational motions in an aqueous LiI solution. *J. Chem. Phys.* **79**, 3467–3473.

Tamura, Y., Tanaka, K., Spohr, E., and Heinzinger, K. (1988). Structural and dynamical properties of an $LiCl \cdot 3H_2O$ solution. *Z. Naturforsch.* **43a**, 1103–1110.

Tanaka, K. (1975). Measurements of self-diffusion coefficients of water in pure water and in aqueous electrolyte solutions. *J. Chem. Soc., Faraday Trans. I* **71**, 1127–1131.

Tanaka, H. and Ohmine, I. (1987). Large local energy fluctuations in water. *J. Chem. Phys.* **87**, 6128–6139.

Tanger, J.C. and Pitzer, K.S. (1989). Calculation of the ionization constant of H_2O to 2273 K and 500 MPa. *AIChE J.* **35**, 1631–1638.

Teleman, O., Jönsson, B., and Engström, S. (1987). A molecular dynamics simulation of a water model with intramolecular degrees of freedom. *Mol. Phys.* **60**, 193–203.

Tivey, M.K., Olson, L.O., Miller, V.W., and Light, R.D. (1990). Temperature measurements during initiation and growth of a black smoker chimney. *Nature* **346**, 51–54.

Tödheide, K. (1982). Hydrothermal solutions. *Ber. Bunsenges. Phys. Chem.* **86**, 1005–1016.

Toukan, K. and Rahman, A. (1985). Molecular dynamics study of atomic motions in water. *Phys. Rev.* **B31**, 2643–2648.

Valyashko, V.M. (1977). Studies of water-salt systems at elevated temperatures and pressures. *Ber. Bunsenges. Phys. Chem.* **81**, 388–396.

Valyashko, V.M. (1990). Sub- and supercritical equilibria in aqueous electrolyte solutions. *Pure & Appl. Chem.* **62**, 2129–2138.

Von Damm, K.L. (1990). Seafloor hydrothermal activity: Black smoker chemistry and chimneys. *Ann. Rev. Earth Planet. Sci.* **18**, 173–204.

Weingärtner, H. (1982). Self-diffusion in liquid water. A reassessment. *Z. Phys. Chem. N.F.* **132**, 129–149.

Wood, D.W. (1979). Computer simulations of water and aqueous solutions, in *Water. A Comprehensive Treatise*, F. Franks, ed., Vol. 6, Plenum, New York, pp. 279–409.

Woolley, H.W. (1980). Thermodynamic properties for H_2O in the ideal gas state, in *Water and Steam*, J. Straub and K. Scheffler, eds., Pergamon, Oxford, pp. 166–175.

Zakirov, I.V. and Kalinichev, A.G. (1980). Dependence of the nonideality of homogeneous gaseous mixtures on critical temperatures of components. *Dokl. Akad. Nauk SSSR* **253**, 1214–1216 (in Russian).

Chapter 2
Estimating Thermodynamic Properties by Molecular Dynamics Simulations: The Properties of Fluids at High Pressures and Temperatures

D.G. Fraser and K. Refson

Introduction

Computer simulations of geochemical reactions are beginning to enable experimentalists to explore processes which are difficult or impossible to reproduce in the laboratory because of either their kinetics or the physical conditions involved. In some cases, it has even been claimed that the calculated data are more reliable than available experimental results (Lie and Clementi, 1986), and in the case of H_2O, such a claim was vindicated by subsequent better experimental data (Soper and Phillips, 1986). The widespread availability of mini-super-computers has made such calculations tractable in most laboratories, and since their first applications to geochemistry in the mid–1970s (Woodcock et al., 1976; Matsui et al., 1981; Angell et al., 1982), very rapid progress has been made. Recent reviews of applications to silicate minerals (Parker and Price, 1990) and diffusion in silicate melts (Kubicki and Lasaga, 1990) have appeared and give good summaries of these areas. It is predictable that such calculations will become increasingly important to experimentalists over the next few years as a prelude to designing new experiments or to help pinpoint crucial experiments in difficult areas so that a good understanding of simulation techniques and their basic assumptions will be essential.

The use of thermodynamics and the application of experimental data obtained from studies of simple systems to geological problems require both accurate standard state data for minerals, melts, and volatile fluids and a detailed knowledge of the effects of mixing. Standard state data are available for many simple systems of geological interest although the range of pressures and temperatures of measurement is often limited. In addition, the wide range of possible mineral compositions in natural rocks makes it necessary to have robust theoretical mixing models for interpolation and extrapolation. Slow reaction kinetics also make some experiments impossible to carry out in reasonable times so that alternative means of investigation are required.

Theoretical studies of the behaviour of geological materials are thus of great importance in geochemistry and petrology. Recently, there has been much interest in applying these methods to study the properties of volatile fluids at high pressure and temperature (e.g., Fraser and Refson, 1990; Belonoshko and Saxena, 1991) and this chapter describes work in this field, with particular emphasis on the properties of H_2O.

Behaviour of Fluids at High Pressure: Intermolecular Forces

The Gibbs energy of a phase of any composition may be calculated at high pressure from a knowledge of its properties at low pressure if the variation in its molar volume with pressure is known

$$G_{P,T} = G_{P^0,T} + \int_{P^0}^{P} V \cdot dP. \tag{1}$$

The variation of V with P and T may be complex and defines an equation of state for the phase. In the case of simple fluids at extremely low pressures, where interatomic distances are large, the variation of molar volume V with P and T may be approximated by the ideal gas equation of state

$$PV = RT \tag{2}$$

so that

$$G_{P,T} = G_{P^0,T} + \int_{P^0}^{P} \frac{RT}{P} \cdot dP. \tag{3}$$

For real gases, the ideal gas equation rapidly becomes invalid at pressures above even a moderate to high vacuum as the fluid density increases. This was interpreted by Van der Waals (1873) as being the result of the existence of *long-range attractive* forces in the fluid opposed by *short-range repulsive* forces acting to define the contribution to the fluid volume of the volumes of molecules themselves and is expressed by the Van der Waals equation of state

$$\left(P + \frac{a}{V^2}\right)(V - b) = RT. \tag{4}$$

The concept that long-range attractive forces and short-range repulsive forces determine the P-V-T properties of fluids is still central to attempts to simulate the behaviour of fluids at high pressures and temperatures.

The Van der Waals equation can be used only over relatively limited ranges of pressure, and many attempts have been made to describe the behaviour of real fluids over more extended pressure ranges. The virial equation, for example, has the form

$$\frac{PV}{RT} = 1 + \frac{B(T)}{V} + \frac{C(T)}{V^2} + \dots, \tag{5}$$

where again V is the molar volume of the fluid and $B(T)$ and $C(T)$ are empirical coefficients which are properties of the fluid in question. The second virial coefficient carries important information about the interactions of pairs of molecules that is important in determining the pair potential $U(r)$ needed for the calculations described below. However, the virial equation does not converge at high pressures and it depends on empirical knowledge of the potentials.

Other empirical approaches to describing the properties of high-pressure fluids include the law of corresponding states and the Redlich–Kwong and modified Redlich–Kwong equations.

In the former, the equation of state is expressed in terms of so-called "reduced" variables P_R, V_R, and T_R, in which the P, V, and T are related to the respective values at the critical point: Pc, Vc, and Tc

$$P_R = \frac{P}{Pc}, \quad V_R = \frac{V}{Vc}, \quad T_R = \frac{T}{Tc}. \tag{6}$$

Using these reduced variables, we see that the properties of many volatile fluids are similar so that the properties of fluids which have not been investigated at high pressures may be estimated from experimental data for other fluids for which data already exist using correlations based on the reduced variables (e.g., Wood and Fraser, 1976).

The Redlich–Kwong equation (Redlich and Kwong, 1949) can be considered an extension to the Van der Waals equation. It also has two empirical variables a and b

$$\left[P + \frac{a}{(V^2 + bV)\sqrt{T}} \right] (V - b) = RT. \tag{7}$$

Holloway (1977, 1981) and Kerrick and Jacobs (1981) have both used further modified Redlich–Kwong (MRK) equations in which Temperature, or Temperature and Pressure dependent terms are introduced into the coefficients a and b to improve the ability of the equations to describe the behaviour of H_2O, CO_2, and H_2O-CO_2 mixtures at high temperatures and pressures, and this approach has been followed by several subsequent authors (Bottinga and Richet, 1981; Halbach and Chatterjee, 1982). A review of equations of state of fluids is given by Saxena and Fei (1987).

Computer Simulations of Fluids

In the approaches described above, the equations themselves are essentially empirical although the coefficients used are related in a general way to intermolecular interactions. An alternative approach is to attempt to calculate the bulk properties of high-pressure fluids from descriptions of these interactions at the molecular level (e.g., Allen and Tildesley, 1987). Such methods have been extensively investigated for H_2O at low pressures (Jorgensen, 1982; Jorgensen et al., 1983; Jorgensen and Madura, 1985; Lie and Clementi, 1986) and the intermolecular interactions under those conditions are quite well known. More recently, calculations based on the work of these authors have been performed to estimate the P-V-T properties of H_2O to 1 Mbar and 2000 K (Fraser and Refson, 1990; Belonoshko and Saxena, 1991) and some calculations over a limited pressure range were reported by Brodholt and Wood (1990). Similar studies of the thermodynamic properties of N_2 by molecular dynamics simulation have been successful in predicting structural phase transformations in solid N_2 up to 100 GPa (1 Mbar) (Belak et al., 1990; Nosé and Klein, 1983).

Statistical mechanics allows bulk thermodynamic properties to be calculated from a knowledge of the motions and energy states of the constituent particles at the atomic or molecular level. The fundamental assumption is the ergodic hypothesis that a time-averaged property of a given system is equivalent to the instantaneous average of the property calculated for many such systems. Two main methods are used. Monte Carlo (MC) methods examine the behaviour of an ensemble of particles moved at random. Molecular dynamics (MD) uses the Newton–Euler equations of motion to calculate the motion of each particle in the system.

In MD, models of the long- and short-range intermolecular forces in a system are used to calculate the accelerations and velocities of individual molecules using the equations of simple mechanics. By knowing the exact positions and velocities of molecules, the pressure and temperature, for example, may be calculated for a given fluid density or volume, thus yielding an equation of state for the fluid. In addition, because the dynamical trajectories of individual particles are calculated, dynamic properties of the fluid such as diffusion rates and viscosity may be calculated.

The potential energy of an assemblage of molecules may be expanded in terms of single-body, two-body, three-body, etc. interactions. Thus,

$$U = \Sigma v_i(\mathbf{r}_i) + \Sigma\Sigma v_2(\mathbf{r}_i, \mathbf{r}_j) + \Sigma\Sigma\Sigma v_3(\mathbf{r}_i, \mathbf{r}_j, \mathbf{r}_k)\ldots, \qquad (8)$$

where the first term represents the effects of external fields and interactions with the container walls and is normally zero, the second term all the pairwise interactions, the third term all three-body interactions, etc., and the terms $\mathbf{r}_i$ or $\mathbf{r}_i$, $\mathbf{r}_j$, etc. give the coordinates of individual atoms, pairs, and so on. Interactions for terms higher than three are very small and are usually ignored in such calculations. In addition, in order to economize on computer time, it is usual to incorporate the three-body interactions into the second term by using an *effective* pair potential $v_i^{\text{eff}}(r_i, r_j)$. Thus, the PE is given by

$$U(r) = \Sigma v_i(\mathbf{r}_i) + \Sigma\Sigma v_2^{\text{eff}}(\mathbf{r}_i, \mathbf{r}_j). \qquad (9)$$

The description of the effective pairwise potential defines the potential energy well for the system. This potential is usually designed to include London or dispersion forces and Coulomb interactions in the case of charged or polar systems. One common expression for these interactions is the 12-6 Lennard–Jones potential which describes the potential energy well in the form

$$U(r) = \frac{\Sigma\Sigma z_i z_j}{4\pi\varepsilon r_{ij}} + \Sigma\Sigma\varepsilon\left[\left(\frac{\sigma}{r_{ij}}\right)^{12} - \left(\frac{\sigma}{r_{ij}}\right)^{6}\right]. \quad (10)$$

(Coulombic) (12-6 Lennard–Jones)

The coefficients in the 12-6 Lennard–Jones term describe the form of the potential energy well as shown in Fig. 1 and can be related to two variables: the collision diameter σ, which is the separation at which $U(r) = 0$, and the value of the separation distance r_m for the energy minimum ε. Values of ε and σ for a number of common gases are given in Table 1. Alternative equations can be used to describe the form of the potential energy well; for example, semiempirical equations using a single-site model fitted to experimental data have been used

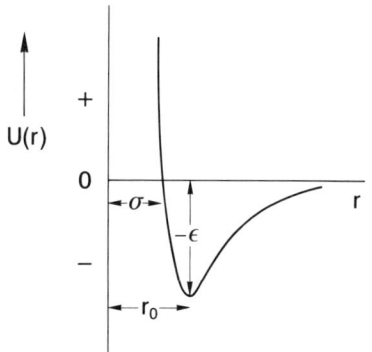

Fig. 1. The intermolecular potential energy well. ε is the well depth and σ, the collision diameter at $U(r) = 0$.

Table 1. Approximate values of maximum well depth (ε) and the collision diameter (σ) for some common gases (modified from Rigby et al., 1986). The value for water is for part of the Lennard–Jones interaction only.

Substance	ε (kJ·mol^{-1})	σ (nm)
He	0.08655	0.2602
Ne	0.3492	0.2755
Ar	1.177	0.3350
Kr	1.661	0.3581
Xe	2.336	0.3790
N_2	0.8663	0.3632
O_2	1.050	0.3382
CO_2	2.039	0.3762
CH_4	1.341	0.3721
H_2O	0.64869	0.3154

by Ross and Ree (1980) and Belonoshko and Saxena (1991). The Lennard-Jones formulation, however, is extensively used in the literature and we have used this in our simulations, together with a description of the geometry and mass and charge distribution of the H_2O molecule based on structural observations. The coefficients used in these simulations thus have the merit of relating to the physical properties of real molecules and should provide the best basis for extrapolation beyond the bounds of existing experimental data. This will be particularly important if fluid mixtures are to be considered successfully.

We have used a 12-6 Lennard-Jones model to investigate the behaviour of H_2O at high pressures and temperatures. A correct description of the intermolecular potentials according to (9) is essential if the calculated thermodynamic quantities of the bulk fluid are to agree with experimental data. Much research has already been done on simulating the properties of H_2O at low temperatures and pressures because of its importance as a solvent and in biological systems (Lie and Clementi, 1986, Jorgensen, 1981, 1982; Jorgensen and Madura, 1985; Jorgensen et al., 1983). In addition to a small number of *ab initio* models of intermolecular potentials for water based on quantum chemistry, of which one of the best known is the MCY potential of Matsuoka et al. (1976), there are dozens of semiempirical potentials. A comparison of results obtained using six such models (SPC, TIP3P, BF, TIPS2, and TIP4P) is given by Jorgensen et al. (1983) who show that most achieve good agreement with experimental data and that TIPS2 and TIP4P give particularly good agreement with experimental data for the O-O radial distribution function obtained by neutron diffraction (Thiessen and Narten, 1982). Simulations of the high-pressure properties of water using MCY water have been reported by Impey et al. (1981) and similar calculations using TIP4P have been used to calculate accurate densities of H_2O at up to 10 kbar at 298 K (Reddy and Berkowitz, 1987; Madura, et al., 1988). Recently, we have investigated the use of this potential at much higher pressures and temperatures and have reported calculated *P-V-T* properties of water up to 300 kbar and 1500 K elsewhere (Fraser and Refson, 1990).

Simulation Conditions

In the TIP4P model, as in MCY, the water molecule is considered to be rigid with the geometry and point mass and charge distribution shown in Fig. 2. The

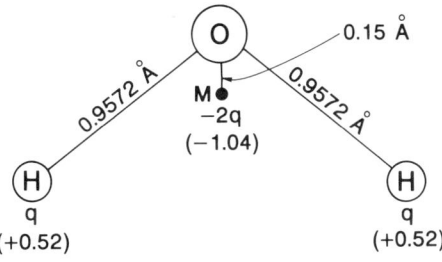

Fig. 2. The TIP4P model of the water molecule.

Table 2. Values of parameters of the TIP4P model used in the MD simulation of H_2O.

Well depth ε (kJ·mol^{-1})	0.64869
Collision diameter σ (nm)	0.31536
Partial charge (H)	0.52
Partial charge (M)	−1.04
O–H length (nm)	0.09572
O–M length (nm) along HOH bisector	0.015
H–O–H angle	104·52°

interatomic angle H–O–H is 104·52° and the O–H bond length 0.9572 Å, which are the values measured for a single water molecule in the gas phase (Benedict et al., 1956). Although the molecules can rotate freely in the MD cell, subject to the constraints of the potential energy equation, the intramolecular geometry is fixed and pressure-independent. In our simulations, a molecular dynamics cell containing 256 H_2O molecules was used as a microcanonical ensemble of fixed energy, volume, and number of particles (N,V,E). The cell satisfied periodic boundary conditions so that molecules leaving the right-hand side of the cell reentered at the left-hand side and so on in other dimensions. Within the box all molecules were subjected to forces as a result of the potential described above. The equations of motion were integrated using a modified Beeman algorithm (Beeman, 1976, Refson and Pawley, 1987) with a time step of 0.2 femto-seconds (1 fs = 10^{-15} s). The size of time steps was chosen so as to keep the total energy constant and give sufficiently accurate measurements of the paths of the molecules as they move with respect to each other's potential energy wells. The long-range Coulombic term was calculated using the Ewald sum technique (Berthault, 1952). After initial trials it was found sufficient for convergence to carry out the calculations to 9 Å in real space and 2 Å^{-1} in reciprocal space, beyond which the calculations produced minimal further changes in the potential energy of the system. The parameters used in conducting the calculation are summarized in Table 2. The positions, velocities, and orientations of all 256 molecules were recorded every 50 time steps (10 fs) so that the dynamics of the system could be observed for each temperature and pressure. The initial configuration was a regular array of randomly orientated H_2O molecule positions, slanted obliquely across the MD cell.

Results

Some typical results for a simulation of TIP4P H_2O at 1000 K for a density of 1.75 are shown in Fig. 3. The calculated translational, rotational, potential, and total energies are shown for 5000 time steps. In order to achieve rapid convergence, the kinetic energy was scaled every 10 time steps during the first 2500

Estimating Thermodynamic Properties by Molecular Dynamics Simulations 67

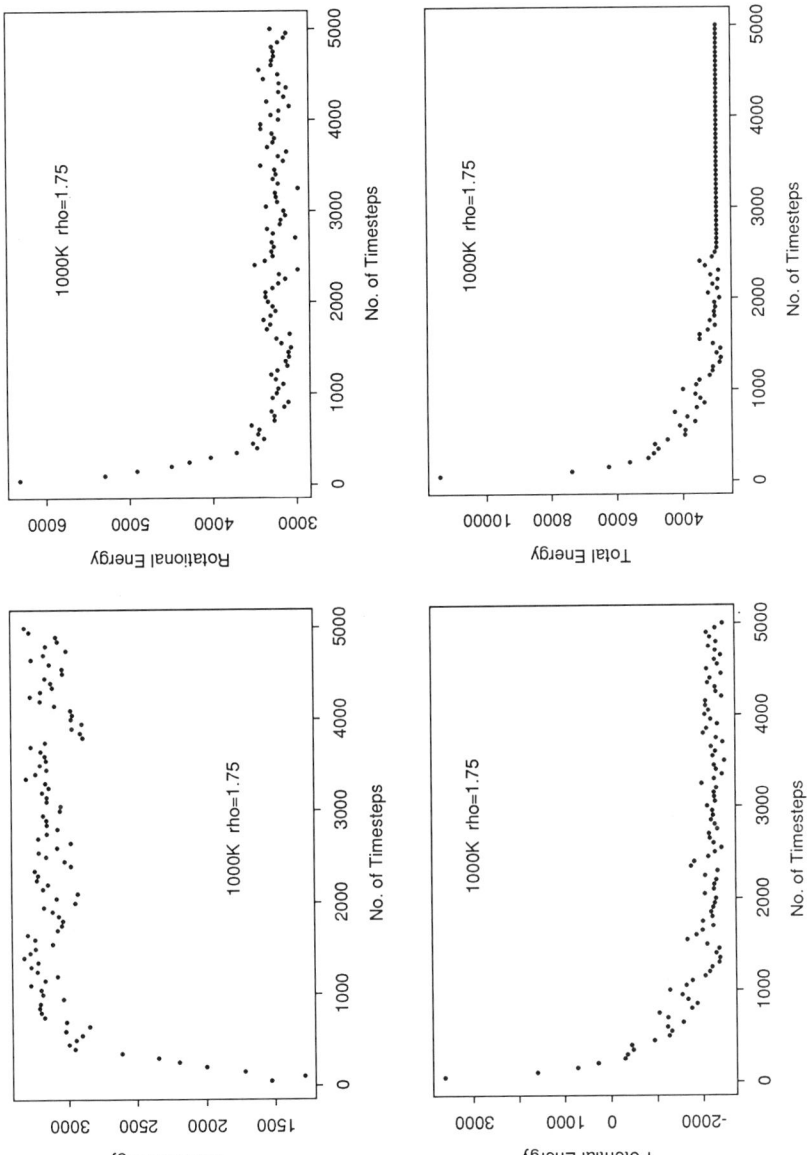

Fig. 3. Evolution of calculated potential, rotational, kinetic, and total energies for TIP4P water at 1000 K and density = 1.75 over 5000 time steps. Velocity Scaling was carried out for the first 2500 timesteps.

Table 3. Thermodynamic results obtained from MD simulations using TIP4P water.

Density (g·cm^{-3})	Temperature (K)	Pressure (kbar)
0.75	789.5	3.269
1.00	789.5	10.69
1.25	787.3	28.35
1.75	794.6	138.2
2.00	812.7	279.9
2.25	833.4	512.9
0.25	999	0.931
0.50	1021	2.41
1.00	1003	15.63
1.25	1013	34.90
1.50	1004	77.53
1.75	995	156.7
2.00	1023	304.0
2.25	1011	548.9
0.50	1117	2.865
0.75	1176	8.631
1.00	1104	18.24
1.25	1127	41.11
1.75	1121	164.5
2.00	1159	308.4
2.25	1181	548.6
0.25	1252	1.478
0.75	1276	9.253
1.00	1241	20.43
1.25	1229	44.92
1.75	1282	175.4
2.00	1278	318.8
2.25	1288	556.2
0.25	1492	1.901
0.50	1496	5.047
0.75	1450	10.19
1.00	1538	24.63
1.25	1528	53.53
1.75	1485	187.8
2.00	1496	339.7
2.25	1532	578.0

steps to give the desired value of temperature. Changes in the calculated values were no longer significant after around 1500 steps and the scaling was turned off after 2500 steps, thus fixing the total energy as shown in Fig. 3. In carrying out runs to determine thermodynamic quantities, measurements were made in this way for 5000 time steps for each change in density, and final data were obtained by accumulating and averaging the values over the last 1000 time steps. Longer

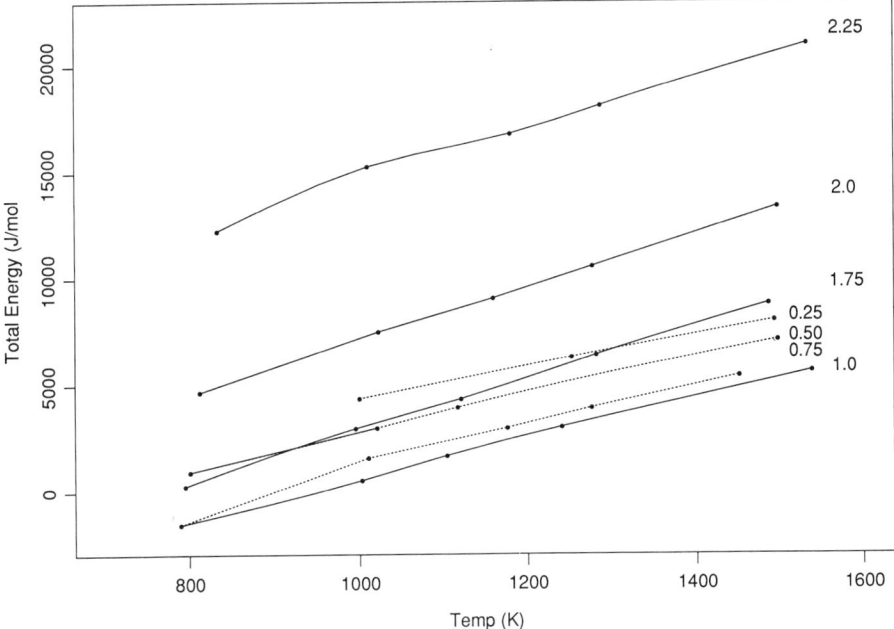

Fig. 4. Variations in energy of TIP4P water with temperature for different densities.

runs of up to 20,000 time steps for sample runs at density 2.0 and $T = 1500$ K produced identical results. The complete configuration was stored at the end of each run and used as a starting configuration for subsequent simulations for the same density at different temperatures. The order in which calculations at the different temperatures were made was randomized. In addition, some points were calculated both up- and down- temperature so as to simulate experimental "reversals" as a check on equilibrium.

Calculations were carried out over a range of temperatures for fluid densities from 0.25 to 2.25, and the *P-V-T* results are given in Table 3. A plot of the total energy of the system for these densities as a function of temperature is shown in Fig. 4. For a given temperature, the total energy varies with density and, with increasing fluid density, first decreases before increasing rapidly beyond a density of around 1.0. These data reflect the shape of the potential energy well.

Discussion

In order to assess how these calculations using the TIP4P model compare with experimental *P-V-T* measurements, data at 1000 K from Table 3 are compared with available experimental measurements in Fig. 5. There are few experimental measurements of the high-pressure and temperature properties of H_2O. Hydrostatic measurements have been made by Burnham et al. (1969) up to 1173 K and

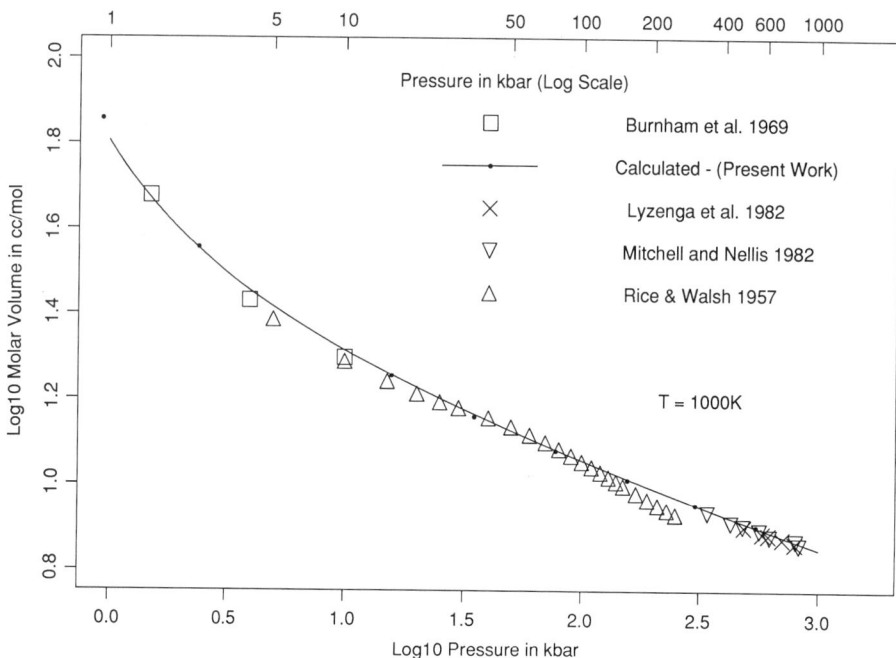

Fig. 5. The *P-V* properties of TIP4P water along the 1000 K isotherm. Experimental data are from Rice and Walsh (1957), Mitchell and Nellis (1982), Lysenga et al. (1982), and Burnham et al. (1969).

8.9 kbar and were extrapolated to 1273 K and 10.0 kbar. At higher pressures, several shock-wave measurements of fluid densities were made without direct measurements of temperature (Rice and Walsh, 1957, Mitchell and Nellis, 1982). In addition, a very limited number of shock-wave data are also available for which temperatures were simultaneously determined (Lysenga et al., 1982). These lie in the *P-T* range 48.9 GPa (489 kbar) to 80.0 GPa (800 kbar) and 3280 to 5270 K and are included in Fig. 5 for comparison.

The results of our MD simulations using TIP4P agree surprisingly well with the available experimental data, despite the rigidity of the model molecule. At low pressures the calculated *P-V-T* data lie close to the isotherms of Burnham et al. (1969). The Rice and Walsh (1957) data appear to agree with both our calculations and the data of Mitchell and Nellis (1982) up to about 50 kbar. Beyond 50 kbar, our calculated *P-V* data lie close to the data of Mitchell and Nellis (1982), but the Rice and Walsh data indicate higher densities. The temperatures measured in the experiments of Lysenga et al. (1982) are much higher, but their observed densities in the range 500 to 800 kbar are similar to those calculated by us using the simple TIP4P potential and the temperature dependence is quite small. Thus, it seems that this simple approach is capable of yielding *P-V-T* data which agree with experiment over a wide range of conditions.

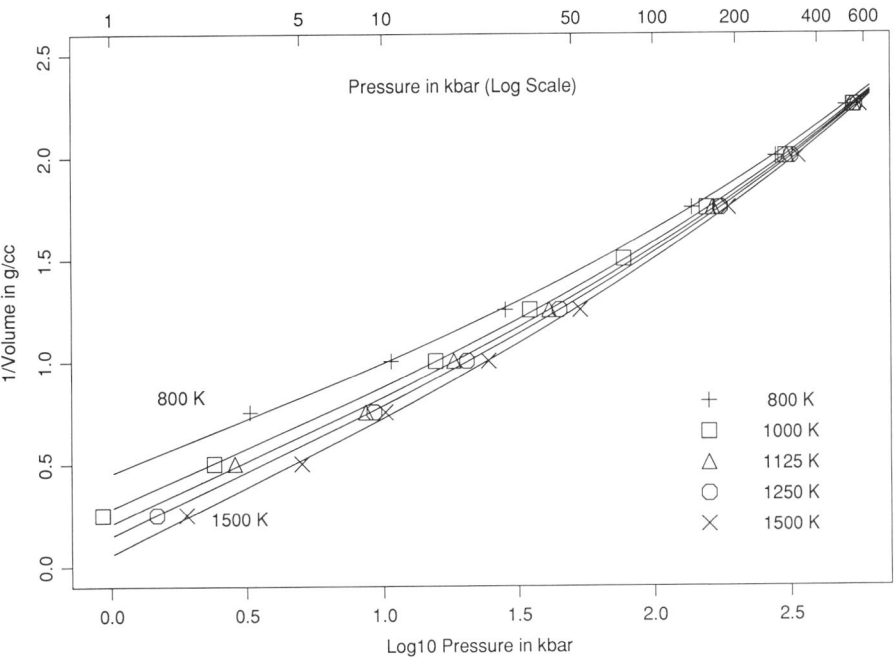

Fig. 6. The *P-V-T* surface of TIP4P water calculated using the TIP4P water model.

The entire *P-V-T* surface calculated for H_2O using the above model is shown in Fig. 6. The data were fitted first to quadratics on each isotherm. Inspection of the residuals indicated significant cubic terms and these were incorporated before the full *P-V-T* fit was attempted. Fitting in this way to the simulated data gives the following equation of state for water:

$$\frac{1}{V} = -0.39675 + 0.8\log_{10}P - 0.013875(\log_{10}P)^2 + 0.023657(\log_{10}P)^3 + \frac{679.202}{T} - \frac{217.76(\log_{10}P)}{T}. \tag{11}$$

Curves calculated using this equation are shown together with the data in Fig. 6. The specific volume of water at 1000 K, 10 kbar using this equation is 1.29 $cm^3 \cdot g^{-1}$, which can be compared with the value of 1.253 $cm^3 \cdot g^{-1}$ tabulated by Burnham et al. (1969), and we expect the agreement to be rather better at higher pressures as deviations related to the flexibility of the H_2O molecule are dominated by the short-range "hard-sphere" repulsions. Comparisons of these data with the results of the simulations of Belonoshko and Saxena (1991) and with the equation of Halbach and Chatterjee (1982) based on a modified Redlich–Kwong model fitted to the data of Burnham et al. and Rice and Walsh (1957) are shown in Fig. 7.

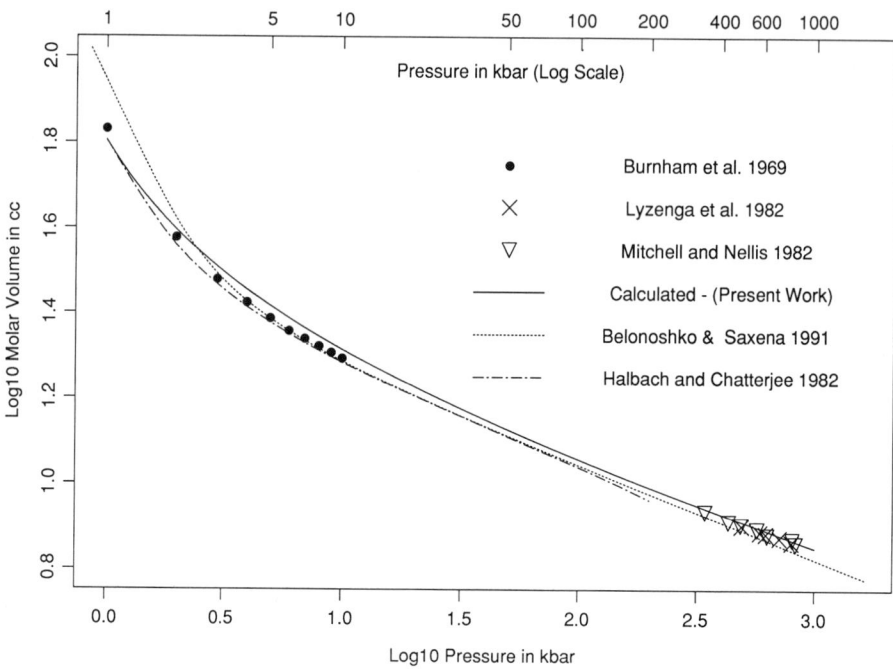

Fig. 7. Comparison of the equation of state [Eq. (12)] of H_2O calculated from the present work with the equation of Belonoshko and Saxena (1991), based on a single-site model and with the extended Redlich–Kwong equation of Halbach and Chatterjee (1982).

Structural Information

Comparisons of the predictions of these models of the structure of H_2O with experimental measurements of structure are also instructive. Soper and Silver (1982) found good agreement between the pairwise radial distribution function (r.d.f.) $g_{HH}(r)$ calculated using MCY and $g_{HH}(r)$ obtained by time-of-flight neutron diffraction. Lie and Clementi (1986) modified MCY to include flexible water–water interactions and internal vibrations of the H_2O molecule and found that still better agreement is obtained using the extended MCYL model. It should be noted that this model is based on *ab initio* calculations and contains no empirical coefficients other than the atomic masses, electronic charge, and Planck's constant.

The TIP4P model is an empirical model, and has the advantage that it is simple and efficient to evaluate. As noted above, it is surprising that apparently good agreement can be obtained between the available experimental *P-V-T* data and the results of simulations based on the rigid TIP4P model since it was originally calibrated at relatively low pressures and allows no molecular flexibility. In Fig. 8 pairwise radial distribution functions $g(r)$ are shown for O–H, H–H, and

O–O for densities ranging from 0.25 to 2.5 g·cm^{-3} at 1000 K. The plot of $g_{OH}(r)$ shows the sharp peak at 0.9572 Å corresponding to the fixed O–H distance of the TIP4P H$_2$O molecule. It also shows a general increase in ordering and a shift of peak positions with increasing density to closer distances of approach. The inflection and decrease in the number of pairs at large distances is a computational artifact caused by the compression of the 256 water molecules into volumes less than the cell volume at high pressure so that the RDF (radial distribution function) falls off. Similar features can be observed in $g_{HH}(r)$. However, it should be noted that at the highest pressures, a significant number of H–H approaches can be seen at distances smaller than the intramolecular fixed H–H distance of TIP4P.

The O–O RDF, $g_{OO}(r)$, measures the distributions of O–O distances between molecules. It shows greatly increased order for relative densities above 1.75. Reddy and Berkowitz (1987) found that for calculations using TIP4P at 298 K, increasing the pressure up to 6 kbar caused a decrease in intensity of the O–O peak at around 2.75 Å and a disappearance of the peak at 4.33 Å, which they interpreted in terms of a disappearance of the tetrahedral low-pressure structure of water with increasing pressure. At the much higher pressures and temperatures studied by us, we confirm the movement of the peak positions of $g_{OO}(r)$ to smaller values with increasing pressure. However, several marked new effects are apparent. The small second peak of Reddy and Berkowitz (1987) is not present at the high pressures and temperatures of these runs and indicates that the low-temperature and -pressure tetrahedral framework of H$_2$O has disappeared. The degree of order increases strongly however with growing pressure. The small broad peak in $g_{OO}(r)$ at around 6 Å at low pressure moves rapidly to smaller distances and clearly splits into two at densities of 2.0 and above. In addition, the first radial distribution peak moves to smaller distances, becomes much sharper, and rapidly increases in intensity at density 2.0. These features strongly suggest crystallization in the system. Although crystallization kinetics may be slow, a number of "reversals" carried out by repeating calculations both up- and down-temperature from configurations previously stored at the end of runs made for other conditions give good agreement and indicate that equilibrium was attained in the runs. Furthermore, as shown in Fig. 3, the translational and rotational energies had clearly converged so that at each point on the curve, we believe equilibrium was achieved. To minimize systematic errors, the data were collected for each density in a randomized order of temperatures.

In order to investigate the apparent phase transition in more detail, we recorded the atomic coordinates in the MD simulation every 50 time steps and followed the trajectories of the H$_2$O molecules as a film sequence. At low pressures, the TIP4P water molecules are seen to move randomly in the disordered fluid. At high pressures, however, while the H-atoms are disordered, the O-atoms define a lattice structure. At low temperatures and high pressures above about 20 kbar (2GPa), ice exists as the proton-ordered BCC (body-centred cubic) ice VIII structure (e.g., Demontis et al., 1988). On heating, this undergoes a phase transition to form the cubic orientationally disordered phase ice VII (Wong and Whalley, 1976). Comparison of $g_{HH}(r)$ and $g_{OO}(r)$ in Fig. 8

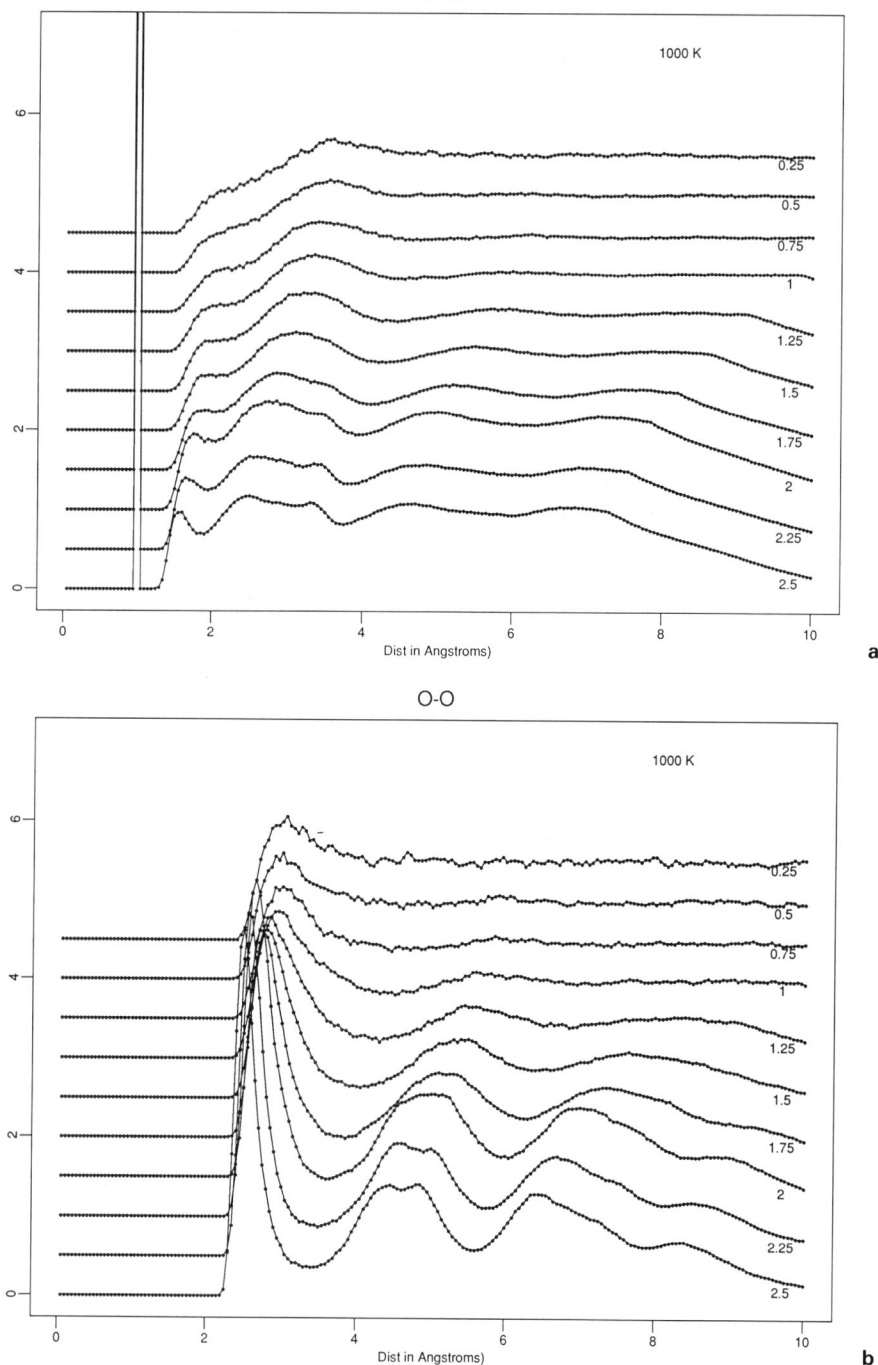

Fig. 8. Pairwise radial distribution functions $g_{OH}(r)$, $g_{HH}(r)$, and $g_{OO}(r)$ at 1000 K. The curves for different densities are each offset by 0.5 units on the y-axis. Note that peak positions move to smaller distances with increasing pressure and the second nearest-neighbor peak around 4.5 Å in $g_{OO}(r)$ becomes resolved above a density of 2.0. Also shown

(continued)

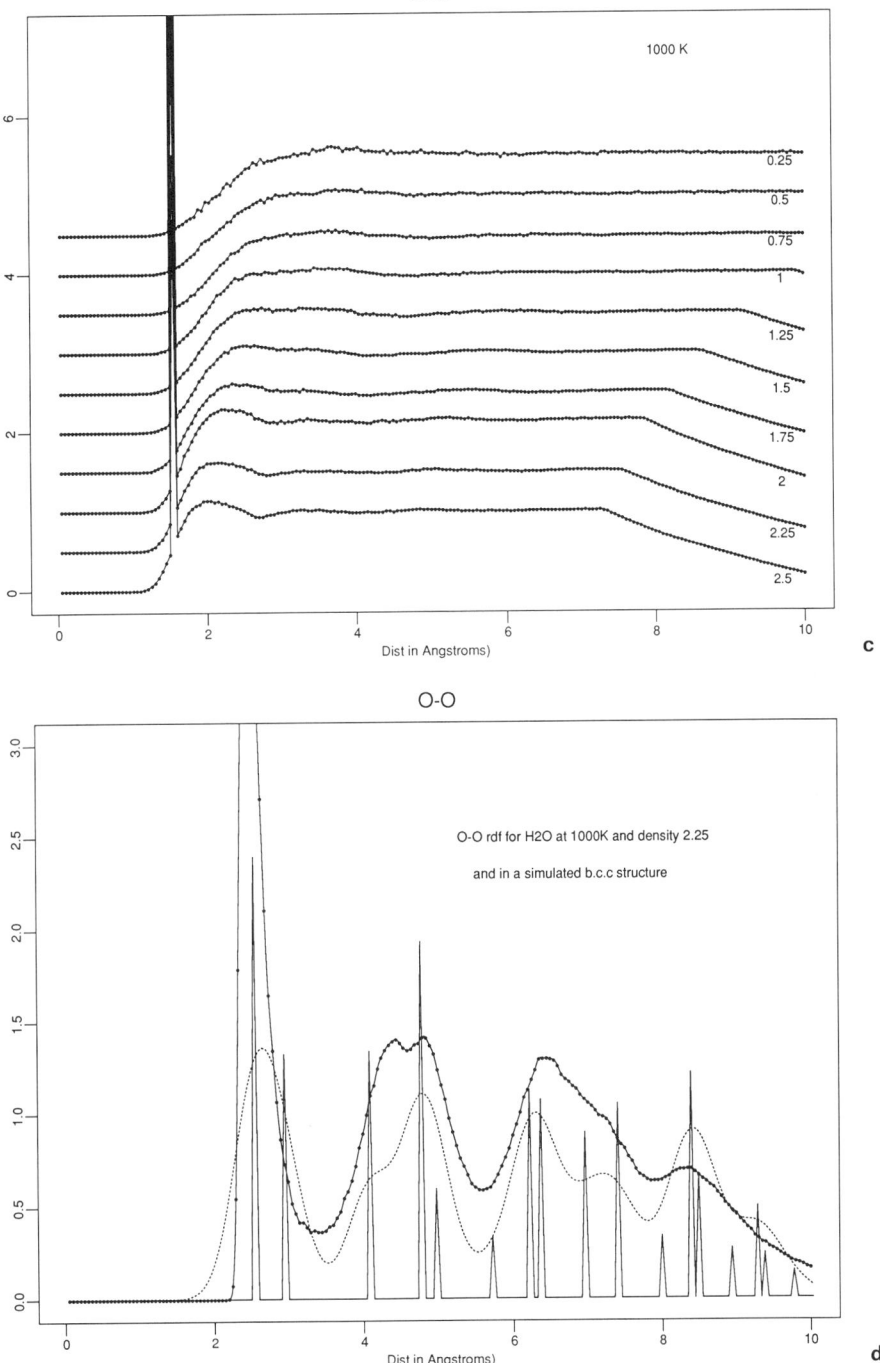

for comparison is the pairwise radial distribution function $g_{oo}(r)$ at 1000 K for density = 2.25 and the sharp peaks of $g_{oo}(r)$ calculated for a crystalline BCC structure like that of ice VII. The fall-off in the rdf at about 8° increasing with density is an artifact of the computational procedure and has no physical meaning.

shows these effects clearly, with $g_{HH}(r)$ showing very little order indeed in contrast to the high degree of order exhibited by $g_{OO}(r)$. We have calculated $g_{OO}(r)$ for a body-centred cubic lattice like ice VII and this is also shown for comparison in Fig. 8. The agreement between the peak positions suggests that TIP4P water at 1000 K, 500 kbar has an ice VII structure. Since no dissociation is possible in this model, ionic contributions could not be investigated.

Summary and Conclusions

The molecular dynamics calculations reported above give calculated P-V-T properties for H_2O up to 1500 K and 100 GPa, which agree remarkably well with the available experimental data. We also observe the phase transition to a crystalline, orientationally disordered cubic ice structure. No account was taken of molecular flexibility in these calculations nor of potential dissociation at high pressures as suggested by Hamman (1981). However, we note that the closest next-nearest-neighbour O–H approach remains significantly greater than the TIP4P fixed O–H bond length within the water molecule for all pressures studied. The equation of state proposed here should be useful for estimating the properties of H_2O at up to 1500 K and 100 GPa (1 Mbar) and is much easier to use in practice than modified Redlich–Kwong equations.

Extension of these methods to the studies of other fluids and of fluid mixtures at high temperatures and pressures will require good potential models for the species involved, and this is likely to involve a combination of good *ab initio* work and semiempirical modelling. Once developed, these models should allow robust predictions of thermodynamic properties beyond the range of the experimental data on the basis of fundamental molecular information.

References

Allen, M.P. and Tildesley, D.J. (1987). *Computer Simulation of Liquids*. Clarendon Press, Oxford.

Angell, C.A., Cheeseman, P.A., and Tammaddon, S. (1982). Pressure enhancement of ion mobilities in liquid silicates from computer simulation studies to 800 kbar. *Science* **218**, 885–887.

Burnham, C.W., Holloway, J.R., and Davis, N.F. (1969). Thermodynamic properties of water to 1000°C and 10,000 bars. Paper 132, Boulder, Colorado. Geological Society America.

Beeman, D. (1976). Some multistep methods for use in molecular dynamics calculations. *J. Comput. Physics* **20**, 130–139.

Belak, J., LeSar, R., and Etters, R.D. (1990). Calculated thermodynamic properties and phase transitions of solid N_2 at temperatures $0 < T < 300$ K and pressures $0 < P < 100$ GPa. *J. Chem. Phys.* **92**, 5430–5441.

Belonoshko, A. and Saxena, S.K. (1991). A molecular dynamics study of the pressure-volume-temperature properties of super-critical fluids: I. H_2O. *Geochim. Cosmochim. Acta* **55**, 381–387.

Benedict, W.S., Gailar, N., and Plyler, E.K. (1956). Rotation-vibration spectra of deuterated water vapor. *J. Phys. Chem.* **24**, 1139–1165.

Bottinga, Y. and Richet, P. (1981). High pressure and temperature equation of state and calculation of the thermodynamic properties of gaseous carbon dioxide. *Amer. J. Sci.* **281**, 615–660.

Berthault, F. (1952). L'energie electrostatique de reseaux ioniques. *J. Phys Radium* **13**, 499–505.

Brodholt, J. and Wood, B.J. (1990). Molecular dynamics of water at high temperatures and pressures. *Geochim. Cosmochim. Acta* **54**, 2611–2616.

Demontis, P., LeSar, R., and Klein, M.L. (1988). New high pressure phases of ice. *Phys. Rev. Lett.* **60**, 2284–2287.

Fraser, D.G. and Refson, K. (1990). The thermodynamic properties of water to 300 kb by molecular dynamics simulation. *Terra Abstracts* **2**, 10.

Halbach, H. and Chatterjee, N.D. (1982). An empirical Redlich–Kwong equation of state for water to 1000°C and 200 kbar. *Contrib. Mineral. Petrol.* **79**, 337–345.

Hamman, S.D. (1981). Properties of electrolyte solutions at high pressures and temperatures, in *Physics and Chemistry of the Earth*, Vol. 13, (D.T. Rickard and F.E. Wickman, eds., Oxford, pp. 89–112.

Holloway, J.R. (1977). Fugacity and activity of molecular species in super-critical fluids, in *Thermodynamics in Geology*, D.G. Fraser, ed., D. Reidel, Dordrecht, pp. 161–181.

Holloway, J.R. (1981). Volatile interactions in magmas, in *Thermodynamics of Minerals and Melts*, R.C. Newton et al., eds., *Advances in Physical Geochemistry*, Vol. I, pp. 273–293. Springer Verlag, New York.

Impey, R.W., Klein, M.L., and McDonald, I.R. (1981). Molecular dynamics study of the structure of water at high temperatures and density. *J. Chem. Phys.* **74**, 647–652.

Jorgensen, W.A. (1981). Transferable intermolecular potential functions for water, alcohol and ethers. Application to liquid water. *J. Amer. Chem. Soc.* **103**, 335–340.

Jorgensen, W.A. (1982). Revised TIPS for simulations of liquid water and aqueous solutions. *J. Chem. Phys.* **77**, 4156–4163.

Jorgensen, W.A. and Madura, J.D. (1985). Temperature and size dependence for Monte-Carlo simulations of TIP4P water. *Mol. Phys.* **56**, 1381–1392.

Jorgensen, W.A., Chandrasekhar, J., Madura, J.D., Impey, R.W., and Klein, M.L. (1983). Comparison of simple potential functions for simulating liquid water. *J. Chem. Phys.* 79 (1983). 926–935

Kerrick, D.M. and Jacobs, G.K. (1981). A modified Redlich–Kwong equation for H_2O, CO_2 and H_2O-CO_2 mixtures at elevated pressures and temperatures. *Amer. J. Sci.* **281**, 735–767.

Kubicki, J.D. and Lasaga, A.C. (1990). Molecular dynamics and diffusion in silicate melts, in *Advances in Geochemistry*, Vol. 8, J. Ganguly, ed., Springer-Verlag, New York, pp. 1–50.

Lie, G.C. and Clementi, E. (1986). Molecular dynamics simulation of liquid water with an ab initio flexible water–water interaction potential. *Phys. Rev. A* **33**, 2679–2693.

Lysenga, G.A., Ahrens, T.J., Nellis, W.J., and Mitchell, A.C. (1982). The temperature of shock-compressed water. *J. Chem. Phys.* **76**, 6282–6286.

Madura, J.D., Pettitt, B.M., and Calef, D.F. (1988). Water under high pressure. *Mol. Phys.* **64**, 325–336.

Matsui, Y., Kawamura, K., and Syono, Y. (1981). Molecular dynamics calculations applied to silicate systems: Molten and vitreous $MgSiO_3$ and Mg_2SiO_4, in *High Pressure Research in Geophysics, Advances in Earth and Planetary Science*, Vol. 12, S. Akimoto and M.H. Manghnani, eds., Reidel, Dordrecht, pp. 551–524.

Matsuoka, O., Clementi, E., and Yoshimine, M. (1976). Study of the water dimer potential surface. *J. Chem. Phys.* **64**, 1351–1361.

Mitchell, A.C. and Nellis, W.J. (1982). Equation of state and electrical conductivity of water shocked to the 100 GPa (1 Mbar) pressure range. *J. Chem. Phys.* **76**, 6273–6281.

Nosé, S. and Klein, M.L. (1983). Structural transformations in solid nitrogen at high pressures. *Phys. Rev. Lett.* **50**, 1207–1210.

Parker, S.C. and Price, G.D. (1990). Computer modelling of the structure and thermodynamic properties of silicate minerals, in *Computer Modelling of Fluids, Polymers and Solids*, C.R.A. Catlow, S.C. Parker, and M.P. Allen, eds., Series C, *Mathematical and Physical Sciences*, Vol. 293, Kluwer, Dordrecht, pp. 405–429.

Reddy, M.R. and Berkowitz, M. (1987). Structure and dynamics of high-pressure water. *J. Chem. Phys.* **87**, 6682–6686.

Redlich, O. and Kwong, J.N.S. (1949). An equation of state. Fugacities of gaseous solutions. *Chem. Rev.* **44**, 233–244.

Refson, K. and Pawley, G.S. (1987). Molecular dynamics studies of the condensed phases of n-butane and their transitions. *Mol. Phys.* **61**, 669–692.

Rice, M.H. and Walsh, J.M. (1957). Equation of state of water to 250 Kilobars. *J. Chem. Phys.* **26**, 824–830.

Rigby, M., Smith, E.B., Wakeham, W.A., and Maitland, G.C. (1986). *The Forces Between Molecules*. Clarendon Press, Oxford.

Ross, M. and Ree, F.H. (1980). Repulsive forces of simple molecules and mixtures at high density temperature. *J. Chem. Phys.* **73**, 6146–6152.

Saxena, S.K. and Fei, Y. (1987). High pressure and high temperature fugacities. *Geochim. Cosmochim. Acta* **51**, 783–791.

Soper, A.K. and Phillips, M.G. (1986). A new determination of the structure of water at 25°C. *Chem. Phys.* **107**, 47–60.

Soper, A.K. and Silver, R.N. (1982). Hydrogen–hydrogen pair correlation function in liquid water. *Phys. Rev. Lett.* **49**, 471–474.

Thiessen, W.E. and Narten, A.H. (1982). Neutron diffraction study of light and heavy water mixtures. *J. Chem. Phys.* **77**, 2656–2662.

Wong, P.T.T. and Whalley, E.H. (1976). Raman spectrum of ice VIII. *J. Chem. Phys.* **64**, 2359–2366.

Wood, B.J. and Fraser, D.G. (1976). *Elementary Thermodynamics for Geologists*. Oxford University Press, Oxford.

Woodcock, L.V., Angell, C.A., and Cheeseman, P. (1976). Molecular dynamics studies of the vitreous state: Ionic systems and silica. *J. Chem. Phys.* **65**, 1565–1567.

Chapter 3
Equations of State of Fluids at High Temperature and Pressure (Water, Carbon Dioxide, Methane, Carbon Monoxide, Oxygen, and Hydrogen)

A.B. Belonoshko and S.K. Saxena

Introduction

The problem of calculating properties of fluids at high temperature (T) and pressure (P) remains one of the main problems of physical chemistry. More than 100 years have passed since the contemporary approach of studying a fluid state was devised (van der Waals, 1881). The comprehensive state of understanding of the fluid state was described in a review of Barker and Henderson (1976).

There are many reasons for the attention this problem has received. First, an understanding of the nature of the fluid state is of basic scientific interest. Second, the experimental (static) measurements of the P-V-T properties of fluids are restricted to comparatively low values of T and P (about 1000 K and 10,000 bars). Investigators in many branches of science including physical geochemistry need to know the properties of fluids at a much higher pressure and temperature than mentioned above. The list of applications of the equation of state (EOS) at extremal T and P is quite extensive.

In addition to the static measurements of P-V-T properties, the shock-wave experiment is a very important source of data. The precision of the latter is much lower as compared to the former. Moreover, temperature in a shock-wave experiment is not actually measured. The two exceptions known to the authors are the experiments of Lysenga et al. (1982) and Radousky et al. (1990). To acquire simultaneously P, V, and T, one has to apply some form of EOS. Therefore, it is understandable that the shock-wave P-V-T data (but only P-V) are not purely experimental, but have a model-dependent character.

There is a very wide group of EOS with polynomial formulation. The coefficients of these polynomials are usually calculated with the least-squares method over a range of experimental (static) data that exist. The extrapolation with these EOSs leads to large errors, because at high-temperature and -pressure conditions, the polynomial term with the highest degree dominates and calculated values of pressure or volume depend practically only on the value of the coefficient of this term. Occasionally, it may not even make any physical sense. It is for this reason that the use of such EOSs is recommended mostly for interpolation and some limited extrapolation (see Saxena and Fei, 1987a). Some polynomial EOSs (Saul and Wagner, 1989; Hill, 1990) have been formulated on the basis of shock-wave data (e.g., Walsh and Rice, 1957). However, as mentioned above, these data are model-dependent and therefore the EOSs are valid only as models to describe initial P-V experimental shock-wave data (e.g., Rice and Walsh, 1957). All other EOSs of fluids, which claim to be valid for extrapolation, could be divided into two large groups. Each of these two groups can be further classified as has seen done by Belonoshko and Saxena (1991b), but here we want to emphasize only one of their main distinctions.

The first group of EOSs consists of equations with parameters calculated directly from experimental data. The ideas of molecular size and interaction potentials are used to obtain a proper formulation of such EOSs. This approach has had a long history starting with the paper of van der Waals (1881) (Holloway, 1977; Bottinga and Richet, 1981; Delany and Helgeson, 1978; Ferry and Baumgartner, 1987; Fuller, 1976; Grevel, 1990; Halbach and Chatterjee, 1982; Kerrick and Jacobs, 1981; Mills et al., 1977; Mitchell and Nellis, 1982; Redlich and Kwong, 1949; Rice and Walsh, 1957; Rimbach and Chatterjee, 1987; Saxena and Fei, 1987a; Saxena and Fei, 1987b; Shmonov and Shmulovich, 1974; Tait, 1889; Tziklis et al., 1975). The list of references is not exhaustive and several important studies for low P-V-T data are not mentioned here.

To the second group belong the equations formulated to a considerable degree on the basis of statistical physics using the methods of computer simulation: molecular dynamics (MD) and Monte Carlo (MC). The main concept in this approach is one of molecular interaction. There exist numerous techniques to calculate the P-V-T properties if the interaction potential (IP) is known (Allen and Tildesley, 1987; Barker and Henderson, 1976; Brown, 1987; Boublik, 1977; Carnahan and Starling, 1969, 1972; Fiorese, 1980; Johnson and Shaw, 1985; Kalinichev and Heinzinger, 1991; Kataoka, 1987; Luckas and Lucas, 1989; Nellis et al., 1983; Ree, 1982; Ross, 1987; Ross and Ree, 1980; Ross et al., 1983; Saager and Fischer, 1990; Saager et al., 1990; Shmulovich et al., 1982; van Waveren et al., 1986; Weeks et al., 1971; Wentorf et al., 1950). The precision of the calculated data with such techniques is about the same as the precision in the estimation of the IP. Two of the most precise methods are MD and MC. The very first attempt to calculate properties of liquid argon with the MD approach resulted in a highly accurate reproduction of the existing experimental P-V-T data (Verlet, 1967, 1968). Since that time, the methods of computer "experiments" have been applied to solve a wide variety of problems in fields as different as molecular biology and aerodynamics.

We consider that the EOS-based MD or MC methods of simulation are better for extrapolation at high temperature and pressure than the EOSs of the first group. The EOSs of the first group are based on experimental data in the range of low temperature and pressure. These regressed coefficients could be wrong at a higher temperature and pressure. Therefore, the authors of these

EOSs must restrict the range of applicability of their equations which in most cases must be adhered to drastically. The EOSs of the second group are based on the fundamental properties of the matter energy of intermolecular interaction. If we can calculate IP from experimental data with a theoretically based method with high precision, we can apply MD and calculate P-V-T within the precision of the IP calculation. The nature of IP is the same under any temperature and pressure, as long as the species remain unchanged. It is evident for monatomic (simple) fluids and should be discussed in more detail for molecular fluids.

In this chapter, we review our recent calculations following the method of MD simulation. We calculated IP in the range of the experimental static P-V-T data. Using such IPs, we then simulated P-V-T data up to 4000 K and 1 Mbar with the MD technique for six species: water, carbon dioxide, methane, carbon monoxide, oxygen, and hydrogen. The simulated data were then fitted with a viriallike (Tait, 1889) equation for use in phase equilibrium computations for chemical systems at high to ultra-high pressures and high temperatures.

Simulation of the P-V-T Data Using MD

The aim of this section is not to present a complete review of the method of MD. It has been done many times before (e.g., Allen and Tidesley, 1987; Boublik, 1977; Kubicki and Lasaga, 1990). We present only the basic idea of the approach. For the sake of simplicity and brevity, many important details are omitted in the following description. We can imagine fluid (as well as solid, glass, melt, plasma, and so on) as a set of N particles (atoms, ions, molecules). If the initial coordinates and velocities are known, we can solve Newtonian equations of motion

$$m\frac{d^2 r_i}{dt^2} = f_i(r_1, r_2, \ldots, r_N), \qquad i = 1, 2, \ldots, N, \tag{1}$$

where $r_i = \{r_{xi}(t), r_{yi}(t), r_{zi}(t)\}$ is the vector of the ith particle, t time, and f_i force acting on the particle. If f_i at all i are known, it is possible to get as long a history of the set of N particles as is needed. The expression for f_i is

$$f_i = -\sum_{j \neq i}^{N} \frac{du_{ij}}{dr_{ij}}, \tag{2}$$

where u_{ij} is the potential energy of interaction between particles i and j, or IP.

Rene Descartes (1644) in "Principia Philosophiae" claimed that if he could know the coordinates and velocities of all bodies in the universe, he could calculate the future of the universe at any given time. This is exactly the idea of MD. Unfortunately, Rene Descartes had no computers to realize his dream. Furthemore, for some reasons it was a "little bit" difficult to know the coordinates and velocities of all bodies. As soon as computers were available, the MD approach became the method of choice. Although we still do not know the initial velocities and coordinates of all particles in a system of interest, it is quite apparent that

1. For fluids, it does not matter what the initial velocities and coordinates are (they influence only the time of "equilibration").
2. As compared to the number of bodies in the universe, a restricted number N of particles is sufficient for a meaningful solution.

There exist a number of techniques on avoiding surface effect and to solve numerically Eqs. (1). We refer the reader again to the several reviews mentioned before. Let us now assume that we know how to calculate the velocities and coordinates (and IP, of course) of N particles in a volume V. The statistical mechanics allow us to calculate in such a case P and T as follows:

$$T = \frac{1}{3Nk} \sum_{i=1}^{N} mv_i^2 \tag{3}$$

$$P = \frac{NkT}{V} - \frac{1}{3V} \sum_{i=1}^{N-1} \sum_{j>i}^{N} \frac{du_{ij}}{dr_{ij}} r_{ij}, \tag{4}$$

where v_i is the velocity of the ith particle, m the mass of the particle, and k the Boltzmann constant.

Calculation of IP in the C–O–H System

Strictly speaking for the exact prediction of the properties of matter, it is necessary to solve the many-body Schrodinger equation describing the motion of all nuclei and electrons. However, the following assumptions help us to simplify the calculations:

1. The Born–Oppenheimer (1927) approximation that the potential energy of a system of N atoms depends only on coordinates of nuclei.
2. The molecules are rigid.
3. The interactions are pair-additive.

These assumptions reduce the problem to that of the calculation of the potential energy of N molecules as follows:

$$U_N = \sum_{i<j} u_{ij}(r_i, \Omega_i, r_j, \Omega_j). \tag{5}$$

Hence, the calculation of configuration energy and the forces in Eq. (1) requires the knowledge of only the pair interactions between molecules, which depend on their positions and coordination r and Ω. However, it should be pointed out that the assumptions 2) and 3) can (and sometimes do) lead to remarkable errors in calculated values. Due to quantum effects, assumption 1) becomes untenable for light atoms like H or He at low temperatures. The necessity of accounting for the orientation of the molecules does not allow us to use simple analytical techniques to calculate thermodynamic properties. However, it is evident that under high enough temperatures any fluid can be imagined as an ideal gas. Of course, this temperature should be sufficiently high for a dense fluid to behave as an ideal gas. In any case, there must exist a low temperature at which we can describe the fluid in terms of a simple liquid; in other words, interaction in such fluids need not depend on the orientation of molecules above a certain "critical" temperature. This is true for monatomic and spherically symmetric molecules. It is enough for a fluid to be orientationally disordered to become a simple liquid. The question is, what is the value of such a critical temperature? The answer to this question one can only obtain posteriori, that is, by trying to calculate the fluid properties by disregarding the orientational effect and comparing the calculated data with the experimental.

Water among the fluids of a C–O–H composition is the most structured species due to hydrogen bonds. If we could show that under some not very high temperature it is possible to disregard the orientational interaction, then we could also ensure that such a temperature would be lower for other species. Gorbaty and Demjanetz (1983) have demonstrated experimentally that the water structure is very similar to that of a simple liquid under high temperature. The temperature and pressure of their experiments were up to 1000 bar and 773 K. Stillinger and Rahman's (1974) MD simulation of water showed that as the density of water increased from 1 to 1.346 g/cm^3, the average coordination number increased from 5.8 to 11.8. The latter value is typical for a simple fluid as a condensed phase. Later it was confirmed by Kalinichev (1986, 1991, and his chapter in this volume) that the high-temperature structure of water is indeed similar to the structure of a simple liquid. Undoubtedly, the contribution of hydrogen bond energy to the energy of interaction must be taken into account even for the simple liquid structure. However, as soon as high temperature leads to orientational disorder, this contribution can be effectively taken into account with the calculation of spherically symmetric IP (Belonoshko, 1989; Belonoshko and Shmulovich, 1986, 1987; Belonoshko and Saxena, 1991a).

An interaction between molecules of simple liquids depends only on the distance between them, which is the distance between their centers of mass. There are two types of IP most widely used: the Lennard–Jones (LJ) potential

$$u(r) = 4\varepsilon\left[\left(\frac{\sigma}{r}\right)^{12} - \left(\frac{\sigma}{r}\right)^{6}\right] \tag{6}$$

and the exponential-6 (exp-6) potential

$$u(r) = \varepsilon\left\{\frac{6}{\alpha - 6}\exp\left[\alpha\left(1 - \frac{r}{r^*}\right)\right] - \frac{\alpha}{\alpha - 6}\left(\frac{r^*}{r}\right)^{6}\right\}. \tag{7}$$

An interaction at small separations depends on r exponentially (Ross, 1987; Ross and Ree, 1980; Ross et al., 1983). The LJ IP becomes too stiff at small r, so it is understandable why the application of exp-6 IP allowed Ross and Ree (1983) and Nellis et al. (1983) to model shock-wave data. On the other hand, LJ IP allows us to reproduce experimental P-V-T data with good precision in the range of moderate densities (V/V_0 about 0.9–1.1) (Verlet, 1967; Barker and Henderson, 1976; Allen and Tildesley, 1987). It is natural to assume that IP such as LJ IP with equivalent r about σ would allow us to reproduce the experimental P-V-T in the range of moderate densities and the IP of the theoretically correct form such as exp-6 IP would be useful over a wide range of P-V-T.

The parameters of r^*, ε in Eq. (7) and σ, ε in (6) have identical sense. The parameter ε represents the depth of the energetic well, i.e., the minimum of the potential, and parameters r^* and σ are responsible for the location of this well. If $r^* = 2^{1/6}\sigma$, Eqs. (6) and (7) will result in the potential minimum. It means that if $\varepsilon_{LJ} = \varepsilon_{exp-6}$, then LJ and exp-6 will be very similar, especially at $\sigma < r < r^*$ exactly in the range of moderate densities. Hence, we can try to calculate parameters σ and ε of LJ IP and afterwards from MD simulation to calculate parameter α. It would be ideal to calculate all three parameters of exp-6 IP from a MD simulation. However, due to purely technical restrictions we are forced to adopt this solution.

The most well developed analytical theory of liquid state is perturbation theory (Weeks et al., 1971). This theory allows one to connect the parameters of the IP and properties of a fluid. Shmulovich et al. (1982) obtained a rather simple analytical expression for the pressure of LJ fluid. This expression was parametrized to coincide with MD simulation results for pressure under given volume, temperature, and parameters σ and ε. As one could expect, the pressure at high density in this method became too high due to the extreme stiffness of LJ IP. Nevertheless, the EOSs for CO_2, CH_4, and Ar up to 10 Kbar and under supercritical temperatures reproduced experimental values of pressure rather well. Their success prompted us to use this method of calculation of r^* as $2^{1/6}\sigma$ and ε for each of the gases and to calculate afterwards with these parameters the parameter α from MD simulations. The calculation was done by minimizing

$$\sum_{i=1}^{K}\left[1 - \frac{Z_{WCA}(T_i, P_i, \varepsilon, r^*)}{Z_{exp}(T_i, P_i)}\right], \tag{8}$$

where Z is the compressibility of a species and K the number of experimental points. The experimental data were chosen from the high-temperature and -pressure experimental measurements of volume (Babb et al., 1968; Burnham et al., 1969; Jusa et al., 1965; Kortbeek et al., 1986; Mel'nik, 1972, 1978a, 1978b; Mills et al., 1977; Presnall, 1969; Robertson and Babb, 1970; Shmonov and Shmulovich, 1974; Shmulovich and Shmonov, 1978; Tziklis, 1977; Tziklis and Kulikova, 1965; Tziklis et al.,1975; van Thiel and Wasley, 1964; Vukalovich et al., 1963). The temperature in (8) was higher than the critical temperature and in all cases not less than the room temperature. The pressure in (8) was higher than 2000 bars (the only exception is hydrogen pressure, 1500 bars). We have assumed that under these conditions the C–H–O fluids are simple fluids as discussed before.

The finally adopted parameters correspond to the fit of the experimental P-V-T data within 2 to 3% errors on compressibility. Therefore, we obtained a two-parameter EOS for supercritical fluids in a C–O–H system above 2 Kbar over the ranges of experimental P and T. This EOS cannot be used at high fluid densities because calculated pressure is too high in comparison to the observed or computed data (Dick, 1972a, 1972b; Fiorese, 1980; Hill, 1990; Johnson and Shaw, 1985; Mills et al., 1977; Mitchell and Nellis, 1982; Nellis et al., 1981, 1983; Rice and Walsh, 1957; Ross et al.,1983; Saul and Wagner, 1989; Walsh and Rice, 1957; Zubarev and Telegin, 1962). Therefore, the third parameter α of exp-6 IP was calculated with the MD method simulating the experimental pressure (Table 1). The calculated parameters of IP (7) are shown in Table 2. Our parameters of effective IP (7) and parameters of effective IP (6) and (7) calculated by others are compared in Table 3. The table shows that our parameters are not at significant variance with the data of others. The only difference is that our set of parameters is self-coordinated.

The present use of effective IP rather than a rigorously calculated many-centered potential may be justified by considering the results of Kataoka (1987) and others. Kataoka's (1987) could only simulate the P-V-T data with "real" (i.e., quantum chemical IP with the real structure of the H_2O molecule) water–water IP over a wide range of density and temperature with significant errors of pressure. Another example may be found in the work on methane. Van Waveren

Table 1. Anchor points to calculate parameter α of potential (7).

	T(K)	V(cm^3/mol)	P(bar)	Ref.
H_2O	800	17.79	10,000	Burnham et al., 1969
CO_2	1000	38.15	8,000	Shmulovich and Shmonov, 1978
CH_4	673	34.17	8,106	Tziklis, 1977
CO	573	29.09	10,000	Babb et al., 1968
O_2	673	25.75	10,132	Tziklis and Koulikova, 1965
H_2	300	10.0	30,200	Mills et al., 1977

Table 2. Calculated parameters of effective IP[a].

	ε/k(K)	σ(Å)	α
H_2O	446.7	2.8431	12.75
CO_2	198.634	3.6963	13.9
CH_4	140.941	3.7165	14.05
CO	90.805	3.6531	14.48
O_2	99.495	3.3688	14.45
H_2	25.144	2.9780	13.34

[a] Parameter $r^* = 2^{1/6}\sigma$.

Table 3. Potential minima location divided $2^{1/6}(\sigma)$ and depth of well ε/k calculated earlier.

	$\varepsilon/k(K)$	$\sigma(\text{Å})$	Potential	Ref.
H_2O	371.0	2.89	LJ	Ben–Amotz and Herschbach, 1990
	809.1	2.641	LJ	Svehla, 1962
CO_2	195.2	3.941	LJ	Svehla, 1962
	245.6	3.716	exp-6	Nellis et al., 1981[a]
	211.3	3.6963	LJ	Shmulovich et al., 1982[b]
	338.3	3.6468	exp-6	Johnson and Shaw, 1985[c]
	247.0	3.69	LJ	Ben–Amotz and Herschbach, 1990
CH_4	136.5	3.882	LJ	Lennard–Jones and Ingham, 1925
	148.6	3.758	LJ	Svehla, 1962
	149.1	3.743	LJ	McDonald and Singer, 1972
	154.1	3.761	exp-6	Ross and Rese, 1980[a]
	139.571	3.725	LJ	Shmulovich et al., 1982
	149.92	3.7327	LJ	Fischer et al., 1984
	147.8	3.73	LJ	Jorgensen et al., 1988
	142.0	3.73	LJ	Ben–Amotz and Herschbach, 1990
CO	91.7	3.690	LJ	Svehla, 1962
	108.3	3.645	exp-6	Nellis et al., 1981[a]
	90.38	3.6534	LJ	Shmulovich et al., 1982[b]
	98.0	3.69	LJ	Ben–Amotz and Herschbach, 1990
O_2	106.7	3.467	LJ	Svehla, 1962
	125.0	3.422	exp-6	Nellis et al., 1981[a]
H_2	33.3	2.968	LJ	Lennard–Jones and Ingham, 1925
	59.7	2.827	LJ	Svehla, 1962
	25.8457	2.9568	LJ	Shmulovich et al., 1982[b]
	36.4	3.0558	exp-6	Ross et al., 1983[d]

[a] Parameter α of exp-6 potential equals 13.0.

[b] Parameters are temperature-dependent in the model; values are given at 500 K.

[c] Parameter α of exp-6 potential equals 14.4.

[d] Parameter α of exp-6 potential equals 11.1.

et al. (1986) used the five-centerd LJ potential to calculate isothermal pressures at room temperature and high densities; their calculated pressure at a molar volume 31.25 cm^3 was about 60% too high. The same P-V-T data were modeled by Saager et al. (1990) and Saager and Fischer (1990) using an effective IP within the experimental errors. It appears that the sophisticated many-centered IPs (e.g., Kalinichev and Heinzinger, 1991) have too many parameters to be determined as a self-coordinated set to reproduce fluid properties with high precision. At the same time, the IP proposed might not allow us to model properties strictly depending on structure such as diffusion or viscosity. Therefore, we may conclude that there are still many uncertainties in the calculation of many-centered IP.

MD-Simulated Data and the EOS

The P-V-T data for six species in the C–O–H system were simulated with the MD approach using potential (7) with the parameters in Table 2. The technical details of simulation were described previously (Belonoshko and Saxena, 1991a, 1991b). The goal was to get the set of P-V-T points more or less uniformly distributed in the temperature and pressure space. For that purpose, pressure under specified T and V has been calculated. The range of pressures, volumes, and temperatures is presented in Table 4.

The statistical errors of calculated T and P are of the same magnitude as those calculated with the commonly adopted methods (Allen and Tildesley, 1987; Hill, 1962). It should be taken into account that the lower the pressure, the larger the error of pressure calculation.

As discussed recently by Belonoshko and Saxena (1991), the SW data are quite consistent with our simulated data on H_2O, CO_2, CH_4, O_2, and H_2. Note that this consistency exists only in a broad sense. In comparing the simulated and the SW data, it is important to note that the estimated temperature on the Hugoniots has a large uncertainty. There are several experimental (SW) data for which the temperatures (as evaluated by Nellis et al., 1981; Ross and Ree, 1980; Ross, 1987) were above the upper limit of our calculations (approximately 4000 K). Our results on CO do not fit the SW data above about 230 Kbar. Nellis et al. (1981) found that the products of the SW experiments above 230 Kbar are solid carbon, O_2 CO, and CO_2. Of course, even a small amount of a solid phase precipitated from a fluid would decrease the total pressure drastically. The comparison of our results with the results of other models and experimental P-V-T data (Babb et al., 1968; Bottinga and Richet, 1981; Grevel, 1990; Halbach and Chatterjee, 1982; Shmonov and Shmulovich, 1974; Shmulovich and Shmonov, 1978; Tziklis and Koulikova, 1965; Tziklis et al., 1975) is discussed below.

The MD-simulated P-V-T data together with experimental P-V-T data in the range above 5 Kbar as referred to before are fitted with the following polynomial in T and V:

$$P = \frac{a}{V} + \frac{b}{V^2} + \frac{c}{V^m}, \qquad (9)$$

Table 4. P-V-T range of MD-simulated data.

Fluid	V(cm^3/mol)	T(K)	P(Kbar)
H_2O	30–7	695–4356	1.35–1044.12
CO_2	40–11	401–5060	2.57–1100.78
CH_4	40–9	409–4293	2.89–1872.04
CO	40–9	393–4286	2.69–1301.81
O_2	40–8	394–4067	1.80–1114.88
H_2	25–3	296–3995	2.36–2271.80

where

$$a = \left(a_1 + a_2 \frac{T}{1000}\right) \times 10^4$$

$$b = \left(b_1 + b_2 \frac{T}{1000}\right) \times 10^6$$

$$c = \left(c_1 + c_2 \frac{T}{1000}\right) \times 10^9.$$

Equation (9) without the third term is very similar to the equation of Tait (1889). Sysoyev (1980) showed with the theory of perturbation of liquid that the Tait equation appears theoreticaly reliable. Vasserman and Rabinovitch (1968) also applied an EOS of that kind to describe the properties of liquid air and its components. Spiridonov and Kvasov (1986) reviewed an EOS for dense fluids and showed that an EOS similar to the EOS (9) is the most reliable and effective for a description of the properties of the dense fluid phase.

The coefficients and some other data on the fits are shown in Table 5. The average error in pressure is between 3.27 to 5.53%, which corresponds to an average error of 1 to 2% in volume. In the range of low temperature and high pressure, i.e., in the range of high compressibility (Z around 100 and higher), the maximum error is about 15 to 20%. Therefore, the error in volume in these points is of the same magnitude, i.e., to 2%. The quality of the fit of the calculated data to the experimental data and a comparison with the high P-T data from other equations are shown in Fig. 1. There are two SW experiments with simultaneously measured temperature. The comparison of calculated and measured volumes for H_2O and CH_4 under experimental T and P is shown in Table 6. The molar volumes and fugacities are given in the appendix. The fugacities of the fluids at T and $P > 5$ Kbar may be accomplished by using the following equation for the experimentally determined fugacities, as reviewed by

Table 5. Parameters of EOS (9) in C–O–H system.

	H_2O	CO_2	CH_4	CO	O_2	H_2
a_1	14.0203	18.3050	21.7694	8.4373	23.5710	5.1381
a_2	−4.1234	−3.2756	0.4731	5.6585	−7.3095	5.7896
b_1	−7.7278	−10.9876	−13.2852	−5.7048	−8.6234	−1.0837
b_2	6.7012	12.5299	9.7013	7.0165	8.1585	1.9852
c_1	6.5201	85.3963	18.4128	13.0116	6.3861	0.1385
c_2	−0.4558	−6.4540	−0.8874	−0.4606	−0.3475	−0.0066
m	4.5860	4.6877	4.1674	4.2415	4.2187	3.6637
N points	141	151	79	84	94	74
Average error, %	3.56	4.57	3.72	3.27	5.53	3.49
Maximum error, %	17.57	22.82	11.41	12.79	16.10	10.76

Equations of State of Fluids at High Temperature and Pressure   89

Fig. 1. The *P-V-T* data as calculated from the existing equations and the experimental data as referred in the text and compared to the data from Eq. (9) together with MD-simulated points (stars): (a) H_2O, (b) CO_2, (c) CH_4, (d) CO, (e) O_2, (f) H_2.

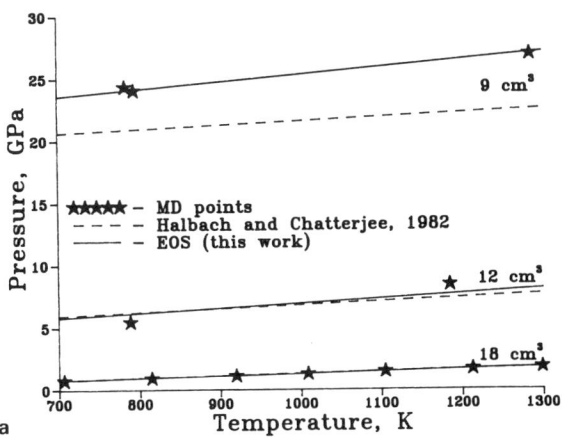

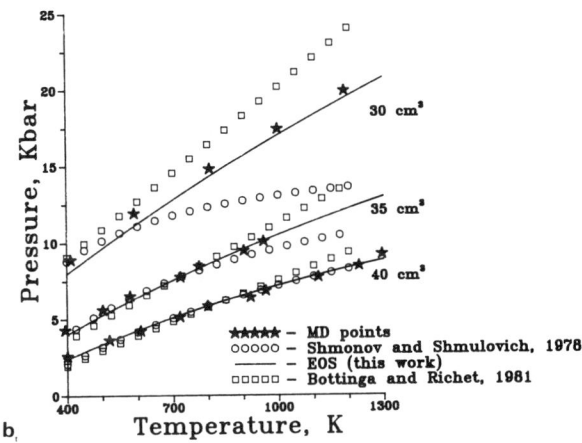

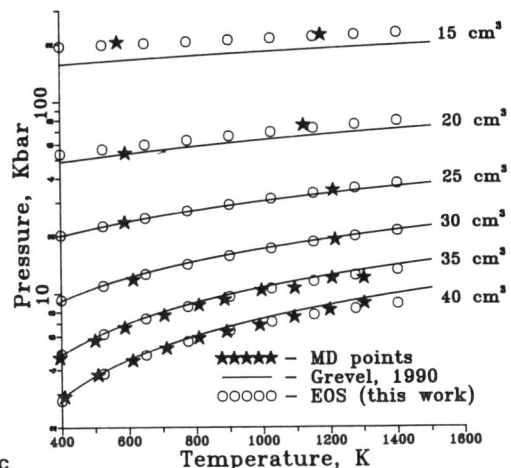

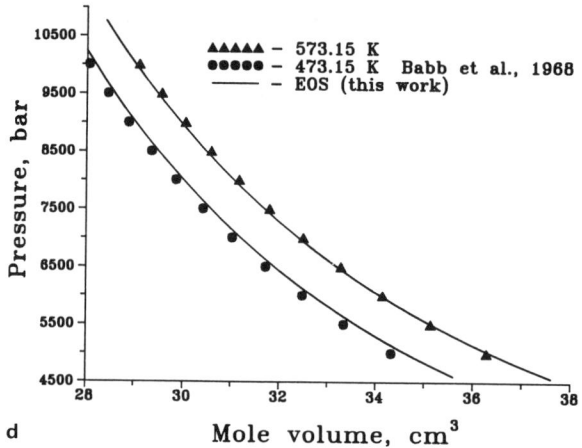
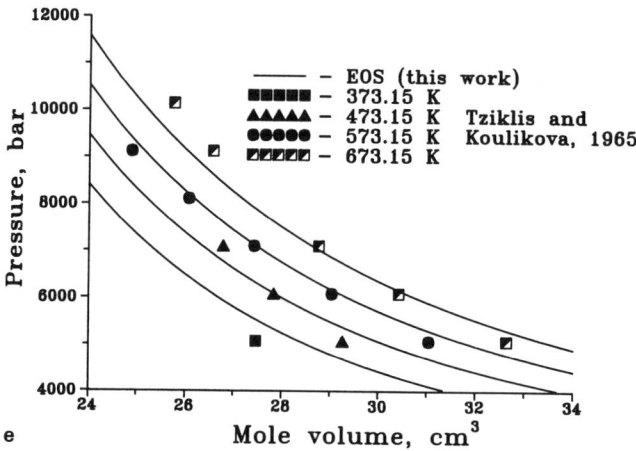
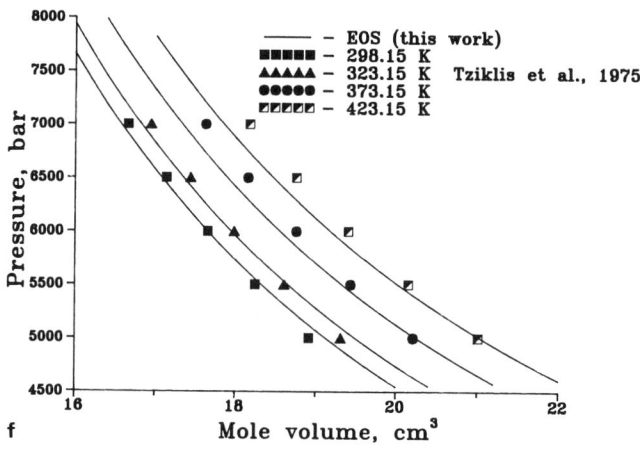

Fig. 1. (Continued.)

Table 6. Experimental and calculated densities at shock-wave experimental P-V-T.

	P(GPa)	T(K)	V(cm³/mol) Exp.	V(cm³/mol) Calc.	Ref.
H_2O	48.9 (0.9)[a]	3280	7.978	8.417	Lysenga et al., 1982
H_2O	58.5 (1.1)	3830	7.705	8.147	Lysenga et al., 1982
H_2O	61.9 (1.2)	4090	7.640	8.078	Lysenga et al., 1982
H_2O	71.0 (3.0)	4480	7.481	7.842	Lysenga et al., 1982
H_2O	80.0 (1.8)	5270	7.300	7.728	Lysenga et al., 1982
CH_4	33.5	3700 (200)	13.76	14.51	Radousky et al., 1990
CH_4	42.0	4330 (200)	12.98	13.80	Radousky et al., 1990

[a] Numbers in parentheses stand for experimental error.

Table 7. Coefficients of Eq. (10).

	H_2O	CO_2	CH_4	CO	O_2	H_2
f_1	−130.517	−40.468	−19.293	−3.382	−10.370	6.803
f_2	0.0650	0.0670	0.0673	0.0696	0.689	0.0701
f_3	19.483	8.348	5.407	2.779	3.450	0.177

Note: CO_2 coefficients from Saxena and Fei (1987a).

Mel'nik (1972, 1978a, 1978b):

$$RT \ln f(5 \text{ Kbar}, T) = 1000 \times (f_1 + f_2 T + f_3 \ln T). \tag{10}$$

The coefficients are listed in Table 7.

Reliability of EOS at Extremal Temperature and Pressure

The characteristic error in P-V-T calculations from the proposed method here is about 5% on volume at the highest T and P and is about 1 to 2%, i.e., about the same as the experimental error in the low range of temperature and pressure. The error of fugacity calculation is proportional to the error in volume. Therefore, we have an error of the same magnitude for fugacity. Of course, all this applies if there is no dissociation of fluids at high temperature and pressure. For example, Ree (1982) discusses the possibility that along the Hugoniot, dissociation may start at approximately 3000 K and 40 GPa. The work of Hamann (1981) is indicative of the dissociation of water into hydrogen and hydroxyl ions between shock densities of 1.6 and 1.9 g/cm^3 at a temperature of 1300 K or more. However, direct structural measurements of shock-wave compressed water with Raman spectrometry did not show the presence of such ions in water (Holmes et al., 1985). Radousky et al. (1990) also indicated the significant stability of the shock-wave compressed methane up to very high pressures and temperatures. Thus, the data on the dissociation and stability of high-TP fluids are controversial. We must wait for experimental determinations.

The fluids in the temperature and pressure range investigated may undergo solid-fluid transformation (Grace and Kennedy, 1967; Stishov, 1974). One cannot apply the theory of liquid state in this case, as was done for high compressed fluid (see Brown, 1987 and Ree, 1982). The advantage of our approach is that the MD technique allows us to calculate properties both in the liquid and solid state.

Conclusions

With the MD simulation, we have calculated the properties of six fluids of C–O–H composition (water, carbon dioxide, methane, carbon monoxide, oxygen, hydrogen), the most commonly occurring species in natural fluids. The theory of perturbation of liquid and direct MD calculation has been used for the calculation of the potential of intermolecular interaction. The IP adopted is of the α-exp-6 type. The data obtained with a MD simulation up to 4000 K and 1 Mbar, together with the low-temperature and -pressure data, were fitted to obtain an EOS for six species. The range of applicability of the EOS obtained is

$$5000 \text{ bar} < P < 1 \text{ Mbar},$$

$$T_{\text{spec}} < T < 4000 \text{ K},$$

where T_{spec} is 400 K for all species except water. T_{spec} for water equals 700 K. A comparison with the available experimental data shows that our EOSs reproduce these data with a maximum error of about 5 to 6% on volume and fugacity. It makes these EOSs useful for geochemical calculations in the specified range of T and P, i.e., under the Earth's mantle conditions.

Acknowledgments

This research was financially supported by the Swedish Natural Science Research Council (NFR).

References

Allen, M.P. and Tildesley, D.J. (1987). *Computer Simulation of Liquids*. Clarendon Press, Oxford.

Babb, S.E., Robertson, S.E., and Scott, G.T. (1968). *PVT* properties of gases at high pressures. Final Rept., Univ. Oklahoma Res. Inst.

Barker, J.A. and Henderson, D. (1976). What is "liquid"? Understanding the states of matter. *Rev. Mod. Phys.* **48**(4), 587–671.
Belonoshko, A.B. (1988). Molecular dynamics simulation of water on β-quartz surface. *Zh. Phis. Khimii* **62**(1), 118–121 (in Russian).
Belonoshko, A.B. (1989). The thermodynamics of the aqueous carbon dioxide fluid within thin pores. *Geoch. Cosm. Acta* **53**(10), 2581–2590.
Belonoshko, A.B. and Shmulovich, K.I. (1986). Molecular dynamics study of dense fluid in micropores. *Geokhimiya* (11), 1523–1534 (in Russian).
Belonoshko, A.B. and Shmulovich, K.I. (1987). Fluid phase in thin porous media under high pressure. *Doklady Akademii Nauk SSSR* **295**(1), 625–629 (in Russian).
Belonoshko, A.B. and Saxena, S.K. (1991a). A molecular dynamics study of the pressure-volume-temperature properties of supercritical fluids: I. H_2O. *Geoch. Cosmochim. Acta* **55**, 381–388.
Belonoshko, A.B. and Saxena, S.K. (1991b). A molecular dynamics study of the pressure-volume-temperature properties of supercritical fluids: II. CO_2, CH_4, CO, O_2, H_2. *Geoch. Cosmochim. Acta*. In press
Ben–Amotz, D. and Herschbach, D.R. (1990). Estimation of effective diameters for molecular fluids. *J. Phys. Chem.* **94**, 1038–1047.
Born, M. and Oppenheimer, J.R. (1927). 1. Zur Quantentheorie der Molekeln. *Ann. Phys. (Leipz)* **84**, 457–484.
Bottinga, Y. and Richet, P. (1981). High pressure and temperature equation of state and calculation of the thermodynamic properties of gaseous carbon dioxide. *Amer. J. Sci.* **281**, 620–659.
Boublik, T. (1977). Progress in statistical thermodynamics applied to fluid phase. *Fluid Phase Equilibria* **1**, 37–87.
Brown, W.B. (1987). Analytical representation of the excess thermodynamic equation of state for classical fluid mixtures of molecules interacting with α-exponential-six pair potentials up to high densities. *J. Chem. Phys.* **87**, 566–577.
Burnham, C.W., Holloway, J.R., and Davis, N.F. (1969). Thermodynamic properties of water to 1000°C and 10,000 bars. Paper 132, Geological Society of America. Washington, D.C.
Carnahan, N.F. and Starling, K.E. (1969). Equation of state for nonattracting rigid spheres. *J. Chem. Phys.* **51**, 635–636.
Carnahan, N.F. and Starling, K.E. (1972). Intermolecular repulsions and the equation of state for fluids. *Amer. Inst. Chem. Eng.* **18**, 1184–1189.
Delany, J.M. and Helgeson, H.C. (1978). Calculation of the thermodynamic consequences of dehydration in subducting oceanic crust to 100 Kb and > 800°C. *Amer. J. Sci.* **278**, 638–686.
Dick, R.D. (1972a). Some Hugoniot data for liquid deuterium and hydrogen. *Bull. Amer. Phys. Soc.* **17**, 1092.
Dick, R.D. (1972b). Shock wave data for liquid hydrogen initially at 20 K. *Bull. Amer. Phys. Soc.*, **17**, 1302.
Ferry, J.M. and Baumgartner, L. (1987). Thermodynamic models of molecular fluids at the elevated pressures and temperatures of crustal metamorphism. *Rev. Mineral.* **17**, 325–365.
Fiorese, G. (1980). Monte-Carlo calculations for molecular H_2 in the fluid phase. *J. Chem. Phys.* **73**, 6308–6315.
Fischer, J., Lustig, R., Breitenfelder-Manske, H. and Lemming, W. (1984). Influence of

intermolecular potential parameters on orthobaric properties of fluids consisting of spherical and linear molecules. *Mol. Phys.* **52**, 485–497.

Fuller, G.G. (1976). A modified Redlich–Kwong–Soave equation of state capable of representing the liquid state. *Ind. Eng. Chem. Fundam.* **15**, 254–257.

Gorbaty, Yu. E. and Demjanetz, Yu. N. (1983). The pair correlation functions of water at a pressure of 1000 bar in the temperature range 25–500°C. *Chem. Phys. Lett.* **100**, 450–453.

Grace, J.D. and Kennedy, G.C. (1967). The melting curve of five gases to 30 Kbar. *J. Phys. Chem. Solids* **28**, 977–981.

Grevel, K.-D. (1990). A modified Redlich–Kwong equation of state for methane at temperatures between 150 K and 1500 K and pressures up to 300 Kbar, in *Thermodynamic Data Systematics*, Uppsala Univ. Symp., Wik, June 10–14, 1990, abstract.

Halbach, H. and Chatterjee, N.D. (1982). An empirical Redlich-Kwong type equation of state for water to 1000°C and 200 Kbar. *Contrib. Mineral. Petrol.* **79**, 337–345.

Hamann, S.D. (1981). Properties of electrolyte solutions at high pressures and temperatures, in *Physics and Chemistry of the Earth*, D.T. Rickard and F.E. Wickman, eds., Vol. 13, Oxford, pp. 89–112.

Hill, P.G. (1990). A unified fundamental equation for the thermodynamic properties of H_2O. *J. Phys. Chem. Data* **19**, 1233–1274.

Hill, T.L. (1962). *An Introduction to Statistical Thermodynamics*. Addison-Wesley, Reading, Mass.

Holloway, J.R. (1977). Fugacity and activity of molecular species in super-critical fluids, in *Thermodynamics in Geology*, D.G. Fraser, ed., Dordrecht-Holland, pp. 161–181.

Holmes, N.C., Nellis, W.J., Graham, W.B., and Walrafen, C.E. (1985). Spontaneous Raman scattering from shocked water. *Phys. Rev. Lett.* **55**, 2433–2436.

Johnson, J.D. and Shaw, M.S. (1985). Thermodynamics using effective spherical potentials for CO_2 anisotropies. *J. Chem. Phys.* **83**, 1271–1275.

Jorgensen, W.L., Buckner, J.K., Boudon, S., and Tirado-Rives, J. (1988) Efficient computation of absolute free energy of binding by computer simulations. Application to the methane dimer in water. *J. Chem. Phys.* **89**, 3742–3746.

Jusa, I., Kmonicek, V., and Sifner, O. (1965). Measurements of the specific volume of carbon dioxide in the range of 700 to 4000 bar and 50 to 475°C. *Physica* **31**, 1735–1744.

Kalinichev, A.G. (1986). Monte Carlo study of the thermodynamics and structure of dense supercritical water. *Internat. J. Thermophys.* **7**, 887–900.

Kalinichev, A.G. and Heinzinger, K. (1991). Computer simulation of aqueous fluids at high temperature and pressure, in *Advance in Physical Geochemistry*, S.K. Saxena, ed., Vol. 10, Springer-Verlag, New York.

Kataoka, J. (1987). Studies of liquid water by computer simulations. V. Equation of state of fluid water with Caravetta–Clementi potential. *J. Chem. Phys.* **87**, 589–596.

Kerrick D.M. and Jacobs G.K. (1981). A modified Redlich–Kwong equation for H_2O, CO_2 and H_2O-CO_2 mixtures at elevated pressures and temperatures. *Amer. J. Sci.* **281**(6), 735–767.

Kortbeek, P.J., Biswas, S.N., and Trappeniers, N.J. (1986). pVT and sound velocity measurements for CH_4 up to 10 kbar. *Physica* **139/140B**, 109–112.

Kubicki, J.D. and Lasaga, A.C. (1990). Molecular dynamics and diffusion in silicate melts, in *Advances in Physical Geochemistry*, J. Ganguly, ed., Vol. 8. Springer-Verlag, New York. 1–50.

Lennard–Jones, J.E., and Ingham, A.E. (1925). On the calculation of certain crystal potential constants, and on the cubic crystal of least potential energy. *Proc. Roy. Soc.* **107A**, 636–653.

Luckas, M. and Lucas, K. (1989). Thermodynamic properties of fluid carbon dioxide from the SSR-MPA potential. *Fluid Phase Equilibria* **45**, 7–23.

Lysenga, G.A., Ahrens, T.J., Nellis, W.J., and Mitchell, A.C. (1982). The temperature of shock-compressed water. *J. Chem. Phys.* **76**, 6282–6286.

Mel'nik, Yu. P. (1972). Thermodynamic parameters of compressed gases and metamorphic reactions involving water and carbon dioxide. *Geokhimiya* (6), 654–662 (in Russian).

Mel'nik, Yu. P. (1978a). Thermodynamic properties of carbon monoxide and methane at high temperatures and pressures—a new correlation based on the principle of corresponding states. *Geokhimiya* (11), 1677–1691 (in Russian).

Mel'nik, Yu. P. (1978b). *Termodinamicheskiye svoistva gazov v usloviyakh glubinnogo petrogenezisa.* Naukova Dumka, Kiev (in Russian).

Mills R.L., Liebenberg J.C., Bronson J.C., and Schmidt L.C. (1977). Equation of state of fluid n-H_2 from P-V-T and sound velocity measurements to 20 Kbar. *J. Chem. Phys.* **66**, 3076–3084.

Mitchell, A.C. and Nellis, W.J. (1982). Equation of state and electrical conductivity of water shocked to the 100 GPa (1 Mbar) pressure range. *J. Chem. Phys.* **76**, 6273–6281.

Nellis, W.J. and Mitchell, A.C. (1980). Shock compression of liquid argon, nitrogen and oxygen to 90 GPa (900 Kbar). *J. Chem. Phys.* **73**, 6137–6145.

Nellis, W.J., Ree, F.H., van Thiel, M., and Mitchell, A.C. (1981). Shock compression of liquid carbon monoxide and methane to 90 GPa (900 kbar). *J. Chem. Phys.* **75**, 3055–3063.

Nellis, W.J., Ross, M., Mitchell, A.C., van Thiel, M., Young, D.A., Ree, F.H., and Trainor, R.J. (1983). Equation of state for molecular hydrogen and deuterium from shock-wave experiments to 760 Kbar. *Phys. Rev.* **A2**, 608–611.

Presnall, D.C. (1969). Pressure-volume-temperature measurements on hydrogen from 200°C to 600°C and up to 1800 atmospheres. *J. Geophys. Res.* **74**, 6026–6033.

Radousky, H.B., Mitchell, A.C., and Nellis, W.J. (1990). Shock temperature measurements of planetary ices: NH_3, CH_4 and "synthetic Uranus." *J. Chem. Phys.* **93**, 8235–8239.

Redlich, O. and Kwong, J.N.S. (1949). On the thermodynamics of solutions: V. An equation of state. Fugacities of gaseous solutions. *Chem. Rev.* **44**, 233–244.

Ree, F.H. (1982). Molecular interaction of dense water at high temperature. *J. Chem. Phys.* **76**, 6287–6302.

Rice, M.H. and Walsh, J.M. (1957). Equation of state of water to 250 Kilobars. *J. Chem. Phys.* **26**, 824–830.

Rimbach, H. and Chatterjee, N.D. (1987). Equations of state for H_2, H_2O, and H_2-H_2O fluid mixtures at temperatures above 0.01°C and at high pressures. *Phys. Chem. Min.* **14**, 560–569.

Robertson, S.L. and Babb, S.E. Jr. (1970). Isotherms of carbon monoxide to 10 Kbar and 300°C. *J. Chem. Phys.* **53**, 1094–1097.

Ross, M. (1987). Physics of dense fluids, in *High Pressure Chemistry and Biochemistry*, R. van Eldik and J. Jonas, eds., Reidel, Dordrecht, pp. 9–49.

Ross, M. and Ree, F.H. (1980). Repulsive forces of simple molecules and mixtures at high density and temperature. *J. Chem. Phys.* **73**, 6146–6152.

Ross, M., Ree, F.H., and Young, D.A. (1983). The equation of state of molecular hydrogen at very high density. *J. Chem. Phys.* **79**, 1487–1494.

Saager, B., Lotfi, A., Bohn, M., Nguyen, V.N., and Fischer, J. (1990). Prediction of gas PVT data with effective intermolecular potentials using the Haar–Shenker–Kohler equation and computer simulations. *Fluid Phase Equilibria* **54**, 237–246.

Saager, B. and Fischer, J. (1990). Predictive power of effective intermolecular pair potentials: MD simulation results for methane up to 1000 MPa. *Fluid Phase Equilibria* **57**, 35–46.

Saul, A. and Wagner, W. (1989). A fundamental equation for water covering the range from the melting line to 1273 K at pressures up to 25,000 MPa. *J. Phys. Chem. Ref. Data* **18**(4), 1537–1563.

Saxena, S.K. and Fei, Y. (1987a). High pressure and high temperature fugacities. *Geochim. Cosmochim. Acta*, **51**, 783–791.

Saxena, S.K. and Fei, Y. (1987b). Fluids at crustal pressures and temperatures. I. Pure species. *Contrib. Mineral. Petrol.* **95**, 370–375.

Shmonov, V.M. and Shmulovich, K.I. (1974). Molal volumes and equations of state of CO_2 at temperatures from 100 to 1,000°C and pressures from 2,000 to 10,000 bars. *Akad Nauk SSSR Doklady* **217**, 205–209 (in Russian).

Shmulovich, K.I. and Shmonov, V.M. (1978). *Tables of Thermodynamic Properties of Gases and Liquids, Carbon Dioxide*. Moscow, Standard Press (in Russian).

Shmulovich, K.I., Tereschenko, E.N., and Kalinichev, A.G. (1982). Equation of state and isochores of nonpolar gases up to 2000 K and 10 GPa. *Geokhimija* (11), 1598–1613 (in Russian).

Spiridonov, G.A. and Kvasov, I.S. (1986). Empirical and semiempirical equations of state for gases and liquids. *Rev. Thermophy. Prop. Matt.* **57**(1), 45–116, (in Russian).

Stillinger, F.H. and Rahman, A. (1974). Improved simulation of liquid water by molecular dynamics. *J. Chem. Phys.* **60**, 1545–1557.

Stishov, S.M. (1974). Thermodynamics of melting of pure species. *Uspekhi Fiz. Nauk* **114**, 3–29 (in Russian).

Svehla, R.A. (1962). Lennard–Jones potential parameters from viscosity data. NASA Tech. R-132, Lewis Res. Ctr., Cleveland, Ohio.

Sysoev, V.A. (1980). Isothermal equation of state for dense gases and liquids. One-component systems. *Ukrainskii Phiz. Zh.* **25**(1), 123–130 (in Russian).

Tait, P.S. (1889). On the virial equation for molecular forces, being Part IV. of a paper on the foundations of the kinetic theory of gases. *Proc. Roy. Soc. Edin.* **16**, 65–72.

Tziklis, D.S. (1977). *Dense Gases*. Moscow, Khimija Press (in Russian).

Tziklis, D.S. and Koulikova, A.I. (1965). Oxygen compressibility determination at pressure to 10,000 atm and temperature to 400°C. *Zh. Phiz. Khimii* **39**, 1752–1756 (in Russian).

Tziklis, D.S., Maslennikova, V.Y., Gavrilov, S.D., Egorov, A.N., and Timofeeva G.V. (1975). Molar volumes and equation of state of molecular hydrogen at high pressures. *Dokl. Akad. Nauk SSSR* **220**, 189–191 (in Russian).

van der Waals, J.H. (1881). *Die Continuitat des Gasformigen und Flussigen Zustcindes.* Leipzig, Barth.

van Thiel, M. and Wasley, M. (1964). Compressibility of liquid hydrogen to 40,000 atm and 1100 K. U.S. Atomic Energy Comm., Univ. Calif., Livermore.

van Waveren, G.M., Michels, J.P.J., and Trappaniers, N.J. (1986). Molecular dynamics simulation of CH_4 in the dense fluid phase. *Physica B* **139/140**, 144–147.

Vasserman, A.A. and Rabinovitch, V.A. (1968). *Thermophysical Properties of Liquid Air and Its Components*. Moscow, Standart Press (in Russian).

Verlet, L. (1967). Computer "experiments" on classical fluids. I. Thermodynamical properties of Lennard–Jones molecules. *Phys. Rev.* **159**, 98–103.

Verlet, L. (1968). Computer "experiments" on classical fluids. II. Equilibrium correlation functions. *Phys. Rev.* **165**, 201–214.

Vukalovich, M.P., Altunin, V.V., and Timoshenko, N.I. (1963). Specific volume of CO_2 at high pressure and temperature. *Teploenergetika* (10), 92–93 (in Russian).

Walsh, J.M. and Rice, M.H. (1957). Dynamic compression of liquids from measurements of strong shock waves. *J. Chem. Phys.* **26**, 815–823.

Weeks, J.D., Chandler, D., and Andersen, H.C. (1971). Role of repulsive forces in determining the equilibrium structure of simple liquids. *J. Chem. Phys.* **54**, 5237–5246.

Zubarev, V.N. and Telegin, G.S. (1962). Shock compressibility of liquid nitrogen and solid carbon dioxide. *Doklady Akad Nauk SSSR* **142**, 309–312 (in Russian).

Chapter 4
Thermodynamic Properties of Minerals: Macroscopic and Microscopic Approaches

P. Richet, P. Gillet, and G. Fiquet

Introduction

Thermodynamic modeling of experimental or natural-phase equilibria has become an integral part of petrology. In this respect, the isobaric heat capacity (C_p) has manifold importance. First, C_p data constitute the basis of third-law determinations of the entropy of minerals. Second, these data are needed to calculate the variation with temperature of the entropy, the enthalpy, and the Gibbs free energy. As a result, it necessary to know accurately heat capacities when retrieving thermodynamic information from phase equilibria data, especially when trying to separate the effects of the enthalpies and entropies of transformation.

In this paper, we broadly review the main empiricial and theoretical aspects of the heat capacity of minerals. We begin with a brief review of the three main techniques that are currently in use for determining heat capacities from 0 to 2000 K, namely, adiabatic, differential scanning (DSC), and drop colarimetry, paying attention to the experimental constraints that limit measurements to certain conditions. When minerals can be subjected at best to limited calorimetric measurements, either because of lack of gram-sized samples or of instability at high temperatures (as if often the case with high-pressure minerals), other ways have to be found for predicting standard entropies and high-temperature properties. The validity of empiricial methods of prediction of the heat capacity as a function of temperature and composition will thus be discussed.

Because these empirical methods have also their own limitations, it is useful to turn to C_p calculations from spectroscopic data for minerals that have not been investigated calorimetrically. We give a few examples to illustrate the usefulness of the well-known model of Kieffer (1979), that is applicable to virtually all minerals thanks to the rather limited spectroscopic information it requires. A shortcoming of this model is its harmonic nature, which prevents reliable calculations at mantle temperatures. Hence, we also discuss the role of anharmonicity and the simple way it can be taken into account when high-pressure and high-temperature spectroscopic measurements are available.

In fact, the property calculated from spectroscopic data is the isochoric heat capacity (C_v), which is related to C_p by

$$C_v = C_p - TV\alpha^2 K_T, \qquad (1)$$

where α is the thermal expansion coefficient and K_T the isothermal bulk modulus. However, thermal expansion coefficients and bulk moduli are generally poorly known at high temperatures. As pointed out by Richet et al. (1982), Fei and Saxena (1987), and Saxena (1989), C_p measurements and C_v calculations enable one to place constraints on these parameters through Eq. (1). We conclude this review with a check of the consistency between high-temperature thermal and volume properties. The temperature dependence of the bulk modulus, the pressure dependence of the thermal expansion coefficient, and a few relationships that have been proposed recently to correlate α, K_T, and C_p will be included in the discussion.

Calorimetry

Measurements

Heat capacities are either measured directly or determined from relative enthalpy measurements through

$$C_p = \left[\partial\left(\frac{H_T - H_{T_0}}{\partial T}\right)\right]_P. \qquad (2)$$

In any case, one actually measures the change in enthalpy resulting from a given temperature increase. Basically, the difference between relative enthalpy measurements by drop calorimetry and direct C_p measurements by adiabatic or differential scanning calorimetry is that in the latter case, only the change in temperature is small enough that the differential of Eq. (2) can be approximated by a finite difference. With most techniques, it is necessary to enclose the sample in a container whose heat capacity or enthalpy must be measured first in order to separate its contribution to the measured property from that of the sample itself. With adiabatic and drop calorimetry, the tight control of the thermal conditions enables one to perform these measurements only once, whereas they have to be made repeatedly with DSC.

Another important point is that heat capacity measurements are needed from the (generally high) temperature of geochemical interest down to very low temperatures. Of course, the reason is that third-law entropies are determined with

$$S = \int_0^T \frac{C_p}{T} dT, \qquad (3)$$

and reliable calculations require measurements down to temperatures of about 20 K, below which C_p/T becomes negligible in the absence of magnetic transitions.

Adiabatic Calorimetry

The most accurate method for C_p determinations is adiabatic calorimetry (e.g., Robie and Hemingway, 1972). With this technique, the sample is enclosed in a small calorimeter that is electrically heated from the inside and enclosed in a shield separating the calorimeter itself from an outer zone that is constantly maintained at the same temperature as the calorimeter. No heat is exchanged between the calorimeter and the exterior in this way; hence, the term *adiabatic* is applied to the shield and the technique. Typically, the electrical energy supplied is that needed to increase the temperature of the calorimeter by a few degrees. This has long been done manually, but measurements are no longer tedious, with automated equipment by which a product is run smoothly from liquid helium to room temperature within a week.

The major difficulty stems from the decrease of C_p at lower temperature, which necessitates samples of 10 to 20 g or more in order to measure significant amounts of heat at the lowest temperatures. This need of big samples is the main constraint on adiabatic calorimetry. As for the temperature limitation of the technique, it results from the increasing difficulty of maintaining adiabatic conditions when radiative heat transfer becomes important at high temperatures. The upper limit of adiabatic calorimetry is thus about 1000 K, even though only a few laboratories succeed in making measurements well above room temperature (i.e., Grønvold, 1967).

The accuracy of the technique depends on the temperature range of the measurements. The relative contribution of the calorimeter itself to the measured heat increases with decreasing temperatures and it can represent 90% of the total measured heat at 10 K. The uncertainties can be greater than 10% below 10 K, but this is not much of a problem because C_p data in this temperature range usually make a negligible contribution to the third-law entropy. In contrast, errors can be as small as 0.2% from about 50 K to room temperature (e.g., Krupka et al., 1985). Standard entropies can thus be known with the same uncertainty, as listed in the thermodynamic tables of Robie et al. (1979).

Differential Scanning Calorimetry

Differential scanning calorimetry (DSC) has rapidly become a popular method for measuring heat capacities from ambient to about 1000 K. Schematically, the sample and a standard are heated under conditions as similar as possible. The heat supplied to both products is assumed to be the same and is determined from the temperature rise of the standard. Then, the heat capacity of the sample can be determined from this amount of heat and its own temperature rise. Temperatures are usually calibrated with measurements of the melting point of metals. For oxide minerals, heat calibrations are made with sapphire whose heat capacity is well known. Even though DSC was originally designed mainly for rapid characterization purposes, one can obtain heat capacities accurate to ±1% to about 1000 K (e.g., Krupka et al., 1979; Hemingway, 1987). This requires careful and repeated calibrations, but measurements between 400 and 1000 K can nevertheless be performed within a week. With laboratory-built equipment, Lange et al. (1991) have recently extended the range of measurements up to 1760 K. The estimated errors on the order of 5% for C_p are large, but they are nonetheless smaller than those obtained with commerically available DSC apparatus designed to attain 2000 K that are inappropriate for quantitative C_p measurements.

The first advantage of DSC is the commercial availability of the apparatus. However, the main geochemical interest of the technique is that only a few tens of mg of product are required for the measurements. This allows measurements on phases synthesized in a piston cylinder apparatus or even in multianvil apparatus (e.g., Watanabe, 1982). One can also note that the temperature limitation of quantitative DSC is not a serious problem for these high-pressure minerals because they generally decompose rapidly to low-pressure forms on moderate heating at room pressure. Another advantage of DSC is the ease with which transitions can be studied and, in particular, the influence of kinetics on transition properties. It must be noted, however, that such measurements have mainly a comparative usefulness. As shown in Fig. 1 for the α-β quartz transition, the most recently published DSC measurements (Hemingway, 1987) have only a

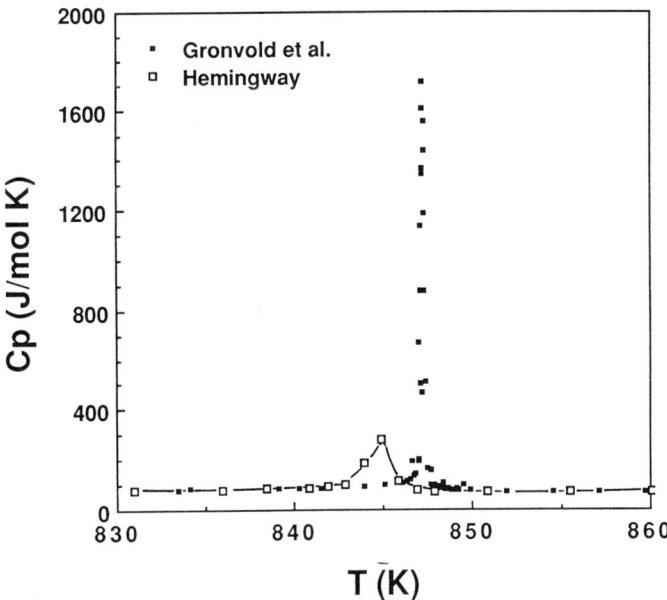

Fig. 1. Heat capacity of quartz at the α-β transition. Adiabatic data of Grønvold et al. (1989) and DSC measurements of Hemingway (1987). The difference in the transition temperature is likely due to the fact that the α-β transition temperature varies slightly from one sample to another.

fraction of the resolution of the outstanding adiabatic measurements of Grønvold et al. (1989).

Drop Calorimetry

Above 1000 K, it becomes more advantageous to make one measurement only at high temperatures, namely, that of the temperature of the sample, and to perform the calorimetric measurement near room temperature where heat exchange with the exterior can be controlled tightly. With this technique, which has long been used for geochemical purposed (e.g., White, 1919; Kelley et al., 1953; Kelley, 1960), a sample is heated to a given temperature T and then rapidly dropped into a calorimeter at a low temperature T_0, where the relative enthalpy $H_T - H_{T_0}$ is measured. Of course, some heat is lost during the drop by conduction and radiation. Experience shows that the heat loss can be accounted for by the measurements on the empty crucible because this loss is the same for the empty and the loaded crucible.

Various kinds of calorimeters can be used to measure relative enthalpies. With copper-block or water calorimeters, the enthalpy of the sample is determined from the temperature rise of the calorimeter once thermal equilibrium is reached with the initially hot sample. The changing temperature difference between the calorimeter and the exterior makes it difficult to control heat leaks to the exterior. These problems are avoided with isothermal calorimeters in which a univariant phase change is monitored, such as the melting of ice at 273 K, or of diphenyl-ether at 300 K. Drop calorimetry with an ice calorimeter constitutes, in fact, the primary high-temperature calorimeteric apparatus since the NBS C_p data for sapphire, which are generally used for thermal calibrations, have been obtained in this way up to 2200 K (Ditmars and Douglas, 1971; Ditmars et al., 1982).

In Fig. 2 we represent schematically the equipment we have recently set up to

Fig. 2. (a) and (b). Schematic representation of drop-calorimetry set-up for measurements between 400 and 1100 K. The ice calorimeter consists essentially of a vessel maintained at 273.15 K and filled with water and mercury. The water is partly crystallized prior to the measurements. Some ice melts when heat is supplied. Since the volume of the calorimeter is constant, mercury then enters the vessel to compensate for the negative volume of melting of ice. This quantity of mercury is proportional to the amount of heat, and it is simply given by the weight change of a beaker from which mercury flows into the calorimeter. Key to letters shown: A, alumina-sheathed Pt-PtRh 10% thermocouples, with the hot junctions shown by the two dots within the crucible; B, to potentiometric equipment for temperature measurements; C, water cooler; D, alumina tubes; E, alumina shield; F, silver heat pipe; G, Kanthal-wound furnace; H, alumina rod holder for the crucible; J, Pt-Rh 15% crucible; K, hook for suspending the crucible; L, hydrolic jack; M, gates; N, to vacuum pump and argon gas cylinder; O, calibrated glass capillary for measuring the continuous slight heat supply from the exterior; P, beaker of mercury; Q, calorimeter well; R, mercury-tempering glass coil; S, evacuated external vessel; T, internal vessel; U, water; V, ice mantle; W, copper block fitting the shape of the crucible for speeding up heat exchange with the calorimeter; X, mercury; Y, ice bath.

complement between 400 and 1100 K the measurements previously made in our laboratory between 800 and 1800 K as described by Richet et al. (1982). The ice calorimeter is similar to that made at NBS (see Ditmars and Douglas, 1971) and has already been described in detail (Deniélou et al., 1971; Richet et al., 1982). With an ice calorimeter, the errors on enthalpies are small enough that the uncertainties of the reported results originate mostly from the measurement of the temperature itself or from temperature gradients within the crucible. Tem-

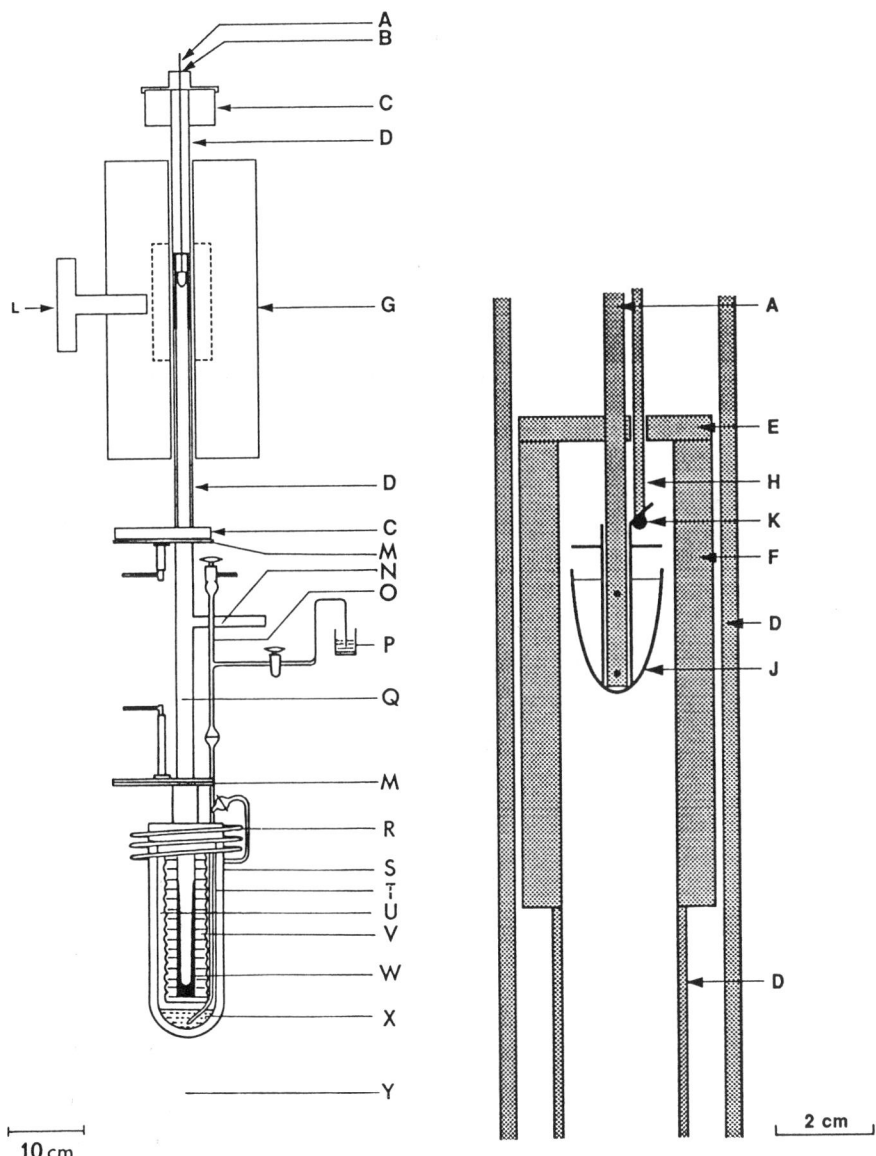

a   b

Table 1. Experimental relative enthalpy results for sapphire (kJ/mol).

Run	T (K)	$H_T - H_{273}$
DG.6	411.6	12.135
DG.10	453.5	16.387
DG.1	507.1	21.953
DG.2	591.1	31.108
DG.7	678.2	41.021
DG.3	760.2	50.586
DG.8	812.0	56.868
DG.4	888.1	65.975
DG.9	956.0	74.405
DG.5	1001.6	80.042

perature gradients are reduced to a fraction of a degree over a few cm with the simple silver heat pipe shown in Fig. 2. In addition, the measurement of temperature with two termocouples within the crucible itself, with the same technique as described in Richet et al. (1982), allows not only precise observations, but also rapid detection of thermal stability. Runs can thus be kept as short as possible, which makes measurements possible on metastable phases that do not transform within less than 1 h.

The example of sapphire will be used to illustrate the capabilities of drop calorimetry at high and intermediate temperatures. Results for $\alpha\text{-}Al_2O_3$ obtained with the apparatus of Fig. 2 are listed in Table 1 and plotted in Fig. 3(a) along with those of Ditmars et al. (1982) and Richet et al. (1982). If we fit the results of Table 2 and Richet et al. (1982) in the form

$$H_T - H_{T_0} = \int_{T_0}^{T} C_p \, dT, \quad (4)$$

a least-squares fit to the data gives the following equation, valid between 400 and 1800 K:

$$C_p = 139.504 + 5.8903 \cdot 10^{-3} T - 24.606 \cdot 10^5 T^{-2} - 589.23 T^{-0.5}. \quad (5)$$

With Eqs. (4) and (5), one reproduces the input data of Table 2 and Richet et al. (1982) with average absolute deviations of 0.09 and 0.03%, respectively, and obtains heat capacitites that agree to within ±0.5% with the adiabatic data of Grønvold (1967) and with the drop-calorimetry values of NBS (Ditmars et al., 1982). Routinely, four to five runs can be performed a day with each of our set-ups, and an absolute and accurate C_p determination over a 1500 K interval can thus be made in only a few days.

As for adiabatic calorimetry, but to a lesser extent, the main drawback of drop calorimetry is the few grams of sample that must be run. In addition, heat capacity determinations from drop-calorimetry measurements are valid only if

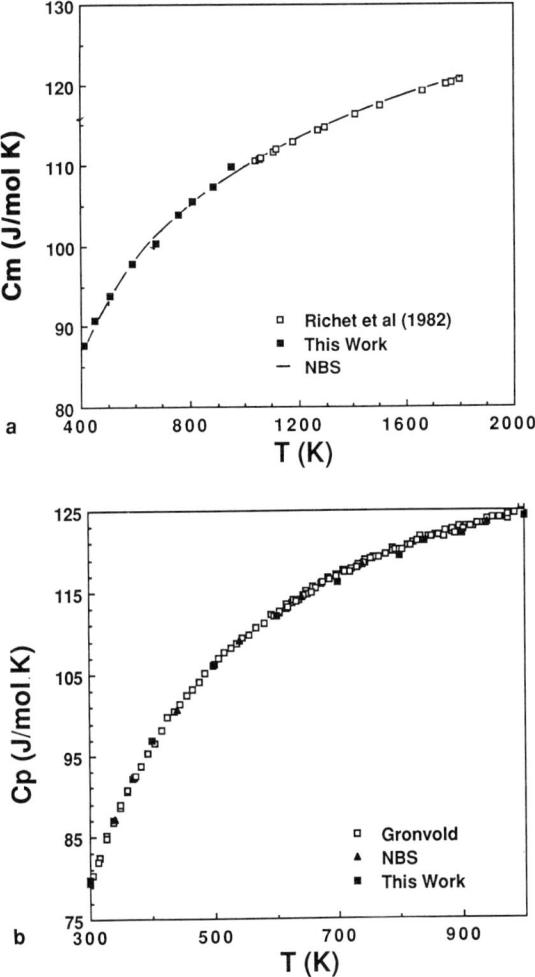

Fig. 3. Properties of α-Al$_2$O$_3$ (sapphire). (a) Comparison of mean heat capacities, $C_m = (H_T - H_{273.15})/(T - 273.15)$, obtained in this work and by Richet et al. (1982) with the NBS-fitted values (Ditmars and Douglas, 1973; Ditmars et al., 1982). (b) Heat capacity as given by the adiabatic measurements of Grønvold (1967) and the drop-calorimetry results of NBS (Ditmars et al., 1982) and our laboratory (Table 1 and Richet et al., 1982).

the sample attains the same final state at the reference temperature T_0, regardless of the initial temperature T from which a given phase is dropped. For kinetic reasons, it may thus be needed to set T_0 higher than T in instances like order–disorder reactions. The problem with such transposed drop-calorimetry experiments is that the heat exchanged cannot be measured as accurately at 1000 K or more as near room temperature. Errors are about 1% for the enthalpies, which is generally acceptable for the enthalpy itself, but they preclude reliable C_p

Table 2. Summary of the merits of C_p equations.

	Eq. (6)	Eq. (8)	Eq. (9)	Eq. (10)	Eq. (11)
Representation of measurements	medio.	excel.	good	excel.	excel.
Low-temperature extrapolation[a]	good	medio.	bad	bad	bad
High-temperature extrapolation of					
DSC measurements	bad	bad	medio.	medio.	medio.
Drop calorimetry data up to					
1800 K	medio.	bad	good	good	excel.

[a] Extrapolation to lower temperatures for phases that are stable at high temperatures only.

determinations since derivative properties are about one order of magnitude less accurate than the original data.

C_p Equations

Theoretical treatments of the thermodynamics of anharmonic oscillators do not lead to analytical expressions for the heat capacity. The C_p-temperature relationships that are used to obtain the representations of the experimental data needed for thermodynamic calculations are thus purely empirical. In addition, none of the rather simple equations used as flexible enough to represent the σ-shape of C_p vs. T curves below room temperature. The well-known consequence is that standard entropies cannot be determined from low-temperature extrapolations of data measured above room temperature. They must be obtained from adiabatic measurements performed up to room temperature that are usually smoothed numerically.

Above 298 K, in contrast, the temperature dependence of C_p is usually slight enough that comparably simple equations allow one to account for the experimental data. Can these equations be extrapolated safely to determine the entropy and other thermodynamic functions of minerals at mantle temperatures? This would be especially useful when only DSC data are available, and thus we will discuss briefly some of the equations that are either used extensively or have been proposed recently for high-temperature extrapolation purposes.

As long as computing facilities have been limited, the expression proposed by Maier and Kelley (1932),

$$C_p = a + bT + \frac{c}{T^2}, \qquad (6)$$

has remained the most popular. The drawbacks of this expression for accounting for measurements from room to high temperature led Chipman and Fontana (1935) to early recommend instead

$$C_p = a + bT + \frac{d}{T^{0.5}}. \qquad (7)$$

However, this equation too cannot reproduced experimental data to within their error margins from 298 to 1200 K or more. Haas and Fisher (1976) thus combined Eqs. (6) and (7) and added yet another term to obtain better qualities of fit

$$C_p = a + bT + \frac{c}{T^2} + \frac{d}{T^{0.5}} + eT^2. \tag{8}$$

The main advantage of this equation is its flexibility, which indeed enables one to reproduce well the experimental data from room temperature to at least 2000 K. This equation has thus been used in the well-known thermodynamic tables of Robie et al. (1979). Its main limitation is that it frequently gives clearly incorrect heat capacities when extrapolated to high temperatures because the bT and eT^2 terms become predominant.

Guessing systematics in C_p variations, Holland (1981) suggested to uses dummy C_p data to constrain Eq. (8) at high temperatures, but the validity of the assumptions made is not readily tested since they rest on data that are already extrapolated. Looking for an equation that would provide instead a correct built-in high-temperature behavior, Berman and Brown (1985) and Berman (1988) proposed a development in reciprocal temperature

$$C_p = k_0 + \frac{k_{0.5}}{T^{0.5}} + \frac{k_2}{T^2} + \frac{k_3}{T^3}. \tag{9}$$

Basically, the same form has been selected by Fei and Saxena (1987) who switched from a $k_{0.5}/T^{0.5}$ to a k_1/T term and also included another term linear in temperature, as justified by the presence of the $TV\alpha^2 K_T$ product in Eq. (1)

$$C_p = k_0 + bT + \frac{k_1}{T} + \frac{k_2}{T^2} + \frac{k_3}{T^3}. \tag{10}$$

Finally, another related equation was recommended by Richet and Fiquet (1991), namely,

$$C_p = k_0 + k_{\ln} \ln T + \frac{k_1}{T} + \frac{k_2}{T^2} + \frac{k_3}{T^3}. \tag{11}$$

The relative merits of these equations have been discussed by Richet and Fiquet (1991) whose conclusions are summarized in Table 2. With Eqs. (9), (10), and especially (11), one can achieve both accuracy of representation of the experimental data and reliability of high-temperature C_p extrapolations. A somewhat obvious result, however, is that a prerequisite for reliable high-temperature extrapolations is the availability of data at temperatures as high as possible. The properties of a number of important minerals, including oxides, are thus known satisfactorily at 2000 K or more when the measurements extend beyond 1500 K. However, a number of mantle minerals can be studied at best by DSC only over narrow temperature intervals. In such a case, no empirical equation guarantees correct C_p extrapolations up to 2000 K.

Empirical Methods of Prediction of the Heat Capacity and Entropy

Standard Entropy

Early systematic measurements of solids have shown that the heat capacity can be expressed approximately as an additive function of composition above room temperature (e.g., Kopp, 1865). Below room temperature, however, this approximation is clearly incorrect, as shown in Fig. 4 by the measurements of Kelley et al. (1953) for sodium aluminosilicate minerals. Below 200 K, the heat capacity of albite ($NaAlSi_3O_8$) agrees closely with values calculated from the heat capacities of quartz (SiO_2) and nepheline ($NaAlSiO_4$). In contrast, that of jadeite ($NaAlSi_2O_6$) is much lower than given by the sum of the heat capacities of quartz and nepheline. As pointed out by Kelley et al. (1953), the reason for this difference is that aluminum is in octahedral coordination in jadeite, whereas it substitutes for Si in albite and nepheline where it is tretrahedrally coordinated with oxygen.

As a general rule, both heat capacities and standard entropies increase when coordination numbers decrease. This is also shown very simply by the data for SiO_2 and GeO_2 polymorphs. Stishovite and tetragonal GeO_2, with six-fold coordinated Si or Ge, have a standard entropy 34 and 43% lower than that of quartz and hexagonal GeO_2, respectively, where these elements are in tetrahedral coordination (see Holm et al., 1967; Counsell and Martin, 1967; Richet et al., 1982; Richet, 1990). To account for the influence of the structure, Helgeson et al. (1978) evaluated standard entropies of minerals from simple additive schemes involving structurally related compounds. A more general procedure has been proposed by Robinson and Haas (1983) who recast chemical compositions in terms of oxide components with specific coordination numbers. They derived in this way a new additive model of calculation of the heat capacity and

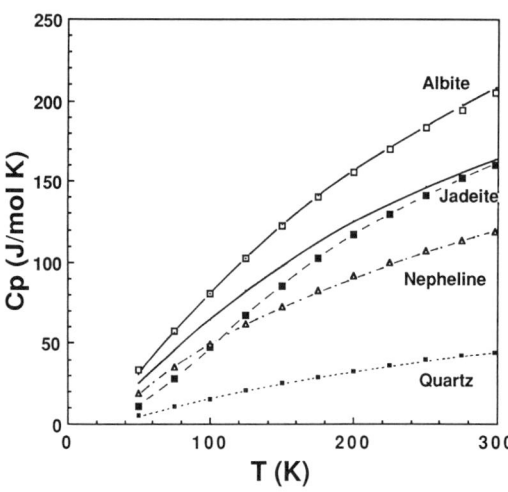

Fig. 4. Low-temperature heat capacity of albite, jadeite, nepheline, and quartz. Symbols: experimental data points of Kelley et al. (1953); solid curves: values calculated for nominal compositions $NaAlSi_2O_6$ and $NaAlSi_3O_8$ from appropriate sums of the heat capacities of quartz and nepheline components; the agreement is good for albite and bad for jadeite.

entropy for oxide and silicate minerals. Following Fyfe et al. (1958), Holland (1989) then modified this procedure to also take into account the slight dependence of the entropy on molar volume through a simple formula based on the Einstein and Debye theories of solids

$$S = kV + \sum n_i S_i'. \tag{12}$$

In this equation, the parameter k is 1 J/K cm^3, n_i is the number of moles of oxide i, and S_i' are the empirically determined coefficients.

In contrast to simpler additive schemes, the models of Robinson and Haas and Holland are very useful in estimating standard entropies. With Holland's model, for instance, the standard entropy of jadeite is predicted to be 134.8 J/mol K. This result is in good agreement with the experimental measurement of 133.5 ± 1.2 J/mol K, whereas the sum of the entropies of quartz and nepheline is clearly discrepant at 165.8 J/mol K. Prediction of the entropy for polymorphic modifications provides a more stringent test since the influence of chemical effects are separated from those of chemical factors. Silicon is four-fold coordinated in both coesite and quartz. The entropy difference between these forms cannot be predicted with the model of Robinson and Haas (1983), whereas the model of Holland (1989) gives 38.1 and 40.1 J/mol K for the standard entropies of coesite and quartz, respectively. This method yields entropies in the right order since the experimental values are 39.4 ± 0.4 and 40.4 ± 0.2 J/mol K, respectively (Holm et al., 1967; Richet et al., 1982). On the other hand, the polymorphs of NaAlSiO$_4$, nepheline and carnegieite, do not seem to show major coordination differences. According to Holland's model, the higher molar volume of carnegieite should translate into an entropy 2 J/mol K higher than for nepheline. This is at variance with the calorimetric standard entropies 124.3 ± 1.3 and 118.7 ± 0.2 J/mol K for nepheline and carnegieite, respectively (Kelley et al., 1953; Richet et al., 1990).

Additional calorimetric and structural data will probably allow further refinement of these models. The main limitation of such empirical approaches for mantle minerals is, in fact, the very lack of data that can be used to obtain and check the validity of their empirical coefficients. Low-temperature calorimetric data for six-fold coordinated Si are limited to the observations of Holm et al. (1967) for stishovite. No mineral with twelve-fold coordinated Ca, Mg, or Fe has been investigated and the trend shown by the coefficients of Holland (1989) for the various coordination states of these oxides is erratic. This shortcoming thus prevents prediction of the entropy of a number of high-pressure minerals, such as those with perovskite structures.

High-Temperature Heat Capacity

One would expect additive models to be followed more closely at high temperatures when the Dulong and Petit limit is approached, as embodied in the simple model of Berman and Brown (1985). To check this commonly made assumption, Richet and Fiquet (1991) considered six minerals for which C_p had been determined in the same way above 1000 K. They observed that additivity is actually obeyed for pyrope, pseudowollastonite, and anorthite, whereas deviations from

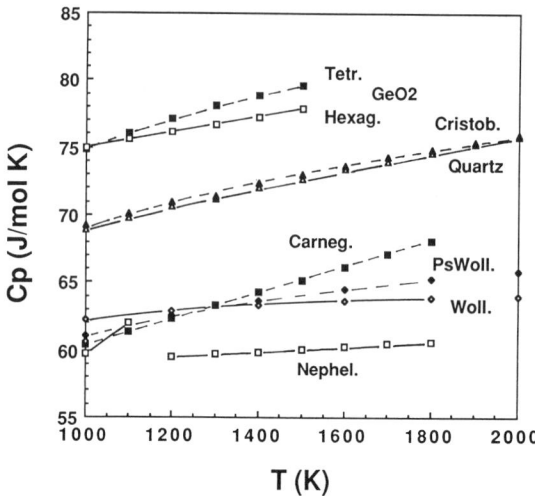

Fig. 5. High-temperature heat capacity of hexagonal and tetragonal GeO_2, cristobalite and quartz (SiO_2), carnegieite and nepheline ($NaAlSiO_4$), and pseudowollastonite and wollastonite ($CaSiO_3$). Solid and open symbols for the high- and low-temperature forms, respectively. Data from Richet (1990) and Richet et al. (1982, 1990, 1991).

additive models increase with temperature for spinel, diopside, and forsterite. For forsterite, an excess C_p of about 6% is observed at 2000 K with respect to the constituting oxides. These deviations are all positive and they likely result from the nonadditivity of the anharmonic contribution to C_p that, as described in the next section, is smaller for the individual oxides than for the other minerals. The puzzling aspect of these deviations, however, is that they are apparently unrelated to the kind of coordination polyhedra present in the minerals.

To look in another way at the relationship between the structure and the high-temperature heat capacity, we plotted recent calorimetric data for pairs of polymorphs in Fig. 5. The main feature is the existence of a general trend, whereby the high-temperature form has higher C_p above 1000 K. The C_p contrast is small between wollastonite and pseudowollastonite, which is consistent with the structural similarities between these $CaSiO_3$ polymorphs. The minor C_p difference between cristobalite and quartz is also expected in view of the tetrahedral coordination of Si in both SiO_2 forms. Along the same lines, the greater difference between the GeO_2 modifications could be related to a major coordination difference for Ge between the four- and six-fold coordination states of the hexagonal and tetragonal forms, respectively. The problem, however, is that the C_p difference does not conform to the expected trend according to which denser phases have a higher entropy at high temperatures (Navrotsky, 1980). On the other hand, the C_p difference reaches more than 10% between nepheline and carnegieite, which are structurally analogous to tridymite and cristobalite, respectively. In summary, SiO_2 and $CaSiO_3$ polymorphs conform to the trends assumed from structural arguments, whereas GeO_2 and $NaAlSiO_4$ show unexpected differences. Hence, it does not seem possible to predict currently the magnitude of C_p differences at high temperatures from simple crystallochemical arguments.

Calculations from Vibrational Spectra

General Remarks

The specific variations of the heat capacity and entropy of minerals discussed in the previous sections show that the empirical estimation of these properties can be fraught with difficulties. Calculation of the heat capacity from statistical methods thus appears to provide an alternative for mantle minerals or other phases for which calorimetric data are lacking. In addition to their practical interest, these calculations also constitute the best way for understanding the specificity of the variations of these properties. Starting with the well-known harmonic modeling, we will then focus on anharmonic factors that are likely at the root of the complex high-temperature variations reviewed in the previous section.

The property evaluated most readily from spectroscopic data is the isochoric heat capacity C_v, from which C_p is obtained through Eq. (1). Strictly speaking, defects and impurities also contribute to the heat capacity, but their contribution is small enough that they can be safely neglected for oxides and silicates. In addition, important crystal-field and magnetic contributions can exist for compounds like iron-bearing minerals, which must be evaluated specifically (see Wood, 1981). Leaving aside these factors, one assumes in the harmonic approximation that the lattice energy of the crystal U_0 is constant at constant volume. Then, C_v is determined only from the derivative of the vibrational internal energy (U_{vib})

$$C_v = \left(\frac{\partial U}{\partial T}\right)_V = \left(\frac{\partial U_{vib}}{\partial T}\right)_V. \tag{13}$$

In a solid, atoms do not vibrate individually, but collectively through vibrational waves called phonons because they have the quantified energy of an Einstein oscillator, $E = hv(n + 1/2)$, where v is the frequency and n the vibrational quantum number (e.g., Kittel, 1971). (Spectroscopic data are usually reported in the form of wave numbers, ratios of frequencies to the speed of light, which are expressed in cm^{-1}.) The vibrational modes are split into acoustic branches, which represent one compressional and two shear waves, and optic branches where the atoms of a given unit cell vibrate along different directions. Because of the symmetry of the lattice, the wave vectors of the vibrational waves are bound to vary within the first Brillouin zone of the crystal, i.e., between 0 and the parameters of the reciprocal lattice. The variation of the frequency with the wave vector constitutes the dispersion relation of a vibrational mode [Fig. 6(a)]. When the wave vector is zero, all unit cells of the crystal vibrate in phase. When the wave vector is maximum, analogous atoms in contiguous unit cells vibrate out of phase.

A mineral with N atoms in its unit cell has $3N$ vibrational modes, of which 3 are acoustic and $3N-3$ optic. Regardless of their nature, however, the contribu-

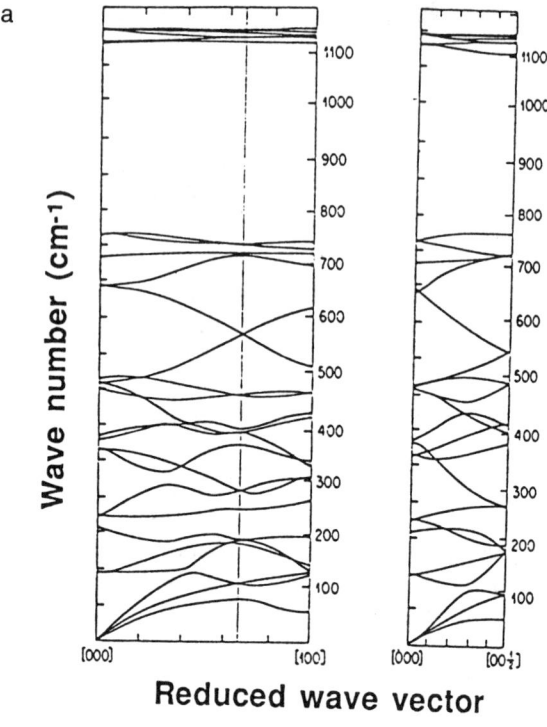

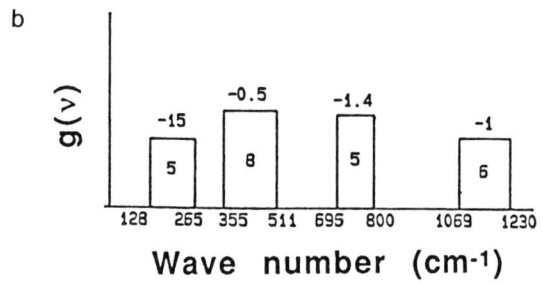

Fig. 6. Dispersion relations and density of states of quartz: (a) Calculated from spectroscopic measurements and inelastic neutron scattering (Elcombe, 1967); (b) simplified with a Kieffer model (Gillet et al., 1990a), sinusoidal relations for the acoustic modes and four continua for the optic modes (whose number of modes is given in the corresponding boxes).

tion of each mode to the heat capacity is that of an Einstein oscillator. For the ith mode, one has

$$C_{vi} = \frac{(hv_i/kT)^2 \exp(hv_i/kT)}{[\exp(hv_i/kT) - 1]^2}, \qquad (14)$$

where h and k are the Planck and Boltzmann constants, respectively. Heat capacity calculations thus reduce to the determination of the density of vibrational states $g(v)$, i.e., the number of modes with a wave number comprised between v and $v + dv$, from $v = 0$ to the highest wave number v_m. When dispersion relations are known for all modes, the density of states is readily evaluated [Fig. 6(b)] and one obtains C_v through

$$C_v = \int_0^{v_m} g(v) \frac{(hv/kT)^2 \exp(hv/kT)}{[\exp(hv/kT) - 1]^2} dv. \tag{15}$$

The high-temperature limit of Eq. (15) is k for one oscillator. For one mole of a crystal with N atoms in the formula unit, the high-temperature limit of C_v is thus $3NR$, which is the so-called limit of Dulong and Petit already mentioned in the previous section.

Densities of states are most rigorously determined from lattice dynamical modeling of inelastic neutron scattering measurements (e.g., Ghose, 1988). Unfortunately, this very time-consuming work with experiments requiring about 50 g of sample is restricted to a few minerals (i.e., Barron et al., 1976; Rao et al., 1988). The density of states has also been calculated for minerals rather simple structurally from interatomic potentials derived from crystallographic and elastic data (e.g., Price et al., 1987). The good results obtained from simpler modeling suggests, however, that these detailed calculations of $g(v)$ are not necessarily needed in heat capacity calculations. As shown in Fig. 6(a), if dispersion is the basic feature of acoustic modes, dispersion effects are generally slight for the optic modes. We will thus restrict ourselves to these simpler models, whereby the variations of the mode frequencies with the wave vector over the first Brillouin zone are neglected for the optic modes.

Kieffer Model

In the widely used model of Kieffer (1979), for instance, the three acoustic modes are treated in a Debye-like fashion with sinusoidal dependences of the frequencies on the wave vectors. And the optic modes are counted either as discrete modes or continua, delimited by low- and high-frequency cutoffs, over which a uniform distribution of models is assumed (Fig. 6). The resulting simplification to the actual density states is shown in Fig. 6(b) for quartz. With proper selection of the number of modes and boundaries of the optic continua, one usually reproduces the low-temperature calorimetric measurements when the relevant acoustic and spectroscopic data are known. Extensive reviews and numerous applications of these calculations have been published (Kieffer, 1979, 1980, 1985; Ross et al., 1986; Hofmeister, 1987; Hofmeister et al., 1987; Chopelas, 1990; Hofmeister and Chopelas, 1991) and thus we will just present a few new or recent examples to illustrate some of the possibilities offered by this model.

Below 50 K, only the acoustic modes and the optic modes with wave numbers lower than about 100 cm^{-1} contribute significantly to C_v. At these temperatures, C_v is thus very sensitive on the lower end of the density of states, i.e., on the assumed low-frequency cutoff. Unfortunately, this cutoff is not always determined readily because of either very low intensity or of Raman or infrared inactivity of the relevant modes. This makes it permissible to adjust this cutoff in order to better reproduce the calorimetric data below 50 K. Of course, this cannot be done when these data are lacking. For the sake of consistency, the low-frequency cutoff has thus deliberately been taken as the lowest observed frequency in all examples discussed below. This can result in errors of 10 to 20% for C_v between 20 and 50 K, but these errors do not have consequences that are too serious because heat capacities in this interval have a relatively minor influence on standard entropies.

A First Example: Glaucophane

The example of glaucophane shows that useful C_p calculations are not restricted to simple minerals, as also shown by Hofmeister et al. (1987) for beryllium-bearing minerals and Hofmeister and Chopelas (1991) for garnets. With 41 atoms in its unit cell, glaucophane [$Na_2Mg_3Al_2Si_8O_{22}(OH)_2$] has 123 vibrational modes. Of the 120 optic modes, only 34 have been observed by Raman spectroscopy and 33 by infrared spectroscopy, half of these bands being visible in both kinds of spectra (Gillet et al., 1989). Even though this might appear to represent rather limited information, these modes are numerous enough to construct a density of states consisting of three optic continua and one high-frequency oscillator in addition to the three acoustic continua (Fig. 7). As shown in this figure, the heat capacities and entropies calculated from this model agree to within 1% with the experimental results obtained up to 1000 K on the same specimen (Gillet et al., 1989; Robie et al., 1991).

Polymorphic Systems: Ortho-Clinozoïsite

Phase relations between polymorphs are frequently difficult to determine experimentally. As illustrated by Salje and Werneke (1982), Akaogi et al. (1984), or Ross et al. (1986), for instance, entropies of transition calculated from simple vibration modeling can give estimates of the slope of equilibrium curves (through $dP/dT = \Delta S/\Delta V$) that are frequently subject to significant experimental uncertainties. As another example, the equilibrium curve of the transition between ortho- and clinozoïsite [$Ca_2Al_3Si_3O_{12}(OH)$] has not been determined unambiguously. Heat capacity and entropy data are available for orthozoïsite (Perkins et al., 1980), whereas sample problems prevent measurements on clinozoïsite whose properties have been determined only from phase-equilibria data (e.g., Newton, 1987). As also observed for the structurally related $MgSiO_3$ polymorphs enstatite and clinoenstatite, there are only slight differences in the Raman and infrared spectra of clino- and orthozoïsite. The spectra of both forms can be modeled in the same way (Fig. 8), yielding standard entropies of 320.7 and 296.5 J/mol K for clino- and orthozoïsite, respectively (Le Cléac'h, 1989). The calculated entropy of transition increases to 30.2 J/mol K at 700 K. Orthozoïsite being denser than clinozoïsite, this indicates a positive P-T slope

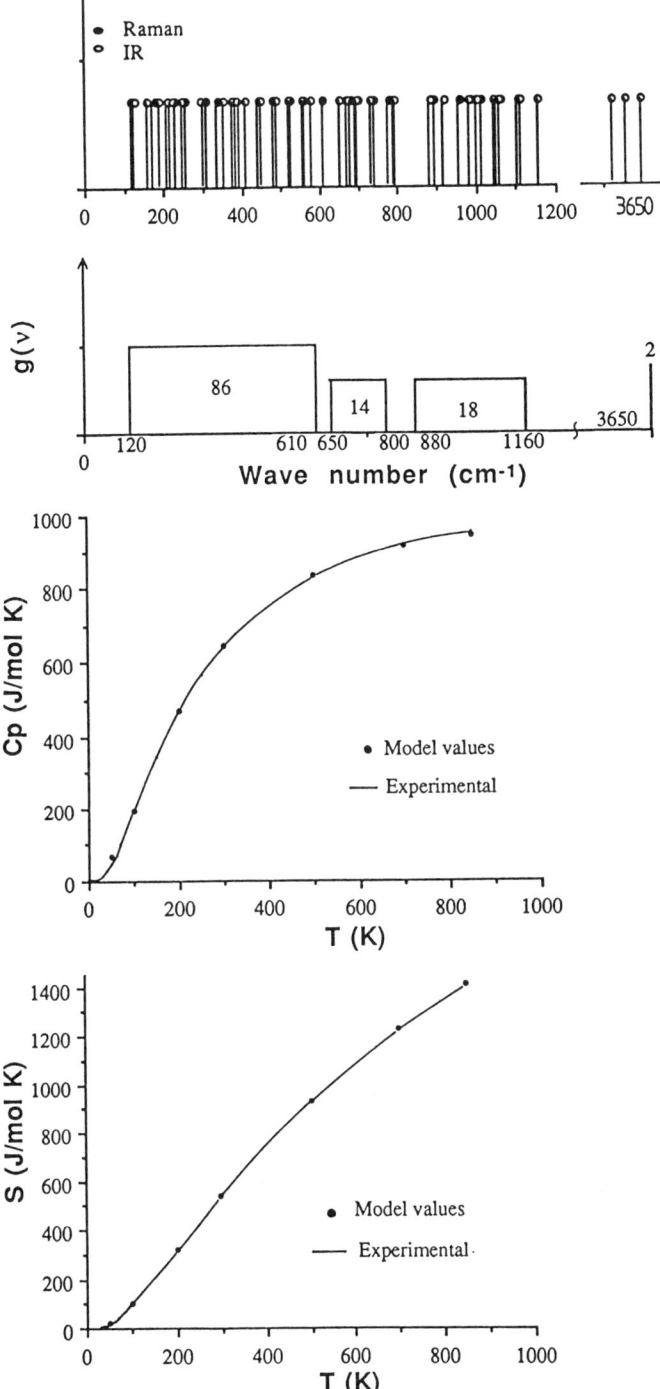

Fig. 7. Spectroscopic and thermodynamic properties of glaucophane. From top to bottom: Observed Raman and infrared frequencies; model density of states, showing the optic modes only, with numbers of modes indicated for continua and discrete modes; and heat capacity and entropy as a function of temperature.

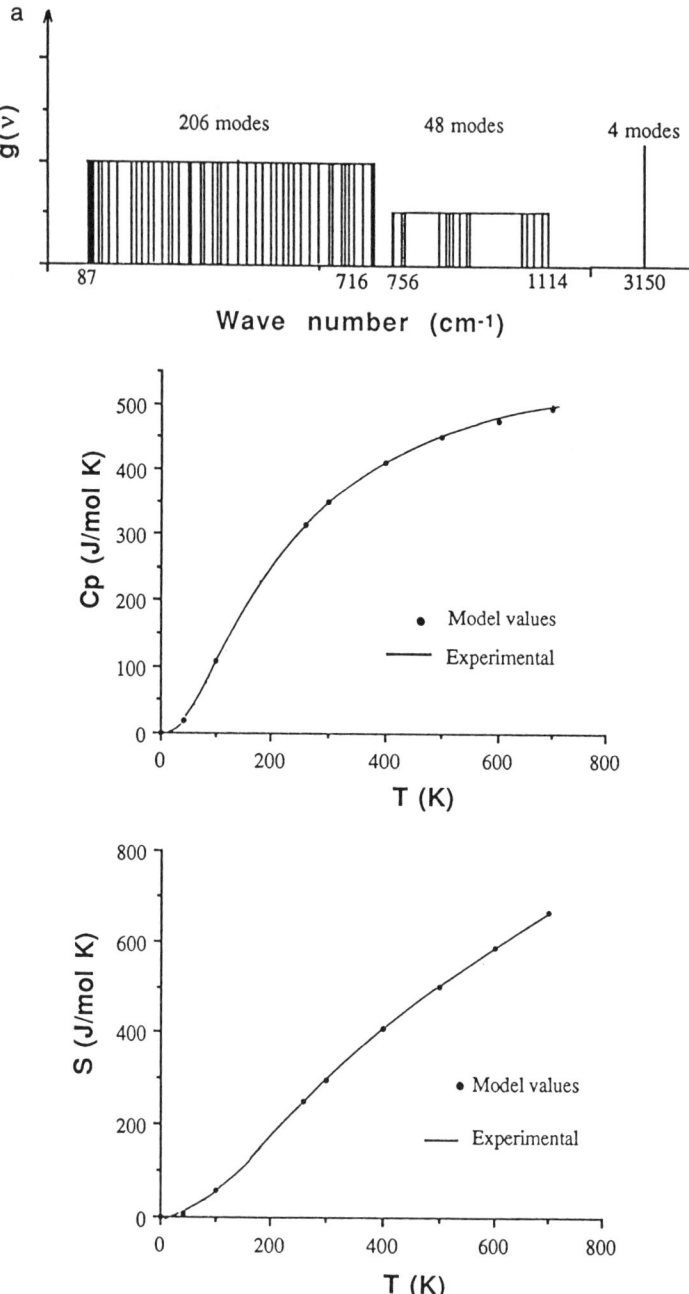

Fig. 8. Observed Raman and infrared spectra and simplified density of states, heat capacity, and entropy of orhtozoïsite (a) and clinozoïsite, the Fe-free end member of epidotes (b). For epidotes, cut-offs of the optic continua are functions of the Fe/(Fe + Al) ratio noted p.

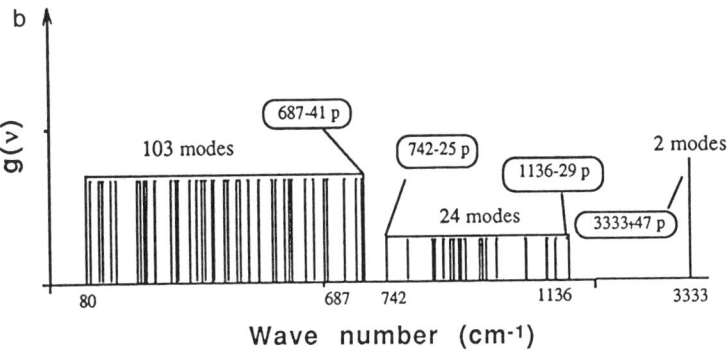
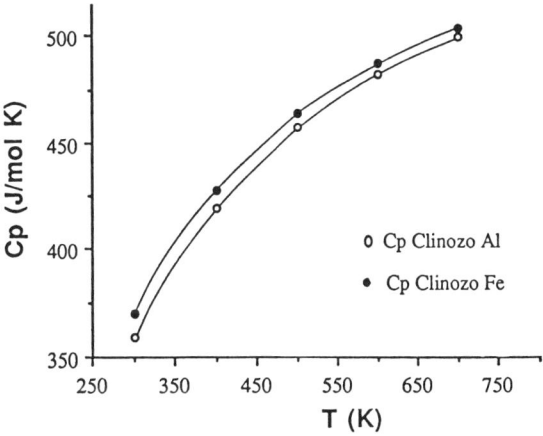
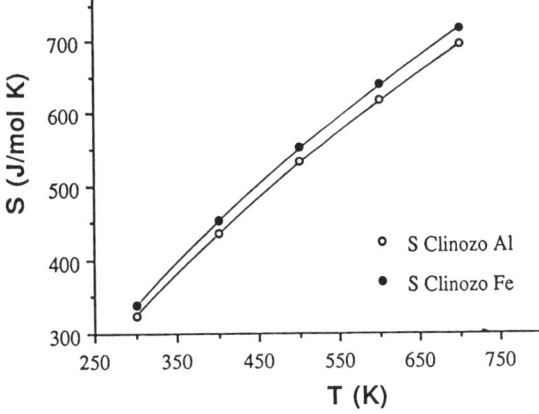

Fig. 8. (Continued.)

for the clino-ortho transition, with the ortho form the high-pressure polymorph. It has been suggested instead that clinozoïsite is the high-pressure and low-temperature form (see Newton, 1987), but the difference of only 0.5 J/mol K between the calorimetric (Perkins et al., 1980) and the calculated standard entropy of orthozoïsite lends independent support to the conclusion drawn from vibrational modeling.

Effect of Substitution: Epidotes

Vibrational spectra depend on the chemical composition along a solid solution. The substitution of a heavier element generally results in systematic changes in vibrational frequencies which can be interpreted in terms of simple models of oscillators (e.g., Jeanloz, 1980; Madon and Price, 1989). Clinozoïsite is also the aluminous end member of epidotes which show general decreases of vibrational frequencies when Al replaced by Fe, with the exception of the OH-related bands whose frequency increase [Fig. 8(b)]. Hence, the composition dependence of the entropy along the join $[Ca_2(Al_{1-p}Fe_p)AlO(SiO_4)(Si_2O_7)]$ can be determined from vibrational modeling. One finds that the heat capacity and entropy vary linearly with composition. Complete substitution results in an entropy increase of 16 and 22 J/mol K at 300 and 700 K, respectively, from the Al to the Fe end member. This is consistent with the numbers predicted by the empirical model of Holland (1989) which gives an entropy difference of 14.4 J/mol K at 298 K.

Use of Mineral Analogues: GeO_2 vs. SiO_2 Polymorphs

In the same way as vibrational spectra vary regularly along a solid solution, the spectra of structurally analogue minerals show correspondences that can be used to determine the thermodynamic properties of minerals for which calorimetric data are lacking and C_p calculations are not straightforward. McMillan and Ross (1987) obtained mixed results in this way for corundum and $MgSiO_3$ ilmenite, using rather comprehensive vibrational information for corundum but only Raman data for ilmenite. Using the same Raman and infrared information for both phases, however, Gillet et al. (1991b) obtained satisfactory results for stishovite, the high-pressure polymorph of SiO_2, and tetragonal GeO_2, which are isomorphous to rutile. In contrast to stishovite, tetragonal GeO_2 can be investigated calorimetrically up to 1300 K, near its 1-bar melting point of 1350 K (Fig. 9). For both compounds, however, C_p modeling is made difficult by the existence of a limited number of 8 measurable bands representing 11 out of 15 optic modes over about a 1000 cm^{-1} interval. However, it is possible to construct and test a density of states with the comprehensive calorimetric data available for tetragonal GeO_2. From the model shown in Fig. 9, one can reproduce the observed heat capacity of tetragonal GeO_2 (Counsell and Martin, 1967; Richet, 1990) to within 5% at 100 K and 1% between 300 and 1300 K, and to within 2 J/mol K the standard entropy. Transferring with the appropriate frequency shifts this density of state to stishovite, Gillet et al. (1990) also found good agreement with the low-temperature calorimetric data of Holm et al. (1967). The

Fig. 9. Model density of states, heat capacity, and entropy of stishovite and tetragonal GeO_2.

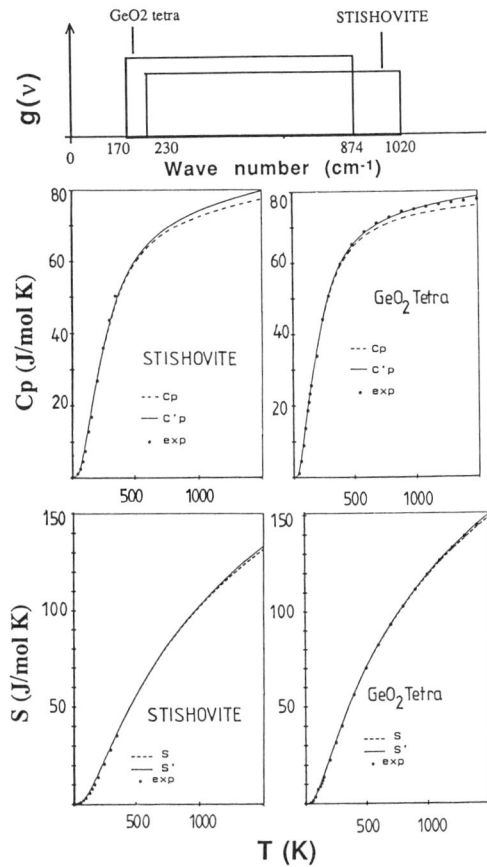

calculated standard entropy of stishovite is 30.5 J/mol K, which confirms the low calorimetric result of 27.8 J/mol K of Holm et al. (1967).

Anharmonic Modeling

Vibrational Anharmonicity

A harmonic solid would not undergo thermal expansion. Hence, the $TV\alpha^2 K_T$ term of Eq. (1) represents an anharmonic contribution that is usually found to be significant above room temperature in C_p calculations. As illustrated by the examples of the previous section, satisfactory results can be obtained in this way up to about 1000 K when the harmonic approximation is used for C_v. At higher temperatures, however, one observes that the difference between C_v and the limit of Dulong and Petit depends specifically on the mineral considered. For periclase and lime, C_v actually tends toward this limit, whereas it exceeds $3NR$ by 5% at 2000 K for forsterite (Fig. 10). As will be described below, this excess C_v is most simply ascribed to the intrinsic anharmonicity of the vibrational modes.

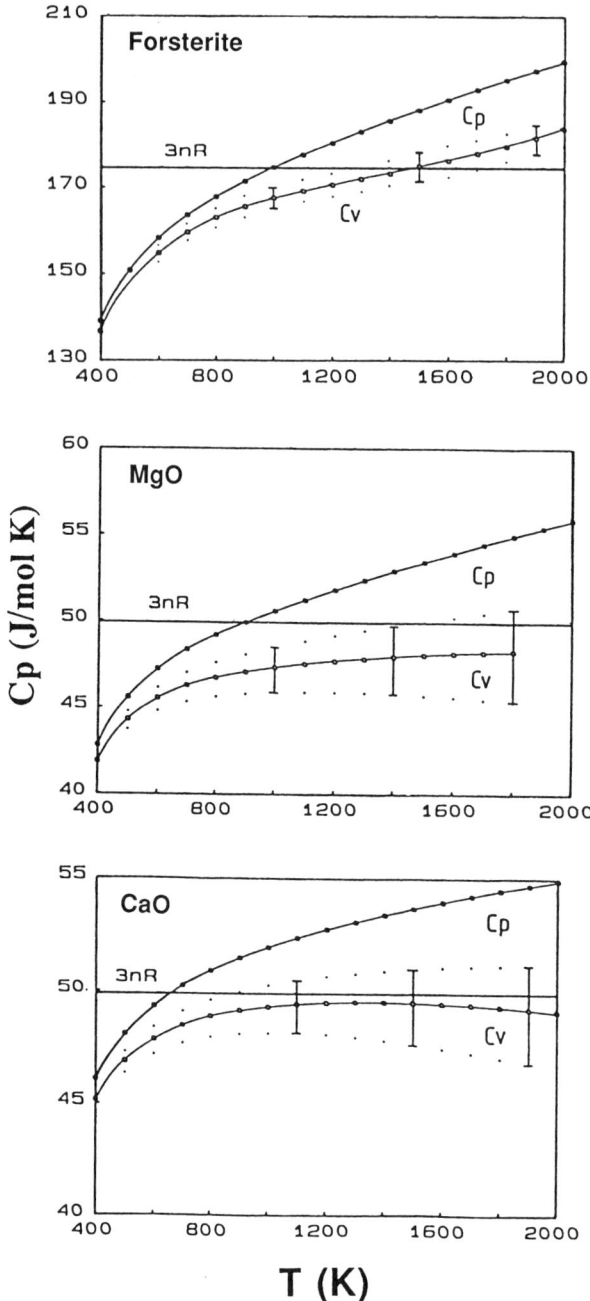

Fig. 10. Heat capacities of forsterite, periclase, and lime: Experimental C_p and C_v as calculated by Gillet et al. (1991a) and Fiquet (1980) with Eq. (1) from the experimental C_p, K_T, and α data of Gillet et al. (1991a), Richet and Fiquet (1991), Isaak et al. (1989a,b), Chang and Graham (1977), and Smith and Leider (1968). The horizontal lines are the Dulong and Petit harmonic limits. Error bars for C_v included as resulting from those on K_T and α.

A basic assumption of the harmonic approximation is that vibrational frequencies do not depend on temperature at constant volume. It might seem difficult to test the validity of this assumption since spectroscopic measurements on solids as a function of temperature are not possible at constant volume. However, the appropriate derivative $(\partial v/\partial T)_V$ can be expressed as a function of measurable properties because vibrational frequencies are thermodynamic state variables, and standard thermodynamic calculations lead to in a logarithmic form

$$\left(\partial \ln \frac{v}{\partial T}\right)_V = \alpha K_T \left(\partial \ln \frac{v}{\partial P}\right)_T + \left(\partial \ln \frac{v}{\partial P}\right)_T. \tag{16}$$

This derivative can thus be obtained from spectroscopic measurements performed at constant pressure and constant temperature. Actually, vibrational modes have their own dependecies, and anharmonic parameters must be defined separately for each mode, as done by Sharma et al. (1989) Mammone and Sharma (1980)

$$a_i = \left(\partial \ln \frac{v_i}{\partial T}\right)_V = \alpha K_T \left(\partial \ln \frac{v_i}{\partial P}\right)_T + \left(\partial \ln \frac{v_i}{\partial P}\right)_T. \tag{17}$$

Only a few minerals have been subjected to high-pressure and high-temperature spectroscopic measurements, and we will use forsterite as a test compound to assess the magnitude of these anharmonic parameters. For orthosilicates and other minerals like garnets, the high-frequency modes describe essentially the internal stretching of the SiO_4 tetrahedra, whereas the low-frequency modes involve lattice displacements through complex coupled translations of all the mineral polyhedra (Hofmeister et al., 1987; Chopelas, 1990; Gillet et al., 1991a,b). For these internal or lattice modes, frequencies generally increase with pressure and decrease with temperature (Fig. 11), although there are some exceptions to these trends (see Hemley, 1987 for stishovite and quartz, e.g.). The net effect of these trends is that the anharmonic parameters are generally negative and of the order of 10^{-5} K^{-1}. As illustrated by the a_i parameters plotted in Fig. 11, anharmonic effects seem more important for the lattice than for the internal modes, and they appear to correlate more with the kind of motion proper to a given mode than with the frequency of this mode.

Anharmonic Calculations

That the a_i parameters are not zero shows that vibrational modes are not strictly harmonic, but it is necessary to assess the influence of anharmonicity on thermodynamic properties. This can be done simply in a quasiharmonic way, by taking into account $(\partial \ln v_i/\partial T)_V$ when differentiating the partition function to derive the internal energy of an anharmonic oscillator. With this approximation, one finds that

$$U_i = U_i^h(1 - a_i T), \tag{18a}$$

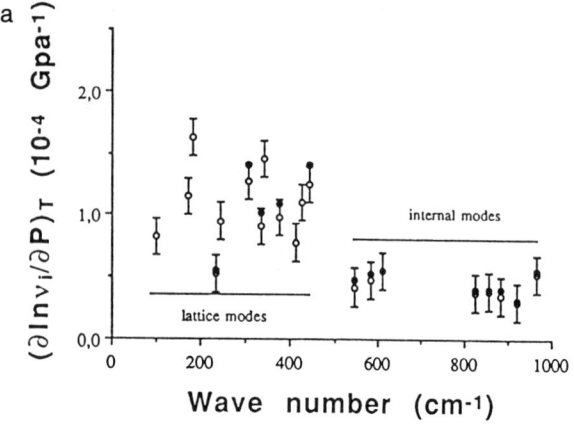

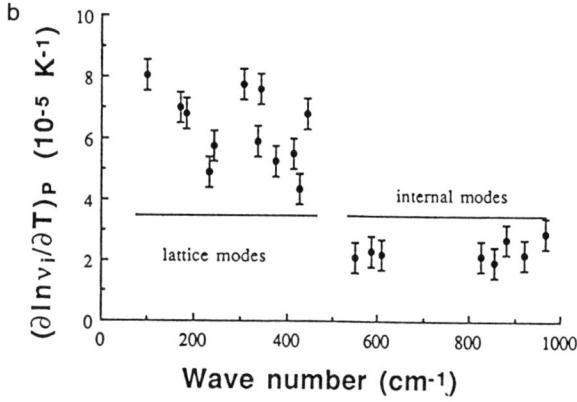

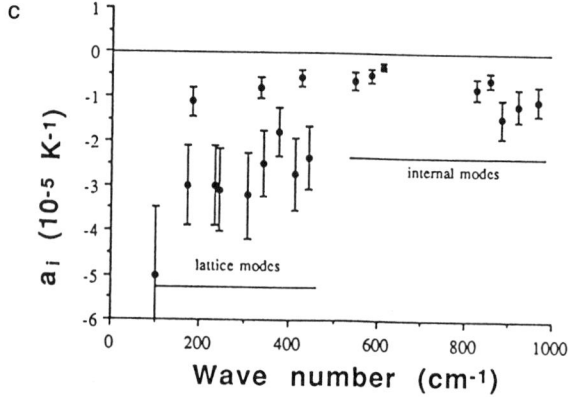

Fig. 11. Relative frequency shifts with pressure (a) and temperature (b) of the Raman modes as a function of wave number and anharmonic parameters (c) of forsterite.

where U_i^h is the harmonic energy. Summing over all the modes with respect to temperature, then one obtains the anharmonic C_v. The calculation is somewhat tedious (see Gillet et al., 1991a), but it gives eventually a very simple result, namely,

$$C_v = 3NR \sum C_{vi}^h (1 - 2a_i T), \tag{18b}$$

where C_{vi}^h is the harmonic C_v, as given by Eq. (13), of the ith continuum or discrete mode.

Of course, all the a_i parameters cannot be determined since only part of the vibrational modes are optically active. When using a Kieffer model, the simplest way to evaluate the anharmonic contribution consists of averaging the a_i's over the same continua as used for calculating the harmonic part of C_v. The anharmonic parameters of the acoustic modes could be determined from acoustic measurements as a function of pressure and temperature. For minerals with a great number of optical modes, however, the contribution of the acoustic modes to the heat capacity is small and their anharmonicity can be neglected.

The results plotted in Fig. 12 for forsterite summarize the main features of these calculations. As already noted, the difference between C_p and C_v is negligible below room temperature, above which the different density of states that can be set up from available spectroscopic measurements leads to the same heat capacities. The $TV\alpha^2 K_T$ and intrinsic anharmonicity terms begin to be significant at 300 and 1300 K, respectively. Without the latter contribution, the difference between the experimental and calculated C_p data would reach 5% at 2000 K. If forsterite is representative of mantle minerals in this respect, then the conclusion is that reliable predictions of heat capacities under mantle conditions will require not only good thermal expansion data, but also comprehensive spectroscopic information in order to evaluate the anharmonic contribution to C_v.

Grüneisen Parameters

Spectroscopic measurements can also be used to calculate the Grüneisen parameter γ that is needed for relating P, T, and V along adiabatic paths. Macroscopically, this parameter is

$$\gamma = \frac{\alpha K_T V}{C_v}. \tag{19}$$

In terms of microscopic quantities, we note that the parameter γ_m can be expressed as (Slater, 1939)

$$\gamma_m = K_T \sum \left(\partial \ln \frac{v_i}{\partial P} \right)_T \frac{C_{vi}}{\sum C_{vi}}. \tag{20}$$

A comparison of Grüneisen parameters evaluated with Eqs. (19) and (20) shows that the latter values are generally too small with respect to the former, and that the discrepancy seems to increase with temperature (Hemley et al., 1989; Gillet et al., 1991a). As an example, one obtains 1.28 at 300 K and 1.06 at 2000 K for

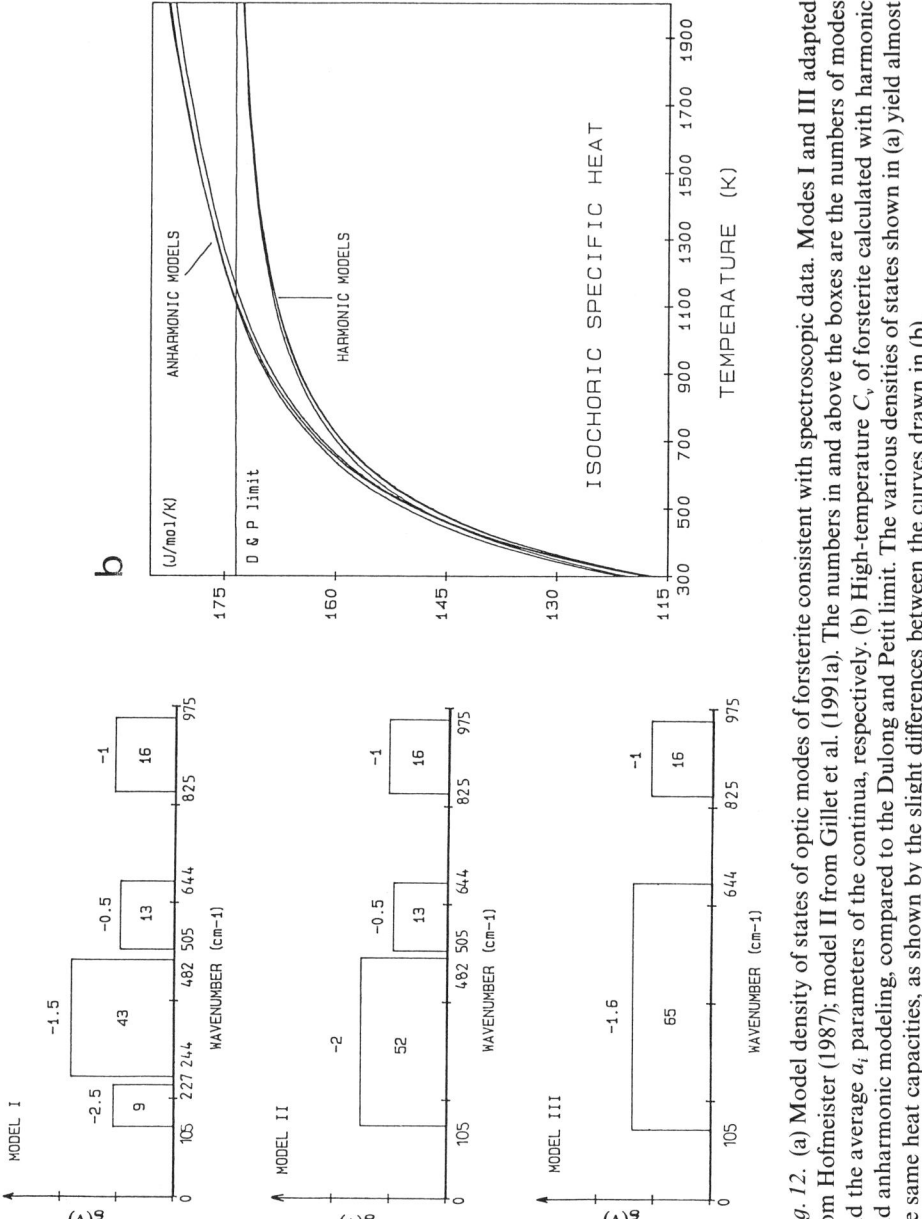

Fig. 12. (a) Model density of states of optic modes of forsterite consistent with spectroscopic data. Modes I and III adapted from Hofmeister (1987); model II from Gillet et al. (1991a). The numbers in and above the boxes are the numbers of modes and the average a_i parameters of the continua, respectively. (b) High-temperature C_v of forsterite calculated with harmonic and anharmonic modeling, compared to the Dulong and Petit limit. The various densities of states shown in (a) yield almost the same heat capacities, as shown by the slight differences between the curves drawn in (b).

forsterite with Eq. (19), whereas Eq. (20) gives 1.22 and 0.83 at the same temperatures (Gillet et al., 1991a).

To determine whether anharmonicity could be responsible for this discrepancy, Gillet et al. (1991a) carried out an anharmonic calculation of γ_m and obtained in the limit of high temperatures

$$\gamma_m = \sum \frac{\gamma_{iT} C_{vi}^h}{C_v} + \sum \frac{\gamma_{iT} a_i (U_i^h - TC_{vi}^h)}{C_v} + \sum \frac{(\partial \gamma_{iT}/\partial T)_V U_i^h}{C_v}. \qquad (21)$$

For forsterite, one finds actually that anharmonicity contributes little to γ_m and another reason for the discrepancy has thus to be found. In fact, the lack of high-temperature data for $(\partial \ln v_i/\partial P)_T$ makes it necessary to use the room-temperature slope of the frequency shifts with pressure for evaluating Eqs. (20) and (21). Contrary to what is often assumed, these slopes could increase with increasing temperature. This is suggested by the simple argument depicted in Fig. 13, which rests on the assumption that frequencies are mainly determined by the volume. If true, such increases with a temperature of $(\partial \ln v_i/\partial P)_T$ could

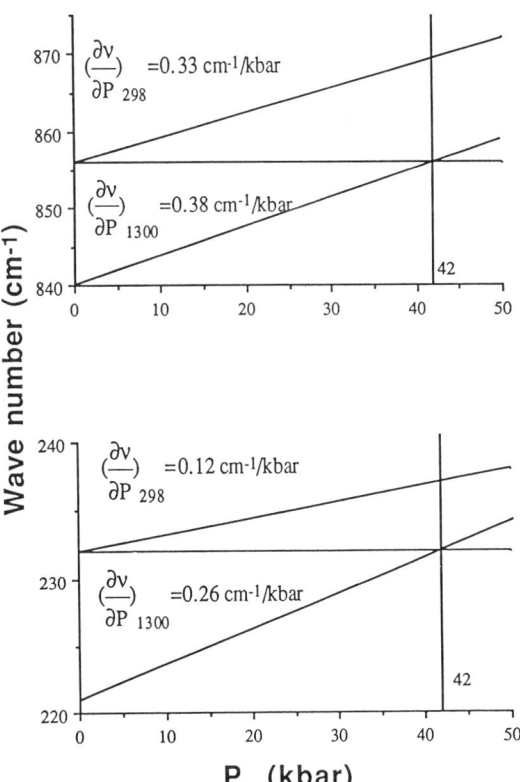

Fig. 13. Effect of temperature on the pressure dependence of the vibrational frequencies of forsterite. The pressure dependence of room temperature frequencies is measured directly. For a temperature of 1300 K, the 1-bar frequencies are also measured values. At this temperature, the molar volume of forsterite is 45.22 cm³ and is back to 43.67 cm³, its 298-K, 1-bar value, at a pressure of 42 kbar according to the elastic data of Isaak et al. (1989a) used with a Birch–Murnaghan equation of state. Hence, frequencies should be similar at 298 K and 1 bar, and at 1300 K and 42 kbar if they depend primarily on volume. With this assumption and a linear pressure dependence of the frequencies, one finds that $(\partial v/\partial P)$ generally increases with temperature in a specific way, as illustrated by the two Raman-active modes considered in this figure.

resolve the discrepancy. High-temperature measurements of the frequency shifts with pressure are thus needed to check this effect, which would be analgous to the decrease to thermal expansion coefficients at high pressure.

Consistency of Thermal and Volume Data

The few minerals for which several determinations of the thermal expansion coefficients are available show values differing by 10 to 20% between 300 and 1400 K, and the differences can reach 50% when extrapolating these results up to 2000 K, as summarized by Goto et al. (1989) and Isaak et al. (1989a) for corundum and forsterite. In addition, high-temperature measurements of the bulk modulus are very difficult to perform and thus restricted to very few minerals, namely, corundum, periclase, and forsterite (Goto et al., 1989; Isaak et al., 1989a,b).

Fei and Saxena (1987) and Saxena (1988, 1989) pointed out that Eq. (1) provides not only a means of ensuring the needed consistency between thermal and volume properties required in thermodynamic calculations, but also a simple way of constraining the pressure and temperature derivatives of the volume at high temperatures. For MgO, e.g., Saxena (1989) used experimental C_p values to optimize simultaneously the values of α, K_T, and C_v.

Within the framework of anharmonic calculations, C_v has no longer to be considered as an adjustable parameter at high temperatures. This allows a simplification of the optimization procedure that has been used by Gillet et al. (1991a) to determine the temperature dependence of α and K_T for forsterite. Comparison of the derived α with available data in Fig. 14 shows very good agreement with the high-temperature observations and extrapolations of Kajiyoshi (1986), which are the recommended values of Isaak et al. (1989a). The

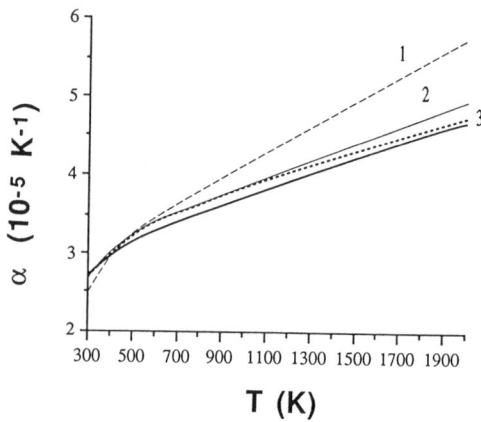

Fig. 14. Thermal expansion of forsterite. Curve 1 from Suzuki et al. (1984); curve 2 from Kajiyoshi (1986); curve 3 from Fei and Saxena (1987); thick curve from $C_p - C_v$ inversion from Gillet et al. (1991a).

inverted values of K_{T0} and (dK_T/dT) are also in excellent agreement with the experimental results of Isaak et al. (1989a). Such results thus suggest that reliable estimates of the high-temperature thermal expansion and bulk modulus at 1 bar can be obtained through anharmonic modeling from the room-temperature properties and the experimental C_p and calculated C_v data.

Acknowledgments

We gratefully thank A. Le Cléac'h, F. Guyot and O. Vidal for their work related to the topics discussed in this paper and Y. Bottinga, M. Madon, and B. Reynard for helpful comments. This work was supported by the DBT program, "Fluides et cinétiques," and grant CNRS-INSU-DBT 241.

References

Akaogi, M., Ross, N.L., McMillan, P., and Navrotsky, A. (1984). The Mg_2SiO_4 polymorphs (olivine, modified spinel and spinel)—thermodynamic properties from oxide-melt solution calorimetry, phase relations, and model of lattice vibrations. *Amer. Mineral.* **69**, 499–512.

Barron, T.H.K., Huang, C.C., and Pasternak, A. (1976). Interatomic forces and lattice dynamics of α-quartz *J. Phys.* **C9**, 3925–3940.

Berman, R.G. (1988). Internally consistent thermodynamic data for minerals in the system Na_2O–K_2O–CaO–MgO–FeO–Fe_2O_3–Al_2O_3–SiO_2–TiO_2–H_2O–CO_2. *J. Petrol* **29**, 445–522.

Berman, R.G. and Brown, T.H. (1985). Heat capacity of minerals in the system Na_2O–K_2O–CaO–MgO–FeO–Fe_2O_3–Al_2O_3–SiO_2–TiO_2–H_2O–CO_2: Representation, estimation, and high-temperature extrapolation. *Contrib. Mineral. Petrol.* **89**, 168–183.

Chang, Z.P. and Graham, E.K. (1977). Elastic properties of oxides in the NaCl structure. *J. Phys. Chem. Solids* **38**, 1355–1362.

Chipman, J. and Fontana, M.G. (1935). A new approximate equation for heat capacities at high temperatures. *J. Amer. Chem. Soc.* **57**, 48–51.

Chopelas, A. (1990). Thermal properties of forsterite at mantle pressures derived from vibrational spectroscopy. *Phys. Chem. Mineral.* **17**, 149–156.

Counsell, J.F. and Martin, J.F. (1967). The entropy of tetragonal germanium dioxide. *J. Chem. Soc. (A)*, 560–561.

Deniélou, L., Fournier, Y., Petitet, J.P., and Téqui, C. (1971). Etude calorimétrique des sels entre 273°K et 1533°K. Technique et application aux sulfates alcalins. *Rev. Int. Hautes Temp. Réfract.* **8**, 119–126.

Ditmars, D.A. and Douglas, T.B. (1971). Measurement of the relative enthalpy of pure α-Al_2O_3 (NBS heat capacity and enthalpy standard reference material No. 720) from 273 to 1173 K. *J. Res. NBS* **75A**, 401–420.

Ditmars, D.A., Ishihara, S., Chang, S.S., and Bernstein, G. (1982). Enthalpy and heat-capacity standard reference material synthetic sapphire (α-Al_2O_3) from 10 to 2250 K. *J. Res. NBS* **87**, 159–163.

Elcombe, M.M. (1967). Some aspects of the lattice dynamics of quartz. *Proc. Roy. Soc.* **91**, 947–958.

Fei, Y. and Saxena, S.K. (1987). An equation for the heat capacity of solids. *Geochim. Cosmochim. Acta.* **51**, 251–254.

Fiquet, G. (1990). Propriétés thermodynamiques des minéraux du manteau supérieur. Calorimétrie à haute température et spectroscopie Raman à haute pression et haute température. Ph.D. Thesis, University of Rennes I.

Fyfe, W.S., Turner, F.J., and Verhoogen, J. (1958). Metamorphic reactions and metamorphic facies. *Geol. Soc. Amer. Mem.* **75**.

Ghose, S. (1988). Inelastic neutron scattering. *Rev. Mineral.* **18**, 161–192.

Gillet, P., Reynard, B., and Téqui, C. (1989). Thermodynamic properties of glaucophane: New data from calorimetric and spectroscopic measurements. *Phys. Chem. Miner.* **16**, 659–667.

Gillet, P., Le Cléac'h, A., and Madon, M. (1990). High-temperature Raman spectroscopy of SiO_2 and GeO_2 polymorphs: Anharmonicity and thermodynamic properties at high temperatures *J. Geophys. Res.* **95**, 21636–21655.

Gillet, P., Richet, P., Guyot, F., and Fiquet, G. (1991a). High-temperature thermodynamic properties of forsterite. *J. Geophys. Res.* **96**, 11805–11816.

Gillet, P., Fiquet, G., Malézieux, J.M., and Geiger, C.A. (1991b). High-pressure and high-temperature Raman spectroscopy of end-member garnets: Pyrope, grossular and andradite. *Eur. J. Mineral.* (in press).

Goto, T., Anderson, O.L., Ohno, I., and Yamamoto, S. (1989). Elastic constants of corundum up to 1825 K. *J. Geophys. Res.* **94**, 7588–7602.

Grønvold, F. (1967). Adiabatic calorimeter for the investigation of reactive substances in the range from 25 to 775°C. Heat capacity of α-aluminum oxide. *Acta Chem. Scand.* **21**, 1695–1713.

Grønvold, F., Stølen, S., and Svendsen, S.R. (1989). Heat capacity of α quartz from 298.15 to 847.3 K, and of β quartz from 847.3 to 1000 K—Transition behaviour and re-evaluation of the thermodynamic properties. *Thermochimica Acta* **139**, 225–243.

Haas, J.L., Jr and Fisher, J.R. (1976). Simultaneous evaluation and correlation of thermodynamic data. *Amer. J. Sci.* **276**, 525–545.

Helgeson, H.C., Delany, J.M., Nesbitt, H.W., and Bird, D.K. (1978). Summary and critique of the thermodynamic properties of rock-forming minerals. *Amer. J. Sci.* **278A**, 1–229.

Hemingway, B.S. (1987). Quartz: Heat capacities from 340 to 1000 K and revised values for the thermodynamic properties. *Amer. Mineral.* **72**, 273–279.

Hemley, R.J. (1987). Pressure dependence of Raman spectra of SiO_2 polymorphs: α-quartz, coesite and stishovite, in *High-Pressure Research in Mineral Physics*, M.H. Manghnani and Y. Syono, eds., A.G.U. Washington, D.C., pp. 347–359.

Hemley, R.J., Cohen, R.E., Yeganeh-Haeri, A., Mao, H.K., Weidner D.J., and Ito, E. (1989). Raman spectroscopy and lattice dynamics of $MgSiO_3$ perovskite at high pressure, in *Perovskite: a Structure of Great Interest to Geophysics and Materials Science*, A. Navrotsky and D.J. Weidner, eds., A.G.U. Geophysical Monograph 45, Washington, D.C., pp. 35–44.

Hofmeister, A.M. (1987). Single-crystal absorption and reflection infrared spectroscopy of forsterite and fayalite. *Phys. Chem. Miner.* **14**, 499–513.

Hofmeister, A.M. and Chopelas, A. (1991). Thermodynamic properties of pyrope and grossular from vibrational spectroscopy. *Amer. Mineral.* **76**, 880–891.

Hofmeister, A.M., Hoering, T.C., and Vigro, D. (1987). Vibrational spectroscopy of beryllium aluminosilicates: Heat capacity calculations from bond assignment. *Phys. Chem. Mineral.* **14**, 205–224.

Holland, T.J.B. (1981). Thermodynamic analysis of simple mineral systems. *Adv. Phys. Geochem.* **1**, 19–34.

Holland, T.J.B. (1989). Dependence of entropy on volume for silicate and oxide minerals: A review and a predictive model. *Amer. Mineral.* **74**, 5–13.

Holm, J.L., Kleppa, O.J., and Westrum, E.F., Jr. (1967). Thermodynamics of polymorphic transformations in silica. Thermal properties from 5 to 1070°K and pressure-temperature stability fields for coesite and stishovite. *Geochim. Cosmochim. Acta* **31**, 2289–2307.

Isaak, D.G., Anderson, O.L., Goto, T., and Suzuki, I. (1989a). Elasticity of single-crystal forsterite measured to 1700 K. *J. Geophys. Res.* **94**, 5895–5906.

Isaak, D.G., Anderson, O.L., and Goto, T. (1989b). Measured elastic moduli of single-crystal MgO up to 1800 K. *Phys. Chem. Mineral.* **16**, 704–713.

Jeanloz, R. (1980). Infrared spectra of olivine polymorphs: α, β phase and spinels. *Phys. Chem. Mineral.* **5**, 327–341.

Kajiyoshi, K. (1986). High-temperature equation of state for mantle minerals and their anharmonic properties. M.S. thesis, Okayama Univ., Okayama, Japan (quoted by Isaak et al., 1989a).

Kelley, K.K. (1960). Contribution to the data on theoretical metallurgy. XIII. High-temperature heat content, heat capacity, and entropy data for the elements and inorganic compounds. U.S. Bureau Mines Bull. 584.

Kelley, K.K., Todd, S.S., Orr, L.R., King, E.G., and Bonnickson, K.R. (1953). Thermodynamic properties of sodium-aluminum silicates. U.S. Bureau Mines Rept. Inv. 4955.

Kieffer, S.W. (1979). Thermodynamics and lattice vibrations of minerals: 3. Lattice dynamics and an approximation for minerals with application to simple substances and framework silicates. *Rev. Geophys. Space Phys.* **17**, 827–849.

Kieffer, S.W. (1980). Thermodynamics and lattice vibrations of minerals: 4. Application to chain and sheet silicates and orthosilicates. *Rev. Geophys. Space Phys.* **18**, 862–886.

Kieffer, S.W. (1985). Heat capacity and entropy: Systematic relations to lattice vibrations. *Rev. Mineral.* **14**, 65–126.

Kittel, C. (1971). *Introduction to Solid State Physics.* Wiley, New York.

Kopp, M. (1865). Investigation of the specific heat of solid bodies. *Phil. Trans. Roy. Soc. Lond.* **155**, 71–202.

Krupka, K.M., Robie, R.A., and Hemingway, B.S. (1979). High-temperature heat capacities of corundum, periclase, anorthite, $CaAl_2Si_2O_8$ glass, muscovite, pyrophillite, $KAlSi_3O_8$ glass, grossular, and $NaAlSi_3O_8$ glass. *Amer. Mineral.* **64**, 86–101.

Krupka, K.M., Robie, R.A., Hemingway, B.S., Kerrick, D.M., and Ito, J. (1985). Low-temperature heat capacities and derived thermodynamic properties of antophyllite, diopside, enstatite, bronzite, and wollastonite. *Amer. Mineral.* **70**, 249–260.

Lange, R.A., De Yoreo, J.J., and Navrotsky, A. (1991). Scanning calorimetric measurement of heat, capacity during incongruent melting of diopside. *Amer. Mineral.* **76**, 904–912.

Le Cléac'h, A. (1989). Contribution à l'étude des propriétés physiques des minéraux de haute pression. Mém. Docum. Centre Arm. et Struct. Socles **36**, Rennes.

McMillan, P.F. and Ross, N.L. (1987). Heat capacity calculations for Al_2O_3 corundum and $MgSiO_3$ ilmenite. *Phys. Chem. Mineral.* **14**, 225–234.

Madon, M. and Price, G.D. (1989). Infrared spectroscopy of the polymorphic series (enstatite, ilmenite, and perovskite) of $MgSiO_3$, $MgGeO_3$, and $MnGeO_3$. *J. Geophys. Res.* **94**, 15687–15701.

Maier, C.G. and Kelley, K.K. (1932). An equation for the representation of high-temperature heat content data. *J. Amer. Chem. Soc.* **54**, 3243–3246.

Mammone, T.F. and Sharma S.K. (1980). Pressure and temperature dependence of the

Raman spectra of rutile structure oxides. *Year Book, Carnegie Inst. Washington*, **79**, 369–373.

Navrotsky, A. (1980). Lower mantle phase-transitions may generally have negative pressure-temperature slopes. *Geophys. Res. Lett.* **7**, 709–711.

Newton, R.C. (1987). Thermodynamic analysis of phase-equilibria in simple mineral systems. *Rev. Mineral.* **17**, 1–33.

Perkins, D., Westrum, E.F. Jr., and Essene, E.J. (1980). The thermodynamic properties and phase relations of some minerals in the system $CaO-Al_2O_3-SiO_2-H_2O$. *Geochim. Cosmochim. Acta* **44**, 61–84.

Price, G.D., Parker, S.C., and Leslie, M. (1987). The lattice dynamics and thermodynamics of the Mg_2SiO_4 polymorphs. *Phys. Chem. Mineral.* **15**, 181–190.

Rao, K.R., Chaplot, S.L., Choudury, N., Ghose, S., Hastings, J.M., Corliss, L.M., and Price, D.L. (1988). Lattice dynamics and inelastic neutron scattering from forsterite, Mg_2SiO_4: Phonon dispersion relation; density of states and specific heat. *Phys. Chem. Mineral.* **16**, 83–97.

Richet, P. (1990). GeO_2 vs SiO_2: Glass transitions and thermodynamic properties of polymorphs. *Phys. Chem. Miner.* **17**, 79–88.

Richet, P. and Fiquet, G. (1991). High-temperature heat capacity and premelting of minerals in the system $CaO-MgO-Al_2O_3-SiO_2$. *J. Geophys. Res.* **96**, 445–456.

Richet, P., Bottinga, Y., Deniélou, L., Petitet, J.P., and Téqui, C. (1982). Thermodynamic properties of quartz, cristobalite and amorphous SiO_2: Drop calorimetry measurements between 1000 and 1800 K and a review from 0 to 2000 K. *Geochim. Cosmochim. Acta* **46**, 2639–2658.

Richet, P., Robie, R.A., Rogez, J., Hemingway, B.S., Courtial, P., and Téqui, C. (1990). Thermodynamics of open networks: Ordering and entropy in $NaAlSiO_4$ glass, liquid, and polymorphs. *Phys. Chem. Miner.* **17**, 385–394.

Richet, P., Robie, R.A., and Hemingway, B.S. (1991a). Thermodynamic properties of wollastonite and $CaSiO_3$ glass and liquid. *Eur. J. Mineral.* **3**, 475–484.

Robie, R.A. and Hemingway, B.S. (1972). Calorimeters for heat of solution and low-temperature heat capacity measurements. *U.S. Geol. Surv. Prof. Paper* 755.

Robie, R.A., Hemingway, B.S., and Fisher, J.R. (1979). Thermodynamic properties of minerals and related substances at 298.15 K and 1 bar (10^5 Pascals) pressure and at higher temperatures. *U.S. Geol. Surv. Bull.* 1452.

Robie, R.A., Hemingway, B.S., Gillet, P., and Reynard, B. (1991). On the entropy of glaucophane $Na_2Mg_3Al_2Si_8O_{22}(OH)_2$. *Contrib. Mineral. Petrol.* **107**, 484–486.

Robinson, G.R. and Haas, J.L., Jr., (1983). Heat capacity, relative enthalpy, and calorimetric entropy of silicate minerals: An empirical method of prediction. *Amer. Mineral.* **68**, 541–553.

Ross, N.L., Akaogi, M., Navrotsky, A., Susaki, J.I., and McMillian, P. (1986). Phase transitions among the $CaGeO_3$ polymorphs (wollastonite, garnet, and perovskite structures): Studies by high-pressure synthesis, high-temperature calorimetry, and vibrational spectroscopy and calculation. *J. Geophys. Res.* **91**, 4685–4696.

Salje, E. and Werneke, C.H. (1982). The phase equilibrium between sillimanite and andalusite as determined from lattice vibrations. *Contrib. Mineral. Petrol.* **79**, 56–67.

Saxena, S.K. (1988). Assessment of thermal expansion, bulk modulus, and heat capacity of enstatite and forsterite. *J. Phys. Chem. Solids* **49**, 1233–1235.

Saxena, S.K. (1989). Assessment of bulk modulus, thermal expansion and heat capacity of minerals. *Geochim. Cosmochim. Acta* **53**, 785–789.

Slater, J.C. (1939). *Introduction to Chemical Physics*. McGraw Hill, New York.

Smith, D.K. and Leider, H.R. (1968). Low-temperature thermal expansion of LiH, MgO and CaO. *J. Appl. Cryst.* **1**, 246–249.

Suzuki, I., Takei, H., and Anderson, O.L. (1984). Thermal expansion of forsterite, Mg_2SiO_4, in *Proceedings of 8th International Thermal Expansion Conference*, T.A. Hahn, ed., Plenum, New York, pp. 79–88.

Watanabe, H. (1982). Thermochemical properties of synthetic high-pressure compounds relevant to the Earth's mantle, *in High-Pressure Research in Geophysics*, S. Akimoto and M.H. Manghnani, eds., Reidel, Dordrecht, pp. 441–464.

White, W.P. (1919). Silicate specific heats. Second series. *Amer. J. Sci.* **47**, 1–43.

Wood, B.J. (1981). Crystal-field electronic effects on the thermodynamic properties of Fe^{2+} minerals. *Adv. Phys. Geochem.* **1**, 63–84.

Chapter 5
Thermodynamics of Silicate Melts: Configurational Properties

P. Richet and D.R. Neuville

Introduction

The ease with which a liquid adjusts to the shape of its container is a well-known consequence of the hallmark of the molten state, atomic mobility. Atomic mobility is the very reason why liquids flow, even though another salient feature evident through daily experience is that the viscosity increases when the temperature decreases. In fact, if crystallization does not occur, the viscosity eventually becomes so high that flow can no longer take place during the timescale of an experiment. The resulting material is a glass, i.e., a solid with the frozen-in disordered atomic arrangement of a liquid. Glasses have been produced for millennia, but the kinetic nature of the liquid-glass transition and its influence on the properties of glasses have long remained elusive. We will not specifically address these aspects, however, because they have already been extensively discussed in the geochemical literature from a relaxational (Dingwell and Webb, 1990) or thermochemical standpoint (Richet and Bottinga, 1983, 1986). In this review, we will focus on features of liquids that are directly related to atomic mobility, namely, the existence of those contributions to physical properties of liquids that have been termed configurational (Simon, 1931; Bernal, 1936).

Configurational properties are, in effect, at the root of the main differences between liquids and solids. To illustrate the importance of configurational aspects, consider the second-order thermodynamic properties of diopside and $CaMgSi_2O_6$ glass and liquid (Table 1), which are representative of silicates in this respect. Above room temperature, the glass and crystalline phases have similar heat capacities (C_p), thermal expansion coefficients (α), and compressibilities ($\beta = 1/K_T$, where K_T is the isothermal bulk modulus). In contrast, the heat capacity of the liquid is about 30% higher than those of the solid forms and the α and K_T differences are greater still. These differences can have far-reaching consequences when integrated as a function of temperature or pressure, as illustrated by two simple examples. Moderate extrapolation of the C_p data for

Table 1. Comparison between second-order properties of diopside and $CaMgSi_2O_6$ glass and liquid.

	T(K)	Diopside[a]	Glass[b]	Liquid[c]
C_p(J/mol K)	298	166.8	168.7	
	1000	247.2	255.6	334.6
	1700	267.8		334.6
$\alpha(K^{-1})$	298	$3.0\ 10^{-5}$	$2.0\ 10^{-5}$	
	1000	$3.4\ 10^{-5}$		$7.3\ 10^{-5}$
K_T(kbar)	298	1144	741	
	1700	907		234

[a] C_p from Richet and Fiquet (1991); α from Finger and Ohashi (1976), to $\pm 10^{-5}$ K^{-1}; and K_T from Levien and Prewitt (1981).
[b] C_p from Richet et al. (1986); α and K_T from Soga et al. (1979).
[c] C_p from Richet and Bottinga (1984b); α from Bottinga et al. (1982), and K_T from Rivers and Carmichael (1987).

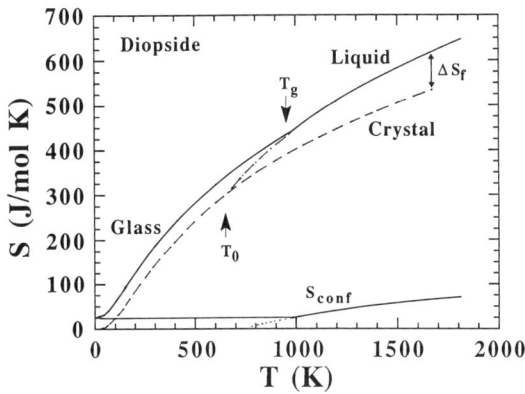

Fig. 1. Entropy of diopside and $CaMgSi_2O_6$ glass and liquid, and configurational entropy of the amorphous phases. The extrapolated entropy of the supercooled liquid, indicated by the dot-dashed curve, is equal to the entropy of the crystal at T_0. (Note that the configurational entropy of the liquid, whose extrapolation is shown as a dotted line, would vanish at a temperature higher than T_0 because the vibrational entropy of the amorphous phase is higher than that of the crystal.) Data from Krupka et al. (1985), Richet and Bottinga (1984b), Richet and Fiquet (1991), and Richet et al. (1986).

supercooled $CaMgSi_2O_6$ liquid toward low temperature would lead to a liquid having a lower entropy than the crystalline form below 640 K (Fig. 1), the temperature of the so-called Kauzmann (1948) entropy catastrophe. In the same way, liquid $CaMgSi_2O_6$ would become denser than diopside at around 130 kbar (Fig. 2).

Such extreme examples show how profound the influence of configurational factors on the thermal and volume properties of liquids can be, and how impor-

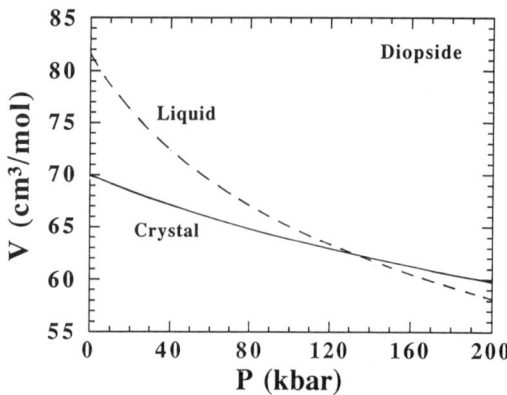

Fig. 2. Volume of diopside and $CaMgSi_2O_6$ liquid as a function of pressure at 1670 K as calculated from the Birch–Murnaghan equations of state reported by Bottinga (1986).

tant the role they play in phase equilibria involving a liquid phase can be. Thermodynamic modeling is a necessary step for interpolating or extrapolating the limited amount of measurements that can be made in view of the considerable ranges of temperature, pressure, and chemical composition of natural magmas (e.g., Bottinga and Weill, 1970, 1972; Carmichael et al., 1977). For this purpose, one thus needs to understand the complex way configurational properties depend on these factors. Even though direct structural information is becoming available through techniques like nuclear magnetic resonance (NMR) spectroscopy (e.g., Stebbins and Farnan, 1989; Farnan and Stebbins, 1990), thermodynamic data are still much more abundant than high-temperature structural information for melts. An essentially phenomenological approach will thus be followed in this review after an introductory, qualitative discussion of configurational changes and their influence on thermodynamic properties.

The configurational heat capacity will be used first to illustrate the specific dependences of configurational properties on temperature and composition. This property is one of the authors' favorite because it has great importance in thermodynamic calculations, and it will also exert strong constraints on the quantitative modeling of the temperature-dependent changes in the relative abundances of the various structural entities present in a melt. In addition, the configurational heat capacity determines the temperature dependence of the configurational entropy that has a major effect on the relaxational properties of melts, of which viscosity is the most important. Quantitative modeling of viscosity will thus be also reviewed briefly within the framework of the configurational entropy theory, which requires understanding the composition and temperature dependencies of configurational heat capacity. In turn, this theory allows determinations of configurational entropies from viscosity measurements, and these data will be valuable for improving or setting up the thermodynamic models of silicate melts that are needed to predict solid–liquid, gas–liquid, or liquid–liquid phase equilibria. Finally, we will conclude this review, admittedly biased by the authors' own interests, with a cursory examination of configurational effects on volume, particularly with respect to high-pressure fusion.

Configurational Properties

Configurational Changes

Atomic positions in solids are determined by local minima of the interatomic potentials. In a crystal, these potentials have long-range symmetry. In a glass, bond angles and distances between next-neighbor atoms are not constant but spread over a range of values. As a consequence, long-range order does not exist and a one-dimensional section of interatomic potentials can be viewed as shown in Fig. 3, where minima are separated by barriers with varying heights and shapes. Now, assume that some heat is delivered instantaneously to a glass. At low temperatures, the subsequent temperature rise is associated with only an increase of the vibrational energy through increasing mean amplitudes of vibration. The heat capacity is only vibrational in nature and the material thus behaves as any solid in this respect.

At some point, however, this energy becomes high enough that atoms can begin to overcome the potential energy barriers separating distinct configurational states. On a very short timescale, i.e., one with a fixed configuration, only the vibrational heat capacity would be measured. But if enough time is left for configurational changes to take place, then the heat capacity shows an excess C_p over the vibrational value. This is the configurational contribution, which represents that part of the heat supplied for a given temperature rise that is used to increase the potential, and not the vibrational energy (Fig. 3). This configurational contribution is necessarily positive (e.g., Davies and Jones, 1953) and the molten and glassy state can thus be distinguished on the basis of the existence

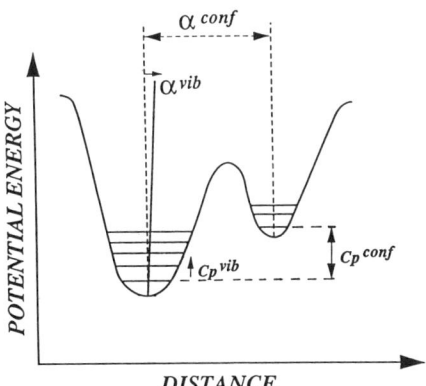

Fig. 3. Schematic one-dimensional representation of interatomic potentials for an amorphous substance. The changes in energy or zero-point energy associated with the vibrational and configurational parts of the heat capacity are shown by arrows. Likewise, arrows point to the vibrational and configurational parts of the (linear) thermal expansion coefficient associated with changes in interatomic distances (r), $\alpha = 1/r(\partial r/\partial T)_P$.

or the lack of this configurational contribution. Hence, the glass transition can be viewed as the onset of the exploration by matter of positions characterized by higher values of interatomic potentials (e.g., Goldstein, 1969). This spreading of configurations over states of higher and higher potential energy is the main feature of atomic mobility, no matter how complex this dynamical process may be at a microscopic level.

Turning now to the volume changes of an amorphous substance, one notes that a general feature of interatomic potentials is their anharmonic nature, i.e., the forces exerted on the vibrating atoms are not strictly proportional to the displacements from their equilibrium positions. Increasing vibrational amplitudes thus results in increases of interatomic distances (Fig. 3). As solids, liquids have also such an anharmonic vibrational expansion, but the configurations of higher energy that begin to be explored above the glass transition are generally associated with increases in interatomic distances. This is the reason why the thermal expansion coefficient usually increases markedly at the glass transition. (Note, however, that a positive configurational thermal expansion coefficient is not a thermodynamic requirement. In a few instances, a densification of the structure is observed with increasing temperatures, as illustrated by the well-known negative α of water below 4°C.)

Consider finally the compressibility. In minerals, the existence of long-range order restricts the compression mechanisms to changes in bond angles or distances if the crystalline structure is to be preserved. As a result, the compressibility of a given phase is somewhat limited. Important compaction takes place through phase transitions characterized generally by increases in coordination numbers. Within a given structure, the effects of pressure can be viewed as inducing only variations in the interatomic potentials characterized by shorter equilibrium distances and steeper slopes around the minima. Because the shape of these potentials determines the vibrational energy levels, this compression will be termed vibrational as long as no significant changes in short-range order take place in the crystal.

In a liquid, this vibrational compaction also exists, but it is considerably enhanced by mechanisms like a progressive switch of cations toward higher oxygen coordination polyhedra, which provides a continuous path for markedly increasing the density (e.g., Waff, 1975; Stolper and Ahrens, 1987). The signature of the availability of configurational states of higher density is the common four- or five-fold increases in compressibility observed on melting or at the glass transition (see Table 1). This configurational part of the compressibility can be positive only (e.g., Davies and Jones, 1953). On the other hand, glass transition temperatures vary with pressure (e.g., Rosenhauer et al., 1979), which would make it useful to describe the glass transition (for a given frequency, see below) as a curve in the P-T plane. However, the effects of pressure and temperature on the properties of glasses are actually of a different nature, for glasses quenched at room temperature from pressures of a few tens of kbar undergo permanent compaction. A high pressure can thus induce irreversible configurational changes at temperatures at which the substance is said to be a glass. Hence, for given frequencies or experimental timescales, the kinetics of pressure- and temperature-induced configurational modifications are markedly different. This dissimilarity essentially originates in that the shape of potential energy wells varies little with temperature, but significantly with pressure. If high kinetic energy is needed to overcome potential barriers at constant pressure, the changes in these barriers with pressure can lead by themselves to new configurational states, at low temperatures, if the pressure is high enough. Without elaborating on these effects already alluded to above, we will just point out that processes taking place in glasses at high pressure thus likely mimic to a significant extent those occurring in liquids, whereas this similarity is clearly nonexistent for the effects of temperature.

Vibrational vs. Configurational Properties

The strong rate dependence of the glass transition shows simply that the kinetics of configurational changes increases tremendously with temperature. In a qualitative way, the time for structural relaxation is a measure of the time required to switch from an initial to a final configuration in order to regain internal equilibrium after a change in temperature or, more generally, in any other state variable. Hence, the glass transition range is usually defined as the temperature interval where time-dependent properties are observed because of the similitude between relaxation times and measurement timescales. For usual calorimetry or volume measurements, experimental timescales are about 1 to 15 min., and this period is similar to the relaxation times of liquids with viscosities the order of 10^{13} poises. The temperature at which this viscosity is attained is thus operationally used as the definition of the glass transition temperature, below which materials cooled at rates the order of 10°/min show solidlike behavior.

Practically speaking, the distinction between vibrational and configurational properties makes sense only if there is some way to determine their relative importance. For this purpose, one must note that atomic vibrations have a period the order of 10^{-12} to 10^{-14} s in solids and in liquids as well. As long as relaxation times are long with respect to 10^{-12} to 10^{-14} s, vibrations in a liquid can thus be considered as taking place instantaneously in a fixed structural environment. Whenever this assumption is met, physical properties can be separated into independent vibrational and configurational contributions. This is actually the case when the viscosity is higher than about 1 poise (0.1 Pa s), as generally found for silicate melts at temperatures of geochemical interest.

In a few instances, configurational and vibrational properties can be determined from measurements made on different timescales. Consider, for example, the measurement of the adiabatic compressibility by an ultrasonic method, whereby the speed of sound (c) in the material is most conveniently measured at MHz frequencies (Fig. 4). The passage of the compression wave will result in configurational changes only if the temperature is high enough that these changes can take place faster than the 10^{-6} s period of the perturbation. Experi-

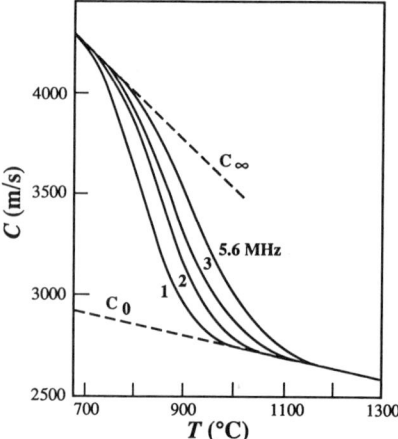

Fig. 4. Speed of sound at various frequencies for amorphous $Na_2Si_2O_5$. The c_0 and c_∞ lines, referring to sound velocities at zero and infinite frequencies, represent the equilibrium and vibrational values, respectively. At these very short timescales, note that liquid-like behavior begins to be observed when the viscosity is as low as 10^6 poises and that a fully relaxed property is measured at viscosities lower than 10^3 poises. The adiabatic compressibility is $\beta_S = -1/V(\partial V/\partial P)_S = 1/\rho c^2$, where ρ is the density. Data from Nikonov et al. (1982).

mentally, this happens only when the viscosity is lower than about 10^3 poises. At higher viscosities, only vibrations relax completely and a rapidly decreasing fraction at most of the configurational part of the compressibility can be measured. Hence, the equilibrium compressibility of the melt can be determined solely at low viscosities and there exists a temperature interval, of the order of a few hundred degrees, where the sound velocity depends on the frequency.

It is generally difficult to measure in such a way the frequency dependence of the heat capacity or the thermal expansion coefficient, especially at the high temperatures at which silicates are molten. The simplest assumption that can be made is that there is no discontinuity in vibrational properties at the glass transition. A priori, this assumption can be justified by data like those discussed above for $CaMgSi_2O_6$ (Table 1), which suggest that second-order thermodynamic properties are mainly determined by the solid or liquid nature and not by the crystalline or amorphous state of a phase. (We will emphasize that this is true above room temperature only. As mentioned in the previous section, vibrational properties depend sensitively on the structure below 200 K). If the vibrational properties of a glass do not differ considerably from those of the isochemical crystal stable at the same pressure, then glasses with different configurational states should exhibit still smaller differences. In other words, the configurational state of a liquid should exert little influence on its vibrational properties, which can be obtained from high-temperature extrapolation of the glass properties.

The most extensive data for second-order properties deal with heat capacity.

For a given composition, one does not observe a significant dependence of C_p on the thermal history of the glass outside the glass transition range (Richet and Bottinga, 1986; Richet et al., 1986), even below 200 K where C_p is a sensitive function of the structure (e.g., Kelley et al., 1953). Likewise, within $\pm 1\%$ the heat capacity of silicate glasses is found above room temperature to be an additive function of composition (Bacon, 1977; Stebbins et al., 1984; Richet, 1987), regardless of the thermal history of the material. Hence, experimental evidence does not point to a significant dependence of the vibrational C_p on the configurational state of a liquid. The validity of this conclusion cannot be ascertained for the thermal expansion coefficient or the compressibility because of the scarcity of relevant experimental data. It is likely, however, that this assumption is valid to a first approximation at least.

Heat Capacity

Configurational Heat Capacity

For silicates, the glass transition takes place when the heat capacity of the glass (C_{pg}) is within a few percent of $3R/g$ atom, with $R =$ gas constant [Fig. 5(a)]. This is the Dulong and Petit harmonic limit for the isochoric heat capacity from which C_p differs little for silicate glasses (Haggerty et al., 1968; Richet and Bottinga, 1986). Under usual cooling conditions, glass transition temperatures of silicates are generally lower than 1200 K. A significant anharmonic contribution to C_{pg} is not expected below 1200 to 1500 K, and the available data discussed below do not show any progressive increase of the heat capacity of the liquid (C_{pl}) beginning above these temperatures. For sodium-silicate liquids, for instance, the heat capacity is constant to within $\pm 1\%$ from 800 to 1800 K. Hence, anharmonicity is unlikely a major feature and the vibrational C_p of silicate melts should not vary significantly with temperature above T_g. This leads immediately to a simple equation

$$C_p^{conf} = C_{pl} - C_{pg}(T_g). \qquad (1)$$

The validity of this equation, which is unlikely to be absolute (Goldstein, 1976), has been justified more rigorously elsewhere for silicates (Richet et al., 1986). This expression provides a simple means of determining calorimetrically the configurational heat capacity, and it also shows that most of the temperature dependence, if any, of C_{pl} can be ascribed to temperature-dependent configurational changes in the liquid. As already noted, the vibrational C_p is a linear function of composition. Deviations of C_{pl} from additive variations are thus attributable to the composition dependence of the configurational heat capacity.

Practically speaking, the only apparent difficulty in using Eq. (1) is that the glass transition of most silicates of geochemical interest is higher than the temperature range at which direct C_p measurements can be made accurately.

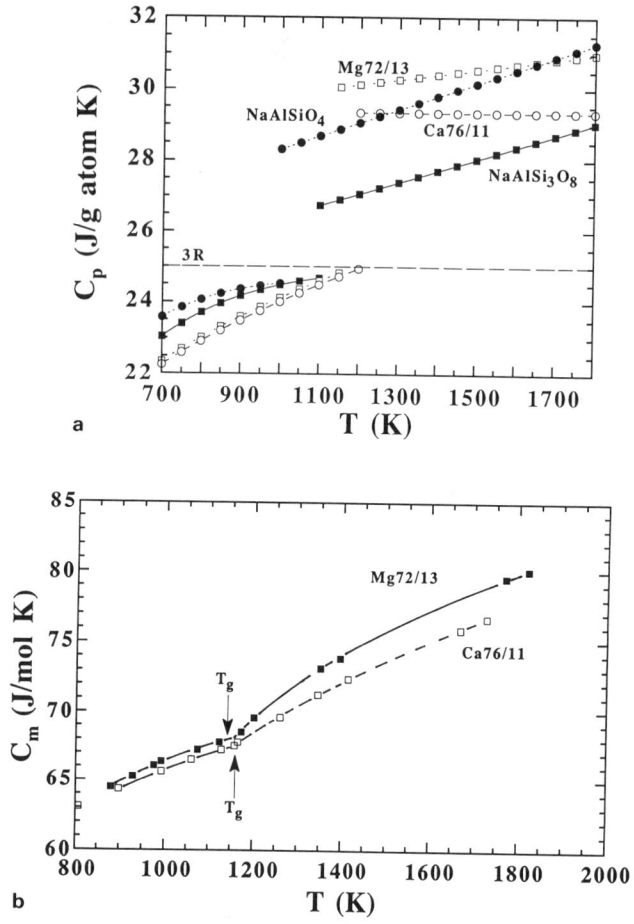

Fig. 5. Calorimetric effects of the glass transition for the calcium and magnesium aluminosilicates of Table 2: (a) true heat capacity, obtained by differentiation of the fitted enthalpy data of Table 2 shown in (b) in the form of mean heat capacities. The heat capacities of $NaAlSiO_4$ and $NaAlSi_3O_8$ (Richet et al., 1990; Richet and Bottinga, 1984a) are included for comparison in (a), where the Dulong and Petit limit of $3R$/g atom is shown as the horizontal dashed line.

Above 1000 K, heat capacities must be determined from relative-enthalpy experiments, $H_T - H_{T_0}$, between a high temperature T and the calorimeter temperature T_0, that is usually around 300 K. Nevertheless, when reported in the form of mean heat capacities, $C_m = (H_T - H_{T_0})/(T - T_0)$, these measurements clearly show the glass transition as an abrupt change in the slope of the C_m curve [Fig. 5(b)]. On both sides of the glass transition, different enthalpy equations can then be fitted and differentiated to yield separately the heat capacities of the glass and liquid phases, as reported in Table 2 for two additional aluminosilicate compositions investigated in this review.

Table 2. Relative enthalpy[a] of two aluminosilicate compositions (kJ/mol).

Run	T(K)	$H_T - H_{273}$	Run	T(K)	$H_T - H_{273}$
	Ca76/11[b]			Mg72/13[c]	
BT.1	807.5	33.689	BU.15	881.0	39.189
BT.14	898.5	40.232	BU.17	929.8	42.854
BT.4	993.1	47.254	BU.19	976.1	46.441
BT.10	1060.8	52.389	BU.13	992.7	47.748
BT.9	1129.2	57.542	BU.1	1074.7	53.880
BT.3	1158.3	59.755	BU.18	1124.1	57.674
BT.8	1165.8	60.495	BU.6	1173.7	61.837
BT.7	1260.8	68.730	BU.11	1201.8	64.594
BT.5	1345.9	76.420	BU.20	1351.6	78.853
BT.12	1414.2	82.594	BU.9	1395.2	82.827
BT.13	1669.1	105.91	BU.10	1770.0	119.07
BT.6	1728.8	111.61	BU.21	1820.1	123.87

[a] Drop-calorimetry measurements made on synthetic samples prepared from oxide (or carbonate) mixes with the ice calorimeter and high-temperature equipment as described by Richet and Bottinga (1984a,b).

[b] Ca76/11: 76.5, 11.75, and 11.75 mol % SiO_2, Al_2O_3, and CaO, respectively. For the liquid phase, $H_T - H_{273} = -46,766 + 91.537\,T$ (J/mol). For the glass phase, $H_T - H_{273} = -24,830 + 64.842\,T + 5.943\,10^{-3}\,T^2 + 18.233\,10^5/T$ (J/mol), with a glass transition temperature of 1199 K.

[c] Mg72/13: 72.5, 13.75, and 13.75 mol % SiO_2, Al_2O_3, and MgO, respectively. For the liquid phase, $H_T - H_{273} = -45,932 + 89.232\,T + 2.24\,10^{-3}\,T^2$ (J/mol). For the glass phase, $H_T - H_{273} = -25,532 + 65.890\,T + 5.879\,10^{-3}\,T^2 + 19.381\,10^5/T$ (J/mol), with a glass transition temperature of 1153 K.

Effects of Temperature and Composition

The main features of the temperature and composition dependences of C_p^{conf} are illustrated in Figs. 6 to 8 for a few simple systems. In all cases, C_p^{conf} increases with decreasing SiO_2 content, from about 10% of C_{pg} for molten silica to more than 30% for metasilicate compositions. At constant SiO_2 content, it seems to increase with the charge of the cation and decreases with its radius (e.g., Stebbins et al., 1984; Richet and Bottinga, 1985), even though the similar effects of Mg and Ca show how rough this trend actually is. In addition, C_p^{conf} does not depend significantly on temperature and practically varies linearly with composition down to pure SiO_2 for Li-, Na-, Ca-, and Ba-bearing liquids [Fig. 6(a)]. Unexpectedly, however, potassium-silicate liquids behave in a quite different way, showing increases of C_p^{conf} with temperature and a variation with composition that is not linear down to pure SiO_2 [Fig. 6(b)]. Quite different behavior is observed when titanium is introduced in alkali silicate melts. For the two potassium or sodium titatanosilicates shown in Fig. 7, the configurational heat capacity is so high that it represents about 50% of 3R/g atom and decreases at higher temperatures to tend approximately toward the values for their Ti-free counterparts.

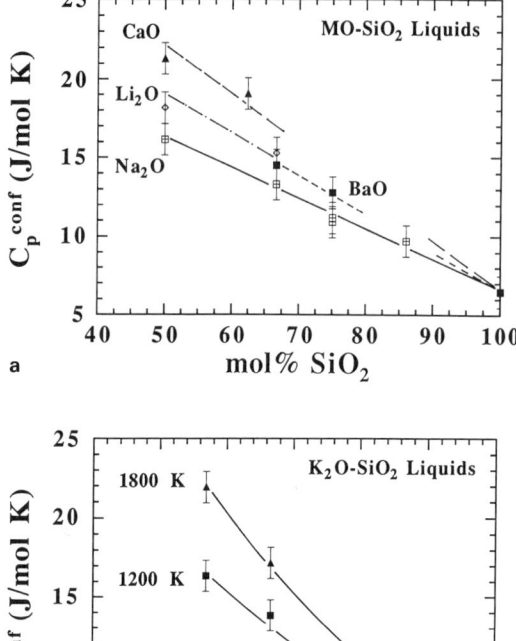

Fig. 6. Configurational heat capacities of MO–SiO$_2$ liquids: (a): MO = Li$_2$O, Na$_2$O, CaO, and BaO, for which measurements yield temperature-independent values; (b) temperature-dependent values for MO = K$_2$O. For reasons of consistency, all the vibrational heat capacities have been assumed to be $3R$/g atom because of the lack of experimental C_p data for some of these glasses at the glass transition. Data from Stebbins et al. (1984) for BaSi$_2$O$_5$ and Li$_2$SiO$_3$, and Richet et al. (1984) and Richet and Bottinga (1985) for the other compositions.

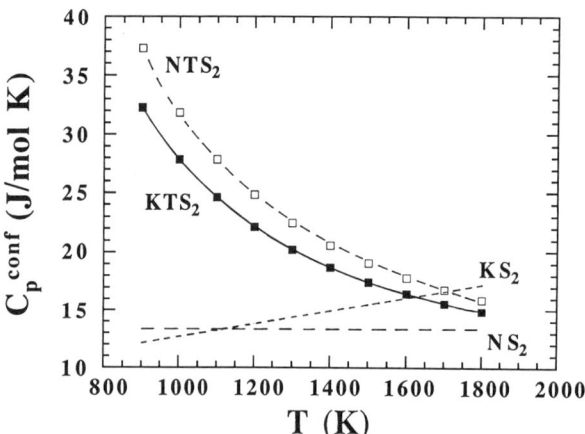

Fig. 7. Configurational heat capacities of two titanium-bearing melts; NTS$_2$ is Na$_2$TiSi$_2$O$_7$ and KTS$_2$ is close to the K$_2$TiSi$_2$O$_7$ composition; NS$_2$ and KS$_2$ are sodium and potassium disilicates. Data from Richet and Bottinga (1985), with vibrational heat capacities assumed to be $3R$/g atom.

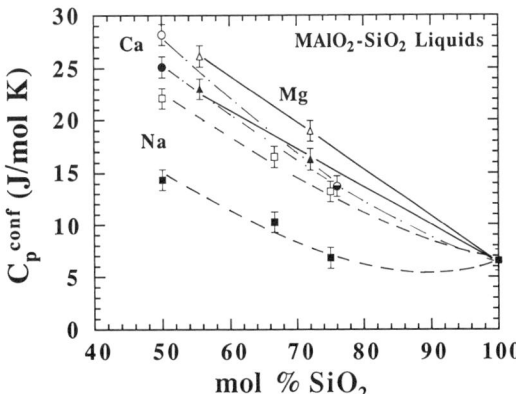

Fig. 8. Configurational heat capacities of aluminosilicate liquids along pseudobinary $MAlO_2$–SiO_2 joins, with M = Na, $Ca_{0.5}$, and $Mg_{0.5}$. The solid and open symbols refer to values at 1200 and 1800 K, respectively. Data from Richet et al. (1982), Richet and Bottinga (1984a, b), and Table 2.

This increase of C_p^{conf} with decreasing SiO_2 content is preserved for aluminosilicate melts when Al is exchanged for Si along $MAlO_2$–SiO_2 joins (Fig. 8), on which lie a number of important mineral compositions. Again, however, the temperature and composition variations depend specifically on the cation M. The configurational heat capacities of Na-bearing melts are the lowest, but they vary most with temperature. In addition, they do not follow a linear variation with composition down to pure SiO_2. The composition of the two products investigated in this study was thus chosen to complement previous observations and determine whether calcium and magnesium aluminosilicates behave in the same way. In both cases, the temperature dependence of C_p^{conf} is significant only for Al-rich, Si-poor liquids. But C_p^{conf} varies linearly with composition from pure SiO_2 to cordierite composition ($Mg_2Al_4Si_5O_{18}$) at least, whereas calcium compositions show deviations from linearity that seem, however, less important than for sodium aluminosilicates.

Even though the structural literature for silicate glasses is extensive, it has limited predictive power to account for these C_p^{conf} data in view of the specificity of interactions in melts that are clearly nonexistent in the glassy phases. Note in this respect that the glass transition of silicates is likely triggered by the onset of oxygen mobility, as suggested by Richet and Bottinga (1983) from an examination of oxygen diffusivity data on both sides of the glass transition (Yinnon and Cooper, 1980) and observed in NMR experiments by Farnan and Stebbins (1990). In addition, the temperature-independent, additive functions of composition found for the configurational heat capacity of the binary systems of Fig. 6(a) suggest, as pointed out by Richet and Bottinga (1985), that C_p^{conf} is mainly determined by the basic structural units which exist from pure SiO_2 to metasilicate compositions at least. The temperature-induced changes would thus affect primarily the short-range order, a conclusion consistent with the thermodynamic interpretation of the viscosity data reviewed in the fifth section.

Specifically, changes in silicon coordination from four to five, or disproportionation reactions of the so-called Q species of the form $2Q_3 = Q_2 + Q_4$ are associated with enthalpies of about 30 kJ/mol (Stebbins, 1988). Over the 1000 K interval that can be generally investigated above the glass transition, the C_p^{conf} data indicate changes in configurational enthalpy of silicate melts ranging from 15 to 25 kJ/g atom, or from 45 to 75 kJ/mol on an oxide formula basis. A comparison of such figures with the above enthalpies of reaction shows at once the importance of practically unquenchable effects such as coordination or Q-species distribution changes. Further work is thus badly needed to relate quantitatively the structural information to the thermodynamic data.

Finally, we will also point out briefly correlations between the variations of the heat capacity and of some other physical properties of aluminosilicate melts. From the high-temperature viscosity data, for instance, Bottinga and Weill (1972) assumed that tetrahedrally coordinated aluminum is more strongly associated with alkali than with alkaline-earth cations, in the order K, Na, Ca, and Mg. On the other hand, molecular orbital calculations have indicated a decrease in the same order of the strength of Al-O bonds (Navrostsky et al., 1985). Finally, enthalpies of mixing also become progressively more negative in the same order along the same joins (Navrotsky et al., 1982; Roy and Navrotsky, 1984). This is in agreement with increasing liquid immiscibility, which is usually viewed as resulting from the competition of cations for oxygen bonding. Not surprisingly, these data thus also point to a central role for oxygen atoms in temperature-induced configurational changes.

Entropy

Residual Entropy of Glasses

Structural relaxation can take place in glasses well below the glass transition range (e.g., Johari, 1976). As shown by available data, however, the calorimetric consequences of these changes are not significant for silicates. Without serious errors, one can assume that the configurational entropy of silicate glasses remains constant from the glass transition down to 0 K. In other words, the residual entropy of a glass at 0 K is the configurational entropy of the liquid frozen in at the glass transition

$$S_g(0) = S^{\mathrm{conf}}(T_g). \qquad (2)$$

This residual entropy can be determined from measurements of the heat capacity and entropy of fusion, as shown in Fig. 1 for $CaMgSi_2O_6$. If the third law is valid for a crystalline form of the material, C_p measurements for this crystal from 0 K to the melting point T_f yield the absolute entropy at T_f. Determination of the entropy of fusion, usually from measurements of the enthalpy of fusion, then gives the absolute entropy of the liquid at T_f. Finally, measurements of the heat capacity from T_f back to 0 K for the supercooled liquid and the glass give $S_g(0)$. In summary, one obtains

$$S^{conf}(T_g) = \int_0^{T_f} \frac{C_{pc}}{T} + dT + \Delta S_f + \int_{T_f}^{T_g} \frac{C_{pl}}{T} dT + \int_{T_g}^0 \frac{C_{pg}}{T} dT, \quad (3)$$

where C_{pc} is the heat capacity of the crystal and ΔS_f its entropy of melting. As illustrated in Fig. 1, all the data needed to evaluate Eq. (3) play an important role. This is the reason why low-temperature C_p data are badly needed for both the crystal and glass. Below room temperature, the heat capacity contrast between the glass and crystal indeed results in important variations temperature the entropy difference between the crystal and the glass for a number of compositions. These trends are shown in Fig. 9 for a few materials, demonstrating the predominantly low-temperature origin of the differences in vibrational entropy between amorphous and crystalline phases.

In fact, glasses are nonequilibrium susbstances because of the kinetic nature of the glass transition, and discussion of their thermodynamic properties is not as straightforward as implicitly assumed in this paper. As discussed in previous reviews, knowledge of the pressure and temperature is insufficient to specify the state of a glass, and the fictive temperature introduced by Tool and Eichlin (1931) represents the simplest one-parameter way to specify the thermal history of a glass (e.g., Richet and Bottinga, 1983). For a glass formed not too slowly by continuous cooling of a liquid (i.e., at rates of 10 K/min or more), the fictive temperature is simply the temperature at which the glass transition took place. Then, one shows that glasses with the same composition, but different thermal histories, have different entropies. As already discussed, available data do not indicate a significant dependence of the heat capacity on the thermal history of

Table 3. Comparisons between residual entropies, $S_g(0)$, obtained from viscosity and calorimetry measurements, and between residual entropies and entropies of (Ca,Mg) or (Si,Al) disordering (S_d).[a]

	J/mol K		J/g atom K	
Composition	$S_g(0)_{Visc}$	$S_g(0)_{Calor}$	$S_g(0)$	S_d
SiO_2		5.1	1.7	
$MgSiO_3$	10.8	11.2 (5)	2.2	
$CaMgSi_2O_6$	28.5	23.0 (4)	2.8	1.1
$CaSiO_3$	10.1	8.8 (2)	2.0	
$NaAlSi_3O_8$	39.5	36.7 (6)	2.8	1.4
$NaAlSiO_4$		9.7	1.4	1.6
$KAlSi_3O_8$	31.8	28.3 (6)	2.4	1.4
$CaAl_2Si_2O_8$	39.2	36.8 (4)	3.0	1.8
$Ca_3Al_2Si_3O_{12}$	49.1		2.5	1.4
$Mg_3Al_2Si_3O_{12}$	56.5	56.3 (13)	2.8	1.4

[a] Calorimetric values from Richet et al. (1982, 1986, 1990, 1991), Richet and Bottinga (1984a, b), and Téqui et al. (1991); viscosity values, to within ±5%, obtained as described by Richet (1984) and Neuville and Richet (1991).

silicate glasses. From Eq. (3), the entropy difference between glasses with different fictive temperatures T_1 and T_2 is thus

$$\Delta S = \int_{T_1}^{T_2} \frac{(C_{pl} - C_{pg})}{T} dT. \quad (4)$$

Hence, comparison between residual entropies can be beset by differences in thermal histories. In the same way, the composition dependence of the glass transition range can be very strong and minor composition errors can result in a significant scatter in the results. All the calorimetrically determined residual entropies listed in Table 3 thus refer to rapidly quenched products for which T_g was chosen in Eq. (3) as the drop-calorimetry glass transition temperature.

Chemical vs. Topological Entropy

The residual entropies have also been listed in Table 3 on a g atom basis to take into account the widely different numbers of atoms in the formula units. At this point, it is useful to separate somewhat arbitrarily the configurational entropy into two parts. In an instance like $NaAlSi_3O_8$, there is an obvious contribution that results from possible (Si,Al) disordering. We will call *chemical* that part of the entropy which originates in the mixing of different elements on structurally equivalent sites. For a glass like SiO_2, the distributions of Si–O–Si and O–Si–O angles and of interatomic distances essentially represent configurational entropy. More generally, these distributions for the various elements of an amorphous phase determine, in principle, the various coordination states, the way bridging and nonbridging oxygens are bonded to Si (Q species distribution), etc., and intermediate-range order as well. We will term *topological* this contribution to the entropy that does not result from element mixing.

In the glassy state, the relative importance of these contributions appears to depend sensitively on composition, as suggested in Table 3 by comparisons of residual entropies with entropies calculated for ideal Ca,Mg and Si,Al disordering. On a g atom basis, pure SiO_2 probably gives a lower bound to the topological entropy of a glass, namely, about 1.7 J/g atom K. The lower value for $NaAlSiO_4$ glass would thus suggest very limited Si,Al disordering, whereas such an aluminum avoidance would not hold for $NaAlSi_3O_8$ whose residual entropy is high enough to suggest complete Si,Al disorder. Indeed, these calorimetric interpretations are consistent with conclusions drawn independently from NMR or Raman spectroscopic observations (Murdoch et al., 1985; Matson et al., 1986).

As pointed out in the next section, however, residual entropies are important only in the neighborhood of the glass transition. At high temperatures, they represent only a small fraction of S^{conf}, which as an integrated form of C_p^{conf}, must be interpreted in the same way as the configurational heat capacity. Indeed, part of the specificity of the C_p^{conf} data described in the third section for aluminosilicates also likely results from disordering effects since the configurational heat capacity

Thermodynamics of Silicate Melts: Configurational Properties

also has chemical and topological parts. For instance, progressive (Si,Al) disordering in the liquid state would be consistent with the higher C_p^{conf} of $NaAlSiO_4$ with regard to that of $NaAlSi_3O_8$ [Fig. 5(a)]. In this respect, it is probably not fortuitous that the data for aluminosilicates suggest a coupling between the temperature dependence of C_p^{conf} and its nonlinear variations with composition.

On the other hand, the distributions of bridging and nonbridging oxygens among the Q species have been investigated extensively in the glassy state (e.g., Maekawa et al., 1991). Unfortunately, their variations with temperature are not known. More generally, attempts to translate quantitatively into thermodynamic terms the available high-temperature structural information are still very few (e.g., Stebbins, 1988). If this subject is well beyond the scope of our review, we will nonetheless emphasize once more the considerable differences between glasses and melts, and the resulting need for a thermodynamic evaluation of the importance of structural factors through their contribution to configurational properties.

Temperature Dependence of Configurational Entropy

It has long been assumed that the configurational entropy of a liquid may be approximated by the entropy difference between the liquid and the crystal. In fact, this assumption is incorrect because the vibrational entropy of a liquid differs from that of the crystal, as shown in particular by the aforementioned differences in low-temperature entropy found between glasses and crystals (Fig. 9). Knowing the residual entropy of a glass with a given fictive temperature, one determines readily the variation temperature of the configurational entropy with

$$S^{\text{conf}}(T) = S^{\text{conf}}(T_g) + \int_{T_g}^{T} \frac{C_p^{\text{conf}}}{T} dT. \tag{5}$$

Since C_p^{conf} is necessarily positive, S^{conf} can only increase with temperature. Less obvious features are the changes in the relative importance of the topological and chemical contributions to the configurational entropy. Especially when

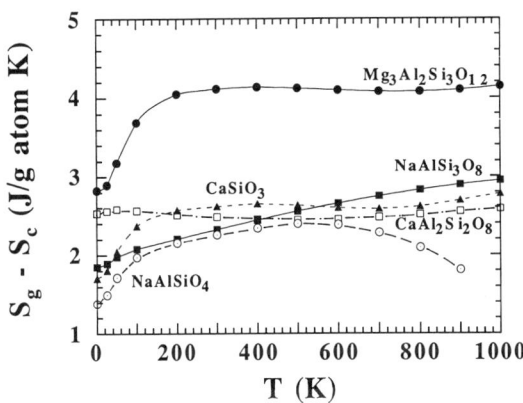

Fig. 9. Low-temperature entropy difference between the glass and crystal forms of pyrope ($Mg_3Al_2Si_3O_{12}$), albite ($NaAlSi_3O_8$), wollastonite ($CaSiO_3$), anorthite ($CaAl_2Si_2O_8$), and carnegieite ($NaAlSiO_4$). Data from Téqui et al. (1991), Haselton et al. (1980, 1983), Robie et al. (1978), Richet and Bottinga (1984a,b), and Richet et al. (1990, 1991).

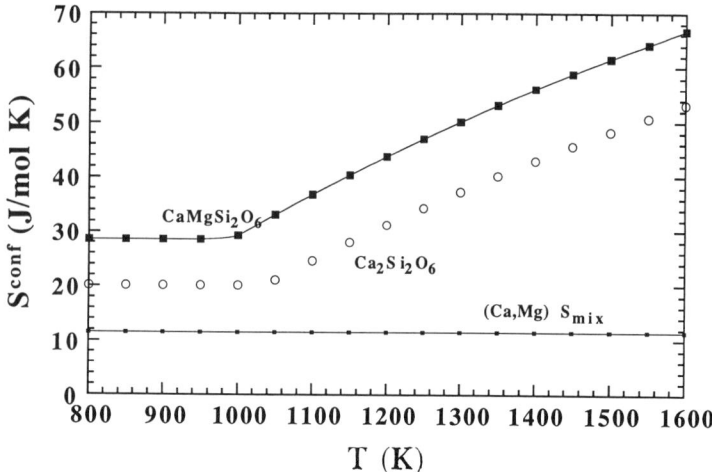

Fig. 10. Chemical and topological entropy for $CaMgSi_2O_6$ glass and liquid. The configurational entropy of $CaSiO_3$ glass and liquid, referred to two formula units, is included for comparison.

mixing can be assumed ideal, there is no evidence for significant variations of the latter contribution, which becomes smaller and smaller with respect to the former when the temperature increases. This is shown in Fig. 10 for configurational entropies of liquids along the join $CaSiO_3$–$MgSiO_3$, and this fact has important consequences regarding the relaxation properties of melts that will be discussed in the next section.

However, these calorimetric determinations of configurational entropies have two serious shortcomings. First, the use of Eq. (2) is restricted to liquids having the same composition as a crystal whose entropy of melting can be measured, i.e., in general, a congruently melting compound. This prevents systematic determinations for intermediate compositions of a solution. Second, as evident in Fig. 1, a calorimetrically determined residual entropy represents a small difference between great numbers. Hence, such determinations can be beset by important errors, especially when some of the data needed to evaluate Eq. (2) are not so well known, as are the entropies of fusion of pyrope ($Mg_3Al_2Si_3O_{12}$) and enstatite ($MgSiO_3$), for example. The possibility of determining more accurately the configurational entropy from viscosity data is thus interesting, especially since any liquid can be investigated, regardless of the complexity of its composition.

Viscosity and Configurational Entropy

Temperature Dependence of the Viscosity

The existence of a link between the thermodynamic and rheological properties of melts might seem puzzling. A striking feature of the viscosity (η) of glass-

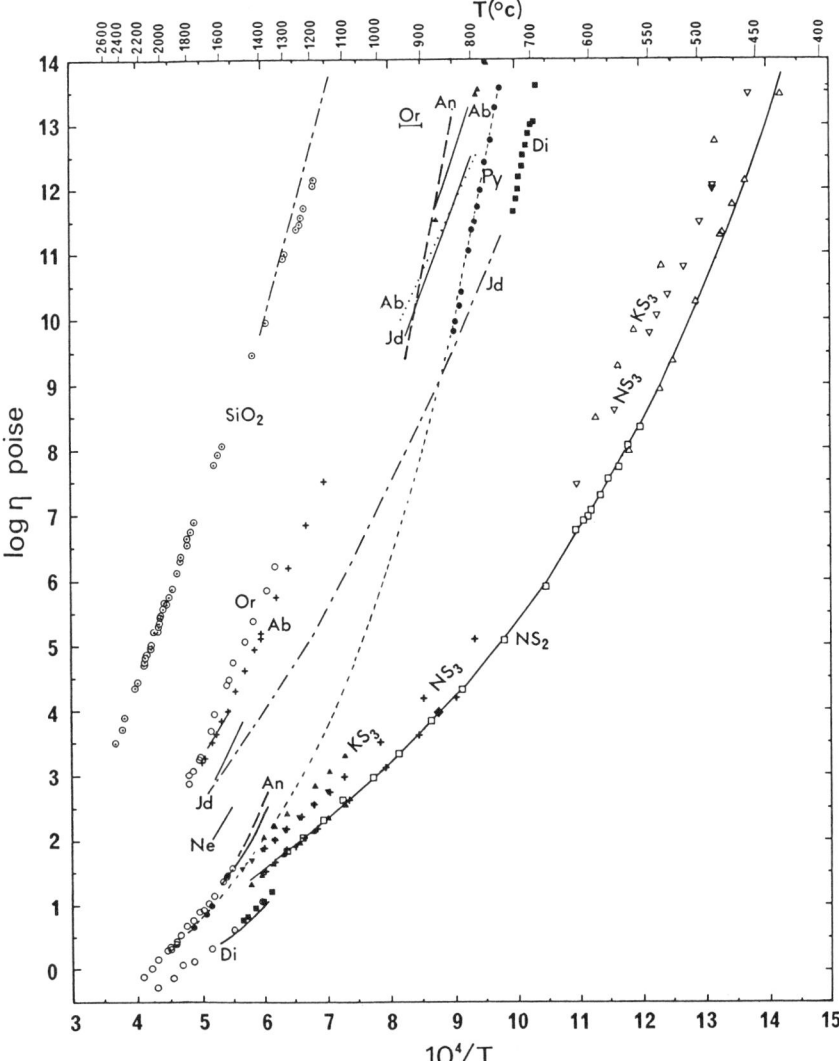

Fig. 11. Viscosity-temperature relationships for a few silicate melts. Abbreviations: Ab ($NaAlSi_3O_8$); An ($CaAl_2Si_2O_8$); Di ($CaMgSi_2O_6$); Jd ($NaAlSi_2O_6$); KS_n ($K_2O.nSiO_2$); NS_n ($Na_2O.nSiO_2$); Ne ($NaAlSiO_4$); Or ($KAlSi_3O_8$); Py ($Mg_3Al_2Si_3O_{12}$). Experimental data as reviewed in Richet (1984) and Neuville and Richet (1991).

forming liquids is that it can span more than 13 orders of magnitude, as shown in Fig. 11 for a few compositions. The viscosity of these melts may differ by many orders of magnitude and the deviations of log η from linear variations of the reciprocal temperature, i.e., from Arrhenius laws, also depend specifically on composition. Regardless of these important features, however, a common trait is that the slopes of the viscosity curves in Fig. 11 increase when the temperature decreases. In many cases, this increase is strong enough that the viscosity would seem to become infinite at a given temperature.

Now, consider a hypothetical liquid with zero configurational entropy. There would be no way to switch any structural entity from one place to another, and the viscosity would thus be infinite. If only two configurations were available for the whole liquid, then mass transfer would require a simultaneous displacement of all entities. The probability of such a cooperative event would be extremely small, but not zero, and the viscosity would be extremely high, but no longer infinite. When the configurational entropy increases, the cooperative rearrangements of the structure required for mass transfer can take place independently in smaller and smaller regions of the liquid. Correlatively, the viscosity thus decreases when the configurational entropy increases. These are the basic ideas of the theory of relaxation processes of Adam and Gibbs (1965), according to which the probability for a cooperative rearrangement is

$$w(T) = A \exp\left(\frac{-B_e}{TS^{\text{conf}}}\right), \qquad (6)$$

where A is a preexponential term and B_e is approximately a constant proportional to the Gibbs free energy barriers hindering the cooperative rearrangements.

The viscosity is proportional to the relaxation times, which are themselves assumed to be inversely proportional to $w(T)$. Thus, one obtains for the viscosity

$$\log \eta = A_e + \frac{B_e}{TS^{\text{conf}}}, \qquad (7)$$

where A_e is a constant. To check the quantitative validity of this theory, Richet (1984) inserted Eqs. (1) and (3) into Eq. (7) and showed that available calorimetrically determined configurational entropies, for liquid $NaAlSi_3O_8$, $KAlSi_3O_8$, and $CaAl_2Si_2O_8$, were consistent with available viscosity data spanning up to 13 orders of magnitude. These examples as well as those of Fig. 11, where the solid curves represent the values calculated with Eq. (7), show the quantitative nature of this relationship, which gives a firm basis to the long-known observation that liquids with higher configurational heat capacities show stronger deviations of viscosity from Arrhenian behavior (e.g., Angell and Sichina, 1976; Angell, 1988).

In turn, one can determine $S^{\text{conf}}(T_g)$ with Eqs. (1) and (7) from viscosity data. The validity of such determinations has been confirmed by subsequent calorimetric measurements, as shown by the good agreement between the calorimetric and viscosity entropies listed in Table 3. In fact, the rheological approach has many practical advantages over the calorimetric determinations. The viscosity measurements are much less tedious to perform than the comprehensive calorimetric experiments required to evaluate Eq. (3). Because they do not represent small differences between large numbers, as do the calorimetric values shown in Fig. 1, they are in addition more accurate. And, especially, they are not restricted to the few compositions for which the whole set of calorimetric measurements can be performed. As a result, they allow determinations of configurational entropies for solutions, i.e., of the thermodynamically important entropies of mixing.

Composition Dependence of the Viscosity

That simple compositional changes can markedly affect viscosity is shown in Fig. 12 by the data for molten (Ca,Mg) pyroxenes. Mixing of Ca and Mg results in a very strong nonlinear variation of viscosity just above the glass transition, with a minimum of about two orders of magnitude. When the temperature increases, the depth of this minimum decreases and η is eventually a linear function of composition at high fluidities.

To interpret such temperature-dependent variations, consider the configurational entropy of an intermediate composition. As a function of the entropies of the endmembers, it can be expressed alternatively as

$$S^{conf}(T) = \sum x_i S_i^{conf}(T) + S_{mix}, \qquad (8)$$

where the temperature dependence of S_i^{conf} is given by Eq. (7). If mixing is ideal, then the entropy of mixing is given by

$$S_{mix} = -nR \sum x_i \ln x_i, \qquad (9)$$

where x_i is the molar fraction of entity i, R the gas constant, and n the number of entities exchanged per formula unit [e.g., $n = 1$ and $x = 0.5$ and 0.5 for (Ca,Mg) exchange in $CaMgSi_2O_6$; $n = 4$ and $x = 0.25$ and 0.75 for (Si,NaAl) exchange in $NaAlSi_3O_8$]. Since configurational heat capacities of mixing are not major, S_{mix} is essentially the chemical entropy referred to in previous sections. And, as already shown in Fig. 10, it is progressively overwhelmed by the topological contribution. The result is a lowering of the viscosity, through the contri-

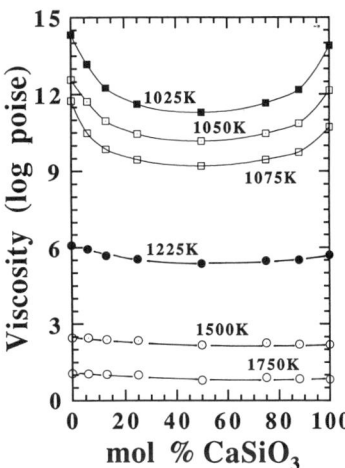

Fig. 12. Viscosity of liquids in the system $CaSiO_3$–$MgSiO_3$ at various temperatures. Low-viscosity data from Bockris and Lowe (1954), Bockris et al. (1955), Urbain et al. (1982), and Licko and Danek (1986). High-viscosity measurements from Neuville and Richet (1991).

bution of the mixing entropy to the total S^{conf} in Eq. (7), which can be effective at lower temperatures only, i.e., at high viscosities.

Entropy Modeling

In Fig. 13, the entropies obtained from the viscosity data of Fig. 12 for molten pyroxenes clearly show the maximum due to mixing. As expected from (Si,Al) mixing, analogous effects on the entropy are also observed along $MAlO_2$–SiO_2 joins, but not along other binary joins of the CaO–MgO–Al_2O_3–SiO_2 system (Neuville, unpublished results). In fact, for both molten pyroxenes and garnets, entropies of mixing are consistent with the ideal mixing of Ca and Mg (Neuville and Richet, 1991). That (Ca,Mg) mixing does not seem to be affected by the presence of aluminum has the noteworthy consequence that the problem of determining entropies of mixing in the quaternary system CaO–MgO–Al_2O_3–SiO_2 reduces to the same problem for the two limiting aluminosilicate ternaries. On the other hand, a preliminary analysis of the viscosity of mixed alkali silicate liquids suggested that Na and K are mixing pairwise (Richet, 1984), and not individually as alkaline-earth cations. This would be consistent with a clustering of the two nonbridging oxygen entities Na–O formed when Na_2O is introduced and reacts with a bridging oxygen. In aluminosilicate liquids, alkali elements are assumed to maintain electrical neutrality around tetrahedrally coordinated Al^{3+} ions, and viscosity data could thus enable one to determine whether the alkali ions mix individually or not.

These examples show not only how entropy models can be set up from viscosity data, but also that models obtained for simple systems should be applicable to systems of more immediate geochemical utility like the joins investigated by Hummel and Arndt (1985) or Tauber and Arndt (1987). The application of these models to phase-equilibria calculations might appear doubtful, however, since only configurational entropies would be taken into account. In

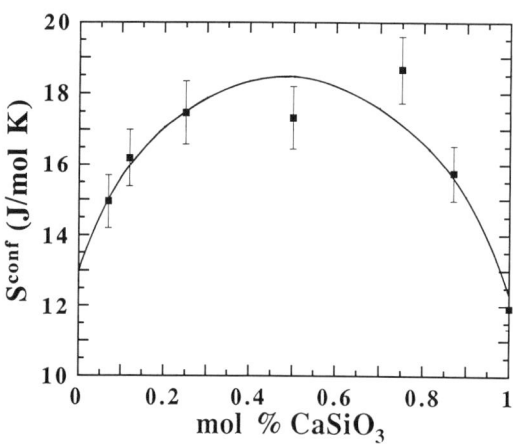

Fig. 13. Configurational entropies at 1100 K of liquids in the system $CaSiO_3$–$MgSiO_3$. The solid squares are values obtained individually for each composition from the viscosity data with Eqs. (5) and (7). The solid curve is obtained from a simultaneous fit of Eqs. (7) and (8), with an ideal (Ca,Mg) mixing model, to all the viscosity data for the binary system. Data from Neuville and Richet (1991).

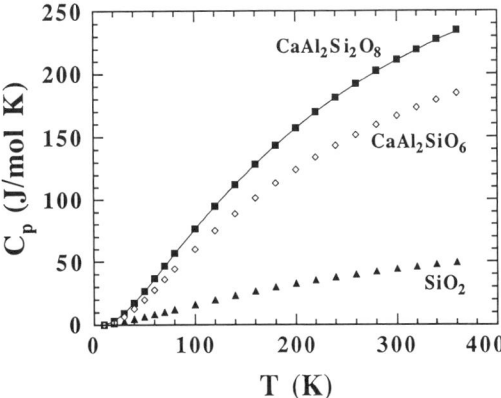

Fig. 14. Low-temperature heat capacity of calcium aluminosilicate glasses. The validity of additive modeling of C_p and entropy is shown by the excellent agreement between the sum of the heat capacities of SiO_2 and $CaAlSi_2O_6$, shown by the solid line for $CaAl_2Si_2O_8$, with the measurements for this glass. Data from Robie et al. (1978), Haselton et al. (1984), and Westrum (as listed in Richet et al., 1982).

fact, recent low-temperature measurements suggest that the heat capacity of silicate glasses does not depart significantly from the additive functions of composition not only above room temperature, as already noted in the second section, but also down to 0 K (Richet, Robie, and Hemingway). The additivity of low-temperature heat capacities over wide composition ranges is shown in Fig. 14 for glasses along the join $Ca_{0.5}AlO_2$–SiO_2. With linear variations of the vibrational C_p from 0 K to T_g, vibrational entropies are also additive. Vibrational entropies of mixing should thus be negligible, and entropies of mixing would reduce to their configurational parts.

Volume

Thermal Expansion

Most of this review has been devoted to the thermal aspects of configurational properties. The reason lies not only in their importance, but also in the wealth of quantitative information that has been recently gathered and which contrasts with the scarce data available for volume properties. However, the configurational contribution to the thermal expansion coefficient of silicate liquids can produce important effects, as shown in Fig. 15 for Na_2O–SiO_2 liquids. In the glassy state, the molar volume increases with the SiO_2 content, whereas the converse holds true above 1200 K in the liquid state. Even though volume is a property more prone to straightforward theoretical evaluation than entropy, the way configurational properties result in such crossovers is still poorly known.

Part of our ignorance originates in the lack or in the uncertainties of thermal expansion data. As noted by Bottinga et al. (1983), one finds discrepancies of more than 40% in the reported thermal expansion coefficients for the most extensively studied melts, namely, binary SiO_2–Na_2O liquids. Indeed, there is no direct way of directly measuring α and this property is usually determined from the differentiation of high-temperature volume measurements having un-

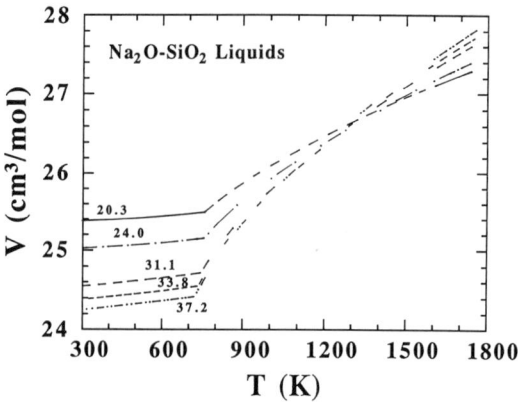

Fig. 15. Volume of sodium silicate glasses and liquids as a function of temperature at 1 bar. Numbers are molar Na_2O contents. Experimental data from Shermer (1956) and Bockris et al. (1956).

Table 4. Thermal expansion of alkali silicate liquids.[a]

mol %	K		cm^3/mol		10^5 K^{-1}			
	T_g	V_{T_g}	V_{1673}	$\alpha_g(T_g)$	α_{lm}	α_{lHT}	α_{lm}/α_g
Li_2O							
32.0	751	21.68	23.64	1.3	9.4	7.9 (0.8)	7.2
34.6	736	21.25	23.36	1.4	10.1	7.7 (0.8)	7.2
37.9	734	20.78	23.06	1.5	11.1	8.4 (0.3)	7.4
39.6	730	20.57	22.91	1.6	11.4	8.9 (0.3)	7.1
Na_2O							
20.3	760	25.49	27.21	1.3	7.1	4.5 (0.8)	5.5
24.0	756	25.15	27.28	1.3	8.8	6.0 (0.8)	6.8
31.1	743	24.71	27.47	1.7	11.4	9.1 (0.9)	6.7
33.8	732	24.55	27.55	1.8	12.2	9.9 (0.9)	6.8
37.2	720	24.42	27.61	2.0	12.9	10.9 (1.4)	6.5
K_2O							
17.3	781	28.05	30.08	1.0	7.8	5.1 (1.1)	7.8
23.2	764	28.53	31.29	1.3	10.2	8.7 (0.7)	7.8
27.8	733	28.82	32.19	1.5	11.8	10.7 (0.5)	7.9
31.8	723	29.09	32.92	1.6	13.0	12.0 (0.5)	8.1

[a] V_{T_g} and V_{1673} are the molar volumes at the glass transition temperature T_g and at 1673 K, respectively. For liquids, α_{lHT} and α_{lm} are the thermal expansion coefficients at 1673 K and mean values between T_g and 1673 K obtained from volume measurements, respectively (Bockris et al., 1956; Shermer, 1956).

certainties of at least 0.5% and covering restricted temperature intervals. As a result, the most reliable data, by Bockris et al. (1956), are accurate to within ±10% at best. In fact, the data of Table 4 show that the α increases at the glass transition are so high as to make such uncertainties acceptable for determinations of the configurational thermal expansion coefficient.

Moreover, as pointed out previously (Bottinga et al., 1982; Richet and Bottinga, 1983), these uncertainties could be reduced further by considering data just above the glass transition, which can be readily obtained by dilatometry, along with those for stable liquids. After all, this is the procedure that has long been used to obtain accurate heat capacities from measurements limited by high liquidus temperatures and crystallization of supercooled liquids [see Fig. 5(b)]. As suggested by Bottinga et al. (1982), the temperature dependence of α could thus also be evidenced in this way. A comparison of average thermal expansion coefficients between T_g and 1673 K with the high-temperature data of Bockris et al. (1956) would indeed suggest a decrease of α with temperature for alkali silicates (Table 4). In contrast, comparisons between the thermal expansion coefficients of Arndt and Häberle (1973) and the values obtained for molten plagioclase by Bottinga et al. (1982) suggest slight increases. These calculated variations could be spurious since they are similar to the uncertainties of the reported numbers. Measurements on supercooled liquids of geochemical interest are still limited (e.g., Taniguchi and Murase, 1987). Fortunately, efforts are currently underway to obtain systematic volume data from the glass transition to the stable liquid field (Knoche et al. 1992). These data will prove extremely useful to model temperature-induced structural changes.

Compressibility and High-Pressure Fusion

The basic difference in the compression mechanisms between a crystal and an amorphous substance has already been described in the second section. The effect to be discussed in this section deals with the decrease in pressure of the volume of melting (ΔV), illustrated in Fig. 2, and it will provide a simple example of the influence of configurational properties on phase equilibria. For a congruently melting compound, the slope of the melting curve is given by the well-known Clausius–Clapeyron equation

$$\frac{dT}{dP} = \frac{\Delta V}{\Delta S_f}, \tag{10}$$

which indicates that the melting temperature decreases with increasing pressure when ΔV is negative (Fig. 16). In general, a crystal will transform at high temperatures to a denser polymorph before the melting volume becomes zero. The dashed line in Fig. 16 thus represents the metastable extension of the melting curve whose negative slope beyond its maximum results from the existence of a configurational contribution to the compressibility of the liquid. (At some point, the glass transition would obtain, but calculations show that the resulting change in the slope of the vitrification curve should not be major, see Richet, 1988).

What would happen if the crystal were brought at pressures beyond the melting curve at low temperatures? In this case, Mishima et al. (1984) guessed that the crystal would bypass a transformation to the stable polymorph in view of the slower and slower kinetics of a reconstructive phase transition. It would amorphize instead, and in this manner Mishima et al. actually obtained at liquid

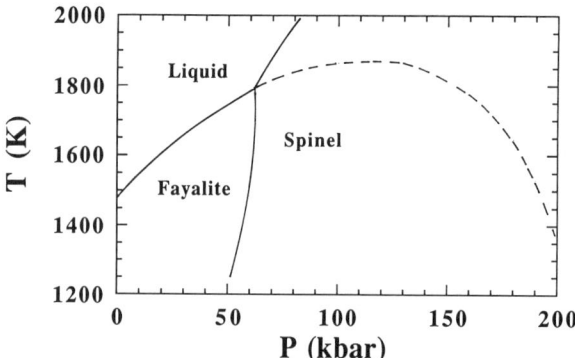

Fig. 16. Phase relationships between fayalite, Fe_2SiO_4 spinel, and Fe_2SiO_4 liquid (experimental data from Akimoto et al., 1967). The dashed line within the spinel (and post-spinel) stability fields is the metastable extension of the fayalite melting curve as calculated by Richard and Richet (1990).

nitrogen temperature the first bulk amorphous H_2O sample. Analogous observations were later made by Hemley et al. (1988) who observed the amorphization of quartz and coesite between 250 and 350 kbar at room temperature. Amorphization through room-temperature compression has also been observed for a variety of inorganic compounds and, in particular, for olivines (Richard and Richet, 1990; Williams et al., 1990). Amorphous silicates with olivine compositions are, in effect, the most difficult to prepare by quenching of a liquid because of the ease with which the isolated SiO_4 tetrahedra of the melt can rearrange to form a crystalline network on cooling. In contrast, room-temperature amorphization thus constitutes an easy route to obtain an amorphous solid whose structural and vibrational properties could be determined.

The scarcity of in-situ measurements for liquids at high pressure, as well as a lack of samples, makes it difficult to ascertain the way pressure-formed amorphous materials are related to glasses quenched from liquids at high pressures. Deviatoric stresses are generally present in such high-pressure experiments. If they could enhance amorphization kinetics, their influence on the bulk thermodynamic properties are unlikely to be significant. Hence, pressure-induced amorphization requires not only that the Gibbs free energy of the amorphous phase be lower than that of the crystal, but also that the amorphous phase be denser than the crystal. Because liquids are the stable, equilibrium phases with the lowest Gibbs free energy, glasses formed at the same high pressure by quenching of liquids should have a Gibbs free energy lower than materials obtained by low-temperature compression. This means that the actual metastable extension of the melting curve should be crossed at pressures lower than those at which high-pressure amorphization becomes thermodynamically possible. In other words, pressure-induced amorphization is a thermodynamically driven phenomenon associated with a bell-shaped melting curve, which can be actually crossed at high pressures because of the possibility of low-temperature configu-

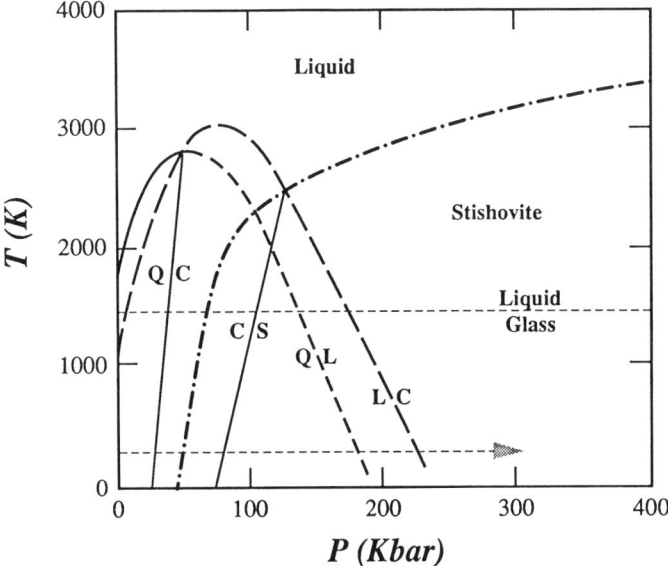

Fig. 17. Schematic melting relationships of SiO_2 polymorphs. The dotted line represents an assumed constant glass transition temperature, below which amorphization takes place. (Adapted from Hemley et al., 1988, with thermodynamic calculations reported by Richet, 1988.)

rational changes described in the second section. This is illustrated in a different way by coesite and other SiO_2 polymorphs (Fig. 17). As already noted, coesite amorphizes at around 300 kbar at room temperature. On the other hand, thermodynamic calculations show that it becomes unstable with respect to glassy SiO_2 at around 875 K at 1 bar, i.e., well below the glass transition temperature of about 1500 K (Richet, 1988). Amorphization of coesite thus takes place on both external sides of the melting curve, either through heating or compression. In contrast, owing to the respective positions of their melting curves, quartz and stishovite amorphize only at low and high pressures, respectively.

Acknowledgments

We thank Y. Bottinga, F. Guyot, and A.M. Lejeune for helpful comments. This research was supported by grant CNRS-INSU-DBT 293.

References

Adam, G. and Gibbs, J.H. (1965). On the temperature dependence of cooperative relaxation properties in glass-forming liquids. *J. Chem. Phys.* **43**, 139–146.

Akimoto, S.I., Komada, E., and Kushiro, I. (1967). Effect of pressure on the melting of olivine and spinel polymorphs of Fe_2SiO_4. *J. Geophys. Res.* **72**, 679–686.

Angell, C.A. (1988). Perspective on the glass transition. *J. Phys. Chem. Solids* **49**, 863–871.

Angell, C.A. and Sichina, W. (1976). Thermodynamics of the glass transition: Empirical aspects. *Ann. N.Y. Acad. Sci.* **279**, 53–67.

Arndt, J. and Häberle, F. (1973). Thermal expansion and glass transition temperatures of synthetic glasses of plagioclase-like compositions. *Contrib. Mineral. Petrol.* **39**, 175–183.

Bacon, C.R. (1977). High-temperature heat content and heat capacity of silicate glasses: Experimental data and a model of calculation. *Amer. J. Sci.* **277**, 109–135.

Bernal, J.D. (1936). An attempt at a molecular theory of liquid structure. *Disc. Farad. Soc.* **336**, 27–40.

Bockris, J.O.'M. and Lowe, D.C. (1954). Viscosity and structure of liquid silicates. *Proc. Roy. Soc., Lond.* **A226**, 423–435.

Bockris, J.O.'M., Mackenzie, J.D., and Kitchener, J.A. (1955). Viscous flow in silica and binary liquid silicates. *Trans. Farad. Soc.* **51**, 1734–1748.

Bockris, J.O.'M., Tomlinson, J.W., and White, J.L. (1956). The structure of liquid silicates: Partial molar volumes and expansivities. *Trans. Farad. Soc.* **53**, 299–310.

Bottinga, Y. (1986). On the isothermal compressibility of silicate liquids at high pressure. *Earth Planet. Sci. Lett.* **74**, 350–360.

Bottinga, Y. and Weill, D.F. (1970). Density of liquid silicate systems calculated from partial molar volumes of oxide components. *Amer. J. Sci.* **269**, 169–182.

Bottinga, Y. and Weill, D.F. (1972). Viscosity of magmatic silicate liquids: A model for calculation. *Amer. J. Sci.* **272**, 438–475.

Bottinga, Y., Weill, D.F., and Richet, P. (1982). Density calculations for silicate liquids. I. Revised method for aluminosilicate compositions. *Geochim. Cosmochim. Acta*, **46**, 909–919.

Bottinga, Y., Richet, P., and Weill, D.F. (1983). Calculation of the density and thermal expansion coefficient of silicate liquids. *Bull. Minéral.* **106**, 129–138.

Carmichael, I.S.E., Nicholls, J., Spera, F.J., Wood, B.J., and Nelson, S.A. (1977). High temperature properties of silicate liquids: Application to the equilibration and ascent of basic magma. *Phil. Trans. Roy. Soc. Lond.* **A286**, 373–431.

Davies, R.O. and G.O., Jones (1953). Thermodynamic and kinetic properties of glasses. *Adv. Phys.* **2**, 370–410.

Dingwell, D.B. and Webb, S.L. (1990). Relaxation in silicate melts. *Eur. J. Mineral.* **2**, 427–449.

Farnan, I. and Stebbins, J.F. (1990). High-temperature ^{29}Si NMR investigation of solid and molten silicates. *J. Amer. Chem. Soc.* **112**, 32–39.

Finger, L.W. and Ohashi, Y. (1976). The thermal expansion of diopside to 800°C and a refinement of the crystal structure at 700°C. *Amer. Mineral.* **61**, 303–310.

Goldstein, M. (1969a). Viscous liquids and the glass transition: A potential energy barrier picture. *J. Chem. Phys.* **51**, 3728–3739.

Goldstein, M. (1976). Viscous liquids and the glass transition. V. Sources of the excess specific heat of the liquid. *J. Chem. Phys.* **64**, 4767–4774.

Haggerty, J.S., Cooper, A.R., and Heasley, J.H. (1968). Heat capacity of three inorganic glasses and supercooled liquids. *Phys. Chem. Glasses* **5**, 130–136.

Haselton H.T. and Westrum, E.F. (1980). Low-temperature heat capacities of synthetic pyrope, grossular, and pyrope$_{60}$ grossular$_{40}$. *Geochim. Cosmochim. Acta* **44**, 701–709.

Haselton, H.T., Hovis, G.L., Hemingway, B.S., and Robie, R.A. (1983). Calorimetric investigation of the excess entropy of mixing in analbite-sanidine solutions: Lack of evidence for Na,K short-range order and implications for two-feldspar thermometry. *Amer. Mineral.* **68**, 398–413.

Haselton, H.T., Hemingway, B.S., and Robie, R.A. (1984). Low-temperature heat capacities of $CaAl_2SiO_6$ glass and pyroxene and thermal expansion of $CaAl_2SiO_6$ pyroxene. *Amer. Mineral.* **69**, 481–489.

Hemley, R.J., Jephcoat, A.P., Mao, H.K., Ming, L.C., and Manghnani, M.H. (1988). Pressure-induced amorphization of crystalline silica *Nature* **334**, 52–54.

Hummel, W. and Arndt, J. (1985). Variation of viscosity with temperature and composition in the plagioclase system. *Contrib. Mineral. Petrol.* **90**, 83–92.

Johari, G.P. (1976). Glass transition and secondary relaxations in molecular liquids and crystals. *Ann. N.Y. Acad. Sci.* **279**, 117–140.

Kauzmann, W. (1948). The nature of the glassy state and the behavior of liquids at low temperatures. *Chem. Rev.* **43**, 219–256.

Kelley, K.K., Todd, S.S., Orr, L.R., King, E.G., and Bonnickson, K.R. (1953). Thermodynamic properties of sodium-aluminum silicates. U.S. Bureau Mines Rept. Inv. 4955.

Knoche, R., Dingwell, D.B., and Webb, S.L. (1992). Temperature-dependent thermal expansivities of silicate melts: The system anorthite-diopside. *Geochim. Cosmochim. Acta,* in press.

Krupka, K.M., Robie, R.A., Hemingway, B.S., Kerrick, D.M., and Ito, J. (1985). Low-temperature heat capacities and derived thermodynamic properties of antophyllite, diopside, enstatite, bronzite, and wollastonite. *Amer. Mineral.* **70**, 249–260.

Levien, L. and Prewitt, C.T. (1981). High-pressure structural study of diopside. *Amer. Mineral.* **66**, 315–323.

Licko, T. and Danek, V. (1986). Viscosity and structure of melts in the system CaO–MgO–SiO_2. *Phys. Chem. Glasses* **27**, 22–29.

Maekawa, H., Maekawa, T., Kawamura, K., and Yokokawa, Y. (1991). The structural groups of alkali silicate glasses determined from ^{29}Si MAS-NMR. *J. Non-Cryst. Solids* **127**, 53–64.

Matson, D.W., Sharma, S.K., and Philpotts, J.A., (1986). Raman spectra of some tectosilicates and of glasses along the orthoclase-anorthite and nepheline-anorthite joins. *Amer. Mineral.* **71**, 694–704.

Mishima, O., Calvert, L.D., and Whalley, E. (1984). "Melting ice" at 77 K and 10 kbar: A new method of making amorphous solids. *Nature* **310**, 393–395.

Murdoch, J.B., Stebbins, J.F., and Carmichael, I.S.E., (1985). High-resolution ^{29}Si NMR study of silicate and aluminosilicate glasses: The effect of network modifying cations. *Amer. Mineral.* **70**, 332–343.

Navrotsky, A., Peraudeau, G., McMillan, P., and Coutures, J.P. (1982). Thermochemical study of glasses and crystals along the joins silica-calcium aluminate and silica-sodium aluminate. *Geochim. Cosmochim. Acta* **46**, 2039–2047.

Navrotsky, A., Geisinger, K.L., McMillan, P., and Gibbs, G.V. (1985). The tetrahedral framework in glasses and melts. Inferences from molecular orbital calculations, and implications for structure, thermodynamics, and physical properties. *Phys. Chem. Mineral.* **11**, 284–298.

Neuville, D.R. and Richet, P. (1991). Vicosity and mixing in molten (Ca,Mg) pyroxenes and garnets. *Geochim. Cosmochim. Acta* **55**, 1011–1019.

Nikonov, A.M., Bogdanov, V.N., Nemilov, S.V., Shono, A.A., and Mikhailov, V.N. (1982). Structural relaxation in binary alkalisilicate melts. *Fyz. Khim. Stekla* **8**, 694–703.

Richard, G. and Richet, P. (1990). Room-temperature amorphization of fayalite and high-pressure properties of Fe_2SiO_4 liquid. *Geophys. Res. Lett.* **17**, 2093–2096.

Richet, P. (1984). Viscosity and configurational entropy of silicate melts. *Geochim. Cosmochim. Acta* **48**, 471–483.

Richet, P. (1987). Heat capacity of silicate glasses. *Chem. Geol.* **62**, 111–124.

Richet, P. (1988). Superheating, melting and vitrification through decompression of high-pressure minerals. *Nature* **331**, 56–58.

Richet, P. and Bottinga, Y. (1983). Verres, liquides, et transition vitreuse. *Bull. Minéral.* **106**, 147–168.

Richet, P. and Bottinga, Y. (1984a). Glass transition and thermodynamic properties of amorphous SiO_2, $NaAlSi_nO_{2n+2}$ and $KAlSi_3O_8$. *Geochim. Cosmochim. Acta* **48**, 453–470.

Richet, P. and Bottinga, Y. (1984b). Anorthite, andesine, diopside, wollastonite, cordierite and pyrope: Thermodynamics of melting, glass transitions, and properties of the amorphous phases. *Earth Planet. Sci. Lett.* **67**, 415–432.

Richet, P. and Bottinga, Y. (1985). Heat capacity of aluminum-free silicate liquids. *Geochim. Cosmochim. Acta* **49**, 471–486.

Richet, P. and Bottinga, Y. (1986). Thermochemical properties of silicate glasses and liquids: A review. *Rev. Geophys.* **24**, 1–25.

Richet, P., Bottinga, Y., and Téqui, C. (1984). Heat capacity of sodium silicate liquids. *J. Amer. Ceram. Soc.* **67**, C6–C8.

Richet, P. and Fiquet, G. (1991). High-temperature heat capacity and premelting of minerals in the system $CaO-MgO-Al_2O_3-SiO_2$. *J. Geophys. Res.* **96**, 445–456.

Richet, P., Bottinga, Y., Deniélou, L., Petitet, J.P., and Téqui, C. (1982). Thermodynamic properties of quartz, cristobalite and amorphous SiO_2: Drop calorimetry measurements between 1000 and 1800 K and a review from 0 to 2000 K. *Geochim. Cosmochim. Acta* **46**, 2639–2658.

Richet, P., Robie, R.A., and Hemingway, B.S. (1986). Low-temperature heat capacity of diopside glass ($CaMgSi_2O_6$): A calorimetric test of the configurational-entropy theory applied to the viscosity of liquid silicates. *Geochim. Cosmochim. Acta* **50**, 1521–1533.

Richet, P., Robie, R.A., Rogez, J., Hemingway, B.S., Courtial, P., and Téqui, C. (1990). Thermodynamics of open networks: Ordering and entropy in $NaAlSiO_4$ glass, liquid, and polymorphs. *Phys. Chem. Mineral.* **17**, 385–394.

Richet, P., Robie, R.A., and Hemingway, B.S. (1991). Thermodynamic properties of wollastonite and $CaSiO_3$ glass and liquid. *Eur. J. Mineral.* **3**, 475–484.

Rivers, M.L. and Carmichael, I.S.E. (1987). Ultrasonic studies of silicate melts. *J. Geophys. Res.* **92**, 9247–9270.

Robie R.A., Hemingway, B.S., and Wilson, W.H. (1978). Low-temperature heat capacities and entropies of feldspar glasses and of anorthite. *Amer. Mineral.* **63**, 109–123.

Rosenhauer, M., Scarfe, C.M., and Virgo, D. (1979). Pressure dependence of the glass transition temperature in glasses of diopside, albite, and sodium trisilicate composition. *Carnegie Inst. Wash. Yearbook* **78**, 547–551.

Roy, B.N. and Navrotsky, A. (1984). Thermochemistry of charge-coupled substitutions in silicate glasses: The systems $M_{1/n}^{n+}AlO_2-SiO_2$ (M = Li, Na, K, Rb, Cs, Mg, Ca, Sr, Ba, Pb). *J. Amer. Ceram. Soc.* **67**, 606–610.

Shermer, H.F. (1956). Thermal expansion of binary alkali silicate glasses. *J. NBS.* **57**, 97–101.

Simon, F. (1931). Uber den Zustand der unterkuhlten Flussigkeiten und Glaser. *Z. Anorg. Allg. Chem.* **203**, 219–227.

Soga N., Yamanaka, H., and Kunugi, M. (1979). Equation of state of metasilicate glasses, in *High-Pressure Science and Technology*, K.D. Timmerhaus and M.S. Barber, eds., Plenum, New York. pp. 200–206.

Stebbins, J.F. (1988). Effects of temperature and composition on silicate glass structure and dynamics: Si-29 NMR results. *J. Non-Cryst. Solids* **106**, 359–369.

Stebbins, J.F. and Farnan, I. (1989). NMR spectroscopy in the earth sciences; structure and dynamics. *Science* **245**, 257–262.

Stebbins, J.F., Carmichael, I.S.E., and Moret, L.K. (1984). Heat capacity and entropies of silicate liquids and glasses. *Contrib. Mineral. Petrol.* **86**, 131–148.

Stolper, E.M. and Ahrens, T.J. (1987). On the nature of pressure-induced coordination changes in silicate melts and glasses. *Geophys. Res. Lett.* **14**, 1231–1233.

Taniguchi, H. and Murase, T. (1987). Some physical properties and melt strutures in the system diopside-anorthite. *J. Volcan. Geoth. Res.* **34**, 51–64.

Tauber, P. and Arndt, J. (1987). The relationship between viscosity and temperature in the system anorthite-diopside. *Chem. Geol.* **62**, 71–81.

Téqui, C., Robie, R.A., Hemingway, B.S., Neuville, D.R., and Richet, P. (1991). Melting and thermodynamic properties of pyrope ($Mg_3Al_2Si_3O_{12}$). *Geochim. Cosmochim. Acta* **55**, 1005–1010.

Tool, A.Q. and Eichlin, C.G. (1931). Variations caused in the heating curves of glass by heat treatment. *J. Amer. Ceram. Soc.* **14**, 276–308.

Urbain, G., Bottinga, Y., and Richet, P. (1982). Viscosity of liquid silica, silicates and aluminosilicates. *Geochim. Cosmochim. Acta* **46**, 1061–1072.

Waff, H.S. (1975). Pressure-induced coordination changes in magmatic liquids. *Geophys. Res. Lett.* **2**, 193–196.

Williams, Q., Knittle, E., Reichlin, R., Martin, S., and Jeanloz, R. (1990). Structural and electronic properties of Fe_2SiO_4 at ultrahigh pressures; amorphization and gap closure. *J. Geophys. Res.* **95**, 21549–21563.

Yinnon, H. and Cooper, A.R., Jr. (1980). Oxygen diffusion in multicomponent glass-forming silicates. *Phys. Chem. Glasses* **21**, 204–211.

Chapter 6
Crystal Chemical and Energetic Characterization of Solid Solution

V.S. Urusov

Introduction

The golden age of classic crystal chemistry (1920s–30s) yielded many well-known empirical rules and generalizations concerning the formation of solid solutions or isomorphous mixtures (mixed crystals). Among them are Vegard's rule of additive dependence of lattice spacings on composition, Goldschmidt and Hume–Roseri's rules of maximal 15% difference of ionic or atomic radii for the existence of wide miscibility, Goldschmidt and Fersman's rules of substitution polarity (in relation to sizes and charges of the ions replacing each other), and the criteria of proximity of polarizabilities or electronegativities of substituents, etc. In the sections that follow we will return to an analysis of these rules from a more sophisticated and modern point of view.

On the other hand, at the same time the development of the thermodynamics of solutions made it clear that the configurational entropy of a random mixture (gaseous, liquid, or solid) could be considered a common cause of miscibility at elevated and high temperatures. If interatomic (intermolecular) forces are attractive or slightly repulsive (as in gases), there is no obstacle for a mixture to be stable at all temperatures or even to form ordered (intermediate) compounds during cooling. Only when repulsive interatomic forces dominate does a solution tend to decompose into phases that are close in their composition to pure components. Such a decomposition is most typical for solid solutions. Therefore, solid solution formation is very often accompanied by the occurrence of repulsive forces, as expressed in a qualitative manner by the classical crystal chemical rules of isomorphism mentioned above.

The description of these forces in a quantitative way is believed to be the main purpose of contemporary crystal chemical and energetic analysis of solid solutions. This survey is devoted to the development and achievements of that theory. The key ideas concern an approach with which one is able to present the energetics of substitution solid solutions by simple and rather accurate

Energetics of Ionic Solid Solutions

Already in the early 1920s Grimm and Herzfeld (1923) had begun the analysis of energetic effects of formation of binary ionic solid solutions and tried to solve the following equation:

$$\Delta H_m = U(x) - x_1 U_1 - x_2 U_2, \tag{1}$$

where ΔH_m is mixing enthalpy, $U(x)$ is lattice energy of a solid solution, U_1 and U_2 are lattice energies of pure components, x_1 and x_2 are their molar fractions. They used the Born–Lande model to express U and to obtain the analytical form of ΔH_m. They concluded that theoretical estimates of ΔH_m for some alkali halide solid solutions had a positive sign and fitted a repulsion parameter in the Born–Lande equation to obtain the correct ΔH_m values. This was a very useful result because ΔH_m is usually a very small portion of the lattice energy U, not more than 0.5%. This quantity is less than the errors in U values, estimated to be about 1 to 2%. Therefore, there is a cancellation of the main errors in Eq. (1).

Later some investigators (Tobolsky, 1942; Wallace, 1949; Fineman, 1950; for references, see Urusov, 1977) developed this approach and found that the mixing enthalpy ΔH_m had to be expressed as a function of the square of the size difference parameter $\delta = \Delta R/R$, where $\Delta R = R_2 - R_1$, R_1 and R_2 are interatomic distances in pure components, and R is the average interatomic distance in the solid solution. However, ΔH_m, calculated by these authors for several systems of alkali halide solid solutions, did not agree very well with calorimetric measurements.

More successful attempts to solve the same problem were performed by Wasastjerna (1949) and his Finnish successors (Hovi, 1950; Hietala, 1963a, b). They used Vegard's rule $R = x_1 R_1 + x_2 R_2$ in the form

$$R_1 = R + x_2 \Delta R, \qquad R_2 = R - x_1 \Delta R, \qquad R_2 > R_1 \tag{2}$$

and the expansion of lattice energy into power series with the small-size parameter $\delta < 1$. The restriction of this series up to quadratic terms again gives ΔH_m as a function of δ^2. Wasastjerna accounted for the local displacements of a common ion C in the solid solution $(A_{x1}, B_{x2})C$ with an NaCl-type structure if this ion is surrounded by ions of various kinds, A and B. Then he divided the total distance 2R between A and B in the following proportion:

$$\bar{R}_1 = \langle A - C \rangle = \left[\frac{R_1}{(R_1 + R_2)} \right] 2R,$$

$$\bar{R}_2 = \langle B - C \rangle = \left[\frac{R_2}{(R_1 + R_2)} \right] 2R. \tag{3}$$

The Wasastjerna–Hovi model was analyzed in detail by Urusov (1977). Here we concentrate somewhat more on Hietala's (1963a) theory that is rather different from the previous treatment. He used Taylor's series expansion for energy

$$U(R) - U(R_0) = U'(R_0)\Delta R + (\tfrac{1}{2})U''(R_0)(\Delta R)^2 + \cdots$$
$$\approx (\tfrac{1}{2})U''(R_0)(\Delta R)^2, \tag{4}$$

where $\Delta R = R - R_0$; considering that the first derivative of the lattice energy $U'(R_0) = 0$ at equilibrium state $(R = R_0, T = 0)$. Hietala also considered the local atomic displacement when the surrounding of a common ion is asymmetric. He presented the change of the interatomic distances $\langle A - C \rangle$ and $\langle B - C \rangle$ in the form

$$\bar{R}_1 = R - u, \qquad \bar{R}_2 = R + u,$$

where u is the displacement of ion C from the center of the linear bonds A–C–B or B–C–A. Solving the system of the two equations

$$\frac{\partial \Delta H_m}{\partial u} = 0 \quad \text{and} \quad \frac{\partial \Delta H_m}{\partial R} = 0,$$

Hietala obtained the displacement $u \approx (1/2)\Delta R$ and the mixing enthalpy

$$\Delta H_m = (\tfrac{1}{4})x_1 x_2 U_1''(R_1)(\Delta R)^2. \tag{5}$$

At $T = 0$,

$$U_1''(R) = \frac{9V_1}{R_1^2 \beta_1}, \tag{6}$$

where V_1 is the molar volume and β_1 the coefficient of isothermal compressibility of pure component 1. Substituting U_1'' from Eq. (5), Hietala obtained

$$\Delta H_m = (\tfrac{9}{4})x_1 x_2 \left(\frac{V_1}{\beta_1}\right)\left(\frac{\Delta R}{R_1}\right)^2. \tag{7}$$

For alkali halides with an NaCl-type structure, V/β is nearly constant and equal to 630 kJ/mol with error bars of a few percent. In other words, the properties of both end-members are very close and $V_1/\beta_1 \approx V_2/\beta_2$. Thus, one finally has

$$\Delta H_m = 1420 x_1 x_2 \left(\frac{\Delta R}{R_1}\right)^2 = x_1 x_2 Q, \tag{8}$$

where interaction parameter $Q = 1420(\Delta R/R_1)^2$ kJ/mol.

A similar, although more elaborate, treatment for solid solutions with a CsCl-type structure (Hietala, 1963b) led to the final equation

$$\Delta H_m = 1.25(\tfrac{9}{4})x_1 x_2 \left(\frac{V_1}{\beta_1}\right)\left(\frac{\Delta R}{R_1}\right)^2 \approx (\tfrac{11}{4})x_1 x_2 \left(\frac{V_1}{\beta_1}\right)\left(\frac{\Delta R}{R_1}\right)^2. \tag{9}$$

Hietala's model takes into account the nonparabolic character of the mixing enthalpy by adding the third-order term in the expansion (4). In this case, one can rewrite Eq. (8) in the following form:

$$\Delta H_m = x_1 x_2 Q[1 + B(x_1 - x_2)], \qquad (10)$$

where $B = 0.6(\text{Å}^{-1})(\Delta R)$. The second term in square brackets in Eq. (10) makes the ΔH_m parabola skew. Due to the asymmetry of ΔH_m (10), its values are relatively higher at $x_1 > x_2$, i.e., when molar fractions of the component 1 with a shorter interatomic distance $R_1 (< R_2)$ are more than 0.5. This conclusion is at least in qualitative agreement with available experimental data.

There exist many extensive calculations of the properties of ionic solid solutions in a nonanalytical way. Durham and Hawkins (1951) and Lister and Meyers (1958) performed detailed analysis of the local atomic displacements and mixing enthalpies for those alkali halide systems that were investigated calorimetrically (Lister and Meyers, 1958).

Many other authors followed Mott–Littleton's theory (Mott and Littleton, 1938) to calculate the local distortions and energetic effects of solution for isolated impurity ion in alkali halides (Brauer, 1953; Fumi and Tosi, 1966; Hardy, 1962; Fukai, 1963; Dick and Das, 1962; Tosi and Doyama, 1966; McDonald, 1966; Douglas, 1966; Kristofel, 1968; see references in Urusov, 1977). The theory was expanded to the full composition range of alkali halide solid solutions by Fancher and Barsch (1969). Their calculations as well as the calculations using Wasastjerna–Hovi's and Hietala's theories are in excellent agreement with the experimental measurements of mixing enthalpies (Lister and Meyers, 1958) for continuous solid solutions in systems KBr–KI, NaI–KI, NaBr–NaI, NaBr–KBr, NaCl–KCl, NaCl–NaBr.

In conclusion, we note that Mott–Littleton's theory was used repeatedly for the prediction of local distortions around isolated bivalent impurity and corresponding solution energies in alkali halides (Bowman, 1973; Stoneham, 1975; Urusov et al., 1980; Bandyopadhyay and Deb, 1988). There are few experimental data to be compared with these theoretical estimates. They come mainly from studies of transport properties, in particular, electrical-conductivity measurements. One can find at least semiquantitative confirmation of the theoretical predictions.

Generalization of Mixing Enthalpy Model

As is known, a majority of inorganic crystals and minerals are interpreted as being not purely ionic, but in fact, the chemical bonding in oxides, chalchogenides, etc. is partly covalent. One can describe the real bonding character by means of the fractional ionicity f_i ($0 \leq f_i \leq 1$) or effective charge of an atom $q = zf_i$. In such a case, the correct measure of cohesion energy of a crystal is the atomization energy E, but not lattice energy U (Urusov, 1975). Therefore, one has to change Eq. (1) with the following:

$$\Delta H_m = E(x) - x_1 E_1 - x_2 E_2, \qquad (11)$$

where $E(x)$ is the atomization energy of a solid solution, E_1 and E_2 are the atomization energies of pure components. It was shown by Urusov (1969a) that when both components are similarly ionic, ΔH_m preserves the general form of the ionic model [Eqs. (7) and (9)] and the ionicity degree f_i is not involved at the first approximation in the final expression. Moreover, in order to include an explicit dependence of interaction parameter Q on the composition it was proposed that the size parameter be used in the form of $\delta = \Delta R/R = \Delta R/(x_1 R_1 + x_2 R_2)$ instead of $\delta_1 = \Delta R/R_1$ in Eqs. (7) and (9). Then we finally have for any isovalent solid mixture of isostructural components an expression very similar to Eq. (7) (Urusov, 1968a, 1974, 1975, 1977):

$$\Delta H_m = \left(\frac{9V}{4\beta}\right) x_1 x_2 \delta^2. \tag{12}$$

The size parameter δ in the last equation varies from $\delta_2 = \Delta R/R_2$ to $\delta_1 = \Delta R/R_1$ within the whole range of compositions, from $x_1 = 0$ and $x_2 = 1$, and vice versa. Hence, simple linear interpolation offers the following useful approximation:

$$\Delta H_m = \left(\frac{9V}{4\beta}\right) x_1 x_2 (x_2 \delta_1^2 + x_1 \delta_2^2) = x_1 x_2 (x_2 Q_1 + x_1 Q_2), \tag{13}$$

which expresses energetic asymmetry in an explicit form. One can compare Eq. (13) with the well-known representation of the excess free energy of mixing in a subregular model (Saxena, 1973)

$$\Delta G_m^{ex} = x_1 x_2 (x_2 W_1 + x_1 W_2), \tag{14}$$

where the Margules parameters W_1 and W_2 are accordingly in close relation to $Q_1 = (9V/4\beta)\delta_1^2$ and $Q_2 = (9V/4\beta)\delta_2^2$.

Let us define an asymmetry parameter η

$$\eta = \frac{Q_1}{Q_2} = \left(\frac{R_2}{R_1}\right)^2. \tag{15}$$

From Hietala's Eq. (10), the asymmetry parameter may be simply presented by the formula

$$\eta_H \approx 1 + 1.2(\text{Å}^{-1})(\Delta R). \tag{16}$$

Table 1 shows a comparison between η and η_H from Eqs. (15) and (16) and the

Table 1. Theoretical and empirical estimates of asymmetry parameter η.

System	η_H	η	η'	Q_1/Q_2, exp.
KI–RbI	1.17	1.08	1.10	1.22
KCl–RbCl	1.17	1.19	1.20	1.21
NaCl–NaBr	1.20	1.13	1.16	1.15
NaBr–KBr	1.37	1.22	1.22	1.38
NaBr–NaI	1.30	1.18	1.22	1.48

available experimental data on the ratio of enthalpy interaction parameters for some alkali halide solutions (Davies and Navrotsky, 1983). One can note satisfactory agreement between theoretical predictions and observations. The predictions by Eq. (15) could be slightly improved by accounting for the individual values of V/β for each end-member of solid solutions (η' in Table 1).

Now it is worthy to emphasize that V/β values are nearly constant not only for alkali halides, but also for various groups of compounds, e.g., halides, oxides, chalcogenides, carbides, and some elements (Urusov, 1970). Similar observations were also made by geophysicists (Anderson and Nafe, 1965; Anderson and Anderson, 1970) for alkali and alkaline-earth halides, oxides, and chalcogenides of the MX type. For instance, V/β values are about 750 kJ/mol for halides of a CsCl-structure type, 630 for halides of the NaCl type, 1950 for oxides $A^{II}B^{VI}$ (the NaCl-structure type), 1200 for chalcogenides $A^{II}B^{VI}$ (the NaCl-structure type), 820 and 1900 for chalcogenides $A^{II}B^{VI}$ and $A^{III}B^{V}$ (the ZnS-structure type). On the other hand, it has long been known (Born, 1923), a relation:

$$\frac{V}{\beta} = \rho E, \qquad (17)$$

where ρ is a constant, dependent only on the parameters in the Mee potential. It is also well known that cohesion energy E is proportional to the product of the number of atoms in molecular unit $A_k B_l$ $m = k + l$, of valences z_A and z_B, and, possibly, of average coordination number v. Hence, $V/\beta \approx m v z_A z_B c$, where c is the only empirical parameter. It was shown by Urusov (1970, 1977) that c occurs to be in close relation to the bonding character f_i, i.e., to the electronegativity difference of constituent atoms A and B: $\Delta\chi = |\chi_A - \chi_B|$. These relations could be written approximately as follows:

$$c(\text{kJ}) \approx 20(2\Delta\chi + 1) \quad \text{or} \quad c(\text{kJ}) \approx 15(10f_i + 1), \qquad (18)$$

if we accept that on the average, for halides $f_i = 0.8$, $\Delta\chi = 3.0$, for oxides $f_i = 0.6$, $\Delta\chi = 2.0$, for chalcogenides $f_i = 0.3$, $\Delta\chi = 1.0$, and for carbides $f_i = 0.2$, $\Delta\chi = 0.5$.

Finally we have the so-called "crystal chemical form" of the formula for the ΔH_m calculations

$$\Delta H_m = x_1 x_2 \, cm \, v z_A z_B \delta^2. \qquad (19)$$

By using Eq. (19), it is possible to predict the values of the mixing enthalpy ΔH_m or interaction parameter $Q = \Delta H_m/(x_1 x_2)$. The calculated Q are given in Table 2 and compared with available experimental estimates (Urusov, 1977; Davies and Navrotsky, 1983). The table includes both rock salt and cesium chloride structures and examples of cubic perovskite and silicate spinel structures. If one takes into account large uncertainties of the empirical data, especially for oxides, then one may observe fair agreement between these two columns of values.

Table 2. Predicted and experimental enthalpy interaction parameter (kJ) for some alkali halide and oxide binary solid solutions.

System	Q, exp.	Q, calc.	System	Q, exp.	Q, calc.
CsCl–CsBr	4.40	3.65	CoO–MgO	0	0.62
CsBr–CsI	8.78	8.04	CoO–NiO	0	0.31
TlCl–TlBr	3.31	2.93	FeO–MgO	5.02	2.49
KI–RbI	2.41	2.17	CoO–MnO	6.01	7.88
KCl–RbCl	3.37	2.97	MgO–MnO	17.08	12.87
KBr–KCl	4.03	3.38	MnO–NiO	9.99	17.74
NaBr–NaCl	5.27	5.07	CaO–SrO	28.80	30.38
KBr–KI	7.44	7.07	SrO–BaO	37.45	33.44
NaBr–KBr	12.27	15.40	$SrTiO_3$–$BaTiO_3$	9.53	8.36
NaI–KI	10.61	11.92	γ-Mg_2SiO_4–γ-Fe_2SiO_4	7.8	6.5
NaBr–NaI	7.98	9.76			
KCl–NaCl	18.48	19.04			

It is obvious that for all cubic crystals, the size parameter δ may be expressed by means of the relative difference of the lattice parameter a

$$\delta_a = \frac{(a_2 - a_1)}{(x_1 a_1 + x_2 a_2)} = \frac{\Delta a}{a} \qquad (20)$$

or of the cubic roots of molar volumes V

$$\delta_V = \frac{(\sqrt[3]{V_2} - \sqrt[3]{V_1})}{(x_1 \sqrt[3]{V_1} + x_2 \sqrt[3]{V_2})}. \qquad (21)$$

That is not the case for crystals of lower symmetry. Then the latter definition of δ_V (21) becomes useful for a rough calculation within a framework of the pseudoisotropic approach. For example, the exact value of the size parameter for spinel-type γ-Mg_2SiO_4 – γ-Fe_2SiO_4 cubic solid solutions $\delta = \Delta a/a = 0.021$. Using this δ value, $m = 3$, $v = 6$, $z_A = 2$, $z_B = 4$ (SiO_4^{4-}), one obtains from Eq. (19) $Q = 6.51$ kJ. That does not stand in contradiction with the experimental estimate: $Q = 7.8$ kJ (Akaogi et al., 1989). Practically speaking, the same value of the "volume" parameter $\delta_V = 0.020$ could be calculated from Eq. (21) for orthorhombic olivine-type α-Mg_2SiO_4 – α-Fe_2SiO_4 solid solutions. However, an experimental estimate of the interaction parameters in this system seems to be about a factor of 2 larger: $Q_1 = 16.8$, $Q_2 = 8.4$ kJ (Wood and Kleppa, 1981). A much better estimation of the size parameter for olivine solid solutions would come from the relative difference of M–Si distances: $\delta_{M1-Si} = 0.018$, $\delta_{M2-Si} = 0.029$. Using these values, one obtains from Eq. (19) $Q = 5.0 - 13.0$ kJ, which is in satisfactory agreement with the experimental asymmetry of mixing enthalpy.

No constraints exist to apply Eq. (19) to sufficiently covalent systems, for instance, semiconductor binary alloys with tetrahedral zincblende and a

Table 3. Comparison between experimental and theoretical predictions of the interaction parameters Q (kJ) in some binary semiconductor systems.

System	Q, exp.	Q^a, Eq. (19)	Q^b, Eq. (22)	Q^c	Q^d	Q^e
ZnSe–ZnTe	6.5	12.4	13.0	8.9	11.1	10.0
ZnTe–HgTe	12.5	6.6	7.6	6.2	5.0	6.9
ZnTe–CdTe	5.6	7.3	8.2	6.8	6.5	—
HgSe–HgTe	2.9	8.0	7.8	6.8	—	7.8
AlAs–GaAs	0.00	0.10	0.00	0.00	0.00	0.00
AlSb–GaSb	0.00	0.01	0.09	0.13	—	0.09
AlAs–InAs	10.5	14.6	9.9	11.7	—	10.03
AlSb–InSb	2.5	9.2	5.8	6.1	—	6.1
GaP–GaAs	1.6; 4.2	3.8	4.1	2.7	4.8	3.0
GaP–InP	13.6; 14.6	14.6	14.9	12.3	16.3	13.3
GaAs–GaSb	16.7; 18.8	18.8	14.0	11.5	17.7	12.3
GaAs–InAs	6.9; 8.3; 12.5	12.5	11.7	10.1	16.3	14.9
GaSb–InSb	6.1; 7.9	7.9	7.5	7.7	—	7.3
InAs–InSb	9.6; 12.1	14.6	9.4	9.1	10.6	9.3
InP–InAs	1.7	2.5	2.4	2.4	2.1	2.3
Si–Ge	5.0	5.0	5.00	3.7	—	3.7
Ge–Sn	31.6	55.0	43.5	32.0	—	36.5
Si–Sn	81.6	90.0	77.4	71.5	—	65.6

[a] Urusov (1977).
[b] Stringfellow (1973).
[c] Fedders and Muller (1984).
[d] Suh and Talwar (1989).
[e] Martins and Zunger (1984).

diamond-type structure. Table 3 contains a comparison of available experimental estimates and theoretical predictions of the interaction parameters Q in such systems.

Another way to calculate these properties of the tetrahedral solid ternary alloys is to use a semiempirical equation, derived by fitting experimental data (Stringfellow, 1973)

$$\Delta H_m(\text{kJ}) = x_1 x_2 7280 \delta^{2.45}. \quad (22)$$

Later it was demonstrated (Fedders and Muller, 1984) that elastic distortion energy is δ^2-dependent, but strongly overestimates (by a factor of about 4) the mixing enthalpy if the crystal structure of an alloy is not relaxed. The tentative value for structure relaxation (Ferreira et al., 1988) allows us to reduce the unrelaxed elastic energy by a factor of 4. That leads to Eq. (7) and, therefore, to Eq. (19).

An alternative approach deals with a semiempirical tight binding method for calculating the energy change and bond length relaxation around isovalent substitutional impurities in $A_{1-x}B_xC$ tetrahedral semiconductor alloys (Talwar et al., 1987; Suh and Talwar, 1989). The interaction parameters Q, calculated from Eqs. (19) and (22) and obtained by the methods of Fedders and Muller (1984) and Suh and Talwar (1989), are compared in Table 3 with some experimental data. As may be seen in Table 3, all these estimates correlate satisfactorily.

If properties of pure components are remarkably different, i.e., $V_1/\beta_1 \neq V_2/\beta_2$, the approximation of Eq. (7) is not completely valid. Then one can use the following correction to Eq. (7) (Hietala, 1963a):

$$\Delta H_m = x_1 x_2 \left[\frac{2s}{(1+s)} \right] Q = x_1 x_2 Q_d, \tag{23}$$

where $s = (R_2 \beta_1)/(R_1 \beta_2)$ and Q_d is the so-called deformation energy. For alkali halide solid solutions, $s \approx 1$ within deviations only of a few percent. Such is not the case for solid solutions like NaCl–AgCl and NaBr–AgBr. Although the end-members in both these systems have very close lattice parameters ($\delta = 0.017$ and 0.035), their elastic properties are significantly different. Thus, the coefficients of compressibility β are equal to 2.26 and 2.48 Mbar^{-1} for AgCl and AgBr and about two times less than 4.26 and 5.08 for NaCl and NaBr. Hence, s values are 0.54 for (Ag,Na)Cl and 0.51 for (Ag,Na)Br systems and the correction factors $2s/(1+s)$ in Eq. (23) will be 0.70 and 0.67, respectively.

Table 4 shows a comparison of theoretical estimations of ΔH_m by means of Eq. (23) and experimental data (Kleppa and Meschel, 1965). Very sharp discrepancy between these two sets of ΔH_m values is evident. It means that other forces in addition to short-range repulsion have to be involved to explain this contradiction.

It is not questionable that bonding characters are rather different in alkali and silver halides. In other words, the previous assumption of proximity of covalency of both components is not valid for Na–Ag substitutions and the approximation $f_1 \approx f_2$ is not fulfilled.

Now suppose the bonding character $f_i(x)$ of a solid solution obeys the simple additivity rule (Urusov, 1968b):

$$f_i(x) = x_1 f_1 + x_2 f_2, \qquad \Delta f_i = f_2 - f_1. \tag{24}$$

Here f_1 and f_2 are the fractional bonding characters of pure components. Using (24), one obtains the following expression for mixing enthalpy:

$$\Delta H_m = x_1 x_2 (Q_d + Q_c), \tag{25}$$

where Q_d is mainly due to elastic forces and Q_c is a chemical term. Q_d coincides with the deformation energy in Eq. (23); Q_c describes a change of long-range electrostatic energy and could be expressed as follows:

$$Q_c = \left(\frac{A z^+ z^-}{R} \right) (\Delta f_i)^2, \tag{26}$$

where A is the Madelung constant, characteristic for a given structure type, and z^+ and z^- are the formal charges of the cation and anion.

Table 4. Comparison between theoretical and experimental mixing enthalpies ΔH_m (J/mol) in the systems NaCl–AgCl and NaBr–AgBr.

	NaCl–AgCl			NaBr–AgBr			
x(NaCl)	Eq. (23)	Eq. (25) $\Delta f = 0.113$	Exp.	x(NaBr)	Eq. (23)	Eq. (25) $\Delta f = 0.101$	Exp.
0.85	41	1450	1465 ± 400	0.87	145	1090	1130 ± 300
0.75	58	2130	2430 ± 330	0.77	230	1720	1845 ± 375
0.60	76	2730	2640 ± 200	0.70	270	2050	2175 ± 375
0.50	82	2850	2720 ± 130	0.65	300	2225	2510 ± 420
0.35	73	2595	2430 ± 200	0.50	330	2460	2135 ± 210
0.15	41	1455	1420 ± 80	0.45	325	2430	2135 ± 170
				0.27	260	1950	1880 ± 125

Equation (25) provides an excellent fit with experimental measurements of ΔH_m for NaCl–AgCl and NaBr–AgBr systems at $\Delta f_i = 0.113 \pm 0.002$ and 0.101 ± 0.005 (see Table 4). One may compare the above-mentioned Δf_i with estimates of the differences of ionicity degrees for these two pairs of the end-members by the well-known Pauling (Pauling, 1960) scale of electronegativities ($\Delta f_i = 0.067$ and 0.068) and Phillips' (Phillips, 1973) spectroscopic definition of ionicity ($\Delta f_i = 0.079$ and 0.084).

The discussion up to now has dealt only with the case of isostructural components of a solid mixture. If the structure types of the end-members are different, theoretical analysis leads to a new form (Urusov, 1969b)

$$\Delta H_m = x_1 x_2 (Q_d + Q_c) + x_2 \Delta H_{tr}, \quad (27)$$

where ΔH_{tr} is the enthalpy of polymorphic transition of end-member 2 to a structure type of end-member 1. The component 2 must undergo the phase transition in order to be soluble in component 1. If the phase transition under consideration is accompanied by a coordination number change, then ΔH_{tr} may be identified with the site preference energy. Thus, the octahedral and tetrahedral site preference energies of some divalent cations were calculated from experimentally determined terminal solubilites in binary oxide and chalcogenide systems with a rock salt or nickel arsenide structures of the end-member 1 and wurtzite or sphalerite structures of the end-member 2 (Urusov, 1969b; Davies and Navrotsky, 1983).

Local Displacements of Atoms in the Solid Solution Structure

As is evident from the previous discussion, the energetic properties of a solid solution strongly depend on the relaxation of its structure. The calculations show that the so-called virtual crystal model in which all of the individual bond lengths in the mixed crystal are equal to their average values according to Vegard's rule [Eq. (2), no relaxation] is not appropriate, even as a first approximation. Under another assumption all the atoms in the solid solution conserve their initial sizes, i.e., with structure relaxation being at a maximum, the individual bond lengths are equal to the bond lengths in the end-members, R_1 and R_2. This model of bond alternation is also not completely adequate, although it is as good as the virtual crystal model that agrees with Vegard's rule for averaged interatomic distances. In other words, Vegard's rule fulfilment is not sensitive to the relaxation of a solid solution structure.

It is clear that the actual changes in the bond lengths of various kinds must lie between both above-mentioned extreme cases. The determination of the degree of structure relaxation, i.e., the local displacements of atoms, requires independent, sufficiently complicated, experimental and theoretical work.

Let us begin with some theoretical arguments to approach a solution of the problem. Wasastjerna's assumption, Eq. (3) (with the second-order terms neglected), leads to the following individual bond lengths A–C and B–C in the asymmetric linear chain A–C–B for the NaCl structure type of solid solution:

$$\bar{R}_1 \approx R_1 - \left(\frac{\Delta R}{2}\right)(1 - 2x_2) + \cdots$$

$$\bar{R}_2 \approx R_2 + \left(\frac{\Delta R}{2}\right)(1 - 2x_1) + \cdots \quad (28)$$

Practically speaking, the same result comes from the minimization of solid solution energy by Hietala (1963a):

$$\bar{R}_1 = R - u = R_1 - \left(\frac{\Delta R}{2}\right)(1 - 2x_2),$$

$$\bar{R}_2 = R + u = R_2 + \left(\frac{\Delta R}{2}\right)(1 - 2x_1).$$

Here it was taken into account that the average interatomic distance R obeys Vegard's rule ($R = R_1 + x_2 \Delta R = R_2 - x_1 \Delta R$) and local displacement u is only very slightly dependent on the composition, $u \approx \Delta R/2$.

At a random distribution of the substitutents, the probability of the bond configuration A–C–A occurring is proportional to x_1^2, the probability of the bond configuration B–C–B–x_2^2, and the probability of the mixed arrangement A–C–B–$2x_1 x_2$. It makes possible the calculation of the average distances of both different types, full numbers of which are proportional to x_1 and x_2 accordingly

$$x_1 R_{A-C} = x_1^2 (R_1 + x_2 \Delta R) + x_1 x_2 \left[R_1 - \left(\frac{\Delta R}{2}\right)(1 - 2x_2)\right]$$

$$= x_1 \left[R_1 + x_2 \left(\frac{\Delta R}{2}\right)\right]; \quad R_{A-C} = R_1 + x_2 \left(\frac{\Delta R}{2}\right), \quad (29)$$

$$x_2 R_{B-C} = x_2^2 (R_2 - x_1 \Delta R) + x_1 x_2 \left[R_2 + \left(\frac{\Delta R}{2}\right)(1 - 2x_1)\right]$$

$$= x_2 \left[R_2 - x_1 \left(\frac{\Delta R}{2}\right)\right]; \quad R_{B-C} = R_2 - x_1 \left(\frac{\Delta R}{2}\right).$$

It means that a maximal increase of the shorter bond A–C as well as maximal decrease of the larger bond B–C in dilute solutions ($x_2 \to 1, x_1 \to 1$) consists of 1/2 of the value of the difference in interatomic distances ΔR. In other words, relaxation parameter λ is equal to 1/2 for the NaCl structure type of solid solutions. It is also clear from Eqs. (29) that the average shortest distance ($R = x_1 R_{A-C} + x_2 R_{B-C} = x_1 R_1 + x_2 R_2$) obeys Vegard's rule at least in the first approximation. This simple consideration is in very good agreement with numerous calculations of lattice distortions around isovalent impurity in an ionic crystal of the NaCl structure, using the Mott-Littleton theory (see "Energetics of Ionic Solid Solutions"). A typical result of such calculations is the fact that the displacement of the nearest neighbors around the impurity ion is nearly half the difference between the interatomic distances: $\delta R \approx (\frac{1}{2})\Delta R$.

A more detailed picture of the atomic local displacements is obtainable by DLS modeling, which consists of the least-square optimization of individual bond lengths, i.e., their fitting some standard distances (which represent the distances in end-members in the case of solid solutions). The application of DLS modeling to isovalent solid solutions of different structures made it possible for Dollase (1980) to introduce the notion of "site compliance." The latter means the actual fraction of increasing (or decreasing) the bond length relative to the difference of the bond lengths in end-members in the limit of infinite dilution (a very small amount of impurity atoms). Dollase concluded (Dollase, 1980) that the compliance parameter c_s is inversely proportional to the coordination number of the nearest neighbors of the impurity ion because they suffer the largest displacements (Fig. 1). Thus, less close-packed structures (ZnS, and ReO$_3$ types) are characterized by larger changes in bond lengths, whereas close-packed ones (NaCl and CsCl types) exhibit relatively smaller changes in bond lengths. It should be noted that the predicted change of bond lengths in the NaCl-structure solid solutions is about 50%, this result being close to the above-mentioned theoretical estimates of the structure relaxation by entirely independent approaches.

It is very difficult to obtain experimental data on the local structure of a solid solution by traditional diffraction methods that carry information on average atomic coordinates. Nevertheless, early X-ray diffraction studies of solid solutions (Wasastjerna, 1945; Iveronova, 1954) detected a decrease in reflection intensities as compared to the case of pure crystals. This fact cannot be explained only by thermal vibrations and requires the assumption of noticeable static displacements of ions in a nonmixed sublattice from their regular positions. An analysis of such effects showed that local displacements are on the order of the difference of interatomic distances of end-members; no strict correlation, however, was observed between them.

About 10 years ago it became possible to determine individual bond lengths in mixed crystals by the EXAFS method (extended X-Ray absorption fine struc-

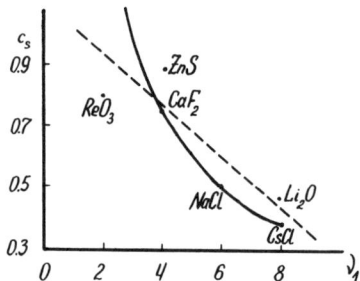

Fig. 1. Correlation between site compliance parameter c_s and first coordination number. Points and dotted line, DLS method, solid line, calculated, Eq. (36).

ture). EXAFS spectroscopy is a very useful tool for characterizing lattice distortions around an impurity atom as it probes one particular element at a time and provides information about the number, distance (with accuracy of 0.005 Å), and chemical identity of the nearest neighbors. In 1979 and 80 this method allowed considerable local displacements of atoms around impurities to be detected in metal alloys and then in mixed halogenides and chalcogenides. For instance, in (K,Rb)Br and Rb(Br,I) solid solutions, maximum change in individual distances amounts to about 40 to 50% (Boyce and Mikkelsen, 1985). Relatively large changes in the shortest distances, about 40%, were also observed for Sr-substituted fluorite $(Ca,Sr)F_2$ (Vernon and Stearns, 1984).

On the other hand, maximum changes in the bond lengths of mixed tetrahedral semiconductors with a ZnS structure are much less, about half that for essentially ionic solid solutions of the NaCl structure. For instance, in a (Ga,In)As system the maximum change in the nearest-neighbor distances Ga–As and In–As corresponding to the dilute solutions amounts only to 20 to 25% of the difference ΔR between the bond lengths in the end-members (Mikkelsen and Boyce, 1983).

It was also observed that the distributions of the Ga–As and In–As distances in the solid solution are nearly the same as in the end-members, which is in accordance with the fact that the nearest environment of both cations is uniform and consists only of arsenic atoms. As distinct from this, there occurs a bimodal distance distribution around As that corresponds to the mixed cation (Ga,In) environment of this atom.

The distances between the second nearest neighbors (cation–cation and anion–anion) considerably differ in their character. EXAFS data indicate the existence of two difference As–As distances in the solid solution: Shorter distances correspond to the As–Ga–As configuration and longer distances to the As–In–As configuration. The weighted average of these two distances corresponds to Vegard's rule. It is clear that the anion packing in the solid solution is strongly distorted, as compared to the regular cubic (closest) packing of anions in the end-members.

A different picture is observed in the case of the distances of the cation–cation second nearest neighbors. All the interatomic distances Ga–Ga, In–In, and In–Ga (Ga–In) vary to obey Vegard's rule within deviations on the order of 0.05 Å, which follows from the virtual crystal model. This means that the atoms in the mixed (cation) sublattice occupy nearly regular positions, and distortions of the ideal packing are relatively slight.

Some other semiconductor alloys were investigated recently by EXAFS: Ga(As,P) (Sasaki et al., 1986), (Ga,In)P (Mikkelsen and Boyce, 1983), etc. All the studies allows us to conclude that real maximum changes of interatomic distances are only about 20 to 30% in such systems, i.e., relaxation parameter λ is approximately 0.7 to 0.8. The latter is sufficiently larger than λ for ionic NaCl-structure solid solutions (~ 0.5), being in accordance with the site compliance prediction by DLS modeling (Dollase, 1980), Fig. 1.

Finally, let us note that there are extensive calculations of lattice distortions associated with substitutional defects in semiconductors, using semiempirical pseudopotential (Srivastava and Weaire, 1987), tight-binding (Talwar et al., 1987), and valence force field (Martins and Zunger, 1984) methods. They lend support to existing EXAFS experimental data: Predicted relaxation parameter λ is in range of 0.6 to 0.8 for most semiconductors. This is also close to the theoretical prediction by DLS modeling (Fig. 1).

Primary Displacements of Atoms in Radial Force Model

As follows from the experimental and theoretical results considered above, the largest displacements in a solid solution structure $(A_{x1}, B_{x2})C$ are experienced by atoms C in that sublattice where no mixing takes place. As for atoms A and B, they form the nearly nondistorted packing. Other consequences of this assumption are as follows.

If the environment of the common atom C is uniform, i.e., if it consists either of atoms A or of atoms B, then all the distances A–C or B–C are the same and equal to the mean $R(x)$; see Eq. (2). If the environment of the atom C is mixed, i.e., consists of some atoms A and B, then bond chains A–C–B appear in the structure and the atom C is displaced from its ideal position on the middle of the bond toward the smaller atom. Let, for example, the atom A be larger than B. In this case, the atom C is displaced from the center of the A–C–B chain toward B and the distance A–C becomes equal to $R + u$, where u is a certain displacement of the atom C.

A change in the distance C–B in the A–C–B chain is dependent on the bond angle $\angle$ A–C–B and can be represented in a first approximation as

$$u_1 = u \cos \alpha, \qquad (30)$$

where $\alpha = \angle$ A–C–B. In the NaCl structure (octahedral arrangement), the A–C–B chain is linear, $\alpha = 180°$, and, consequently, $u_1 = -u$ (Fig. 2). In the ZnS structure (tetrahedral coordination), $\alpha = 109°28'$ and, therefore, $u_1 = (-1/3)u$

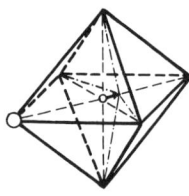

$\delta R = 2 \dfrac{u^2}{2R}$

$P = 6(x_1 x_2^5 + x_2 x_1^5)$

Fig. 2. Octahedral coordination with one substituted vertex. u denotes the arrow length (primary displacement), δR the change in the distances between second nearest neighbors, and P the probability of the configuration.

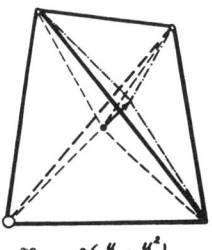

Fig. 3. Tetrahedral coordination with one substituted vertex, corresponding δR and P.

$$\delta R = u + 3\left(-\tfrac{u}{3} + \tfrac{u^2}{3R}\right)$$
$$P = 4(x_1^3 x_2 + x_1 x_2^3)$$

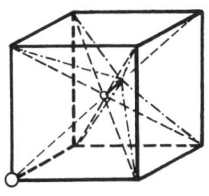

Fig. 4. Cubic coordination with one substituted vertex, corresponding δR and P.

$$\delta R = u + 3\left(-\tfrac{u}{3} + \tfrac{u^2}{3R}\right)$$
$$P = 8(x_1^7 x_2 + x_2 x_1^7)$$

(Fig. 3). In the CsCl structure with a cubic environment of the central atom, there exist three types of chains with the angles $180°$, $109°28'$, and $70°32'$ (Fig. 4). From Eq. (30) it follows that in this case, $u_1 = -u$, $u_1' = (\pm 1/3)u$.

Now let us try to estimate primary displacements using the simple model of radial forces. Denote by $\varepsilon(R)$ the energy of a certain pair of bound atoms separated by a distance R from each other. The energy change in the formation of the solid solution may be represented as

$$\Delta E = Nv\{x_1^2[\varepsilon_1(R) - \varepsilon_1(R_1)] + x_2^2[\varepsilon_2(R) - \varepsilon_2(R_2)] \\ + x_1 x_2[\varepsilon_1(\overline{R}_1) - \varepsilon_1(R_1)] + x_2 x_1[\varepsilon_2(\overline{R}_2) - \varepsilon_2(R_2)]\} \quad (31)$$

Here N is Avogadro's number, v is the coordination number, $\varepsilon_1(R_1)$ and $\varepsilon_2(R_2)$ are the energies of pairwise interactions in the end-members, $\varepsilon_1(R)$ and $\varepsilon_2(R)$ are the energies of two bond types (A–C and B–C) at a mean distance R in symmetric bond chains, and $\varepsilon_1(\overline{R}_1)$ and $\varepsilon_2(\overline{R}_2)$ are the energies of these bond types in nonsymmetric bond chains. The distances $\overline{R}_1$ and $\overline{R}_2$ depend on displacements of C atoms in accordance with the bond angles as shown by the previous Eq. (30). For example, for the NaCl structure $\overline{R}_2 = R + u$, $\overline{R}_1 = R - u$.

Expand the energy ΔE into a Taylor series. Leaving the first- and second-order terms and taking into account that $\varepsilon'(R) = 0$ (at $T = 0$ K) in the equilibrium state, we have $\varepsilon(R) - \varepsilon(R_0) = (\tfrac{1}{2})\varepsilon''(R_0)(R - R_0)^2$. If we assume that the properties of the components are close, i.e., $\varepsilon_1''(R_1) \approx \varepsilon_2''(R_2) = \varepsilon''(R)$, we obtain from (31)

$$\Delta E = (\tfrac{1}{2})N v \varepsilon''(R)[x_1^2(R - R_1)^2 + x_2^2(R - R_2)^2 + x_1 x_2 (R - R_1 - u)^2$$
$$+ x_1 x_2(R - R_2 + u)^2]. \tag{32}$$

Using Vegard's rule, we can rewrite expression (32) as follows:

$$\Delta E = (\tfrac{1}{2})x_1 x_2 N v \varepsilon''(R)[2x_1 x_2 (\Delta R)^2 + (x_2 \Delta R - u)^2 + (-x_1 \Delta R + u)^2]. \tag{33}$$

Minimizing ΔE as a function of the displacement u, we arrive at the condition

$$\frac{d\Delta E}{du} = \left(\frac{x_1 x_2}{2}\right) N v \varepsilon''(R)(-2x_2 \Delta R + 2u - 2x_1 \Delta R + 2u) = 0,$$

from which $4u = 2\Delta R(x_1 + x_2)$. As $x_1 + x_2 = 1$, finally we have $u = (\tfrac{1}{2})\Delta R$.

This result means that the displacement of the common atom in the NaCl-structure solid solution amounts to half the difference of the interatomic distances in the end-members. This estimate is close to the experimental data reported in Mikkelsen and Boyce (1985) and to the compliance parameter of the DLS method (Dollase, 1980).

In solid solutions of the CsCl type, each primary displacement $\pm u$ is accompanied by three displacements $(\pm 1/3)u$. Therefore, the total energy change in the approximation of radial force field may be expressed as

$$\Delta E = \left(\frac{x_1 x_2}{2}\right) N v \varepsilon''(R) \left[2x_1 x_2 (\Delta R)^2 + (x_1 \Delta R - u)^2 + (-x_2 \Delta R + u)^2 + 6\left(\frac{u}{3}\right)^2 \right].$$
$$\tag{34}$$

Minimizing (34) as a function of u yields

$$\frac{d\Delta E}{du} = \left(\frac{x_1 x_2}{2}\right) N v \varepsilon''(R) \left[-2x_1 \Delta R + 2u - 2x_2 \Delta R + 2u + \left(\frac{4}{3}\right) u \right] = 0$$

and $u = (\tfrac{3}{8})\Delta R$. This result is also in good agreement with the estimate of the site compliance in the CsCl structure: $c_s = 0.38$ (Dollase, 1980) (3/8 = 0.375).

For the ZnS-structure solid solutions, a simple radial force model was used earlier (Shih et al., 1985). The obtained result, $u = (\tfrac{3}{4})\Delta R$, is in agreement with the atomic displacements observed by EXAFS (20–25% of the bond length difference in the end-members) and with the predictions of the DLS method (Dollase, 1980) and more rigorous calculations (see the preceding section).

Generalizing the above estimates of primary displacements of the central atom in the coordination polyhedron in the case of its nonsymmetric environment, we may write the expression

$$u = \left(\frac{3}{v_1}\right) \Delta R, \tag{35}$$

where v_1 is the first coordination number. This is in agreement with Dollase's conclusion that the site compliance is inversely proportional to the number of ligands. In other words, we obtain the following simple relation between c_s and v_1:

$$c_s = \frac{3}{v_1}, \tag{36}$$

which is compared in Fig. 1 with the DLS results.

One can easily see from the previous discussion that all primary displacements in a solid solution structure completely compensate for one another and cannot be the reason for deviations from Vegard's rule [Eq. (2)]. Nevertheless, the deviations from this additivity rule may be observed very often through accurate X-ray measurements. Usually, at least in simple binary systems, they could be described by a second-degree parabola

$$R - R_V = \delta R = x_1 x_2 d, \tag{37}$$

where d is positive in a majority of systems. Let us construct a possible geometric model of the deviations from Vegard's rule.

Secondary Displacements and Deviations from Vegard's Rule

Primary displacements of the neighbors of the central atom in mixed coordination polyhedron entail displacements of the second and higher orders. Let us consider an octahedral arrangement with one substituted atom in the vertex of a regular octahedron if the assumption is made of the regular undistorted packing of the mixed sublattice (Fig. 2). As can be seen, the displacement of the atom C from the octahedron center along the bond causes the changes not only of the two bond lengths by amounts $+u$ and $-u$, respectively, but also of the other four bond lengths. It can easily be found that the change in the symmetric bond length is $\sigma_1 = \sqrt{(R^2 + u^2)} - R \approx u^2/2R$ (to second-order terms). In the case of two substituted atoms in the octahedron, all the bonds experience additional secondary displacements by $\sigma_2 = u^2/R$, and in the case of three substituted atoms by $\sigma_3 = 3u^2/2R$.

The general expression for secondary displacements is as follows: $\sigma_i = (i/2)(u^2/R)$ ($i = 1, 2, 3$). The doubled and tripled displacements for $i = 2$ and $i = 3$, respectively, are connected with the fact that all of the ligands become the second nearest neighbors of two ($i = 2$) or three ($i = 3$) substituents.

The proportion of various configurations for different compositions can be calculated from the binomial formula

$$P = (x_1 + x_2)^v. \tag{38}$$

The probabilities of individual configurations are determined from the expression

$$P_i = m_i x_1^{v-\omega} x_2^\omega, \tag{39}$$

where ω is the amount of atoms of the other sort in the polyhedron vertices, and m_i is the multiplicity of a given configuration (the corresponding coefficient in the binomial theorem). The multiplicities m_i can be found from the relation

$$m_i = \frac{n_i}{n}, \tag{40}$$

where n is the symmetry order of the regular coordination polyhedron, and n_i is the symmetry order of the ith configuration ("substituted" polyhedron). Thus, for instance, the multiplicity of the configuration in Fig. 2 is $m_1 = 48/8 = 6$, because 48 is the symmetry (m3m) order of the regular octahedron, 8 is the symmetry (4mm) order of the one-substituted octahedron.

It should be noted that secondary displacements refer to the distances between the central atom and second neighbors of the substitutional defect, and that Eqs. (38) and (39) determine the relative probabilities of primary displacements. The number of secondary displacements is to the number of primary displacements as $l = v_2/v_1$, where v_1 and v_2 are the first and second coordination numbers, respectively. Therefore, the relative probabilities of secondary displacements can be calculated from the formula

$$P_i' = m_i l x_1^{v-\omega} x_2^{\omega}. \tag{41}$$

Now it is possible to obtain the final equation

$$\sigma R = \Sigma_i P_i' \delta_i. \tag{42}$$

Taking into account that $l = 12/6 = 2$ for NaCl (B1) and using the above quantities $\sigma_i = iu^2/2R$ ($i = 1, 2, 3$), we may reduce sum (42) to the form

$$\sigma R_{B1} = 6x_1 x_2 \left(\frac{u^2}{R}\right). \tag{43}$$

It was found in the preceding section that $u = \Delta R/2$ for the NaCl structural type. Accordingly,

$$\sigma R_{B1} = (\tfrac{3}{2}) x_1 \frac{(\Delta R)^2}{R}. \tag{44}$$

In the case of a tetrahedral arrangement, there occur two configurations: with one and with two substituents. The ratio of the numbers of secondary and primary displacements $l = 3$ is in accordance with the ratio $v_2/v_1 = 12/4 = 3$. Summation (42) yields in this case

$$\sigma R_{B3} = 4x_1 x_2 \left(\frac{u^2}{R}\right), \tag{45}$$

and, subject to $u = (\tfrac{3}{4})\Delta R$ for the ZnS (B3) structure, we observe that

$$\sigma R_{B3} = (\tfrac{9}{4}) x_1 x_2 \frac{(\Delta R)^2}{R}. \tag{46}$$

The cubic arrangement in the CsCl (B2) structure allows 13 configurations, with substitutions in one, two, three and four vertices of the cube. For B2 the ratio of the number of secondary displacements to that of primary displacements is $l = 3.25$, in accordance with the fact that the second ($v_2 = 6$), third ($v_3 = 12$), and fourth ($v_4 = 8$) neighbors of substitutional defects take part in these displacements. In this case, summation (42) yields

$$\sigma R_{B2} = 8 x_1 x_2 \left(\frac{u^2}{R}\right). \tag{47}$$

With $u = (\frac{3}{8})\Delta R$ for structure B2 in mind, we arrive at the expression

$$\sigma R_{B2} = (\tfrac{9}{8}) x_1 x_2 \frac{(\Delta R)^2}{R}. \tag{48}$$

Equations (43), (45), and (47) can be written in general form as

$$\sigma R = v_1 x_1 x_2 \left(\frac{u^2}{R}\right), \tag{49}$$

whereas Eqs. (44), (46), and (48) can be written in the form

$$\sigma R = \left(\frac{9}{v_1}\right) x_1 x_2 \frac{(\Delta R)^2}{R}. \tag{50}$$

Let us compare our predictions to available experimental data. Few measurements of the composition dependence of the lattice parameter are reported for solid solutions of the CsCl structural type. The deviations from Vegard's rule observed in CsCl–CsBr did not exceed ± 0.002 Å for all compositions (Hietala, 1963b). The estimate by Eq. (48) yields the following result: The maximum deviation for the composition of $x_1 = x_2 = 0.5$ amounts to $+0.0017$ Å, which does not contradict the experimental evidence.

Callahan and Smith (1966) reported the experimental results of the composition dependence of the unit cell parameters for low-temperature (structure B2) and high-temperature (structure B1) solid solutions NH_4Cl–NH_4Br. Positive deviations from Vegard's rule for intermediate compositions ($x \approx 0.5$) of the low-temperature series reach about 0.004 to 0.002 Å (high-angle measurements). Equation (48) gives 0.002 Å, this being in agreement with experiment.

A considerably large number of measurements have been made for solid solutions of the NaCl-structure type. Table 5 compares the deviations from Vegard's rule measured for four systems with those estimated by Eq. (44). As can be seen, the agreement between theory and experiment is very good.

Extensive investigations were made on the composition dependence of the unit cell parameters for solid solutions of the sphalerite (B3) and wurtzite (B4) structures. The results of these measurements made with different degrees of accuracy are sometimes contradictory. Thus, for instance, both positive and negative deviations from Vegard's rule were observed for the ZnS–CdS (B4) system, however, most of the measurements agreed with Vegard's rule within the limits of experimental errors ($\pm n \cdot 10^{-3}$ Å). The calculations by means of the geometric model [Eq. (46)] show that the maximum positive deviation of the

Table 5. Measured and calculated deviations ($\text{Å} \cdot 10^4$) from Vegard's rule.

x_1	NaCl–KCl		NaCl–NaBr		KCl–KBr		RbI–RbBr	
	exp.[a]	calc.	exp.[b]	calc.	exp.[c]	calc.	exp.[d]	calc.
0.1	45	46	20	13	10	10	—	20
0.3	111	110	35	31	16	23	50	50
0.5	127	133	40	37	23	27	80	60
0.7	105	115	27	31	—	23	40	50
0.9	53	50	12	13	8	10	—	20

[a] Barrett and Wallace, 1954.
[b] Nickels et al., 1949.
[c] Slagle and McKinstry, 1966.
[d] Ahtee, 1969.

parameters a and c from linearity for intermediate compositions does not exceed 0.012 to 0.015 Å.

It is possible that the accuracy of predictions by the geometric model worsens with decreasing packing density. This model is based on the assumption that a mixed sublattice forms a regular packing, i.e., the coordination polyhedrons around the central atom (in a nonmixed sublattice) are undistorted. In fact, this is not the case, as can be seen from the EXAFS data for the InAs–GaAs system (Mikkelsen and Boyce, 1983). Thus, the In–In distances are systematically 0.08 Å larger than the Ga–Ga distances for the same compositions of the solid solution, whereas the Ga–In and In–Ga distances are of intermediate values close to the additive ones. Therefore, solid solution relaxation occurs in both sublattices, although to different degrees, and both primary and secondary changes in bond lengths with respect to those in the end-members could be less than in the geometric model.

In general, due to positive deviations from the additivity of bond lengths and lattice parameters, Retger's rule of molar volume additivity

$$V(x) = x_1 V_1 + x_2 V_2 = V_1 + x_2 \Delta V = V_2 - x_1 \Delta V, \qquad \Delta V = V_2 - V_1 \qquad (51)$$

provides normally better approximation than Vegard's rule for most available measurements (Slagle and McKinstry, 1966; Ahtee, 1969; Urusov, 1977).

In addition, as it became clear many years ago (Zen, 1956), Retger's rule indeed corresponds to the slight positive departures from Vegard's rule. At the same time, the validity of the former means that the mixing volume ΔV_m appears to be close to zero in most simple cases.

Vibrational Entropy of a Random Solid Solution

The nonconfigurational (excess) entropy of a random solid solution of insulators or semiconductors arises chiefly from vibrational contribution (possible electronic and magnetic contributions can be neglected). As is shown by the experimental measurements of heat capacities and Debye temperatures of some solid solutions, the excess entropy is usually positive (see, e.q., Urusov, 1977). Moreover, the following empirical correlation was found between the vibrational entropy ΔS_{vib} and the mixing enthalpy:

$$\Delta S_{vib} = \frac{\Delta H_m}{t}, \quad (52)$$

where the empirical parameter t is equal to 2800 ± 500 K (Urusov and Kravchuk, 1976).

Let us look for the theoretical foundation for the correlation of such a type (Urusov and Kravchuk, 1983). Consider the nonconfigurational entropy as a function of volume $S(V)$. Besides that, let us assume that Retger's rule, Eq. (51), is valid and expand $S(V)$ into a Taylor series with respect to ΔV

$$S(V) = S(V_1) + x_2 \left(\frac{\partial S}{\partial V_1}\right)_T \Delta V + \left(\frac{x_2^2}{2}\right)\left(\frac{\partial^2 S}{\partial V_1^2}\right)_T (\Delta V)^2 + \cdots$$

$$= S(V_2) - x_1 \left(\frac{\partial S}{\partial V_2}\right)_T \Delta V + \left(\frac{x_1^2}{2}\right)\left(\frac{\partial^2 S}{\partial V_2^2}\right)_T (\Delta V)^2 + \cdots.$$

Multiplying the first equation by x_1 and the second one by x_2 and summing, we obtain

$$S(V) = x_1 S(V_1) + x_2 S(V_2) + x_1 x_2 \left[\left(\frac{\partial S}{\partial V_1}\right)_T - \left(\frac{\partial S}{\partial V_2}\right)_T\right]\Delta V$$
$$+ (\tfrac{1}{2}) x_1 x_2 \left[x_2 \left(\frac{\partial^2 S}{\partial V_1^2}\right)_T + x_1 \left(\frac{\partial^2 S}{\partial V_2^2}\right)_T\right](\Delta V)^2. \quad (53)$$

Taking into account that $(\partial S/\partial V_1)_T = \alpha_1/\beta_1$ and $(\partial S/\partial V_2)_T = \alpha_2/\beta_2$, where α_1 and α_2 are the volume thermal expansion coefficients and β_1 and β_2 are the coefficients of isothermal compressibility and assuming approximately $\alpha_1/\beta_1 \approx \alpha_2/\beta_2$, we notice that the third term in Eq. (53) is close to zero, while the fourth one is the nonconfigurational entropy of mixing. Remember the well-known Maxwell relationship

$$\left(\frac{\partial S}{\partial V}\right)_T = \left(\frac{\partial P}{\partial T}\right)_V.$$

Differentiation of that gives

$$\left(\frac{\partial^2 S}{\partial V^2}\right)_T = \frac{\partial^2 P}{[(\partial V)_T (\partial T)_V]} = \frac{\theta}{\beta V}, \quad (54)$$

where $\theta = (\partial \ln \beta / \partial T)_V$. If we assume now that $\theta_1/\beta_1 V_1 \approx \theta_2/\beta_2 V_2 = \theta/\beta V$, then finally we obtain

$$S_{vib} = (\tfrac{1}{2}) x_1 x_2 \left(\frac{V\theta}{\beta}\right)\left(\frac{\Delta V}{V}\right)^2. \quad (55)$$

By comparing Eq. (55) with Eq. (7), and keeping in mind that $(\Delta R/R)^2 \simeq \frac{1}{9}(\Delta V/V)^2$, one can easily see the following relation:

$$S_{vib} = 2\theta \Delta H_m \quad \text{or} \quad t = \frac{1}{2\theta}. \tag{56}$$

The average value of θ for the alkali halides is about $1.8 (\pm 0.3) \cdot 10^{-4}$ K^{-1}. Hence, $t = 2800 \pm 500$ K, this being coincident with the above-mentioned empirical estimation, Eq. (52).

Finally, one can write the useful parabolic expression:

$$\Delta S_{vib} = \frac{x_i x_2 Q}{t}. \tag{57}$$

It means that, in fact, there exists a correlation between the vibrational entropy and the size parameter δ, which was first found out empirically for alkali halide solid solutions by Ahtee and Inkinen (1970). The common value of t of about 2800 K could be used not only for alkali halide or metallic solid solutions, however the universality of this parameter is now open to question (Urusov and Kravchuk, 1976, 1983).

Prediction of Solid Solubility and Partition Coefficients

Immiscibility occurs when positive mixing enthalpy ΔH_m outweighs the configurational ΔS_c and nonconfigurational contributions in entropy terms: $T(\Delta S_c + \Delta S_{vib})$. At higher temperatures, the latter may be larger than ΔH_m and the mutual solubility of components is complete, at low temperatures $\Delta H_m > T\Delta S_m$, and a result is a miscibility gap. Under these conditions, when the composition dependence of ΔH_m and ΔS_{vib} can be represented by a second-degree parabola, Eqs. (8) and (57), the critical solution temperature T_{cr} can be calculated from the following equation (Ahtee and Inkinen, 1970; Urusov, 1977):

$$T_{cr} = \frac{Q}{(2kN + Q/t)}, \tag{58}$$

where k is Boltzmann's constant. Inasmuch as the interaction parameter Q is δ^2-dependent, it is evident that T_{cr} is also a function of the size parameter δ. The calculated T_{cr} are compared with the experimental estimates in Table 6 for some simple halide Na–K systems.

First, one can see the obvious decrease of T_{cr} as δ values decrease. Second, one can easily note that the theoretical calculations of T'_{cr} without involving the correction of ΔS_{vib} strongly overestimate the critical temperatures, while inclusion of the effect of ΔS_{vib} by means of Eq. (58) draws hearer the calculated and the experimental T_{cr}.

In fact, the miscibility gaps are asymmetric because of the energetic asymmetry, i.e., the dependence of Q on composition. The greater the asymmetry

Table 6. Comparison between the calculated and experimental estimates of critical solution temperatures (K).

System	δ_1	T_{cr} O, exp.	T'_{cr}, calc.	T_{cr}, calc.
NaF–KF	0.155	(1500)	1800	1400
NaCl–KCl	0.116	773	1060	800
NaBr–KBr	0.104	668	860	690
NaI–KI	0.092	518	620	500

Table 7. Predicted reduced critical temperatures and compositions of mixing for various energetic asymmetry parameters.

η	T_{cr}/T_{max}	$x_2(cr)$	$x_1(cr)$
1.00	1.00	0.50	0.50
1.05	0.98	0.48	0.52
1.10	0.95	0.47	0.53
1.25	0.87	0.43	0.57
1.50	0.83	0.38	0.62

parameter η [see Eq. (15)], the greater the deviations of T_{cr}/T_{max} and x_{cr} from their values, predicted by the simple regular model ($T_{cr}/T_{max} = 1$, $x_{cr} = 0.5$). Here T_{cr}/T_{max} is the reduced critical solution temperature, T_{max} being the value of T_{cr}, corresponding to the maximal value of Q_1 (or maximal size parameter δ_1) for a system under consideration. Table 7 shows predicted critical temperatures and compositions of mixing for various asymmetry parameters.

The asymmetric solvi as a function of the size parameters were calculated for various binary systems (oxides, chalcogenides, etc.) by some investigators (Urusov, 1974, 1975, 1977; Davies and Navrotsky, 1983). The predicted miscibility gaps are in very reasonable agreement with the available experimental data, in particular, for Al_2O_3–Fe_2O_3, TiO_2–SnO_2, CaS–MgS, HgS–HgTe, etc.

Later some useful approximate correlations were established between properties of simple binary melts and solid solutions (Urusov and Kravchuk, 1976; Kravchuk and Urusov, 1978). For instance, the mixing enthalpy of melts very simply anticorrelates with that of solid solutions

$$\Delta H_m^s \approx -0.2 \Delta H_m^l. \tag{59}$$

Here s denotes solid and l the liquid phases. Moreover, the universal correlation between mixing enthalpy and excess (vibrational?) entropy, discussed in the preceding section, is also valid for ionic melts

$$\Delta S_{vib}^l = \frac{\Delta H_m^l}{t}, \tag{60}$$

where t has the same empirical value, being equal to about 2800 K.

It follows from the above that a close relationship exists between all energetic parameters governing the distribution of the isovalent impurity during crystallization from melt. Taking into account previously established correlations, one can write an expression for the partition coefficient K_i of the i-component between solid and liquid phases in the following form:

$$\ln K_i = \left(\frac{\Delta H_{ml}}{kN}\right)\left(\frac{1}{T} - \frac{1}{T_{ml}}\right) - \left(\frac{1.2\Delta \overline{H}_i^s}{kN}\right)\left(\frac{1}{T} - \frac{1}{2800}\right). \tag{61}$$

Here ΔH_{ml} and T_{ml} are the melting enthalpy and melting temperatures, and $\Delta \overline{H}_i^s$ is the partial mixing enthalpy of the i-component in the solid phase.

For the simplest case of infinite dilution (i.e., without regard for the K_i dependence on composition), Eq. (61) is written as

$$\ln K_i = \left(\frac{\Delta H_{ml}}{kN}\right)\left(\frac{1}{T} - \frac{1}{T_{ml}}\right) - \left(\frac{1.2Q}{kN}\right)\left(\frac{1}{T} - \frac{1}{2800}\right). \tag{62}$$

If one remembers that the interaction parameter Q is mainly δ^2-dependent and takes into account the fact that the first term in Eq. (62) is relatively insignificant, one can conclude that $\ln K_i$ is also δ^2-dependent. This result offers a satisfactory explanation of the observed facts that are described by the so-called Onuma's diagrams (Onuma et al., 1968). As known, they represent $\ln K$ as a parabolic function of the radii difference Δr, which is in reality $\ln K \approx f(\delta^2)$.

It should be noted that Eq. (62) makes it possible to calculate K_i for some binary halide, oxide, silicate, etc. systems at least in semiquantative and qualitative agreement with experimental observations (Urusov and Kravchuk, 1976; Kravchuk and Urusov, 1978). Some nontrivial modifications of the theory are necessary for predictions of equilibrium cocrystallization coefficients from water solution (Urusov, 1980).

Finally, it is worth noting that a sharp rise of K_i at trace amounts of i-impurity often observed by geochemists and experimentors could be explained with the help of a mechanism of interaction between impurity atoms and thermal point defects of a crystal lattice, named as the effect of "microimpurity catching by crystal lattice defects" (Urusov and Kravchuk, 1978). These just-mentioned problems necessitate, however, a separate discussion.

Crystal Chemical Rules of Isomorphous Substitutions from an Energetic Point of View

Now we have an opportunity to explain the empirical rules of isomorphism (see the Introduction) from the point of view of the modern quantitative theory described briefly above. As may be inferred from Eqs. (19), (25), and (27), four main factors affect enthalpy of mixing and therefore the boundaries and stability field of a solid solution.

The first and, possibly, the most important is the size difference, i.e., the value of the size parameter δ. Let us write δ in the conventional form

$$\delta = \frac{\Delta R}{R} = \frac{(r_2 + r - r_1 - r)}{(x_1 r_1 + x_2 r_2 + r)} = \frac{\Delta r}{(x_1 r_1 + x_2 r_2 + r)}, \tag{63}$$

using the standard crystal chemical representation of interatomic distances as sums of the cation and anion radii: $R_1 = r_1 + r$, $R_2 = r_2 + r$, r being the radius of a common (nonmixed) structural unit.

From the identity (63), it is clear that the value of δ is chiefly determined by the difference of radii of the ions or atoms being mixed: $\Delta r = r_2 - r_1$. Therefore, too great a difference $\Delta r/r_1$, usually more than 15%, results in considerable strain (or deformation) energy and limitation of solubility (Goldschmidt and Hume–Roseri's rules).

Besides that, as Eq. (63) indicates, the size parameter δ is a function of composition x. The greatest value of δ is reached with minor amounts of the large-size component 2: $x_2 \to 0$, $x_1 \to 1$, $\delta \to \delta_1 = \Delta r/(r_1 + r)$. Indeed, the opposite is true and the least value of δ is realized with the trace contents of the small-size component 1: $x_1 \to 0$, $x_2 \to 1$, $\delta \to \delta_2 = \Delta r/(r_2 + r)$. For this reason, the asymmetry of the interaction parameter Q and solvus occurs so that, for a given size difference Δr, it is practically always easier to put a smaller atom (ion) into a large site than vice versa. It is precisely this fact that explains the well-known Goldschmidt's "polarity rule" of isomorphism.

In addition to that, we should concentrate our attention on the role of the size of a common structural unit, namely, r in Eq. (63). It occurs only in the denominator of the identity, i.e., the larger the size of a common unit, the smaller the size parameter δ and, hence, the interaction parameter Q and critical solution temperature T_{cr}. The examples of Na–K substitutions, given in Table 6, demonstrate the fact that T_{cr} decreases as the radius of anions increases, from F^- ($r = 1.33$ Å) to I^- ($r = 2.20$ Å). It is also worthy noting that only limited solubility at high temperature is the case for the fluoride system NaF–KF and $T_{cr}(exp)$ could be approximately estimated from the eutectic-type phase diagram.

Many other observations confirm this deduction (Urusov, 1977; Navrotsky, 1982, 1985, 1987). In particular, if one compares solid solubility in CaO–MgO (very limited), $CaCO_3$–$MgCO_3$ (extensive), and $Ca_3Al_2Si_3O_{12}$–$Mg_3Al_2Si_3O_{12}$ (complete), one is drawn to the same conclusion. In the examples above the size parameter δ sharply decreases from 0.14 (oxides) to 0.08 (carbonates) and 0.03 (silicates) because of a rise in the size of the unmixed structural unit (O^{2-}, CO_3^{2-}, $Al_2Si_3O_{12}^{6-}$). In other words, the critical solution temperature T_{cr}, being δ^2-dependent, falls nearly 20 times from oxides to silicates. Such solid solution behavior is described by so-named "assistance rule" (Urusov, 1977), if we bear in mind that a size mismatch can be tolerated by a structure with a larger size (volume) of a common structural unit that plays the so-called "assistant role."

The second main factor is the charge or valence z of substituents, as follows from Eq. (19). The higher the charge when isovalent species are mixed, the more positive the mixing enthalpy ΔH_m and the higher T_{cr} for comparable size difference δ. For instance, at a high temperature complete solubility exists in the system CaO–SrO ($z = 2$, $\delta = 0.07$), limited solubilities occur in the systems Al_2O_3–Fe_2O_3 ($z = 3, \delta = 0.04$) and TiO_2–SnO_2 ($z = 4, \delta = 0.05$), and very limited solid solution boundaries are observed in the system MoO_3–WO_3 ($z = 6, \delta = 0.01$), although the size parameter δ declines remarkably in this range. The difference in Zn^{2+} and Mn^{2+} radii ($\Delta r = 0.08$ Å) is much more than in the Mo^{6+} and W^{6+} radii ($\Delta r = 0.02$ Å), but there is complete solubility in $ZnWO_4$–$MnWO_4$ at $T > 800°C$ and limited solubility in $ZnWO_4$–$ZnMoO_4$ even at higher temperatures.

The third factor is the difference in bonding characters or ionicity degrees Δf_i of end-members. This conclusion is a straight consequence of Eq. (25). In cases where the ionicity of the species being mixed differs significantly, $\Delta f_i \geq 0.05$, solubility must be limited, even when the species are similar in size. Examples were NaCl–AgCl and NaBr–AgBr, which show large positive enthalpies of mixing (Table 4) and decompose ($T_{cr} \approx 200$ and $300°C$) in spite of a very small size mismatch.

One can use the electronegativity difference $\Delta \chi$ as a good measure of the ionicity degree difference Δf_i. In the above examples of Na–Ag substitutions, $\Delta \chi$ is equal to 1.0 and Δf_i is equal to 0.10 to 0.11. Thus, one can easily explain the absence or limitation of isomorphous substitutions between K^+ and Ag^+ ($\Delta r = 18\%, \Delta \chi = 1.1$), Ca^{2+} and Cd^{2+} ($\Delta r = 5\%, \Delta \chi = 0.6$) or Hg^{2+} ($\Delta r = 8\%, \Delta \chi = 0.9$), Mg^{2+} and Zn^{2+} ($\Delta r = 12\%, \Delta \chi = 0.4$), although there are extensive or even complete substitutions between Na^+ and K^+ ($\Delta r = 36\%, \Delta \chi = 0.1$), K^+ and Rb^+ ($\Delta r = 12\%, \Delta \chi = 0.0$), Ca^{2+} and Sr^{2+} ($\Delta r = 16\%, \Delta \chi = 0.1$), Sr^{2+} and Ba^{2+} ($\Delta r = 15\%, \Delta \chi = 0.1$), Zn^{2+} and Cd^{2+} ($\Delta r = 19\%, \Delta \chi = 0.1$) with similar or larger Δr, but smaller $\Delta \chi$. In other words, a $\Delta \chi$ value of about 0.4 represents an upper limit, so that larger values of $\Delta \chi$ keep solid solubility from being anywhere efficient.

At last, the fourth main factor is structure difference, as seen from Eq. (27). It concerns nonisostructural systems, which usually show only limited solubility. It should be borne in mind that the enthalpy ΔH_{tr} of the "forced" structure transition of "guest" end-member 2, on dissolving in the "host" end-member 1 with another structure, is positive. The higher the positive ΔH_{tr}, the more limited the solubility of the guest end-member in the host one. Let us consider only one example: The solubility of alabandite MnS (the NaCl structure) in sphalerite ZnS reaches 40 mol % at 600°C, although the size mismatch is significant ($\Delta r = 0.08$ Å), but the solubility of CoS (the NiAs structure) in ZnS is remarkably less, about 30 mol % at 850°C, although the sizes of Zn^{2+} and Co^{2+} are practically alike in octahedral as well as tetrahedral positions. The explanation of these facts consists of the difference in octahedral site preference energies for Mn and Co in sulfides: -5 and -20 kJ, respectively (Urusov, 1977). Therefore, it is more difficult to embed Co in a tetrahedral crystal matrix than Mn. It is also known that there are unstable red modifications of MnS with sphalerite and wurtzite structures, but any analogous modifications of CoS do not exist.

The effect of temperature on solid solubility is unambiguous: due to the mixing entropy contribution, the higher the temperature the more extensive the solubility. Only a few examples of a slight temperature influence on solid solution boundaries are known. One of the most important is FeS solubility in sphalerite ZnS, because this fact forms the basis for well-known sphalerite geo- and cosmobarometer.

As for the pressure influence on solid solubility, the situation is simple only when components have different structures. Then the pressure increase expands the stability field of a denser structure, usually a structure with larger coordination numbers. And the opposite is true: the close-packed component is pushed out of the loose matrix. It is the case for Fe in sphalerite; the terminal solubility of FeS in ZnS at 500°C falls from about 50 mol % at 1 bar to nearly 10% at 2 GPa due to a volume gain of about 6 cm^3/mol FeS.

When components are isostructural, the prediction of pressure influence is somewhat complicated. At first glance, the problem reduces to the determination of the sign of deviations from Retger's rule, Eq. (51). If the deviations are positive (positive mixing volume ΔV_m), the pressure increase corresponds to a solubility decrease, and vice versa. This is true, but in the author's opinion, another factor is most important. It is the dependence of the interaction parameter Q on pressure. It was demonstrated (Fancher and Barsch, 1971) by detailed calculations for alkali halide solid solutions that they have to decompose under high pressure in good agreement with experimental evidence.

Indeed, as follows from the previous discussion, Q is proportional to V/β, which is growing when pressure builds up. The calculations of V/β values at various pressures and corresponding changes of Q were first undertaken by the author (Urusov, 1977) and later refined (Kirkinsky and Fursenko, 1980) by accounting for the different compressibility of end-members. The definite result of these calculations is a strong increase of interaction parameter Q, as well as the critical solution temperature T_{cr} at a pressure rise. This conclusion is consistent with most available experimental data. For instance, the predicted increase of T_{cr} for the NaCl–KCl miscibility gap is equal to about 200°C at 20 kbar, being very close to the experimental estimates (Bhardway and Roy, 1971). The following mineral systems also reveal the effect of solid solubility depression under high pressure: $NaAlSi_3O_8$–$KAlSi_3O_8$, Mg_2SiO_4–$CaMgSiO_4$, $Mg_2Si_2O_6$–$CaMgSi_2O_6$ (Urusov, 1977).

It is probable that one has to take into account the above-mentioned effect if one intends to predict the mixing properties of a solid solution at high pressures. For instance, let us compare the size parameters δ and the values of V/β (kJ) for the series of $MgSiO_3$–$FeSiO_3$ solid solutions of various structures: pyroxene ($\delta = 0.016$; $V/\beta = 3350$), garnet (0.010; 4400), ilmenite (0.006; 5530), perovskite (0.013; 6040). One can see that, despite decreasing δ in the direction from low- to high-pressure phases, the product $(V/\beta)\delta^2$ and, hence, the interaction parameter Q are likely to be largest for the densest polymorphs of the perovskite structure.

It means that, as a rule, high pressure could suppress the stability field of a solid solution. However, this assumption needs further refinement.

Concluding Remarks

The author well understands that this chapter severely suffers from incompleteness; in particular, this contribution deals only with an isovalent type of solid solution. There is no discussion of the crystal chemistry and energetics of heterovalent solutions, which, in fact, represent many varieties. Our knowledge of this field is limited, but some useful semiquantative formulations have been established by Urusov (1977 and elsewhere). The complications concerning the multicomponent and multisite solid solutions are also amenable to similar theoretical treatment. The simplest case of isovalent and isostructural solid solution was chosen as the primary theme in this chapter to demonstrate the fundamentals of such an approach. The author believes that in the near future there will be a rapid development of the crystal energetic theory of solid solutions.

Acknowledgments

The author thanks I.P. Deineko for great help in the preparation of this manuscript, V.P. Volkov, A.S. Marfunin, and N.R. Khisina took part in numerous useful discussions.

References

Ahtee, M. (1969). Lattice constants of some binary alkali halide solid solutions. *Ann. Acad. Sci. Fenn.* **AVI**, N313, 1–14.

Ahtee, M. and Inkinen, O. (1970). Critical solution temperatures of binary alkali halide solid solutions. *Ann. Acad. Sci. Fenn.* **AVI**, N355, 1–14.

Akaogi, M., Ito, E., and Navrotsky, A. (1989). Olivine-modified spinel-spinel transitions in the system Mg_2SiO_4–Fe_2SiO_4: Calorimetric measurements, thermo-chemical calculations and geophisical application. *J. Geophys. Res.* **94**, 15671–15685.

Anderson, O.L. and Nafe, J.E. (1965). The bulk modulus-volume relationship for oxide compounds and related geophysical problems. *J. Geophys. Res.* **70**, 3951–3953.

Anderson, D.L. and Anderson, O.L. (1970). The bulk modulus-volume relationship for oxides. *J. Geophys. Res.* **75**, 3494–3500.

Bandyopadhyay, S. and Deb, S.K. (1988). Divalent defects in alkali halides. *Ind. J. Phys.* **A62**, 298–302.

Barrett, W.T. and Wallace, W.E. (1954). Studies of NaCl–KCl solid solutions. I. Heats of formation, lattice spacings, Schotky defects and mutual solubility. *J. Amer. Chem. Soc.* **76**, 366–380.

Bhardway, M.C. and Roy, R. (1971). Effect of high pressure on crystal solubility in the system NaCl–KCl. *J. Phys. Chem. Solid* **32**, 1603–1607.

Born, M. (1923). Atomtheorie des festen Zustandes. Berlin.

Bowman, R.C. (1973). Theoretical solution enthalpies of divalent impurities in the alkali halides. *J. Chem. Phys.* **59**, 2215–2223.

Boyce, J.B. and Mikkelsen, J.C. (1985). Local structure of ionic solid solutions: Extended X-ray absorption fine-structure study. *Phys. Rev.* **B31**, 6903–6905.

Callahan, S.J.E. and Smith, N.O. (1966). Crystallographic studies of NH_4Cl-NH_4Br solid solutions. *Adv. X-ray Anal.* **9**, 156–169.

Davies, P.K. and Navrotsky, A. (1983). Quantitative correlations of deviations from ideality in binary and pseudobinary solid solutions. *J. Solid State Chem.* **46**, 1–22.

Dollase, W.A. (1980). Optimum distance model of relaxation around substitutional defects. *Phys. Chem. Miner.* **6**, 295–304.

Durham, G.S. and Hawkins, J.A. (1951). Solid solutions of the alkali halides. II. The theoretical calculation of lattice constants, heats of mixing and distributions between solid and aqueous phases. *J. Chem. Phys.* **19**, 149–156.

Fancher, D.L. and Barsch, G.R. (1969). Lattice theory of alkali halide solid solutions. I. Heat of formation. *J. Phys. Chem. Solids* **30**, 2503–2510.

Fancher, D.L. and Barsch, G.R. (1971). Lattice theory of alkali halide solid solutions. III. Pressure dependence of solid solubility and spinodal decomposition. *J. Phys. Chem. Solids* **32**, 1303–1313.

Fedders, P.A. and Muller, M.V. (1984). Mixing enthalpy and composition fluctuation in ternary III–V semiconductor alloys. *J. Phys. Chem. Solids* **45**, 685–688.

Ferreira, L.G., Mbaye, A.A., and Zunger, A. (1988). Chemical and elastic effects on isostructural phase diagrams: The ε-G approach. *Phys. Rev.* **B37**, 10547–10570.

Grimm, H.G. and Herzfeld, K.F. (1923). Uber Gitterenergie und Gitterabstand von Mischkristallen. *Z. Physik*, **16**, 77–83.

Hietala, J. (1963a). Alkali halide solid solutions. II. Heat of formation of the sodium chloride type. *Ann. Acad. Sci. Fenn.* **AVI**, N122, 1–31.

Hietala, J. (1963b). Alkali halide solid solutions. III. Heat of formation of the cesium chloride type. **AVI**, N123, 1–18.

Hovi, V. (1950). On Wasastjerna theory of the heat of formation of solid solutions. *Soc. Sci. Fenn., Comment. Phys.-Math.* **XV**, N12, 1–20.

Iveronova, V.I. (1954). Deformations du reseau cristallique dans les solutions solides. *Trudy Inst. Krist. Akad. Nauk SSSR* **10**, 339–347.

Kirkinsky, V.A. and Fursenko, B.A. (1980). Estimation of thermodynamic functions of minerals of variable composition at high pressure using equations of state for end members. *Phys. Earth Planet Inter.* **22**, 262–266.

Kleppa, O.I. and Meschel, S.V. (1965). Heats of formation of solid solutions in the system (Na–Ag)Cl and (Na–Ag)Br. *J. Phys. Chem.* **69**, 3531–3536.

Kravchuk, I.F. and Urusov, V.S. (1978). Theoretical calculations of partition coefficients of isovalent impurities with due regard for excess entropy of mixing in the solid solution and the melt. *Krist. und Techn.* **13**, 1195–1202.

Lister, M.V. and Meyers, N.F. (1958). Heats of formation of some solid solutions of alkali halides. *J. Phys. Chem.* **62**, 145–150.

Martins, J.L. and Zunger, A. (1984). Bond length around isovalent impurities in semiconductor solid solutions. *Phys. Rev.* **B30**, 6217–6220.

Mikkelsen, J.C. and Boyce, J.B. (1983). Extended X-ray absorption fine-structure study of $Ga_{1-x}In_xAs$ random solid solutions. *Phys. Rev.* **B28**, 7130–7140.

Mott, N.F. and Littleton, M.J. (1938). Conduction in polar crystals. I. Electrolytic conduction in solid salts. *Trans. Farad. Soc.* **34**, part 3, 485–495.

Navrotsky, A. (1982). Trends and systematics in mineral thermodynamics. *Ber. Bunsenges. Phys. Chem.* **86**, 994–1001.

Navrotsky, A. (1985). Crystal chemical constraints on thermochemistry of minerals. *Rev. Mineral.* **13**, 225–276.

Navrotsky, A. (1987). Models of crystalline solutions. *Rev. Mineral.* **17**, 35–67.

Nickels, I.E., Fineman, M.A., and Wallace, W.E. (1949). X-ray diffraction studies of sodium chloride—sodium bromide solid solutions. *J. Phys. Colloid. Chem.* **53**, 625–630.

Onuma, N., Higuchi, H., Wakita, H., and Nagasawa, H. (1968). Trace element partition between two pyroxenes and the host lava. *Earth Planet. Sci. Lett.* **5**, 47–51.

Pauling, L. (1960). *The Nature of the Chemical Bond*. 3rd ed., Cornell Univ. Press, Ithaca, N.Y.

Phillips, J.C. (1973). *Bonds and Bands in Semiconductors*, Academic Press, New York.

Sasaki, T., Onda, T., Ito, R., and Ogasawa, N. (1986). An extended X-ray absorption fine-structure study of bond lengths in $GaAs_{1-x}P_x$. *Jap. J. Appl. Phys.* **25**, 231–233.

Saxena, S.K. (1973). *Thermodynamics of Rock-Forming Crystalline Solutions*. Springer-Verlag, Heidelberg-New York.

Shih, C.K., Spicer, W.E., Harrison, W.A., and Sher, A. (1985). Bond-length relaxation in pseudobinary alloys. *Phys. Rev.* **B31**, 1139–1144.

Slagle, O.D. and McKinstry, H.A. (1966). The lattice parameters in the solid solution KCl–KBr. *Acta Cryst.* **21**, 1013–1015.

Srivastava, G.P. and Weaire, D. (1987). The theory of cohesive energy of solids. *Adv. Phys.* **36**, 463–517.

Stoneham, A.M. (1975). *Theory of Defects in Solids*. Clarendon Press, Oxford.

Stringfellow, G.B. (1973). Calculation of regular solution interaction parameters in semiconductor solid solution. *J. Phys. Chem. Solids* **34**, 1749–1751.

Suh, K.S. and Talwar, D.N. (1989). Bond length relaxation and thermodynamic parameter in $A_{1-x}B_xC$ alloy semiconductors. *Cryst. Latt. Def. Amorph. Matt.* **18**, 503–509.

Talwar, D.N., Suh, K.S., and Ting, C.S. (1987). Lattice distortions associated with isovalent defects in semiconductors. *Phil. Mag.* **B56**, 593–609.

Urusov, V.S. (1968a). Effect of size difference on limits of isovalent substitutions. *Geochimiya* **N9**, 1033–1043.

Urusov, V.S. (1968b). Covalency effects in heats of formation of inorganic solid solutions. *Dokl. Akad. Nauk SSSR* **181**, 1185–1188.

Urusov, V.S. (1969a). Application of Wasastjerna–Hovi theory for partly covalent solid solutions. *Zh. Fiz. Chim.* **43**, 537–540.

Urusov, V.S. (1969b). Asymmetry of solid solution boundaries as a consequence of structure and bonding type difference of end members. *Zh. Fiz. Chim.* **43**, 3030–3033.

Urusov, V.S. (1970). Calculations of miscibility gaps of isovalent solid solutions. *Izv. Akad. Nauk SSSR, Neorg. Mater.* **6**, 1209–1214.

Urusov, V.S. (1974). Energetic theory of miscribility gaps in mineral solid solutions. *Fortschr. Miner.* **52**, 141–150.

Urusov, V.S. (1975). *Energetic Crystal Chemistry*. Nauka, Moscow (in Russian).

Urusov, V.S. (1977). *Theory of Isomorphous Miscibility*. Nauka, Moscow (in Russian).

Urusov, V.S. and Kravchuk, I.F. (1976). Energetic analysis and calculations of partition coefficients of isovalent impurities during crystallization of melt. *Geochimiya* **N8**, 1204–1223.

Urusov, V.S. and Kravchuk, I.F. (1978). Effect of microimpurity catching by crystal lattice defects and its geochemical role. *Geochimiya* **N7**, 963–978.

Urusov, V.S. (1980). Energetic formulation of the problem of equilibrium cocrystallization from water solution. *Geochimiya* **N5**, 627–644.

Urusov, V.S. and Kravchuk, I.F. (1983). Vibrational entropy of substitutional solid solutions. *Cryst. Res. Technol.* **18**, 629–636.

Urusov, V.S., Dudnikova, V.B., and Garanin, A.V. (1980). Calculations of solution energies of alkaline-earth ions in alkali halide crystals. *Phys. Stat. Sol.* **102**b, 695–703.

Vernon, S.P. and Stearns, M.B. (1984). Extended X-ray absorption fine-structure study of Y^{3+} and Sr^{2+} impurities in CaF_2. *Phys. Rev.* **B29**, 6968–6971.

Wasastjerna, J.A. (1945). Some experimental values of the atomic scattering factor. Thermal vibrations and lattice distortions in pure and mixed crystals. *Soc. Sci. Fenn., Comment. Phys.-Math.* **XIII**, N5, 1–24.

Wasastjerna, J.A. (1949). On the theory of the heat of formation of solid solutions. *Soc. Sci. Fenn., Comment. Phys.-Math.* **XV**, N3, 1–30.

Wood, B.J. and Kleppa, O.J. (1981). Thermochemistry of forsterite-fayalite olivine solutions. *Geochim. Cosmochim. Acta* **45**, 529–534.

Zen, E-an. (1956). Validity of "Vegard's law." *Amer. Miner.* **41**, 523–525.

Chapter 7
A Structure Energy Model for C2/c Pyroxenes in the System Na–Mg–Ca–Mn–Fe–Al–Cr–Ti–Si–O

G. Ottonello, A. Della Giusta, A. Dal Negro, and F. Baccarin

Introduction

Pyroxenes, like feldspars, occupy a position that is "chemically central to the composition realm of rocks" (Robinson, 1980) and occur ubiquitously in most part of igneous and metamorphic terrains. Understanding their crystal-chemical and thermodynamic properties is thus of primary importance in the earth sciences. Due to their importance in earth sciences, pyroxenes have been the object of various petrologic and thermodynamic investigations. Most published experimental data concern the pyroxene quadrilateral and have been restricted to the binary joins. The work of modeling multicomponent pyroxene mixtures using the binary solution data has begun only recently, and we wish to contribute to its development by presenting a series of structure-energy calculation procedures for the various phases of interest. This first work concerns the structural class C2/c that is the most representative of pyroxenes in nature and for which most crystal-chemical and thermodynamic data are available.

Modeling the C2/c Structure

Although pyroxenes exhibit a wide compositional variability, the bulk of their composition is constituted by magnesian, ferroan, and calcic components ($Mg_2Si_2O_6$–$Fe_2Si_2O_6$–$Ca_2Si_2O_6$). As the maximum molar miscibility of the calcic component (wollastonitic molecule) with the other two is 50%, the pyroxene composition in the compositional space $(Mg,Fe,Ca)_2Si_2O_6$ is often recast in terms of the four end-members: $Mg_2Si_2O_6$, $CaMgSi_2O_6$, $CaFeSi_2O_6$, $Fe_2Si_2O_6$. These are not real end-members in the Gibbs space and their adoption, although useful from a petrologic point of view, can be a source of ambiguity in the thermodynamic reduction of experimental data. Three main phases form in the pyroxene quadrilateral: the monoclinic form C2/c (clinopyroxene, cpx), the orthorombic form Pbca (orthopyroxene, opx), and the structural class P2$_1$/c (pigeonite, clinohypersthene). Although other crystalline phases may form in the (Mg,Fe,Ca)$_2$Si$_2$O$_6$ compositional space, as Pbcn (Smith, 1959) or P2$_1$ca (Steele, 1975), these can be regarded as symmetry subgroups of either C2/c or Pbca (Cameron and Papike, 1980) and their occurrence is limited by unusual f_{O_2}, P, T, X conditions of stability. The presence of sodic terms in a mixture as NaAlSi$_2$O$_6$ (jadeite), NaFeSi$_2$O$_6$ (acmite), and NaCrSi$_2$O$_6$ (ureyite) results in the further complexity of the system at a low temperature ($T \leqslant 860°C$) with the appearance of P2/n omphacites (Holland, 1983).

The C2/c cpx structure is characterized by single tetrahedral chains that form layers parallel to (100) and octahedral layers containing cations in coordination 6-8 with oxygen. The six-fold coordinated sites (M1) are quite regular, whereas the eight-fold coordinated sites (M2) exhibit important distortion. In Fig. 1 we see the structure of C2/c pyroxene projected onto the (100) plane. Figure 1 differs

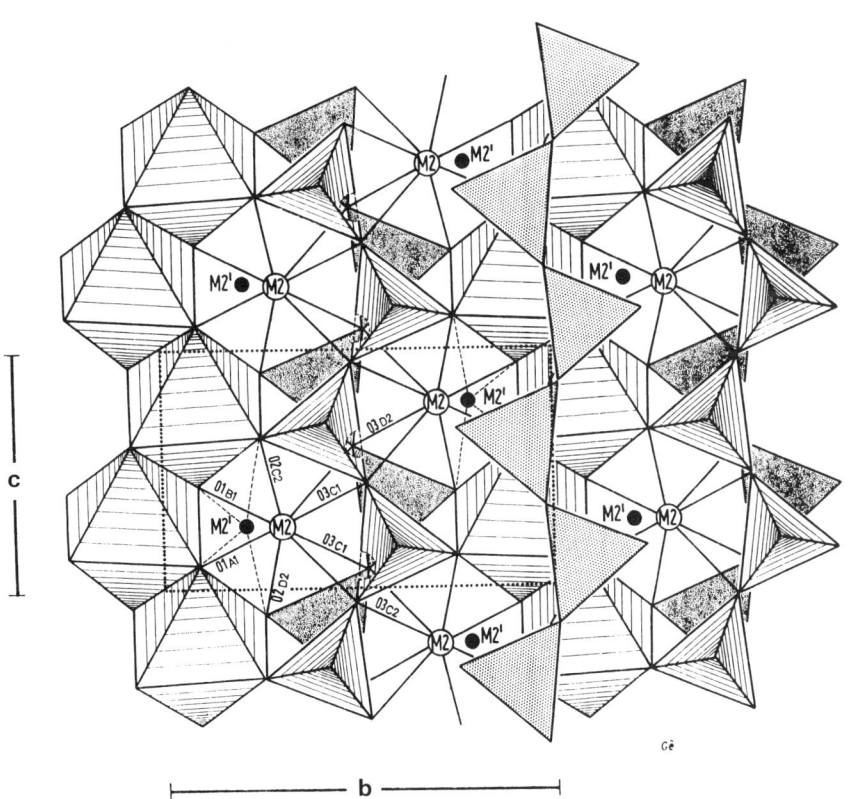

Fig. 1. The crystal structure of C2/c pyroxene projected onto the (100) plane. The geometry of the new M2' site as determined by Dal Negro et al. (1982) has been added to the diagram.

from the original drawing of Cameron and Papike (1980) by the identification of a nonequivalent M2′ site, made by Dal Negro et al. (1982); the atom nomenclature is after Burnham et al. (1967). Although the M1 polyhedron is quite regular, a notable variation is shown by the M1–O2 bond distances (2.021–2.051 Å), which reflect the occupancy by ions smaller than Mg and Fe^{2+}, i.e., Ti^{4+}, Fe^{3+}, Al^{3+}, and Cr^{3+}. The occupancy by the latter ions produces a strong angular distortion of the polyhedron. The crystal-chemical characteristics of the M2 polyhedron largely depend on the Ca occupancy, but in a nonlinear and rather complex fashion (Dal Negro et al., 1982); moreover, the reconaissance of an M2′ site, nonequivalent to M2, by Fourier difference synthesis over a large number of natural samples adds further complexity to the M2 geometry (Dal Negro et al., 1982).

The detailed crystallographic studies existing on a large number of natural and synthetic samples (Clark et al., 1969; Bruno et al., 1982; Carbonin et al., 1991; and references therein) made it possible to treat all nonequivalent interionic distances in the asymmetric unit of C2/c pyroxenes in the same manner as proposed by Ottonello et al. (1989) and Della Giusta et al. (1990) for Pbnm silicate olivines. This leads to simple proportionality expressions involving atomic proportions on sites

$$D_{\text{calc } j} = \sum_s \sum_i \omega_{ij,s} X_{i,s} + \xi_i, \tag{1}$$

where j refers to nonequivalent distances in the asymmetric unit, $D_{\text{calc } j}$ is the calculated jth distance $X_{i,s}$ is the atomic fraction of the ith species on site s, $\omega_{ij,s}$ are adjustable parameters for each ion on each site, and ξ_J is the fitting constant for the jth distance. The treatment involved minimization of the squared reduced residuals

$$\sum \chi_{j,l}^2 = \sum \left[\frac{(D_{\text{calc } j,l} - D_{\text{obs } j,l})}{D_{\text{obs } j,l}} \right]^2, \tag{2}$$

where $D_{\text{obs } j,l}$ is the experimentally "observed" interionic distance conducted over $l = 209$ samples. The complete set of structural coefficients is shown in Table 1.

Thirty independent interatomic distances were considered plus cell edge, cell volume, and mean polyhedral distances. Table 2 compares the simulation of cell edges and mean polyhedral distances for the 209 natural and synthetic pyroxenes. The error in the simulation is in most cases restricted to a few units on the third decimal place in an angstrom scale. Although the degree of accuracy achieved is sufficient for most crystal-chemical purposes, it is still low for structure-energy calculations for the estimation of heat of mixing in chemically complex phases. Interaction energies are, in fact, a small fraction of the bulk energy of the phase (typically less than one per mil) and extreme accuracy is needed in reproducing the actual interionic distances if one wishes sound results in terms of energy. The best way to refine the structure simulation is the distance least-square method (DLS). The DLS procedure is an iteration technique that adjusts the positional parameters of the ions and the lattice parameters in

Energy Model for C2/c Pyroxenes

Table 1. Coefficients to be used in Eq. (1) to calculate interatomic distances, cell edges, β, and volume for C2/c clinopyroxenes.

	T:Si	Al	Ti	M1:Mg	Fe^{2+}	Fe^{3+}	Al	Ti	Cr	Mn^{2+}	M2:Ca	Na	Mg	Fe^{2+}	Mn^{2+}	cost.
T-O1	0.685	0.729	0.772	0.013	0.014	0.052	0.061	0.098	0.053	0.016	0.049	0.035	0.062	0.072	0.046	0.170
T-O2	0.679	0.717	0.782	0.064	0.065	0.110	0.105	0.163	0.103	0.072	0.044	0.007	0.043	0.056	0.057	0.120
T-O3A1	0.709	0.715	0.764	0.023	0.021	0.041	0.032	0.073	0.045	0.041	0.060	0.007	0.026	0.036	0.004	0.169
T-O3A2	0.734	0.731	0.847	−0.011	−0.009	0.020	0.008	0.093	0.018	−0.001	0.071	0.000	0.050	0.020	0.020	0.160
O1A1-O2A1	0.923	0.993	1.100	0.353	0.343	0.411	0.447	0.464	0.413	0.344	0.258	0.202	0.255	0.275	0.261	0.277
O1A1-O3A1	0.735	0.779	0.878	0.079	0.084	0.129	0.117	0.189	0.129	0.111	0.767	0.678	0.724	0.728	0.694	0.366
O2A1-O3A1	0.755	0.791	0.907	0.555	0.553	0.616	0.608	0.706	0.625	0.569	0.288	0.212	0.260	0.271	0.253	0.310
O1A1-O3A2	0.864	0.897	1.015	0.203	0.208	0.260	0.240	0.391	0.242	0.227	0.578	0.482	0.553	0.535	0.473	0.182
O2A1-O3A2	0.773	0.803	0.915	0.213	0.217	0.254	0.250	0.310	0.258	0.226	0.572	0.539	0.578	0.567	0.591	0.243
O3A1-O3A2	1.282	1.289	1.373	−0.029	−0.025	0.025	−0.017	0.141	0.018	−0.001	−0.047	−0.097	−0.035	−0.036	−0.063	0.160
M1-O1A2	0.657	0.620	0.723	0.557	0.633	0.581	0.481	0.569	0.558	0.660	0.047	−0.005	0.020	0.030	−0.023	0.142
M1-O1A1	0.664	0.668	0.689	0.474	0.510	0.473	0.364	0.459	0.401	0.581	0.116	0.101	0.145	0.175	0.157	0.206
M1-O2	0.700	0.663	0.701	0.400	0.427	0.338	0.254	0.233	0.341	0.478	0.096	0.041	0.065	0.088	0.093	0.162
O1A1-O1B1	0.793	0.780	0.873	0.034	0.045	0.040	−0.026	0.034	0.021	0.090	0.671	0.669	0.696	0.749	0.692	0.498
O1B1-O1A2	0.005	−0.010	0.044	4.004	4.148	3.906	3.713	3.714	3.870	4.242	−1.483	−1.552	−1.530	−1.460	−1.567	0.282
O1A1-O1A2	1.445	1.456	1.547	−0.005	0.023	0.008	−0.069	0.094	−0.007	0.092	0.028	−0.054	0.017	0.032	−0.032	0.145
O1A1-O2C1	1.414	1.377	1.411	0.008	0.081	−0.086	−0.235	−0.276	−0.141	0.196	0.048	−0.022	0.010	0.086	0.034	0.143
O1B2-O2C1	1.068	0.906	1.223	0.101	0.188	0.183	0.045	0.183	0.126	0.219	0.597	0.502	0.459	0.436	0.369	0.140
O1A2-O2C1	0.753	0.793	0.666	2.431	2.479	2.364	2.218	2.242	2.330	2.536	−0.671	−0.663	−0.560	−0.520	−0.433	−0.379
O2C1-O2D1	1.302	1.268	1.344	0.116	0.124	0.147	−0.001	0.232	0.096	0.186	0.099	0.025	0.110	0.060	0.163	0.171
M2-O1	1.140	1.082	1.320	−0.025	−0.022	0.095	0.069	0.245	0.088	0.014	−0.013	−0.110	−0.108	−0.254	−0.251	0.124
M2-O2	0.672	0.551	1.002	0.782	0.797	0.981	0.985	1.265	0.968	0.789	−0.049	−0.187	−0.248	−0.370	−0.476	0.267
M2-O3C1	0.968	1.025	0.963	0.119	0.167	0.030	−0.042	−0.222	0.015	0.187	0.323	0.282	0.357	0.554	0.564	0.187
M2-O3C2	0.850	0.779	0.865	0.290	0.294	0.271	0.187	0.218	0.202	0.323	0.532	0.662	0.642	0.740	0.772	0.199
O1A1-O2C2	1.594	1.485	1.850	−0.037	−0.064	0.130	0.147	0.354	0.107	−0.069	−0.089	−0.154	−0.199	−0.242	−0.341	0.072
O1A1-O3C2	1.535	1.403	1.697	0.281	0.297	0.372	0.307	0.385	0.312	0.315	0.110	0.142	0.083	0.064	0.090	0.165
O2C2-O3C2	1.892	1.727	2.200	−2.622	−2.638	−2.404	−2.425	−2.036	−2.452	−2.632	1.748	1.753	1.712	1.618	1.550	1.225
O2C2-O3D2	1.522	1.396	1.831	0.111	0.135	0.257	0.201	0.437	0.211	0.169	0.079	−0.035	−0.077	−0.150	−0.173	0.176
O3C1-O3D2	0.785	0.722	0.902	0.386	0.423	0.389	0.274	0.363	0.315	0.511	0.658	0.640	0.645	0.648	0.681	0.304
O3C1-O3D1	1.517	1.491	1.808	0.070	0.140	0.161	0.102	0.202	0.151	0.195	0.118	−0.157	−0.111	−0.094	−0.192	0.155
a	3.990	3.967	4.410	1.396	1.491	1.486	1.264	1.528	1.385	1.640	0.154	−0.041	0.065	0.105	0.033	0.221
b	3.343	3.290	3.586	3.379	3.476	3.356	3.141	3.151	3.274	3.613	−0.713	−0.827	−0.769	−0.700	−0.673	−0.428
c	1.465	1.460	1.672	1.586	1.588	1.708	1.629	1.975	1.689	1.628	0.576	0.491	0.590	0.559	0.516	0.164
β	30.42	31.19	28.73	25.97	24.94	25.09	25.39	24.68	25.15	25.23	15.29	17.67	17.50	17.70	19.30	3.79
Volume	101.76	95.80	153.85	266.47	277.89	281.21	253.97	296.37	270.94	294.57	1.89	−24.50	−9.19	−6.57	−14.99	−31.86

Table 2. Discrepancies in calculated and observed cell parameters and mean polyhedral distances for 209 C2/c pyroxenes. The listed numbers are frequencies within each calc-obs interval (Å).

$D_{calc}-D_{obs}$	a	b	c	$\langle T-O \rangle$	$\langle M1-O \rangle$	$\langle M2-O \rangle$
0.000–0.003	102	112	106	200	196	180
0.003–0.006	51	68	61	9	13	29
0.006–0.009	35	18	29	0	0	0
0.009–0.012	15	6	9	0	0	0
0.012–0.015	5	2	3	0	0	0
0.015–0.020	1	3	0	0	0	0
0.020–0.030	0	0	1	0	0	0

Table 3. DLS optimization of a chemically complex C2/c pyroxene after four iteration cycles.

Composition	M2	$(Ca_{0.762}, Na_{0.096}, Fe_{0.024}, Mg_{0.111}, Mn_{0.007})$			
	M1	$(Mg_{0.781}, Fe_{0.053}, Al_{0.116}, Cr_{0.037}, Ti_{0.013})$			
	T	$(Si_{1.917}, Al_{0.083})$			

Atom 1	Atom 2	D_{calc}	D_{obs}	$D_o - D_c$	Weight
T	O1A1	1.6131	1.6131	0.0000	0.9793
T	O2A1	1.5929	1.5929	0.0000	0.9793
T	O3A1	1.6626	1.6626	0.0000	0.9793
T	O3A2	1.6814	1.6814	0.0000	0.9793
O1A1	O2A1	2.7473	2.7473	0.0000	0.1660
O1A1	O3A1	2.6796	2.6796	0.0000	0.1660
O2A1	O3A1	2.6663	2.6663	0.0000	0.1660
O1A1	O3A2	2.6895	2.6895	0.0000	0.1660
O2A1	O3A2	2.5804	2.5804	0.0000	0.1660
O3A1	O3A2	2.6506	2.6506	0.0000	0.1660
M1	O1A2	2.0439	2.0439	0.0000	0.3629
M1	O1A1*	2.1139	2.1138	−0.0001	0.3629
M1	O2C1	2.0256	2.0256	0.0000	0.3629
O1A1*	O1B1	2.7856	2.7856	0.0000	0.1660
O1B1	O1A2	2.7654	2.7654	0.0000	0.1660
O1A1*	O1A2	3.0446	3.0446	0.0000	0.1660
O1A1*	O2C1	2.9795	2.9796	0.0001	0.1660
O1B2	O2C1	2.9305	2.9305	0.0000	0.1660
O1A2	O2C1	2.8807	2.8807	0.0000	0.1660
O2C1	O2D1	2.9686	2.9686	0.0000	0.1660
M2	O1A1	2.3524	2.3524	0.0000	0.2537
M2	O2C2	2.3250	2.3250	0.0000	0.2537
M2	O3C1	2.5515	2.5515	0.0000	0.2537
M2	O3C2	2.7293	2.7292	−0.0001	0.2537
O1A1	O2C2	3.1329	3.1329	0.0000	0.1660
O1A1	O3C2	3.6202	3.6202	0.0000	0.1660

Table 3. (Continued.)

Composition	M2 M1 T	$(Ca_{0.762}, Na_{0.096}, Fe_{0.024}, Mg_{0.111}, Mn_{0.007})$ $(Mg_{0.781}, Fe_{0.053}, Al_{0.116}, Cr_{0.037}, Ti_{0.013})$ $(Si_{1.917}, Al_{0.083})$			
Atom 1	Atom 2	D_{calc}	D_{obs}	$D_o - D_c$	Weight
O2C2	O3C2	4.1502	4.1501	−0.0001	0.1660
O2C2	O3D2	3.3839	3.3839	0.0000	0.1660
O3C1	O3D2	2.8951	2.8951	0.0000	0.1660
O3C1	O3D1	3.3281	3.3281	0.0000	0.1660

	A	B	C	β
calc	9.709	8.874	5.263	106.34
obs	9.711(1)	8.874(2)	5.263(1)	106.33(1)

Coordinates		cal			obs	
Atom	X	Y	Z	X	Y	Z
T	0.2879	0.0930	0.2311	0.2878	0.0930	0.2309
M1	0.0	0.9071	0.25	0.0	0.9070	0.25
M2	0.0	0.2999	0.25	0.0	0.2999	0.25
O1A1	0.1149	0.0863	0.1403	0.1148	0.0862	0.1402
O2A1	0.3623	0.2518	0.3205	0.3623	0.2520	0.3204
O3A1	0.3514	0.0179	−0.0026	0.3515	0.0181	−0.0026

the structure until the discrepancy between the predicted distances and the simulated ones is minimum. This method was first developed by Meier and Villiger (1969); we adopt here an updated version, DLS-76 by Baerlocher et al. (1977). The minimized function is expressed by:

$$\sum \chi_j^2 = \sum W_j^2 (D_{calc\,j} - D_{ref\,j})^2, \quad (3)$$

where W_j is the weight assigned to distance j and $D_{ref\,j}$ the refined distance. The summation occurs over all nonequivalent distances in the asymmetric unit. The weight factor is proportional to the Pauling's strength of the bond: the higher the bond strength, the higher the weighting factor. By a trial-and-error procedure, we found that appropriate weighting factors are the following: $W_j = 0.166$ for Na–O and all O–O distances; $W_j = 0.250$ for Ca–O; $W_j = 0.333$ for Mg–O, Fe^{2+}–O, Mn–O; $W_j = 0.5$ for Al–O, Fe^{3+}–O, Cr–O; $W_j = 0.666$ for Ti–O and $W_j = 1.0$ for Si–O.

In Table 3 we see the results of the DLS procedure on a chemically complex C2/c pyroxene. In this case, convergency was achieved in four steps. Although the refined distances do not differ by more than 0.0001 Å from the calculated ones, the refinement modifies cell edges on the third decimal place. The resulting atom positional parameters, which are the first essential step in building the interionic potential model, are strictly representative of C2/c geometry.

From Structure to Static Bulk Lattice Energy

The static lattice energy $U(P, T)$ of a crystalline solid can be considered to consist of three main terms: Coulombic (E_C), repulsive (E_R), and dispersive (E_D). A crystal field stabilization energy term (E_{CFS}) can be evaluated eventually, but its contribution to the bulk static energy is rather limited for the phases of interest here

$$U_{(P,T)} = E_C + E_R + E_D + E_{CFS}. \tag{4}$$

The Coulombic, repulsive, and dispersive terms of energy are the results of pair interactions among all atoms in the crystal, i.e.,

$$E_C = \sum_{i=1}^{n} \sum_{j=i}^{n} Z_i Z_j C_{ij}^c, \tag{5}$$

$$E_R = \sum_{i=1}^{n} \sum_{j=i}^{n} b_{ij} C_{ij}^r, \tag{6}$$

$$E_D = \sum_{i=1}^{n} \sum_{j=i}^{n} dd_{ij} C_{ij}^{dd} + \sum_{i=1}^{n} \sum_{j=i}^{n} dq_{ij} C_{ij}^{dq}, \tag{7}$$

where, Z_i, Z_j are the formal charges of the i, j ions; b_{ij}, dd_{ij}, dq_{ij} are the repulsive and dispersive (dipole–dipole and dipole–quadrupole) coefficients between the i, j ions. C_{ij}^c, C_{ij}^r, C_{ij}^{dd}, and C_{ij}^{dq} are structure coefficients that depend on atomic fractional coordinates and cell constants.

As is explicit in Eq. (4), the bulk lattice energy of a crystalline substance is a function of the intensive variables P, T. Although there are no experimental methods that give exact measurement of the lattice energy, we know from theory that at zero point the bulk lattice energy of the substance is related to its enthalpy of formation from the elements by the Born–Haber–Fayans thermochemical cycle; i.e., for compound $Mg_2Si_2O_6$

$$\begin{array}{ccc}
Mg_2Si_2O_6 & \xleftarrow{U_{(0,0)}} & 2\,Mg^{2+}_{(g)} \quad 2\,Si^{4+}_{(g)} \quad 6\,O^{2-}_{(g)} \\
\uparrow \bar{H}^0_{f(0,0)} & & \uparrow 2\,I_{Mg} \quad \uparrow 2\,I_{Si} \quad \uparrow 6\,E_{a_o} \\
2\,Mg_{(s)} + 2\,Si_{(s)} + 3\,O_{2(g)} & \xrightarrow{2E_{s,Mg} + 2E_{s,Si} + 3E_{d,O_2}} & 2\,Mg_{(g)} + 2\,Si_{(g)} + 6\,O_{(g)}
\end{array} \tag{8}$$

where I_{Mg} is the ionization energy for the metal Mg to its formal valency state in the crystal (2+), E_{a_o} is the electron affinity for oxygen, E_s are the sublimation energies of metals, and $E_{d_{O_2}}$ is the dissociation energy of molecular oxygen.

There is a main difficulty in applying the Hess additivity rule to the energy balance (8), namely, that most of the quantities involved are not specifically available at zero-point conditions. Hence, the Born–Haber–Fayans cycle is commonly referred to with 298.15 K, $P = 1$ bar conditions, is neglected, and the effect of P change the thermal expansion of gaseous (ideal) ions from 0 to

298.15 K is accounted for by the equation-of-state term nRT

$$U_{(298.15\,K,\,1\,bar)} = \sum_M (E_{s_M} + I_M) + 3\,E_{d_{O_2}} + 6\,E_{a_O} + 10\,RT - \overline{H}^\circ_{f(298.15\,K,\,1\,bar)}. \tag{9}$$

The lattice energy at this point is no longer simply static but includes a vibrational term.

Let us now consider the system constituted by Eqs. (4–7) and 9; we have four structural coefficients ($C_{ij}^c, C_{ij}^r, C_{ij}^{dd}, C_{ij}^{dq}$) that describe the dependency of energy terms on atom spacings in the crystal. An accurate simulation of *all* interionic distances in the asymmetric unit through type-(1) equations, followed by DLS treatment of the resulting values leads, as we have seen, to a structural simulation whose accuracy is virtually unaffected by the chemical complexity of the sample. The resulting atomic fractional coordinates and cell parameters are then reconverted to C_{ij} structural factors by standard crystallographic routines (Catti, 1981a,b). Among the energy coefficients in Eqs. (5–7) only b_{ij} terms can be parameterized as all the others are fixed by theory; i.e., Z_i, Z_j are "formal" ionic charges and the dispersive coefficients are derived from the atomistic properties of the intervening ions

$$dd_{ij} = \frac{3}{2}\alpha_i\alpha_j \frac{\overline{E}_i\overline{E}_j}{\overline{E}_i + \overline{E}_j}, \tag{10}$$

$$dq_{ij} = \frac{27}{8} \frac{\alpha_i\alpha_j}{e^2(\overline{E}_i + \overline{E}_j)}\left(\frac{\alpha_i\overline{E}_i}{n_i} + \frac{\alpha_j\overline{E}_j}{n_j}\right), \tag{11}$$

where

α_i, α_j = free ion polarizabilities for i, j ions

$\overline{E}_i, \overline{E}_j$ = mean excitation energy

n_i, n_j = effective electrons

$$n_i = \sqrt{n'_i n''_i} \tag{12}$$

$$n''_i = \frac{\alpha_i \overline{E}_i^2 4\pi m}{e^2 h^2} \tag{13}$$

with

n'_i, n'_j = "outer" electrons for i, j ions

e = electronic charge

m = rest mass of electron

h = Planck's constant

(see Ottonello, 1987, and Ottonello et al., 1989, 1990, for an extended discussion of the energy terms involved and the resulting interionic coefficients). The b_{ij}

repulsive factor is given by

$$b_{ij} = b\beta_{ij} \exp\left(\frac{r_i + r_j}{\rho}\right), \tag{14}$$

where

r_i, r_j = "repulsive" radii for i, j ions

ρ = the hardness factor of the substance

b = its repulsive factor

β_{ij} = the Pauling number (Heitler and London, 1927; Pauling, 1960)

$$\beta_{ij} = 1 + \frac{Z_i}{n'_i} + \frac{Z_j}{n'_j}. \tag{15}$$

In the Born–Mayer approach, b has a constant value in the family of salts and p is variable from salt to salt, whereas in the Huggins–Mayer treatment, b is variable from salt to salt and p is constant (Tosi, 1964).

Adopting the generalized Huggins–Mayer approach b can be regarded as an arbitrary constant since the single ion repulsive contributions

$$b_i = b^{1/2} \exp\left(\frac{r_i}{\rho}\right) \tag{16}$$

become characteristic parameters of ions (Tosi and Fumi, 1964). Stemming from Eq. (16), the evaluation of b_{ij} repulsive coefficients can be readily generalized for chemically complex mixtures as shown by Ottonello (1987). The generalization implies that the b factor of common ions be linearly related to the molar proportions of end-member components in the mixture, i.e., for a binary $(A_x B_{1-x})(A_y B_{1-y})Si_2 O_6$ clinopyroxene

$$b_{A,M1} = b^{1/2}_{A_2 Si_2 O_6} \exp\left(\frac{r_{A,M1}}{\rho}\right), \tag{17}$$

$$b_{A,M2} = b^{1/2}_{A_2 Si_2 O_6} \exp\left(\frac{r_{A,M2}}{\rho}\right), \tag{18}$$

$$b_{Si} = (x+y) b_{Si, A_2 Si_2 O_6} + (2 - x - y) b_{Si, B_2 Si_2 O_6}, \tag{19}$$

$$b_O = (x+y) b_{O, A_2 Si_2 O_6} + (2 - x - y) b_{O, B_2 Si_2 O_6}. \tag{20}$$

The b_{ij} factors for nonequivalent sites will then be dependent on site population

$$b_{M1-M2} = xy b_{A,M1-A,M2} + x(1-y) b_{A,M1-BM2}$$
$$+ (1-x)y b_{B,M1-A,M2} + (1-x)(1-y) b_{B,M1-B,M2}. \tag{21}$$

Although in the cpx structure, there are 10 b_{ij} to be evaluated (b_{M1-M1}; b_{M1-M2}; b_{M1-Si}; b_{M1-O}; b_{M2-M2}; b_{M2-Si}; b_{M2-O}; b_{Si-Si}; b_{Si-O}; b_{O-O}), the real variables can be much less. For instance, in the cpx quadrilateral, if we assume the repulsive radii of ions to be coincident with ionic radii, this will reduce the number of variables to four ($b_{Mg_2Si_2O_6}$; $b_{CaMgSi_2O_6}$; $b_{CaFeSi_2O_6}$; $b_{Fe_2Si_2O_6}$). Obviously, as we know the enthalpy of the four end-members at 298.15 K, 1 bar and we have four b factors to evaluate, the system can be solved for whatever hardness factor is in the family of salts. However, we wish for an energy model that predicts interaction energies in chemically complex mixtures. Therefore, we are forced to a nonlinear minimization procedure where the input data are *all* the existing energy data in chemically complex cpx and the solving set of factors is the *minimum* number of parameters necessary to achieve a *sufficiently* precise energy simulation.

From Gibbs Free Energy of Mixtures to Bulk Lattice Energy and Vice-Versa

In the "mixture" notation, i.e., "solid phases containing more than one substance when all substances are treated in the same way" (see IUPAC, 1979), the standard state is identical for all components in the mixture and is normally that one of "pure component at P, T of interest." The molar Gibbs free energy of a mixture is moreover constituted by three terms

$$\overline{G}_{mix} = \sum_i \mu_i^0 X_i + \overline{G}_{mix\,id} + \overline{G}_{mix\,exc}, \qquad (22)$$

where

X_i = molar fraction of component i

μ_i^0 = standard-state chemical potential of component i

$G_{mix\,id}$ = ideal molar Gibbs free energy of mixing

$G_{mix\,exc}$ = excess molar Gibbs free energy of mixing.

The ideal mixing contribution to the Gibbs free energy of the mixture is given by the permutations at maximum disorder. Through the Stirling's approximation, we have

$$\overline{G}_{mix\,id} = RT \sum_S \sum_M X_{M,S} \ln X_{M,S}. \qquad (23)$$

$X_{M,S}$ terms in (23) are atomic fractions of the various metals on sites where mixing takes place (i.e., M1, M2, T for C2/c pyroxenes). The excess molar Gibbs free energy of mixing is usually expressed in terms of macroscopic interaction parameters, i.e., for a binary A-B mixture

$$\overline{G}_{mix\,exc} = X_A X_B W_{(P,T)}. \qquad (24)$$

The Margules parameter $W_{(P,T)}$ in a regular mixture varies with P, T conditions and it can easily be shown by partial derivation over the two intensive variables that it is constituted by three contributions: enthalpic, entropic, and volumetric (analogous expressions can be derived for subregular excess terms)

$$W_{(P,T)} = W_H - TW_{S_{(0,T)}} + \int_0^P W_{V_{(P,0)}} dP. \quad (25)$$

As in our notation,

$$\mu_i^0 = \overline{G}_i^0 = \overline{H}_{i(P,T)}^0 - T\overline{S}_{i(P,T)}^0 \quad (26)$$

if we consider simply the enthalpy of a given binary mixture at P, T conditions, we will have

$$\overline{H}_{\text{mix}(P,T)} = \sum_i \overline{H}_i^0 X_i + X_i(1-X_i) W_H. \quad (27)$$

Moreover, for all components in the mixture

$$\overline{H}_{i(P,T)} = \overline{H}_{P_r,T_r}^0 + \int_{T_r}^T Cp/dT + \int_{P_r}^P \left[\overline{V}_i - T\left(\frac{\partial V_i}{\partial T}\right)\right] dP. \quad (28)$$

Let us now consider solely the effect of T, i.e.,

$$\overline{H}_{i(1,T)} = \overline{H}_{i(1,T_r)}^0 + \int_{T_r}^T Cp_i dT. \quad (29)$$

For all components, the specific heat at constant P is expressed normally as a polynomial Maier–Kelley-type function of T

$$Cp_i = A_i + B_i T + C_i T^{-2} + \cdots. \quad (30)$$

Combining 27, 29, and 30, we get

$$\overline{H}_{\text{mix}_{P_r,T_r}}^0 = \sum_i \overline{H}_{i_{P_r,T_r}}^0 X_i + X_i(1-X_i) W_H$$

$$= \overline{H}_{\text{mix}_{P_r,T}} - \int_{T_r}^T \left(\sum_i A_i X_i + T\sum_i B_i X_i + T^{-2}\sum_i C_i X_i + \cdots\right) dT. \quad (31)$$

We must then introduce the following considerations: (1) Based on (25), the interaction parameter W_H represents the virtual macroscopic interaction of the substance at zero-point conditions (actually, in most cases, W_H is no more than a fitting parameter relating the measured excess energy terms at various P, T conditions to composition).

(2) The internal disorder (long-range plus short-range terms) for all components of interest here does not vary with temperature, thus, their Cp functions (30) *do not carry any configurational information.*

(3) Short-range ordering is implicitly excluded by interionic potential models, which are based on *fixed* atomic positional parameters, valid in the whole crystal.

With the above provisos, $\overline{H}_{\text{mix},P_r,T_r}^0$ is the enthalpy of the mixture at P_r, T_r and *at the same state of configurational disorder observed at P_r, T.*

From Eq. (31) or its analogous forms, we can reduce the experimental information on the high P, T enthalpy of the various binary cpx mixtures existing in the literature to data that which can be reconverted to lattice energy through the thermochemical cycle (9). Once the repulsive energy is parameterized with a nonlinear minimization of the obtained U values, the reversed operation will lead to $\bar{H}_{\text{mix}, P, T}$ estimates in the whole compositional range of interest.

We wish to stress here that there is no empiricism involved in the reduction of experimental data. Both the internal energy U and enthalpy H are, in fact, state functions and their relations are commonly described by Legendre transforms involving the state parameters. The most familiar of the Legendre transforms relating enthalpy to internal energy is

$$dH = dU + P\,dV + V\,dP. \quad (32)$$

Relation (32) is, however, not particularly interesting to us as it relates U and H through volume and pressure variations. For our purposes, it is better to express both U and H as functions of the intensive variables P, T

$$dU = \left(\frac{\partial U}{\partial T}\right)_P dT + \left(\frac{\partial U}{\partial P}\right)_T dP, \quad (33)$$

$$dH = \left(\frac{\partial H}{\partial T}\right)_P dT + \left(\frac{\partial H}{\partial P}\right)_T dP, \quad (34)$$

$$\left(\frac{\partial U}{\partial T}\right)_P = T\left(\frac{\partial S}{\partial T}\right)_P - P\left(\frac{\partial V}{\partial T}\right)_P = Cp - PV\alpha, \quad (35)$$

$$\left(\frac{\partial U}{\partial P}\right)_T = \left(\frac{\partial U}{\partial V}\right)_T \left(\frac{\partial V}{\partial P}\right)_T = \left[T\left(\frac{\partial P}{\partial T}\right)_V - P\right]\left(\frac{\partial V}{\partial P}\right)_T$$

$$= \beta V \left[P - T\left(\frac{\partial P}{\partial T}\right)_V\right], \quad (36)$$

$$\left(\frac{\partial H}{\partial T}\right)_P = Cp, \quad (37)$$

$$\left(\frac{\partial H}{\partial P}\right)_T = V - T\left(\frac{\partial V}{\partial T}\right)_P = V - TV\alpha, \quad (38)$$

with

$\alpha =$ isobaric thermal expansion coefficient

$\beta =$ isothermal compressibility coefficient.

Let us now consider the partial derivatives (35) and (37). Cp is mainly vibrational and, if we consider substances that do not change their configurational disorder with T, then Cp is simply and solely vibrational. The $PV\alpha$ contribution in (35) reflects the anharmonicity of the vibrational terms, so $(\partial U/\partial T)_P$ is intrinsically vibrational and cannot in any case be evaluated by a static potential model (this fact that seems trivial is often underestimated in interionic potential calculations).

The "trick" involved in relation (31) (or analogous forms) is simply the choice of an appropriate reference temperature *at which* to carry the interionic (static) potential calculation and *from which* to start evaluating the vibrational contribution. Although from a strict theoretical point of view, this temperature is 0 K, three practical considerations suggest the adoption of 298.15 K:

1. Most Cp functions are known and tabulated starting from 298.15 K.
2. Iron-bearing compounds have complex electronic transitions at low T (i.e., below 298.15 K), reflected in complex λ-type Cp functions, which are not readily translated in chemically complex systems.
3. The Born–Haber–Fayans energy terms are not specifically known at 0 K.
4. The PV term is embodied in the Born–Haber–Fayans cycle.

The uncertainty involved in neglecting the effect of α from 0 to 298.15 K for the various compositions is energetically insignificant.

End-Member Energy Data

In Table 4 enthalpy and lattice energy constituents for various end-member components of interest in the C2/c pyroxene family are listed. As can be seen, the bias in the $\bar{H}^0_{T_r,P_r}$ values from the various sources is around 10 kJ/mol for end-members in the cpx quadrilateral and even larger for Ca-Tschermak (~ 26 kJ/mol) and jadeiitic (~ 20 kJ/mol) components. The energies for the last two components in the table were estimated in this work in a manner that will be outlined later on and should be considered preliminary. Also clinoferrosilite ($Fe_2Si_2O_6$, C2/c) was estimated in this work, but the proposed value is strictly consistent with the choice of H values for $Mg_2Si_2O_6$, $CaMgSi_2O_6$, and $CaFeSi_2O_6$ and with the mixture notation. Let us, in fact, consider the formation of the mixture $Ca_{2/3}Fe_{2/3}Mg_{2/3}Si_2O_6$ from end-member components at the opposite sides of the cpx quadrilateral

$$\frac{2}{3}CaMgSi_2O_6 + \frac{1}{3}Fe_2Si_2O_6 \rightleftarrows Ca_{2/3}Mg_{2/3}Fe_{2/3}Si_2O_6, \quad (39)$$

$$\frac{2}{3}CaFeSi_2O_6 + \frac{1}{3}Mg_2Si_2O_6 \rightleftarrows Ca_{2/3}Mg_{2/3}Fe_{2/3}Si_2O_6. \quad (40)$$

If the mixture is ideal, then by definition, the enthalpy change involved in (39) and (40) is also zero; hence,

$$\frac{1}{3}\bar{H}^0_{Fe_2Si_2O_6, T_r, P_r} = \frac{1}{3}\bar{H}^0_{Mg_2Si_2O_6, T_r, P_r} + \frac{2}{3}\bar{H}^0_{CaFeSi_2O_6, T_r, P_r}$$

$$- \frac{2}{3}\bar{H}^0_{CaMgSi_2O_6, T_r, P_r}. \quad (41)$$

Table 4. Energy values for C2/c pyroxene. E_{BH} = Born–Haber–Fayans thermochemical cycle. All values are in kJ/mol. Asterisks mark adopted values.

Component	E_{BH}	$\bar{H}^0_{T_r,P_r}$	U	E_C	E_{DD}	E_{DQ}	E_R
$Mg_2Si_2O_6$	30543.6	−3095.5[a]	−33639.1	−39215.7	−398.0	−122.5	6097.1
		−3094.2[i]	−33637.8				6098.4
		−3088.0[b]	−33631.6				6104.6
$Fe_2Si_2O_6$	31355.9	−2387.2[c]	−33743.1	−39254.6	−590.0	−203.8	6305.4
		−2361.9[d]*	−33717.8				6330.7
$CaMgSi_2O_6$	30119.6	−3210.8[a]	−33330.4	−38878.6	−437.7	−132.6	6118.5
		−3205.5[e,f]	−33325.1				6123.8
		−3203.2[g]	−33322.8				6126.1
		−3201.8[h,i]*	−33321.4				6127.5
$CaFeSi_2O_6$	30525.8	−2849.2[j]	−33375.0	−38766.0	−533.9	−147.5	6099.3
		−2838.2[g,h]*	−33364.6				6109.7
$CaAl_2Si_2O_6$	28301.7	−3306.0[n]	−31604.4	−37152.9	−467.4	−141.9	6157.8
		−3280.2[a,g]*	−31578.6				6183.6
$NaAlSi_2O_6$	31928.5	−3030.7[c]	−35280.3	−41115.0	−459.0	−145.5	6439.2
		−3029.9[k]*	−35279.5				6440.0
		−3029.4[a]	−35279.0				6440.5
		−3023.8[i]	−35272.9				6446.1
		−3011.5[g]	−35261.1				6458.4
		−3010.8[f]	−35260.4				6459.0
$NaFeSi_2O_6$	32174.8	−2593.7[l]*	−34768.5	−40588.7	−468.3	−150.2	6438.6
$CaMnSi_2O_6$	30304.2	−2947.2[m]*	−33251.4	−38472.7	−466.5	−144.8	5832.7
$NaCrSi_2O_6$	32108.1	−2900.7[d]*	−35008.7	−40728.4	−470.4	−150.2	6340.2
$CaTiAl_2O_6$	27244.1	−3553.1[d]*	−30797.2	−36066.6	−490.1	−153.7	5913.1

[a] Robie et al. (1978).
[b] Saxena and Chatterjee (1986).
[c] Saxena et al. (1986).
[d] Present work.
[e] Wagman et al. (1981).
[f] Naumov et al. (1971).
[g] Helgeson et al. (1978).
[h] Saxena (1990).
[i] Berman (1988).
[j] Stull and Prophet (1971).
[k] Hemingway et al. (1981).
[l] Calculated from the enthalpy of formation from oxides reported in Viellard (1982).
[m] Newton and McCready (1948).
[n] Chatterjee (1989).

Table 5. Entropy at $T_r = 298.15$ K, $P_r = 1$ bar and coefficients of the C_p function for C2/c pyroxene end-members. Limits of validity of the C_p functions are also given. $C_p = A + BT + CT^{-2} + ET^{-3} + FT^{-1/2} + GT^{-1}$. Data are in J/mol × K.

Compound	$\bar{S}^0_{T_r,P_r}$	$A \times 10^{-2}$	$B \times 10^3$	$C \times 10^{-6}$	$E \times 10^{-8}$	F	$G \times 10^{-4}$	T_{lim}
$Mg_2Si_2O_6$	134.85	2.889	3.764	−2.70	9.224	0	−3.876	1830[a]
$Fe_2Si_2O_6$	188.84	3.0514	6.524	−2.088	9.224	0	−3.876	—[a,b,c]
$CaMgSi_2O_6$	143.093	2.2121	32.80	−6.586	0	0	0	1664[a,d]
$CaFeSi_2O_6$	170.289	2.2933	34.18	−6.28	0	0	0	—[a,e]
$CaAl_2SiO_6$	156.0	2.324	21.845	−7.2788	0	0	0	1600[d]
$NaAlSi_2O_6$	134.72	2.6566	13.86	3.294	−4.930	0	0	—[a,d]
$NaFeSi_2O_6$	153.55	1.9313	80.793	−3.9581	0	0	0	950[f]
		2.1932	41.882	−3.2133	0	0	0	950–1050[f]
		2.1033	45.564	−3.2133	0	0	0	1050–1400[f]
$CaMnSi_2O_6$	173.31	2.2667	27.93	−7.1065	0	−104.59	0	1664[g]
$NaCrSi_2O_6$	149.64	2.4649	18.249	3.5215	−4.93	492.317	0	—[h]
$CaTiAl_2O_6$	175.80	2.5088	−2.839	−7.2633	0	−5.616	0	844[i]
		2.3655	24.884	−8.2651	0	−5.616	0	844–1800[i]

[a] Saxena (1990).
[b] Saxena and Chatterjee (1986).
[c] Entropy value calculated to be consistent with energy and interactions in the cpx quadrilateral (see text).
[d] Robie et al. (1978).
[e] Saxena et al. (1986).
[f] Helgeson et al. (1978).
[g] Calculated with the exchange reaction (42).
[h] Calculated with the exchange reaction (43).
[i] Calculated with the exchange reaction (44).

If $Ca_{2/3}Fe_{2/3}Mg_{2/3}Si_2O_6$ is not ideal, then as by definition, the excess enthalpy of a mixture is the difference in the enthalpy of the actual substance and an ideal substance of the same composition; the two excess enthalpy terms in (39) and (40) will cancel out by subtraction so that (41) will still be valid in the actual case (note that this is true for all P, T conditions).

To each value of $\bar{H}^0_{T_r,P_r}$ corresponds (through the Born–Haber–Fayans cycle) a discrete value of the bulk lattice energy U. As we see in Table 4, this energy is mainly constituted by the coulombic and short-range repulsive terms, whereas the dispersive terms are greatly subordinate. As the lattice energy is about 10 times larger in magnitude than enthalpy, the uncertainty in the adoption of reference $\bar{H}^0_{T_r,P_r}$ values for the various end-members does not appreciably affect the model, provided the choice remains consistent throughout calculations.

In Table 5 we see the $\bar{S}^0_{T_r,P_r}$ and Cp polynomials adopted in the model. As for enthalpy, the $\bar{S}^0_{T_r,P_r}$ and Cp of clinoferrosilite were calculated to be internally consistent with the other end-members in the cpx quadrilateral. $\bar{S}^0_{T_r,P_r}$ and Cp polynomials for the components $CaMnSi_2O_6$, $NaCrSi_2O_6$, and $CaTiSi_2O_6$ were calculated with the algorithms proposed by Helgeson et al. (1978) [their Eqs. (62), and (80)] based on the exchange reactions

$$CaMnSi_2O_6 + MgO \rightleftarrows CaMgSi_2O_6 + MnO, \tag{42}$$

Table 6. Volume of C2/c pyroxene end-members and volume coefficients adopted in the model. $\bar{V}^0_{T_r,P_r}$ is in cm³/mol; α_0 in K^{-1}; α_2 in K, and $k0$ in Mbar.

Compound	$\bar{V}^0_{T_r,P_r}$	$\alpha_0 \times 10^4$	$\alpha_1 \times 10^8$	α_2	$k0$	$k1$	Ref.
$Mg_2Si_2O_6$	64.353	0.20303	1.28049	0.001	1.1	4.2	a,b,c
$Fe_2Si_2O_6$	66.439	0.2216	0.0	0.0	1.0	4.2	a,d,e
$CaMgSi_2O_6$	66.234	0.18314	1.41273	0.0011	1.0	4.2	a,b
$CaFeSi_2O_6$	67.976	0.18314	1.41273	0.0011	1.0	4.2	a,i
$CaAl_2Si_2O_6$	63.390	0.247	0.0	0.0	1.2	4.2	a,f,b
$NaAlSi_2O_6$	60.367	0.13096	2.23308	0.17442	1.33	4.2	a,d,g,b
$NaFeSi_2O_6$	64.446	0.13096	2.23308	0.17442	1.33	4.2	a,h
$CaMnSi_2O_6$	70.525	0.18314	1.41273	0.0011	1.0	4.2	a,i
$NaCrSi_2O_6$	62.931	0.13096	2.23308	0.17442	1.33	4.2	a,h
$CaTiAl_2O_6$	67.626	0.247	0.0	0.0	1.2	4.2	a,j

[a] Present work.
[b] Saxena and Eriksson (1983).
[c] Saxena (1990).
[d] Skinner (1966).
[e] Akimoto (1972).
[f] Haselton et al. (1984).
[g] Birch (1966).
[h] Volume coefficients assumed to be identical to jadeitite.
[i] Volume coefficients assumed to be identical to diopside.
[j] Volume coefficients assumed to be identical to Tschermak's molecule.

$$NaCrSi_2O_6 + \frac{1}{2}Al_2O_3 \rightleftarrows NaAlSi_2O_6 + \frac{1}{2}Cr_2O_3, \qquad (43)$$

$$CaTiAl_2O_6 + SiO_2 \rightleftarrows CaAl_2SiO_6 + TiO_2 \qquad (44)$$

[note that the values adopted for $NaFeSi_2O_6$ are those derived by Helgeson et al., 1978 with the same method and the same reference component as in (43)].

The molar volumes and volume coefficients adopted in the model for pure components are reported in Table 6. The molar volumes are those obtained by the structural calculations previously outlined. They differ slightly from the values reported in the literature which are based mostly on synthetic materials. The observed differences are unimportant in model calculations for mixtures, provided that the structural model is capable of describing deviations from ideal mixing as far as the volume terms are concerned (this implication will be outlined later). The volume coefficients for thermal expansion and compressibility are essentially those adopted by Saxena (1990).

Structure and Energy of C2/c Phases in the Pyroxene Quadrilateral

The system $Mg_2Si_2O_6$–$CaMgSi_2O_6$–$CaFeSi_2O_6$–$Fe_2Si_2O_6$ has been studied by various authors, either for its structural or energetic aspects. These studies have been complicated by the fact that various phases form in the system (C2/c, Pbca, $P2_1/c$, $P2_1ca$, Pbcn) and each phase is stable only within a limited compositional range.

As far as the structural class C2/c is concerned, the components $Mg_2Si_2O_6$ (clinoenstatite) and $Fe_2Si_2O_6$ (clinoferrosilite) are merely fictive as they never form in nature. Attempts to define the structure of $Mg_2Si_2O_6$ and $Fe_2Si_2O_6$ at room conditions are generally based on energy arguments and not crystallographic considerations. The adopted energy models consider the heterogeneous equilibria

$$Mg_2Si_2O_6 \rightleftarrows Mg_2Si_2O_6, \qquad (45)$$
$$\text{Pbca} \qquad \text{C2/c}$$

$$Fe_2Si_2O_6 \rightleftarrows Fe_2Si_2O_6, \qquad (46)$$
$$\text{Pbca} \qquad \text{C2/c}$$

and by applying the concept of equality of the chemical potential of the components in the two phases (C2/c and Pbca) at equilibrium to experimentally observed miscibility gaps, parameterize the ΔH, ΔS, and ΔV terms of the heterogeneous reactions (45) and (46) (see, e.g., Lindsley et al., 1981; Lindsley, 1981; Saxena, 1981; Holland et al., 1979; Nickel and Brey, 1984). The limits of the models in predicting the volume are due to the large dependence of ΔV on the interaction parameters adopted to describe mixing within each class of substances. Estimates of ΔV for reaction (45) range thus from 0 (Holland et al., 1979) to 0.355 (Lindsley et al., 1981), 0.6188 (Nickel and Brey, 1984), 0.9754 (Davidson and Lindsley, 1989), and 2.0 cm^3/mol (Saxena, 1981).

The structural predictions of our model for the four limiting binary joins in the quadrilateral are compared in Figs. 2 a through d with experimental evidence and theoretical estimates of the C2/c molar volumes. In comparing the results, one should keep in mind that the differences in molar volumes bear the maximum bias as three distinct distances and one angle are involved in the calculation. Discrepancies in the observed and calculated interionic distances are generally lower by a factor of 10 (see, e.g., Table 3). In Fig. 2(a) we see that there is good agreement between model predictions and experimental observations on the join $Mg_2Si_2O_6$–$CaMgSi_2O_6$ in the range $0.6 < X(CaMgSi_2O_6) < 1.0$. Notably, the model reproduces fairly well the convex upward inflection observed by Clark et al. (1962). For $X(CaMgSi_2O_6) < 0.6$, the model predicts molar volumes substantially higher than those observed by Newton et al. (1979). However, the two extreme compositions of Newton et al. (1979) [i.e., $X(CaMgSi_2O_6) = 0.2$ and 0.33, respectively] should be discarded as their diffraction patterns exhibit some complexities not in line with a simple C2/c symmetry (see Newton et al., 1979). Moreover, the model estimate for pure C2/c $Mg_2Si_2O_6$ falls within the volumetric range obtained by various authors from energy considerations.

For the join $CaMgSi_2O_6$–$CaFeSi_2O_6$, the model predictions are very close to the experimental observations of Turnock et al. (1973), with the exception of the small inflection observed by these authors for the single composition $X(CaFeSi_2O_6) = 0.6$, which is not reproduced at all [see Fig 2(b)]. The model volumes along the join $Fe_2Si_2O_6$–$CaFeSi_2O_6$ are slightly larger than the values obtained by Lindsley et al. (1969) and in closer agreement with the results of Davidson and Lindsley (1989) and Syono et al. (1971) for C2/c $Fe_2Si_2O_6$ [Fig. 2(c)]. Along the join $Mg_2Si_2O_6$–$Fe_2Si_2O_6$, there is no experimental information and the model predicts a slight positive excess volume of mixing, which has, as we will see later, some significant consequences on the intrinsic stability of C2/c Ca-poor phases at high P. Due to the impressive experimental work of Lindsley and his coworkers, the miscibility of C2/c and Pbca substances along the joins $Mg_2Si_2O_6$–$CaMgSi_2O_6$ and $Fe_2Si_2O_6$–$CaFeSi_2O_6$ is well known at various P, T conditions (Lindsley, 1981; Lindsley et al., 1981; Davidson and Lindsley, 1989; and the references therein). The mixing properties within the C2/c class, based on Lindsley's data, are usually reconverted to a macroscopic subregular Margules-type formulation (see Table 7). For the join $Mg_2Si_2O_6$–$Fe_2Si_2O_6$, there is general consensus about an essentially ideal mixing behavior within the C2/c structural class (Sack, 1980; Saxena, 1983; Saxena et al., 1986) as far as enthalpic and entropic terms are concerned. One can now parameterize the lattice energy of C2/c pyroxenes in the cpx quadrilateral stemming from the energies calculated for the three joins and the four end-members of interest involved in the calculations.

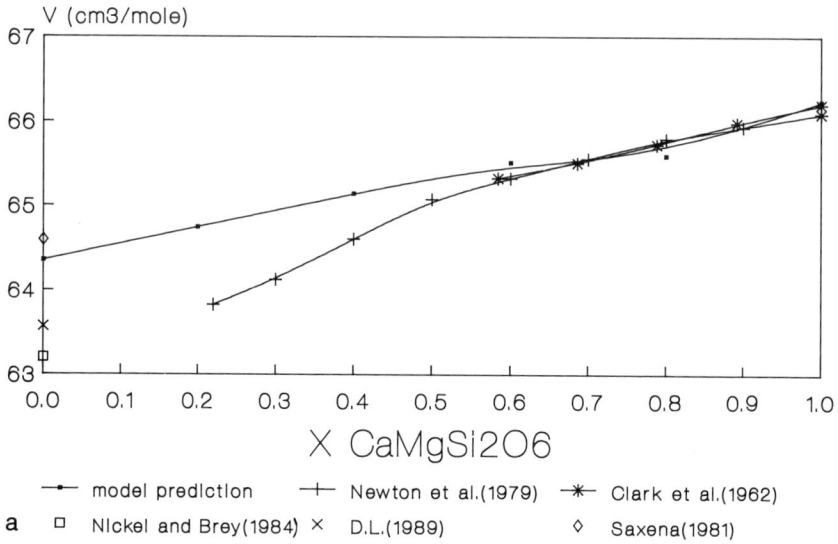

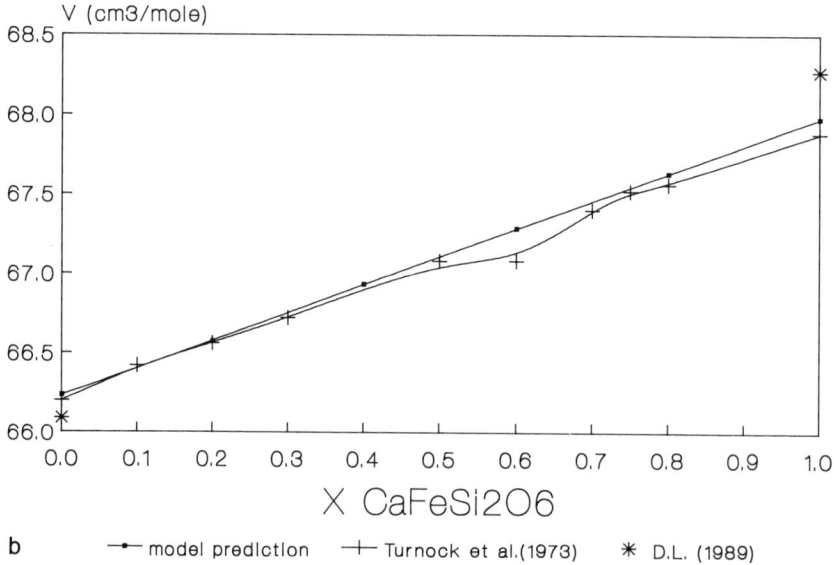

Fig. 2. Structure model predictions (a–d) for the four limiting C2/c binary joins in the pyroxene quadrilateral, compared with experimental values and theoretical estimates. D.L. (1989) = Davidson and Lindsley (1989).

Energy Model for C2/c Pyroxenes

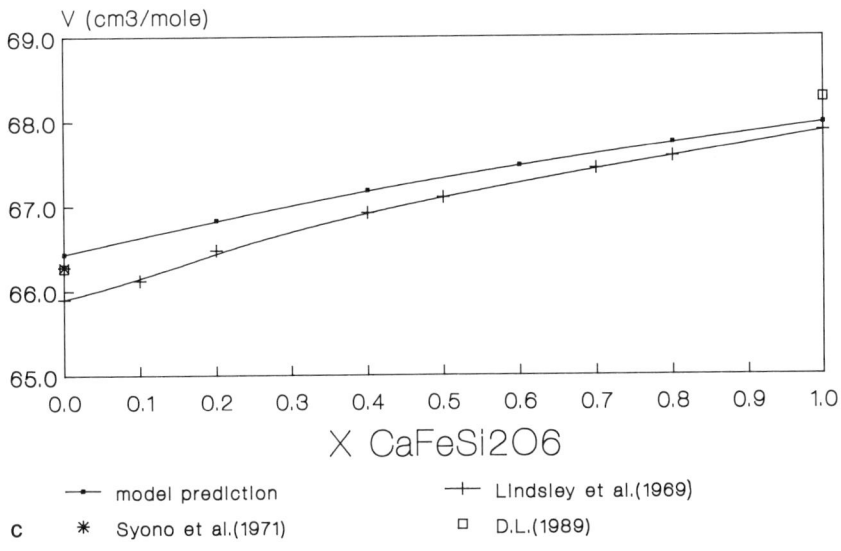

c

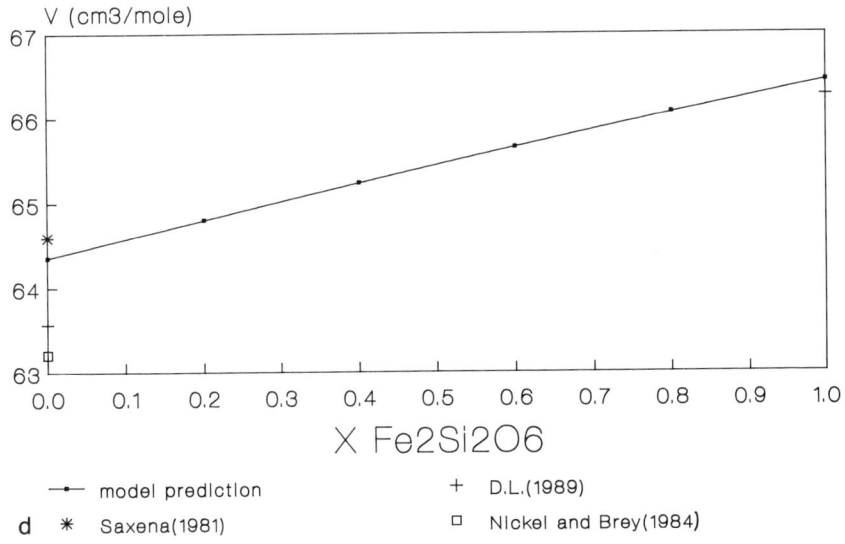

d

Fig. 2. (Continued.)

Table 7. Macroscopic subregular interaction parameters for C2/c pyroxenes. Data are in J/mol (W_H), J/mol × K (W_S), J/mol × bar (W_V). 1 = $Mg_2Si_2O_6$, 2 = $CaMgSi_2O_6$, 3 = $CaFeSi_2O_6$, 4 = $Fe_2Si_2O_6$, 5 = $CaAl_2SiO_6$, 6 = $NaAlSi_2O_6$.

	W_H 12	W_H 21	W_H 13	W_H 31	W_H 14	W_H 41	W_H 23	W_H 32
a								
b	23,663	32,467	52,971	29,085	−960.5	1190.1	13,984	17,958
d	25,484	31,216	50,774	27,841	−1558.6	2518.6		
e			93,300	−20,000	0	0	12,000	12,000

	W_H 24	W_H 42	W_H 34	W_H 43	W_H 25	W_H 52	W_H 26	W_H 62
a	22,590	46,604			25,462	31,693	28,243	44.063
b	17,327	49,128	20,099	16,109	25,397	29,175		
c			20,697	16,941				
e	24,000	15,000						
f						41,214		

	W_S 12	W_S 21	W_S 13	W_S 31	W_S 14	W_S 41	W_S 23	W_S 32
a			15.887	3.538	0.041	−0.081	−0.001	0.004
b	0.016	−0.007	38.907	24.609	24.389	24.651		
d	0.0	0.0						
e			45.0	−28.0	0.0	0.0	0.0	0.0

	W_S 24	W_S 42	W_S 34	W_S 43	W_S 25	W_S 52	W_S 26	W_S 62
a	3.940	14.996	−0.034	0.033	−15.036	−3.879	−0.003	−0.008
b	23.620	41.017	0.0	0.0	9.046	19.027		
c	0.0	0.0						
e					−12.259	30.920		

Table 7. (Continued.)

	$W_V\,12$	$W_V\,21$	$W_V\,13$	$W_V\,31$	$W_V\,14$	$W_V\,41$	$W_V\,23$	$W_V\,32$
a	0.0208	0.0066	0.0416	0.0294	0.0557	−0.0180	0.0068	−0.0084
b	0.0812	−0.0061	−0.0771	0.0510	0.0282	0.0226		
d			0.0	0.0	0.0	0.0		
e							0. 0	0.0

	$W_V\,24$	$W_V\,42$	$W_V\,34$	$W_V\,43$	$W_V\,25$	$W_V\,52$	$W_V\,26$	$W_V\,62$
a	0.0720	0.0190	0.0530	0.0453	0.0108	0.0204	0.0336	0.0301
b	0.0137	−0.004	−0.0023	0.0059	0.0285	−0.0079		
c	0.0	0.0						
e					0.0	0.0		
f								

[a] Structure-energy model, mixing over two sites.
[b] Structure-energy model, mixing on one site.
[c] Lindsley (1981).
[d] Lindsley et al. (1981).
[e] Saxena et al. (1986) (3-oxygen basis).
[f] Ganguly and Saxena (1987).

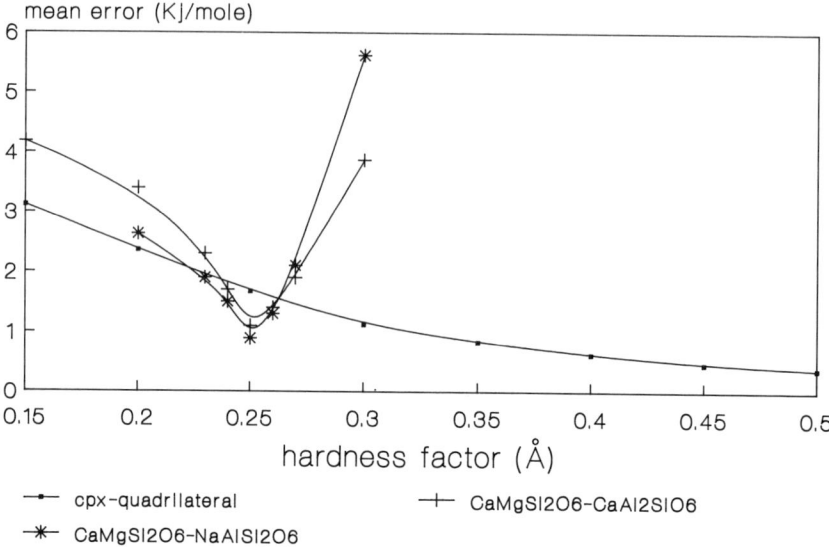

Fig. 3. Fit of the repulsive energy model in the various investigated systems at different conditions. Results are relative to model (2), i.e., $U = E_C + E_R + E_D$.

The parameterization of the lattice energy was performed with the method previously outlined with a nonlinear minimization procedure (James and Roos, 1977) involving 72 distinct compositions in the P, T range of 1 to 20000 bar and 298.15 to 1700 K, respectively. Calculations were done with different values of the hardness parameter for the class of substances (the Huggins–Mayer approach). As shown in Fig. 3, the best results were achieved with a hardness factor of 0.5 Å. The variables involved in the calculations are the repulsive radii of cations on sites M1 and M2 (five values as Ca is assumed to be stable on the M2 site at all P, T, X conditions) and the four repulsive factors b of the end-members. The best fit of the interionic potential calculations is obtained if we consider all energy constituents (i.e., $U = E_C + E_R + E_D + E_{CFS}$; see Table 8). Disregarding the crystal field stabilization energy terms has, however, a very limited effect on the model as the mean error rises only from 268 to 311 J/mol. The following arguments indeed suggest that we disregard the E_{CFS} contribution to the lattice energy of C2/c pyroxenes:

1. No precise CFS estimates are available for the C2/c structure (case 3 in Table 8 was evaluated adopting the canonical CFS term of Dunitz and Orgel, 1957, i.e., 49.7 kJ/mol for Fe^{2+} in both M1 and M2 sites.
2. As the M1 and M2 sites are differently distorted, one would, however, expect slightly different E_{CFS} for M1 and M2 sites; moreover, the different degeneracy states for 3d electrons in the M1 and M2 sites would result in electronic entropy contributions not readily quantifiable in chemically complex mixtures (see Ottonello et al., 1990).

Energy Model for C2/c Pyroxenes

Table 8. Parameters of the repulsive energy obtained by nonlinear minimization of binary-join energetics. ρ, r are in (Å); b factors are in J/mol × 10^{19}.

(a) Optimization on the joins $Mg_2Si_2O_6$–$CaMgSi_2O_6$, $Fe_2Si_2O_6$–$CaFeSi_2O_6$, $Mg_2Si_2O_6$–$Fe_2Si_2O_6$

	$\rho = 0.25$			$\rho = 0.50$	
	$U = E_C + E_R + E_D$	$U = E_C + E_R$	$U = E_C + E_R + E_D$	$U = E_C + E_R + E_D$	$U = E_C + E_R + E_D + E_{CFS}$
b $Mg_2Si_2O_6$	3.23953	3.07047		3.39063	3.39038
b $Fe_2Si_2O_6$	3.117605	2.87426		3.02727	3.08709
b $CaMgSi_2O_6$	3.39273	3.11362		3.34256	3.34270
b $CaFeSi_2O_6$	3.23925	2.92203		3.12258	3.14985
r Mg, M1	0.76363	0.50023		0.54599	0.54735
r Mg, M2	0.78276	0.70205		0.62340	0.62187
r Fe^{2+}, M1	0.87501	0.71456		0.81996	0.81964
r Fe^{2+}, M2	0.78288	0.74346		0.84855	0.83934
r Ca, M2	0.91566	0.81855		0.85120	0.85005
$\Sigma\chi^2$	0.0550	0.0285		0.0020	0.0015
Mean error (kJ/mol)	1.660	1.195		0.311	0.268

(b) Optimization on the join $CaMgSi_2O_6$–$NaAlSi_2O_6$ ($\rho = 0.25$)
$U = E_C + E_R + E_D$

b $NaAlSi_2O_6$	1.80598
b $CaMgSi_2O_6$	2.77368
r Mg, M1	0.5118
r Ca, M2	1.2455
r Na, M2	1.2383
r Al, M1	0.85329
$\Sigma\chi^2$	0.0025
Mean error (kJ/mol)	0.894

(c) Optimization on the join $CaMgSi_2O_6$–$CaAl_2Si_2O_6$ ($\rho = 0.25$)
$U = E_C + E_R + E_D$

b $CaAl_2Si_2O_6$	2.48612
b $CaMgSi_2O_6$	3.18989
r Mg, M1	0.54196
r Ca, M2	1.1475
r Al, M1	0.71426
r Al, T	0.20268
$\Sigma\chi^2$	0.0060
Mean error (kJ/mol)	1.095

Table 8. (Continued.)

(d) Extension of (a) to minor components* ($\rho = 0.50$)
$U = E_C + E_R + E_D$

b $Mg_2Si_2O_6 = 3.39063$
b $CaFeSi_2O_6 = 3.12258$
b $CaTiAl_2O_6 = 2.18952$
b $NaCrSi_2O_6 = 1.63984$
r Fe^{2+}, M1 = 0.81996
r Mg, M2 = 0.62340
r Al, M1 = 0.83292
r Fe^{3+}, M1 = 0.01587

b $Fe_2Si_2O_6 = 3.02727$
b $CaMnSi_2O_6 = 2.56038$
b $NaAlSi_2O_6 = 2.66178$
r Fe^{2+}, M2 = 0.84855
r Ca, M2 = 0.85120
r Al, T = 0.05071
r Cr^{3+}, M1 = 1.34900

b $CaMgSi_2O_6 = 3.34256$
b $CaAl_2Si_2O_6 = 2.41396$
b $NaFeSi_2O_6 = 3.89261$
r Mg, M1 = 0.54599
r Na, M2 = 0.0
r Mn, M1 = 1.12930
r Ti^{4+}, M1 = 0.05143

*Concentration limits of validity: $Mg_2Si_2O_6 = 1.0$, $Fe_2Si_2O_6 = 1.0$, $CaMgSi_2O_6 = 1.0$, $CaFeSi_2O_6 = 1.0$, $NaAlSi_2O_6 = 0.1$, $CaMnSi_2O_6 = 0.1$, $CaAl_2SiO_6 = 0.1$, $CaTiAl_2O_6 = 0.1$, $NaFeSi_2O_6 = 0.05$, $NaCrSi_2O_6 = 0.1$.

3. According to the experimental evidences Fe^{2+}, Mg atoms are randomly distributed over the M1 and M2 sites. In spite of point 2, the CFS energy differences between Fe^{2+} atoms in the M1 and M2 positions would be virtually negligible.

In line with the above arguments, we consider an interionic potential model based on coulombic + repulsive + dispersive terms adequate (at present) at the level of precision offered by the existing experimental data. Static lattice energy terms calculated for various compositions within the cpx quadrilateral are listed in Table 9. All the listed values are relative to the C2/c structure. As we see, the static lattice energy is largely dominated by coulombic interactions, followed by short-range repulsive terms, while the dispersive contributions (E_{DD}, E_{DQ}) are greatly subordinate. The eventual crystal field stabilization terms (when accounted for) would reach at $X = 2.0$, $Y = 0.0$ a maximum value of -99.4 kJ/mol, lower than the lowest E_{DQ} interaction.

$\bar{H}^0_{T_r, P_r}$ in Table 9 is the enthalpy of the various substances derived from U by the Born–Haber cycle (8) at $T_r = 298.15$ K, $P_r = 1$ bar. $\bar{S}^0_{T_r, P_r}$ and $\bar{S}_{T, P_r}$ are the entropies of the substances calculated (for the "mixture" notation), adding to the entropy of components at standard state (the pure component at the T, P of interest) a configurational term arising from the mixing on site of the nonidentical atoms

$$\bar{S}^0_{T_r, P_r} = \sum_i \bar{S}^0_i X_i - R \left(\sum_{M1} X_{M1} \ln X_{M1} + \sum_{M2} X_{M2} \ln X_{M2} \right). \tag{47}$$

As Ca is always stable on the M2 site, this configurational term does not coincide with the maximum disorder within the quadrilateral, but only along the edge $Mg_2Si_2O_6$–$Fe_2Si_2O_6$.

$\bar{G}_{T, P}$ in Table (9) is the Gibbs free energy of the substance at $T = 1773$ K, $P = 20$ Kbar. $\bar{G}_{T, P}$ values were obtained by the model (and analogous values calculated at different T, P conditions) through the relation (22) for the calculation of the Gibbs free energy of mixing terms. These terms were then reconverted to the usual subregular macroscopic formulation and are shown in Table 7.

The macroscopic interaction factors W were calculated with two distinct ideal mixing contributions, i.e.,

$$\bar{G}_{\text{mix id}} = RT \sum_i X_i \ln X_i, \tag{48}$$

$$\bar{G}_{\text{mix id}} = RT \sum_S \sum_M X_{M,S} \ln X_{M,S}. \tag{49}$$

Equation (48) is the usual macroscopic formulation adopted by authors (irregardless of the nature of atoms interacting on sites). Formulation (49) is apparently more appropriate to C2/c pyroxenes that have two energetically distinct sites where mixing takes place (three when Tschermak's terms are involved).

Table 9. Energy values for C2/c pyroxenes in the system $Mg_2Si_2O_6-CaMgSi_2O_6-CaFeSi_2O_6-Fe_2Si_2O_6$. $X = 2Fe/(Mg + Fe)$; $Y = 2Ca/(Ca + Mg + Fe)$. All data are in kJ/mol except $S^0_{T_r,P_r}$ and S^0_{T,P_r} (J/mol K) and $V^0_{T_r,P_r}$ (cm³/mol). $T_r = 298.15$ K; $P_r = 1$ bar; $T = 1773$ K; $P = 20$ Kbar.

X	Y	U	E_C	E_R	E_{DD}	E_{DQ}	$H^0_{T_r,P_r}$	$S^0_{T_r,P_r}$	$V^0_{T_r,P_r}$	H_{T,P_r}	S_{T,P_r}	$\int VdP$	$G_{T,P}$
0.000	1.000	−33,321.4	−38,878.6	6127.5	−437.7	−132.6	−3201.8	143.1	66.234	−2843.8	549.9	137.7	−3681.1
0.200	1.000	−33,324.1	−38,867.0	6129.2	−448.5	−137.9	−3163.9	148.5	66.404	−2804.3	557.1	138.1	−3654.1
0.400	1.000	−33,327.5	−38,855.9	6130.4	−458.9	−143.0	−3126.6	152.7	66.572	−2765.6	563.1	138.4	−3625.7
0.600	1.000	−33,330.8	−38,844.5	6130.7	−469.1	−147.9	−3089.3	156.3	66.761	−2726.8	568.6	138.8	−3596.2
0.800	1.000	−33,334.4	−38,832.9	6130.2	−479.1	−152.6	−3052.3	159.6	66.931	−2688.3	573.6	139.2	−3566.3
1.000	1.000	−33,338.8	−38,821.9	6129.1	−488.8	−157.2	−3016.1	162.5	67.108	−2650.5	578.3	139.5	−3536.5
1.200	1.000	−33,343.6	−38,811.0	6127.2	−498.2	−161.6	−2980.3	165.0	67.284	−2613.3	582.7	139.9	−3506.6
1.400	1.000	−33,348.1	−38,799.4	6124.7	−507.5	−165.9	−2944.1	167.2	67.454	−2575.7	586.7	140.2	−3475.8
1.600	1.000	−33,353.6	−38,788.6	6121.4	−516.5	−170.0	−2909.1	169.0	67.629	−2539.1	590.3	140.6	−3445.3
1.800	1.000	−33,359.1	−38,777.4	6117.5	−525.3	−173.9	−2873.9	170.3	67.806	−2502.5	593.4	141.0	−3413.7
2.000	1.000	−33,364.6	−38,766.0	6113.0	−533.8	−177.7	−2838.8	170.3	67.976	−2465.9	595.3	141.3	−3380.0
0.000	0.800	−33,379.4	−38,939.7	6122.3	−430.8	−131.1	−3175.0	145.5	65.885	−2816.5	552.8	137.0	−3659.7
0.332	0.800	−33,385.5	−38,922.6	6128.8	−451.0	−140.7	−3099.8	155.5	66.191	−2738.4	566.4	137.6	−3605.0
0.668	0.800	−33,393.6	−38,905.6	6132.2	−470.2	−149.7	−3026.8	162.8	66.509	−2662.3	577.3	138.2	−3547.8
1.000	0.800	−33,401.3	−38,888.6	6134.3	−488.8	−158.3	−2953.2	168.8	66.822	−2585.8	587.0	138.7	−3487.8
1.332	0.800	−33,410.7	−38,872.3	6134.5	−506.6	−166.3	−2881.4	173.6	67.129	−2511.0	595.5	139.3	−3427.5
1.668	0.800	−33,421.8	−38,855.8	6131.6	−523.8	−173.8	−2811.3	177.2	67.441	−2437.9	602.7	139.9	−3366.6
2.000	0.800	−33,432.5	−38,839.1	6128.0	−540.4	−181.0	−2740.7	178.2	67.749	−2364.3	607.3	140.4	−3300.6
0.000	0.600	−33,439.0	−39,003.8	6118.0	−423.6	−129.5	−3149.7	145.2	65.506	−2790.8	553.0	136.3	−3635.1
0.286	0.600	−33,445.8	−38,991.8	6127.6	−443.0	−138.6	−3075.4	155.4	65.796	−2713.5	566.9	136.8	−3581.9
0.570	0.600	−33,453.1	−38,980.0	6136.1	−461.9	−147.3	−3001.4	163.1	66.085	−2636.5	578.1	137.2	−3524.4
0.856	0.600	−33,461.1	−38,968.1	6142.8	−480.2	−155.6	−2928.2	169.5	66.368	−2560.3	588.2	137.7	−3465.6
1.144	0.600	−33,470.8	−38,956.9	6147.7	−498.1	−163.5	−2856.7	174.9	66.651	−2485.8	597.3	138.1	−3406.7
1.430	0.600	−33,480.9	−38,946.2	6151.9	−515.4	−171.1	−2785.6	179.4	66.935	−2411.7	605.4	138.6	−3346.5
1.714	0.600	−33,490.9	−38,935.3	6155.1	−532.4	−178.4	−2714.3	182.6	67.212	−2337.5	612.2	139.0	−3284.1
2.000	0.600	−33,501.7	−38,924.3	6157.0	−548.9	−185.4	−2643.9	183.3	67.476	−2264.0	616.5	139.4	−3217.9

Energy Model for C2/c Pyroxenes

Table 9. (Continued.)

0.000	0.400	−33,500.6	−39,071.0	6113.9	−415.9	−127.6	−3126.6	143.5	65.138	−2767.2	551.8	135.6	−3610.1
0.250	0.400	−33,507.9	−39,063.6	6127.0	−434.9	−136.4	−3052.7	154.0	65.405	−2690.3	565.9	136.0	−3557.7
0.500	0.400	−33,515.8	−39,056.7	6139.5	−453.6	−144.9	−2979.4	161.9	65.661	−2614.0	577.4	136.3	−3501.6
0.750	0.400	−33,524.0	−39,050.0	6151.2	−472.0	−153.2	−2906.3	168.6	65.930	−2537.9	587.8	136.7	−3443.5
1.000	0.400	−33,533.0	−39,044.0	6162.2	−490.0	−161.2	−2834.1	174.5	66.186	−2462.8	597.3	137.7	−3384.9
1.250	0.400	−33,541.6	−39,037.4	6172.6	−507.8	−169.0	−2761.4	179.5	66.443	−2387.1	606.0	137.4	−3324.2
1.500	0.400	−33,551.6	−39,032.2	6182.5	−525.3	−176.7	−2690.3	183.6	66.685	−2313.0	613.7	137.7	−3263.5
1.750	0.400	−33,561.3	−39,026.5	6191.8	−542.6	−184.1	−2618.7	186.6	66.942	−2238.5	620.3	138.0	−3200.4
2.000	0.400	−33,572.2	−39,021.8	6200.7	−559.7	−191.4	−2548.4	187.0	67.184	−2165.1	624.4	138.4	−3133.9
0.000	0.200	−33,564.6	−39,141.5	6109.5	−407.4	−125.3	−3105.9	140.3	64.740	−2746.0	549.1	134.8	−3584.9
0.222	0.200	−33,572.7	−39,138.9	6126.5	−426.3	−134.0	−3032.7	151.0	65.002	−2669.8	563.4	135.1	−3533.7
0.444	0.200	−33,580.8	−39,136.5	6143.2	−445.1	−142.5	−2959.6	159.1	65.238	−2593.8	575.2	135.4	−3478.3
0.668	0.200	−33,589.4	−39,134.4	6159.5	−463.6	−150.8	−2886.9	166.2	65.488	−2518.1	585.9	135.7	−3421.2
0.890	0.200	−33,598.4	−39,133.4	6176.2	−482.0	−159.1	−2814.7	172.4	65.711	−2442.9	595.7	136.0	−3363.2
1.110	0.200	−33,606.4	−39,132.1	6193.2	−500.3	−167.2	−2741.5	177.8	65.948	−2366.8	604.8	136.2	−3302.9
1.332	0.200	−33,615.3	−39,131.3	6209.8	−518.6	−175.3	−2669.2	182.5	66.185	−2291.4	613.1	136.5	−3242.0
1.556	0.200	−33,624.6	−39,130.7	6226.2	−536.8	−183.3	−2597.3	186.3	66.401	−2216.5	620.6	136.7	−3180.2
1.778	0.200	−33,634.1	−39,130.9	6243.1	−555.0	−191.3	−2525.5	189.1	66.618	−2141.7	627.0	137.0	−3116.4
2.000	0.200	−33,643.9	−39,131.6	6260.2	−573.2	−199.4	−2454.1	189.3	66.834	−2067.4	630.8	137.2	−3048.7
0.000	0.000	−33,631.6	−39,215.7	6104.6	−398.0	−122.5	−3088.0	134.4	64.353	−2727.7	543.7	134.0	−3557.8
0.200	0.000	−33,640.2	−39,217.4	6125.3	−417.0	−131.1	−3015.4	145.3	64.584	−2652.1	558.2	134.3	−3507.7
0.400	0.000	−33,648.5	−39,219.2	6146.3	−435.9	−139.7	−2942.5	153.6	64.802	−2576.2	570.2	134.5	−3452.8
0.800	0.000	−33,665.9	−39,225.3	6189.8	−473.6	−156.8	−2797.4	167.4	65.238	−2425.2	591.2	134.9	−3338.6
1.000	0.000	−33,674.6	−39,229.1	6212.5	−492.6	−165.4	−2724.9	173.2	65.449	−2349.7	600.6	135.1	−3279.6
1.200	0.000	−33,683.3	−39,233.4	6235.7	−511.7	−174.0	−2652.4	178.3	65.654	−2274.2	609.4	135.3	−3219.4
1.400	0.000	−33,692.1	−39,238.2	6259.7	−530.9	−182.7	−2579.9	182.7	65.866	−2198.7	617.4	135.5	−3158.0
1.600	0.000	−33,700.6	−39,243.2	6284.6	−550.4	−191.6	−2507.2	186.3	66.070	−2123.0	624.7	135.7	−3094.9
1.800	0.000	−33,709.5	−39,249.0	6310.0	−570.0	−200.6	−2434.9	188.8	66.262	−2047.7	630.8	135.8	−3030.4
2.000	0.000	−33,717.8	−39,254.6	6336.6	−590.0	−209.7	−2361.9	188.8	66.439	−1971.7	634.5	136.0	−2960.8

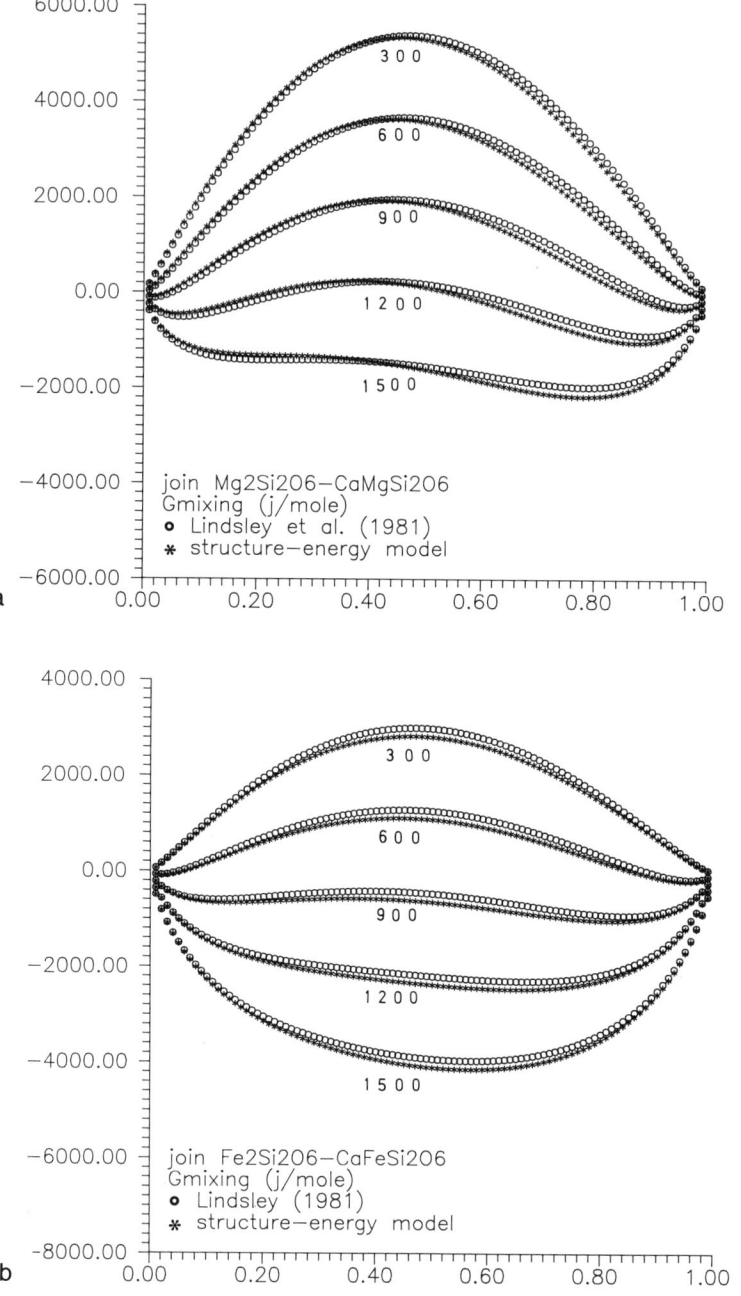

Fig. 4. Molar Gibbs free energy of mixing terms predicted by interionic potential calculations for the binary joins (a) $Mg_2Si_2O_6$–$CaMgSi_2O_6$, (b) $Fe_2Si_2O_6$–$CaFeSi_2O_6$, (c) $Mg_2Si_2O_6$–$Fe_2Si_2O_6$, compared with experimental evidences (a, b) and theoretical estimates (c). Values are relative to different T (K) conditions and $P = 1$ bar.

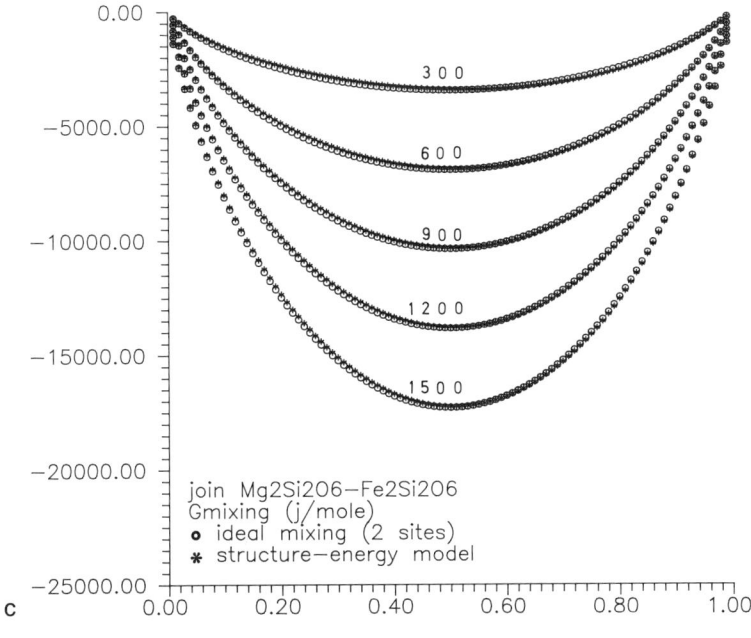

Fig. 4. (Continued.)

We see in Table 7 that the adoption of forms (48) or (49) does not imply any substantial modification of the W_H or W_V terms, but results in different W_S. Moreover, we report that attempts to fit Eq. (49) to the binary joins $Mg_2Si_2O_6$–$CaMgSi_2O_6$ and $Fe_2Si_2O_6$–$CaFeSi_2O_6$ gave unsatisfactory results.

The Gibbs free energy of mixing terms obtained by the model on three binary joins of the cpx quadrilateral are compared in Fig. 4 with experimentally derived and/or estimated values. The agreement between the predicted and observed values for the joins $Fe_2Si_2O_6$–$CaFeSi_2O_6$ and $Mg_2Si_2O_6$–$CaMgSi_2O_6$ is rather impressive [see Figs. 4(a) and (b)]. The molar Gibbs free energy of the mixing terms depicted by the model for the join $Mg_2Si_2O_6$–$Fe_2Si_2O_6$ at $P = 1$ bar conditions is essentially that of an ideal mixture with negligible asymmetry [Fig. 4(c)]. For the remaining three binary joins in the pyroxene quadrilateral, there is no direct experimental information, only estimates (Saxena et al., 1986). The highest bias between model and estimated values is observed for the join $Mg_2Si_2O_6$–$CaFeSi_2O_6$ at low T (see Table 7). In Fig. 5 we see stereographic representations of the Gibbs free energy of mixing terms within the quadrilateral at various P, T conditions based on the structure-energy calculations. As we see, the Gibbs free energy of a mixing surface is not regularly upward-concave, as would be expected from a Kohler-type combination of binary interaction

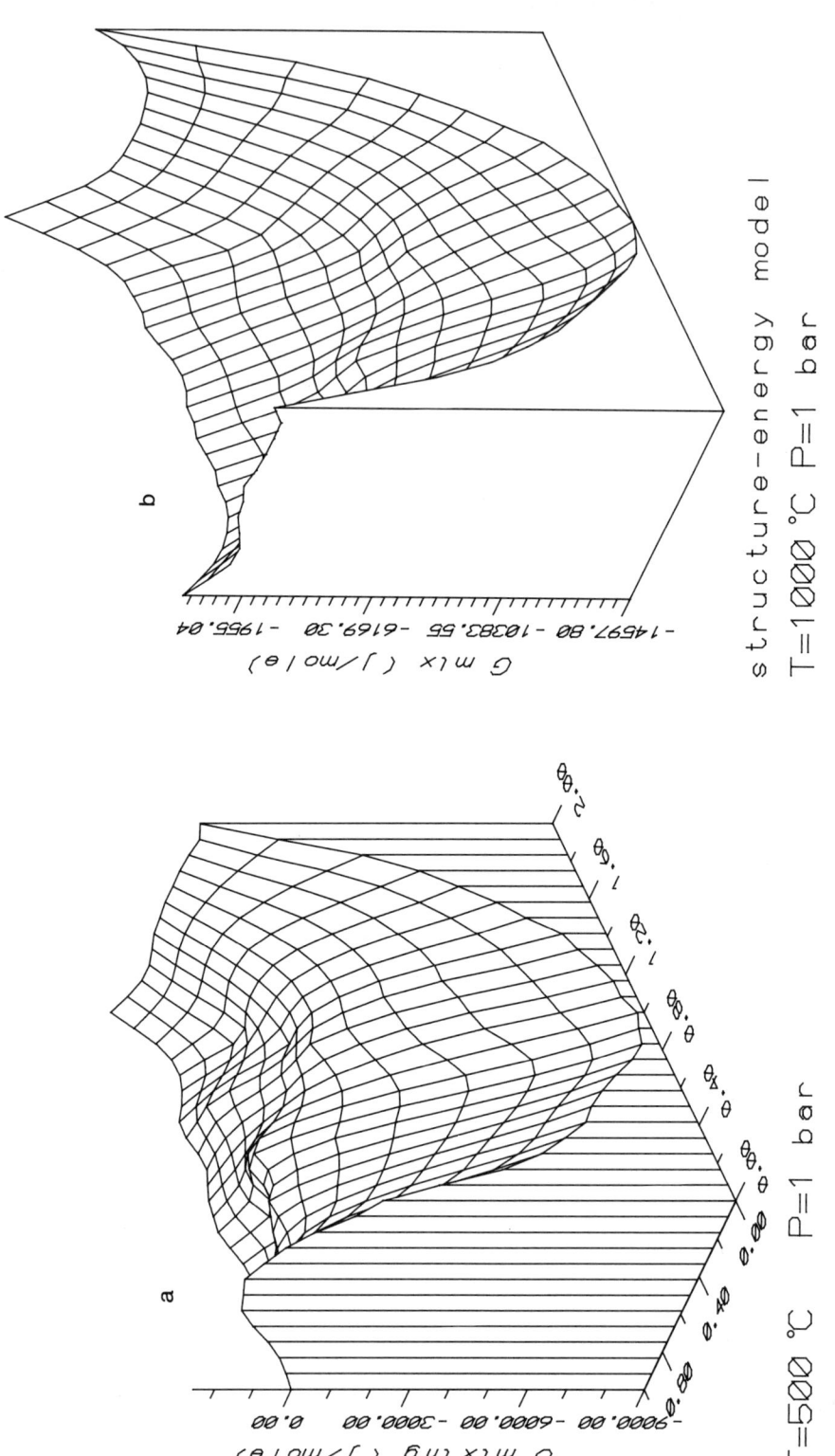

Fig. 5. Molar Gibbs free energy of mixing terms depicted by the model within the pyroxene quadrilateral at various T, P conditions (a–c). All data are relative to the C2/c structure. In (d) the results of the Kohler model at $T = 1000°C$, $P = 1$ bar, obtained by adopting the W_H, W_S, W_V binary interaction terms of Table 7 and disregarding ternary and quaternary terms, are represented for comparative purposes. The compositional coordinates are those of Table 9.

Energy Model for C2/c Pyroxenes

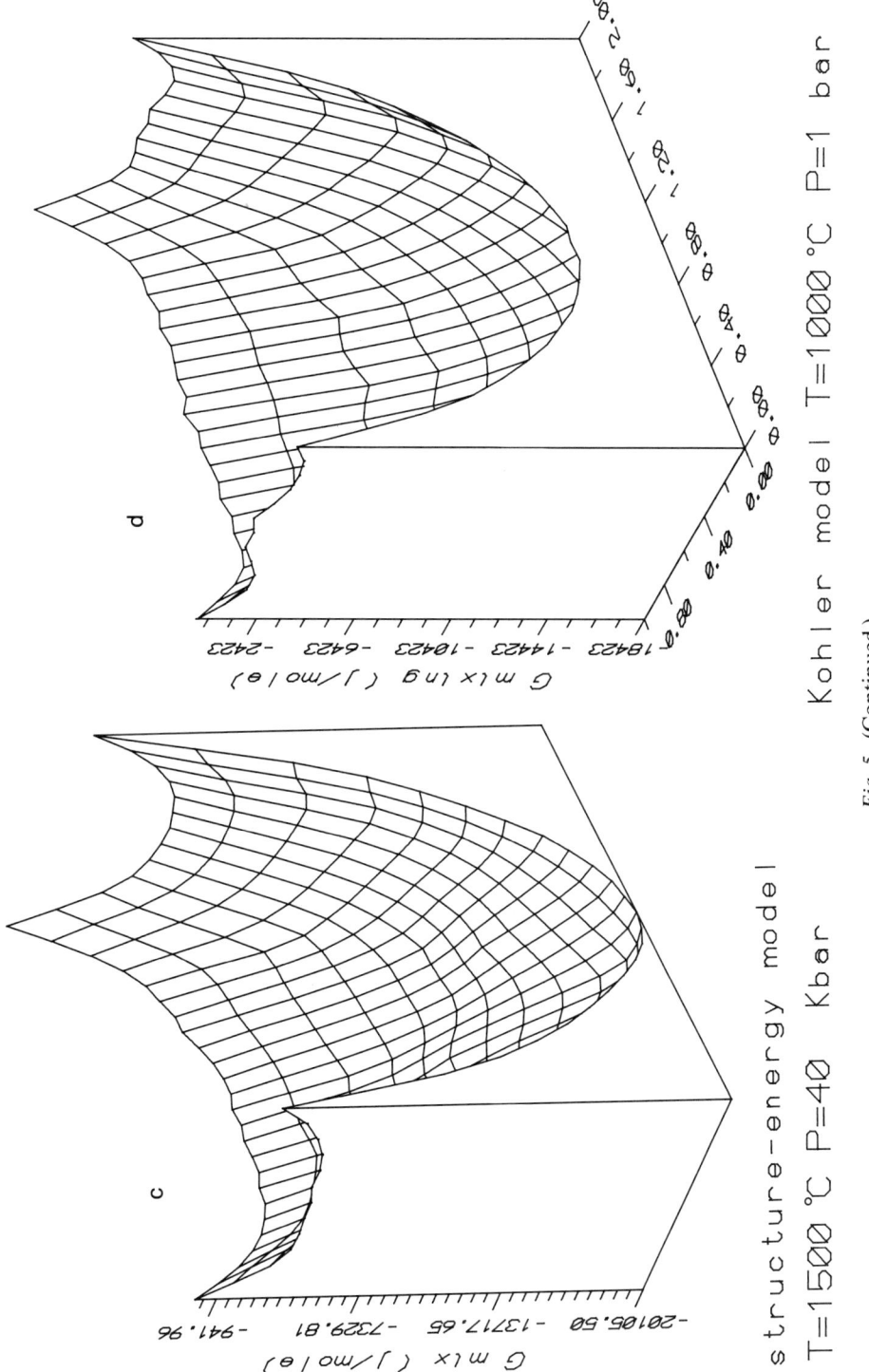

Fig. 5. (Continued.)

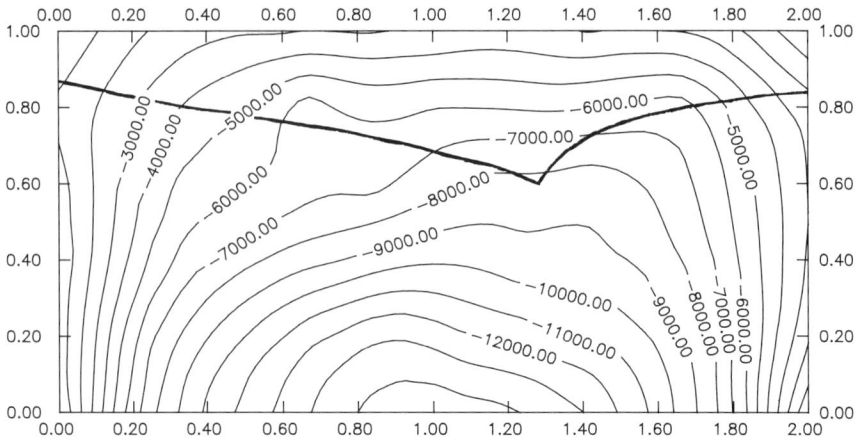

Fig. 6. Topographic representation of the molar Gibbs free energy of mixing terms depicted by the model at $T = 1000°C$, $P = 1$ bar. Compositional coordinates as in Fig. 5. The heavy line is the miscibility gap limb observed at the same T, P conditions by Davidson and Lindsley (1989).

parameters [Fig. 5(d)], but exhibits irregular inflections in the intermediate compositional regions that are indicative of the *intrinsic* instability of the C2/c phase. In Fig. 6 it is shown in more detail that this intrinsic instability is indicated by fluctuations of the isoenergetic lines that occur in the vicinity of the C2/c–Pbca miscibility gap limb of Davidson and Lindsley (1989). Figure 7 shows the effects of pressure on the stability of the C2/c mixtures. To evaluate the compressional work operated on the phase at $P > P_r$

$$\int_{P_r}^{P} \bar{V} dp = \bar{V}_{T_r,P_r}^{0} \exp\left[1 - \alpha_0(T - T_r) + \frac{\alpha_1}{2}(T^2 - T_r^2) - \alpha_2(T^{-1} - T_r^{-1})\right]$$
$$\cdot \left[\frac{k0}{(k1-1)}\right]\left[\left(1 + \frac{k1 \cdot P}{k0}\right)^{(k1-1)/k1} - \left(1 + \frac{k1 \cdot P_r}{k0}\right)^{(k1-1)/k1}\right], \quad (50)$$

we assumed that thermal expansion and compressibility factors of chemically complex mixtures (i.e., α_0, α_1, α_2 and $k0$, $k1$, respectively) may be obtained by linear summation of the end-member values or, in other words, that the excess volume of mixing term is constant over P, T. As we see, decompressing the substance at parity, of T conditions widens the (negative) isoenergetic regions; hence, decompression (at parity of T) stabilizes the C2/c mixture. This effect is more evident along the $Mg_2Si_2O_6$–$Fe_2Si_2O_6$ edge of the cpx quadrilateral, while it vanishes toward the calcic components. The increase of pressure appears thus to enhance the extrinsic instability of Ca-poor C2/c pyroxenes with respect to their Pbca counterparts.

Energy Model for C2/c Pyroxenes

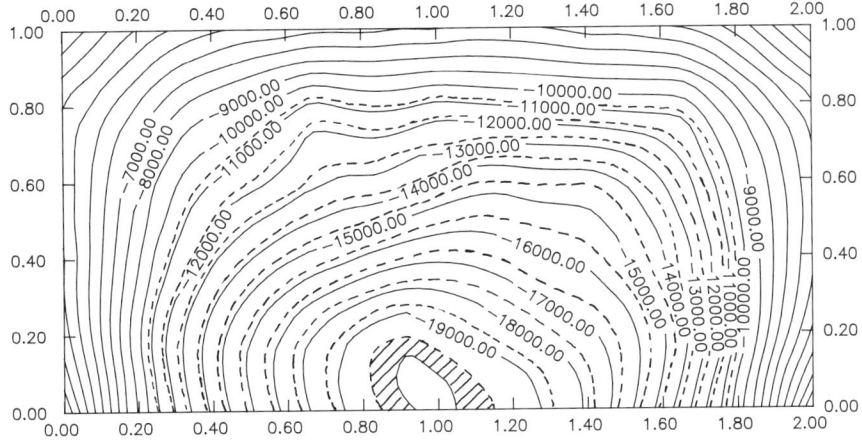

Fig. 7. Effect of pressure on the molar Gibbs free energy of mixing of C2/c pyroxenes in the quadrilateral. Dashed lines are relative to $T = 1500°C$, $P = 1$ bar conditions; solid lines are relative to $T = 1500°C$, $P = 40$ Kbar. The dashed area encompasses the -20 kJ/mol isoenergetic area.

Mixing of the Aluminiferous and Sodic Terms with the Diopsidic Component

The joins $CaMgSi_2O_6$–$CaAl_2SiO_6$ and $CaMgSi_2O_6$–$NaAlSi_2O_6$ have been carefully investigated by various authors due to the importance of Tschermak's and jadeiitic molecules in natural C2/c pyroxenes. Figure 8 shows calculated volumes along the join $CaAl_2SiO_6$–$CaMgSi_2O_6$ compared with various experimental results. According to the structural simulation, the excess volume of mixing is virtually negligible, whereas the experimental data seem to indicate a certain negative deviation from ideal in the $CaMgSi_2O_6$-rich region. The discrepancies between the various experimental data at parity of composition are essentially identical in magnitude to the discrepancy between model and experiment, and no definite conclusion can be reached. In Fig. 9 we see that the agreement between structural predictions and experiments is quite satisfactory for the join $NaAlSi_2O_6$–$CaMgSi_2O_6$. This fact is remarkable due to the volumetric change involved in the binary. The mixing properties of $CaMgSi_2O_6$–$CaAl_2SiO_6$ based on the calorimetric data of Newton et al. (1977) and the analysis of Gasparik and Lindsley (1980) of the phase equilibrium data were converted by Ganguly and Saxena (1987) to a subregular Margules formulation (Table 7). Adopting their formulation and the end-member parameters listed in Tables 4 through 6, we calculated 18 distinct compositions to constrain the interionic potential calculations. The repulsive energy was parametrized at various ρ factors. The best results were obtained with $\rho = 0.25$. As we see in Fig. 3, the value of ρ is tightly constrained and the mean error is higher than the corresponding error for the

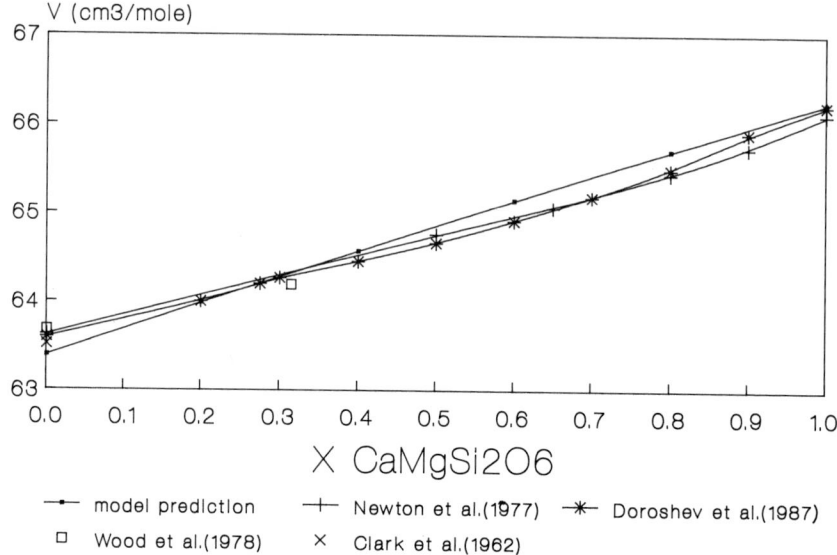

Fig. 8. Structure model predictions for the join $CaAl_2SiO_6$–$CaMgSi_2O_6$, compared with various experimental sources.

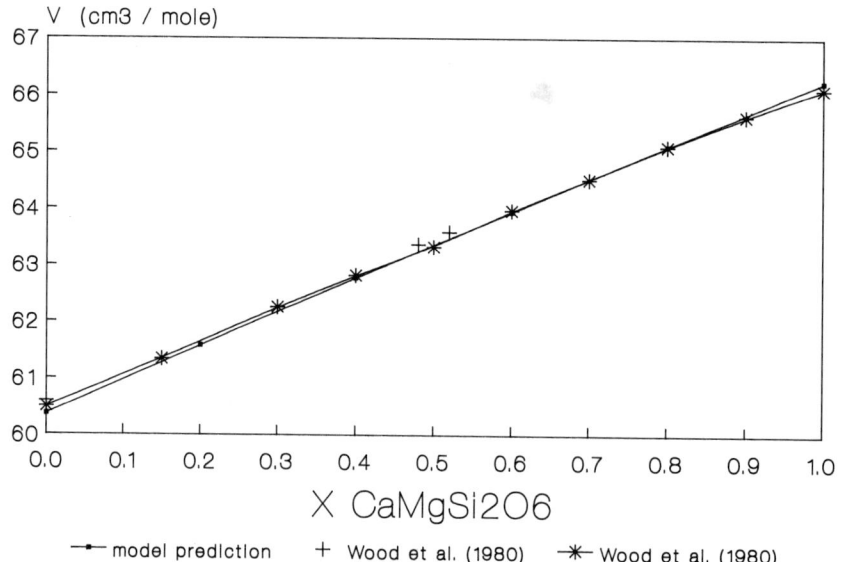

Fig. 9. Structure model predictions for the join $NaAlSi_2O_6$–$CaMgSi_2O_6$, compared with the results of Wood et al. (1980). Crosses indicate two natural samples.

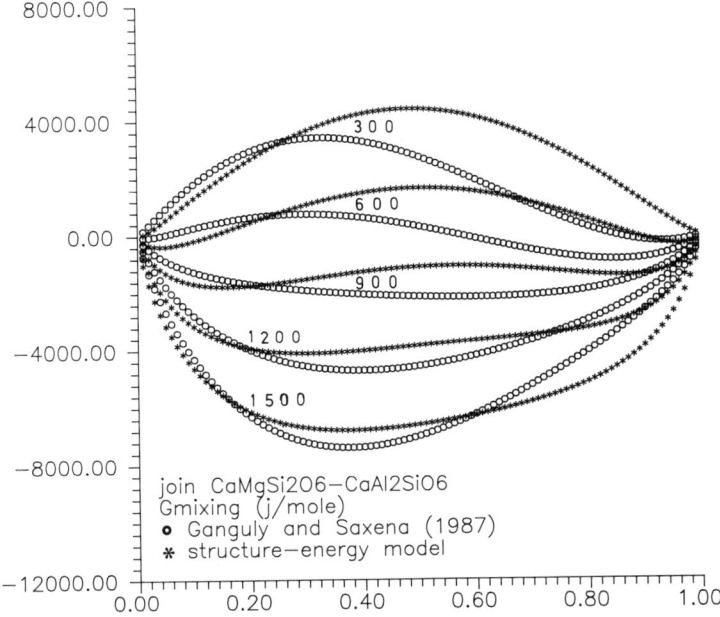

Fig. 10. Molar Gibbs free energy of mixing terms predicted by the model for the join $CaMgSi_2O_6$–$CaAl_2SiO_6$ at various T (K) conditions ($P = 1$ bar), compared with the subregular Margules formulation of Ganguly and Saxena (1987).

cpx quadrilateral at the best-fit condition. The energies obtained via interionic potential calculations were then converted to a subregular macroscopic model, as was previously done for the cpx quadrilateral (Table 7).

The calculated Gibbs free energy of mixing terms is shown in Fig. 10, where it is compared with the subregular formulation of Ganguly and Saxena (1987). As we see, there are marked discrepancies at low T in the $CaAl_2SiO_6$-rich region, while agreement is better at a high temperature.

The mixing properties in the binary $CaMgSi_2O_6$–$NaAlSi_2O_6$ have been investigated by various authors (Ganguly, 1973; Wood et al., 1980; Holland, 1983; Gasparik, 1985; Cohen, 1986). Gasparik (1985) based on his own phase-equilibrium study and on the data of Holland (1983) proposed a Redlich–Kister formulation for macroscopic interactions as follows:

$$A0 = 12,600 - 9.45\, T \text{ (J/mol)}, \tag{51a}$$

$$A1 = 12,600 - 7.60\, T, \tag{51b}$$

$$A2 = -21,400 + 16.2\, T, \tag{51c}$$

$$\bar{G}_{\text{mix exc}} = X(1-X)[A0 + A1(2x-1) + A2(2x-1)^2], \tag{52}$$

with $X = X(NaAlSi_2O_6)$.

Cohen (1986), based on the calorimetric data of Wood et al. (1980) and phase-

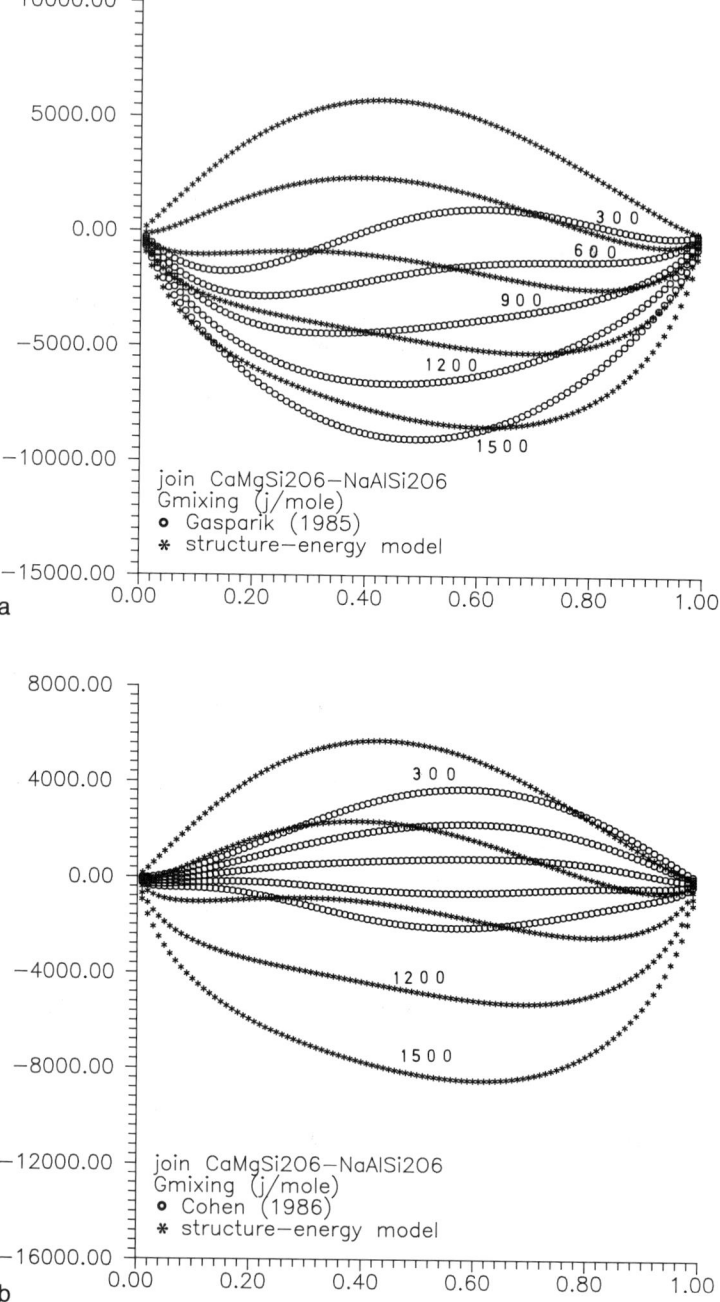

Fig. 11. Molar Gibbs free energy of mixing terms predicted by the model for the join $CaMgSi_2O_6$–$NaAlSi_2O_6$ at various T (K) and $P = 1$ bar, compared with the (a) Van Laar formulations of Gasparik (1985) and (b) Cohen (1986).

equilibrium data, obtained somewhat different Redlich–Kister parameters

$$A0 = 19{,}800 - 4.4\,T \text{ (J/mol)}, \tag{53a}$$

$$A1 = 9600 - 9.1\,T, \tag{53b}$$

$$A2 = -8200 + 16.4\,T, \tag{53c}$$

Both formulations disregard the effect of pressure, which is indeed negligible due to an almost ideal volume of mixing (Fig. 9). The interionic potential model parametrized with the formulation (51) gave the best-fit results at $\rho = 0.25$ analogously to the system $CaMgSi_2O_6$–$CaAl_2SiO_6$ (Table 8 and Fig. 3). The results of the interionic potential calculation were converted to a subregular Margules formulation shown in Table 7. Incidentally, a strictly macroscopic formulation (i.e., one-site mixing) is inadequate to describe the mixing energies in the $CaMgSi_2O_6$–$NaAlSi_2O_6$ join, as Ca and Na are randomly distributed over M2 sites and Al, Mg over M1 sites. Accordingly, attempts to fit the subregular Margules macroscopic formulation to the Gibbs free energy of mixing terms obtained via interionic potential calculations gave bad results involving a mean approximation of several hundreds of joules ($\Sigma\chi^2 = 1.81$, 24 values). The difference was sensibly reduced by accounting for mixing over two sites (formulation 49, $\Sigma\chi^2 = 0.0076$, 24 values). A comparison between the Gibbs free energy of mixing terms arising from the model and the macroscopic formulations of Gasparik (1985) and Cohen (1986) is shown in Figs. 11(a) and (b). As we see, the difference is quite marked at low T between the model predictions and the formulation of Gasparik (1986) (51a–c), but it is less in contrast to the formulation of Cohen (1986) (53a–c). On the other hand, at high T the agreement is more marked with the formulation of Gasparik (1985), whereas the formulation of Cohen (1986) indicates higher Gibbs free energy of mixing terms.

Extension of Interionic Potential Calculations to the Multicomponent Mixture (Na–Mg–Ca–Mn–Fe–Al–Cr–Ti–Si–O)

The ultimate goal of the model is the prediction of energy of the C2/c pyroxenes in a chemical system representative of the actual chemical complexity of the natural phases. The model must then include ferric ($NaFeSi_2O_6$), chromian ($NaCrSi_2O_6$), manganoan ($CaMnSi_2O_6$), and titanian ($CaTiAl_2O_6$) components, that normally occur as "minor" components in natural C2/c pyroxenes. Only for two of the above listed components (i.e., acmite $NaFeSi_2O_6$ and johannsenite $CaMnSi_2O_6$) are $\overline{H}^0_{T_r,P_r}$ enthalpy data available (and we still have some doubts about the value assigned to acmite; see Table 4), whereas the $\overline{H}^0_{T_r,P_r}$ values of ureyite ($NaCrSi_2O_6$) and Ca–Ti–Tschermak ($CaTiAl_2O_6$) are unknown. Reasonable estimates of repulsive energies in $NaCrSi_2O_6$ and $CaTiAl_2O_6$ components can be attempted, based on the relationships existing between repulsive parameters, ionic radii, and other atomistic properties of the substituting ions. It has, in fact, been evident since 1931 that the "short-range" interactions derived through quantum-mechanical calculations for closed-shell atoms can be expressed as exponential functions of the distance, i.e.,

$$\Phi_{REP} = b \exp\left(\frac{-\gamma_{ij} r_{ij}}{a_0}\right), \tag{54}$$

where a_0 is the Bohr's radius and γ_{ij} is a short-range interaction constant that was shown by Zener (1931) to be related to the ionization potentials of the substituting ions (I_1, I_2) through

$$\gamma_{ij} = \sqrt{\frac{2a_0}{e}} (\sqrt{I_1} + \sqrt{I_2}). \tag{55}$$

Moreover, as shown by Della Giusta et al. (1990), the radius of the ion in the crystal (r_i) is related to its ionization potential (I_i) through screen constants (C1,C2) as follows:

$$r_i = \left[\frac{C1/(C2^{-1/2} \cdot n)}{\sqrt{I_i}}\right], \tag{56}$$

where n is the principal quantic number, or to the free ion polarizability as follows:

$$r_i = \left[\frac{C1}{(C3 \cdot n \cdot a_0^{3/4})}\right] \alpha_i^{1/4}. \tag{57}$$

Bokreta and Ottonello (1987) have shown that in garnets the b factors of end-member components are linearly related to ionic radii and free ion polarizabilities on sites X and Y

$$b = \eta_0 + \eta_1 r_{ix} + \eta_2 r_{iy} + \eta_3 \alpha_{ix} + \eta_4 \alpha_{iy}. \tag{58}$$

As for Eq. (57), relation (58) can be converted to a simple linear dependency on r_i. Indeed we see in Fig. 12 that such a simple relationship does exist for C2/c pyroxenes and is analytically represented by

$$b = 3.0044 + 0.5686 r_{i_{M1}} - 0.6326 r_{i_{M2}} + 0.1589 r_{i_T}. \tag{59}$$

Relation (59) that is based on repulsive radii coincident with the ionic radii of Shannon (1976) (and not on the optimized radii of Table 8) leads to a $\Sigma\chi^2$ of 0.0158 on eight values. The maximum error is observed for $NaFeSi_2O_6$ (210 kJ/mol). The mean error is 76 kJ/mol. Because of these large approximations, the values listed in Table 4 for $NaCrSi_2O_6$ and $CaTiAl_2O_6$ should be considered preliminary at present.

Besides the immediate utilization of relation (59) for predictive purposes, we emphasize that the system of Eqs. (54–59) indicates that the short-range interactions can be reconverted to a discrete set of parameters, each one singularly dependent on the interionic distance, has great heuristic validity, and should be further investigated to assess the energetics of trace components.

Energy Model for C2/c Pyroxenes

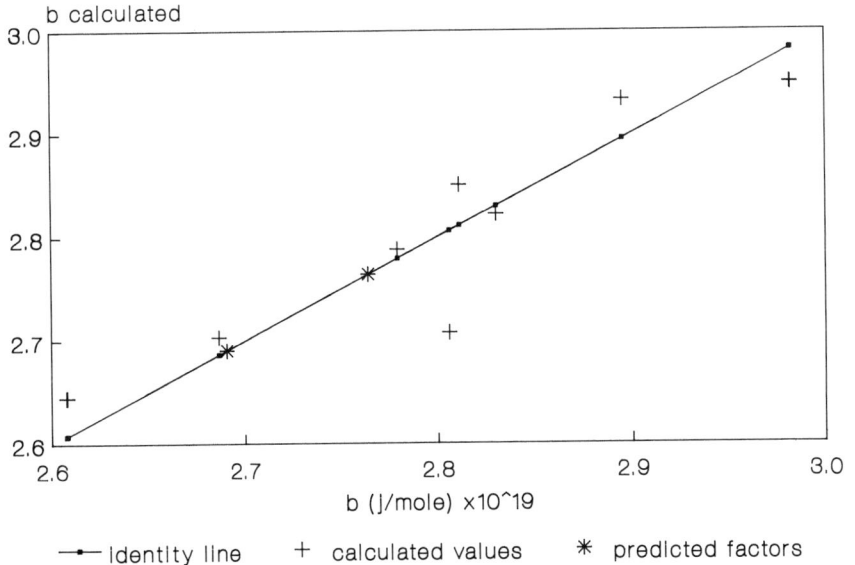

Fig. 12. Repulsive factors (*b*) of end-member C2/c components, compared with values calculated by Eq. (59). The largest bias is observed for NaFeSi$_2$O$_6$. Asterisks denote *b* factors calculated for NaCrSi$_2$O$_6$ and CaTiAl$_2$O$_6$, for which there are no energy data available. Calculations assume that repulsive radii are coincident with Shannon's (1976) ionic radii and cannot thus be extended to mixing properties that require a more accurate parameterization.

Let us assume that the predicted $\bar{H}^0_{T_r, P_r}$ values of NaCrSi$_2$O$_6$ and CaTiAl$_2$O$_6$ are sufficiently accurate. We want now a set of internally consistent ρ, b, r factors capable of reproducing the energetics of C2/c pyroxenes in the Na–Mg–Ca–Mn–Fe–Al–Cr–Ti–Si–O space. We have already seen that the cpx quadrilateral is optimized at $\rho = 0.5$, while the binaries CaMgSi$_2$O$_6$–NaAlSi$_2$O$_6$ and CaMgSi$_2$O$_6$–CaTiAl$_2$O$_6$ need a hardness factor of 0.25. One would be tempted to follow the suggestions of Gilbert (1968) and assign to each atom its own hardness factor (see, e.g., Miyamoto et al., 1982). In this way, the number of variables would become sufficiently large to fit whatever energetic term is in the whole compositional space (and even more !). We prefer, however, to retain the Huggins–Mayer formulation that proved so satisfactory for the cpx quadrilateral.

For this purpose, it is sufficient to limit the compositional space for minor components within the molar proportions observed in the natural phases. The results of the extended calculation and new compositional limits are shown in Table 8(d) and in Table 9 are listed energy values obtained within the pyroxene quadrilateral (all relative to the C2/c structural form). In evaluating the extended model, we assumed that acmite and ureyite behave similarly to jadeite in their mixing properties with the diopsidic components (according to Popp and Gilbert, 1972; acmite and jadeite mix ideally). We assumed, moreover, that

Ca–Ti Tschermak behaves similarly to the Ca-Tschermak molecule when mixing with diopside and finally that diopside and johannsenite mix ideally. The model tolerates the presence in the mixture of minor components within molar concentration limits of $X \leqslant 0.1$ ($NaAlSi_2O_6$, $NaFeSi_2O_6$, $NaCrSi_2O_6$, $CaAl_2SiO_6$, $CaMnSi_2O_6$) and $X \leqslant 0.05$ ($CaTiAl_2O_6$). The mean error observed in the concentration range $0 \leqslant X \leqslant 0.1$ is 1.3 kJ/mol. At the concentration levels commonly observed in natural magmatic C2/c pyroxenes, the error would be much lower and virtually negligible.

Summary

The structure of C2/c pyroxenes in the Na–Mg–Ca–Fe–Al–Cr–Ti–Si–O system can be accurately simulated by the calculation of all interionic distances in the asymmetric unit, followed by a DLS treatment that optimizes cell parameters and atomic fractional coordinates.

The precise structural simulation allows an accurate treatment of the static bulk lattice energy of the phase through two-body interaction potential calculations extended to all atoms within the asymmetric unit.

The parametrization of the repulsive energy is achieved with the Huggins–Mayer formulation that assumes the hardness factor of substances to be constant for all salts in the family and the repulsive factors of components to be variable from salt to salt.

The parametrization of the repulsive energy allows the investigation of bulk phase energetics within the cpx quadrilateral. The results indicate that the Gibbs free energy of mixing of C2/c pyroxenes does not decrease monotonically within the quadrilateral, as would be expected from a simple combinatory approach based on binary macroscopic interaction parameters, but exhibits inflections that are indicative of the intrinsic instability of the phase. This instability is increased by pressure and would favor the extrinsic instability observed in Ca-poor clinopyroxenes with respect to their Pbca counterparts.

The linear relationship observed between repulsive factors of components and ionic radii of substituting ions on sites allows a rough estimate of lattice energy (hence of enthalpy, through the Born–Haber treatment) of unknown components.

The structure-energy model developed for C2/c pyroxenes in the quadrilateral can be extended to the Na–Mg–Ca–Mn–Fe–Al–Cr–Ti–Si–O system when the molar abundances of "minor" components $CaAl_2SiO_6$, $NaAlSi_2O_6$, $NaCrSi_2O_6$, $CaMnSi_2O_6$, $CaTiAl_2O_6$ are limited to $X \leqslant 0.1$ and the molar abundance of $CaTiAl_2O_6$ is restricted to $X \leqslant 0.05$.

The precision of the results rests on the precision of the enthalpy values selected at T_r, P_r reference conditions, as the uncertainty involved in the static lattice energy calculations is virtually negligible. Once new experimental data are available for the end-member components of interest, the model can be easily recast to obey the new limiting conditions.

The package of computer programs necessary to perform structure-energy calculations in C2/c pyroxenes is available from the authors on request.

Acknowledgments

This work was supported by MURST project "Crystal chemistry and Thermodynamics of Minerals."

References

Akimoto, S. (1972). The system MgO–FeO–SiO$_2$ at high pressures and temperatures. Phase equilibria and elastic properties. *Tectonophys* **13**, 161–187.

Baerlocher, C., Hepp, A., and Meier, W.M. (1977). DLS-76. *A program for the Simulation of Crystal Structures by Geometric Refinement.* Institute of Crystallography and Petrography, ETH Zurich, Switzerland.

Berman, R.G. (1988). Internally-consistent thermodynamic data for minerals in the system Na$_2$O–K$_2$O–CaO–MgO–FeO–Fe$_2$O$_3$–Al$_2$O$_3$–SiO$_2$–TiO$_2$–H$_2$O–CO$_2$. J. Petrol., **29**, 445–522.

Birch, F. (1966). Compressibility; elastic constants, *Geology Society American Mem.*, Vol. 97, pp. 97–174.

Bokreta, M. and Ottonello, G. (1987). Enthalpy of formation of end-member garnets. EOS **68**, 448.

Bruno, E., Carbonin, S., and Molin, G.M. (1982). Crystal structures of Ca-rich clinopyroxenes on the CaMgSi$_2$O$_6$–Mg$_2$Si$_2$O$_6$ join. *Tsch. Min. Petr. Mitt.* **29**, 223–240.

Burnham, C.W., Clark, J.R., Papike, J.J., and Prewitt, C.T. (1967). A proposed crystallographic nomenclature for clinopyroxene structures. *Zeit. Krist.* **125**, 1–6.

Cameron, M. and Papike, J.J. (1980). Crystal chemistry of silicate pyroxenes. *Rev. Mineral.* **7**, 5–87.

Carbonin, S., Dal Negro, A., Ganeo, S., and Piccirillo, E.M. (1991). Influence of magma composition and oxygen fugacity on the crystal structure of C2/c clinopyroxenes from a basalt-pantellerite suite. *Contrib. Mineral. Petrol.* **108**, 34–42.

Catti, M. (1981a). The lattice energy of forsterite. Charge distribution and formation enthalpy of the SiO$_4^{4-}$ ion. *Phys. Chem. Mineral* **7**, 20–25.

Catti, M. (1981b). A generalized Born–Mayer parametrization of the lattice energy in orthorombic ionic crystals. *Acta Cryst.* **A37**, 72–76.

Chatterjee, N. (1989). An internally consistent thermodynamic data base on minerals: Applications to the earth's crust and upper mantle. Ph.D. Thesis, City University, New York.

Clark, S.P., Jr., Schairer, J.F., and De Neufville, J. (1962). Phase relations in the system CaMgSi$_2$O$_6$–CaAl$_2$SiO$_6$–SiO$_2$ at low and high pressure. *Carnegie Inst. Wash. Yearbook* **61**, 59–68.

Clark, J.R., Appleman, D.E., and Papike, J.J. (1969). Crystal-chemical characterization of clinopyroxenes based on eight new structure refinements. *Miner. Soc. Amer. Spec. Papers* **3**, 31–50.

Cohen, R.E. (1986). Thermodynamic solution properties of aluminous clinopyroxenes: nonlinear least squares refinements. *Geochim. Cosmochim. Acta.* **50**, 563–575.

Dal Negro, A., Carbonin, S., Molin, G.M., Cundari, A., and Piccirillo, E.M. (1982). Intracrystalline cation distribution in natural clinopyroxenes of tholeiitic, transitional and alkaline basaltic rocks. *Adv. Physical Geochem.* **2**, 117–150.

Davidson, P.M. and Lindsley, D.H. (1989). Thermodynamic analysis of pyroxene-olivine-quartz equilibria in the system $CaO-MgO-FeO-SiO_2$. *Amer. Mineral.* **74**, 18–30.

Della Giusta, A., Ottonello, G., and Secco, L. (1990). Precision estimates of interatomic distances using site occupancies, ionization potentials and polarizability in Pbnm silicate olivines. *Acta Cryst.* **B46**, 160–165.

Doroshev, A.M., Malinovskaya, Ye. K., Surkov, N.V., and Bulakov, V.K. (1987). Synthesis and unit cell parameters of $CaMgSi_2O_6-CaAl_2SiO_6$ clinopyroxenes. *Geochem. Internat.* **16**, 83–92.

Dunitz, J.D. and Orgel, L.E. (1957). Electronic properties of transition metal oxides. II. Cation distribution among octahedral and tetrahedral sites. *J. Phys. Chem. Solids* **3**, 318–323.

Ganguly, J. (1973). Activity-composition relation of jadeite in omphacite pyroxene: Theoretical deductions. *Earth Planet. Sci. Lett.* **19**, 145–153.

Ganguly, J. and Saxena, S.K. (1987). *Mixtures and Mineral Reactions.* Springer-Verlag, Berlin-Heidelberg-New York.

Gasparik, T. (1985). Experimentally determined composition of diopside-jadeite pyroxenes in equilibrium with albite and quartz at 1200–1350°C and 15–34 Kbar. *Geochim. Cosmochim. Acta* **49**, 865–870.

Gasparik, T. and Lindsley, D.H. (1980). Phase equilibria at high pressure of pyroxenes containing monovalent and trivalent ions, in *Reviews in Mineralogy*, C.T. Prewitt, ed., Vol. 7, Mineralogy Society of America Washington, D.C.

Gilbert, T.L. (1968). Soft sphere model for closed-shell atoms and ions. *J. Chem. Phys.* **49**, 2640–2642.

Haselton, H.T., Jr., Hemingway, B.S., and Robie, R.A. (1984). Low-temperatute heat capacities of $CaAl_2SiO_6$ glass and pyroxene and thermal expansion of $CaAl_2SiO_6$. *Amer. Mineral.* **69**, 481–489.

Heitler, W. and London, F. (1927). Wechselwirkung neutraler Atome und homöopolare Bindung nach der Quantenmechanik. *B. Physik.* **44**, 455–472.

Helgeson, H.C., Delany, J.M., Nesbitt, H.W., and Bird, D.K. (1978). Summary and critique of thermodynamic properties of rock-forming minerals. *Amer. J. Sci.* **278A**, 1–229.

Hemingway, B.S., Krupka, K.M., and Robie, R.A. (1981). Heat capacities of the alkali feldspars between 350 and 1000 K from differential scanning calorimetry, the thermodynamic functions of alkali feldspars from 298.15 to 1400 K, and the reaction quartz + jadeite = albite. *Amer. Mineral.* **66**, 1202–1215.

Holland, T.J.B. (1983). The experimental determination of activities in disordered and short-range ordered jadeitic pyroxenes. *Contrib. Mineral. Petrol.* **82**, 214–220.

Holland, T.J.B., Navrotsky, A., and Newton, R.C. (1979). Thermodynamic parameters of $CaMgSi_2O_6-Mg_2Si_2O_6$ pyroxenes based on regular solution and cooperative disordering models. *Contrib. Mineral. Petrol.* **69**, 337–344.

IUPAC (1979). Manual of symbols and therminology for physicochemical quantities and units. *Pure & Appl. Chem.* **51**, 1–41.

James, F. and Roos, M. (1977). MINUIT. A system for function minimization and analysis of the parameter errors and correlations. CERN Computer Ctr., Geneva, Switzerland.

Lindsley, D.H. (1981). The formation of pigeonite on the join hedembergite-ferrosilite at 11.5 and 15 Kbar: Experiments and a solution model. *Am. Mineral.* **66**, 1175–1182.

Lindsley, D.H., Munoz, J.L., and Finger, L.W. (1969). Unit-cell parameters of clinopyroxenes along the join hedenbergite-ferrosilite. *Carnegie Inst. Wash. Yearbook* **67**, 91–92.

Lindsley, D.H., Grover, J.E., and Davidson, P.M. (1981). The thermodynamics of the $Mg_2Si_2O_6$–$CaMgSi_2O_6$ join: A review and a new model, in *Advances in Physical Geochemistry*, R.C. Newton, A., Navrotsky, B.J. and Wood, eds., Springer-Verlag, Berlin-Heidelberg-New York, Vol. 1, pp. 149–175.

Meier, W.M. and Villiger, H. (1969). Die Methode der Abstandsverfeinerung zur Bestimmung der Atomkoordinaten idealisierter Gerustsruckturen *Zeits. Kristallogr.* **129**, 411–423.

Miyamoto, M., Takeda, H., Fujino, K., and Takeuchi, Y. (1982). The ionic compressibilities and radii estimates for some transition metals in olivine structures. *Min. J.* **11**, 172–179.

Naumov, G.B., Ryzhenko, B., and Khodakovsky, I.L. (1971). *Handbook of Thermodynamic Data*. Atomizdat, Moscow.

Newton, R.C., Charlu, T.V., and Kleppa, O.J. (1977). Thermochemistry of high pressure garnets and clinopyroxenes in the system CaO–MgO–Al_2O_3–SiO_2. *Geochim. Cosmochim. Acta* **41**, 369–377.

Newton, R.C., Charlu, T.V., Anderson, P.A.M., and Kleppa, O.J. (1979). Thermochemistry of synthetic clinopyroxenes on the join $CaMgSi_2O_6$–$Mg_2Si_2O_6$. *Geochim. Cosmochim. Acta* **42**, 55–60.

Newton, W. and McCready, N. (1948). Thermodynamic properties of sodium silicates. *C. Phys. Coll. Chem.* **52**, 1277–1283.

Nickel, K.G. and Brey, G. (1984). Subsolidus orthopyroxene-clinopyroxene systematics in the system CaO–MgO–SiO_2 to 60 Kbar: A re-evaluation of the regular solution model. *Contrib. Mineral. Petrol.* **87**, 35–42.

Ottonello, G. (1987). Energies and interactions in binary (Pbnm) orthosilicates: A Born parametrization. *Geochim. Cosmochim. Acta* **51**, 3119–3135.

Ottonello, G., Della Giusta, A., and Molin, G.M. (1989). Cation ordering in Ni–Mg olivines. *Amer. Mineral.* **74**, 411–421.

Ottonello, G., Princivalle, F., and Della Giusta, A. (1990). Temperature, composition and f_{O2} effects on intersite distribution of Mg and Fe^{2+} in olivines. *Phys. Chem. Miner.* **17**, 301–312.

Pauling, L. (1960). *The Nature of the Chemical Bond*, 3rd ed. Cornell University Press, Ithaca, New York.

Popp, R.K. and Gilbert, M.C. (1972). Stability of acmite-jadeite pyroxenes at low pressures. *Amer. Mineral.* **57**, 1210–1231.

Robie, R.A., Hemingway, B.S., and Fisher, J.R. (1978). Thermodynamic properties of minerals and related substances at 298.15 K and 1 bar (10^5 Pascals) pressure and at higher temperatures. *U.S. Geol. Surv. Bull.* 1452, pp. 1–452.

Robinson, P. (1980). The composition space of terrestrial pyroxenes—internal and expernal limits. *Rev. Mineral.* **7**, 419–494.

Sack, R.O. (1980). Some constraints on the thermodynamic mixing properties of Fe–Mg orthopyroxenes and olivines. *Contrib. Mineral. Petrol.* **71**, 257–269.

Saxena, S.K. (1981). Fictive component model of pyroxenes and multicomponent phase equilibria. *Contrib. Mineral. Petrol.* **78**, 245–251.

Saxena, S.K. (1983). Exsolution and Fe^{2+}–Mg order-disorder in pyroxenes, in *Advances in Physical Geochemistry*, S.K. Saxena, ed., Springer-Verlag, Berlin-Heidelberg-New York, Vol. 2, pp. 61–80.

Saxena, S.K. (1990). Programs INSP and THERMO and corresponding data-base (unpublished).

Saxena, S.K. and Chatterjee, N. (1986). Thermochemical data on mineral phases. I. The system CaO–MgO–Al_2O_3–SiO_2. *J. Petrol.* **27**, 827–842.

Saxena, S.K. and Eriksson, G.E. (1983). High temperature phase equilibria in a solar-composition gas. *Geochim. Cosmochim. Acta* **47**, 1865–1874.

Saxena, S.K., Sykes, J., and Eriksson G. (1986). Phase equilibria in the pyroxene quadrilateral. *J. Petrol.*, **27**, 843–852.

Shannon, R.D. (1976). Revised effective ionic radii and systematic studies of interatomic distances in halides and chalcogenides. *Acta Cryst.* **A32**, 751–767.

Skinner, B.J. (1966). Thermal expansion, in *Handbook of Physical Constants*, S.P. Clark, Jr., ed., Geology Society of Amer. Mem. Vol. 97, pp. 75–95.

Smith, J.V. (1959). The crystal structure of proto-enstatite. *Acta Cryst.* **12**, 515–519.

Steele, I.M. (1975). Mineralogy of lunar norite 78235; second lunar occurrence of $P2_1ca$ pyroxene from Apollo 17 soils. *Amer. Mineral.* **60**, 1086–1091.

Stull, D.R. and Prophet, H. (1971). *JANAF Thermochemical Tables*. Data Series, Washington, D.C., Vol. 37, pp. 1–1141.

Syono, Y., Akimoto, S., and Matsui, Y. (1971). High pressure transformations in zinc silicates. *Solid State Chem.* **3**, 369–380.

Tosi, M. (1964). Cohesion of ionic solids in the Born model. *Solid State Phys.* **16**, 1–120.

Tosi, M. and Fumi, F.G. (1964). Ionic sizes and Born repulsive parameters in the NaCl-type alkali halides-II. *Phys. Chem. Solids.* **25**, 45–52.

Turnock, A.C., Lindsley, D.H., and Grover, J.E. (1973). Synthesis and unit cell parameters of Ca–Mg–Fe pyroxenes. *Amer. Mineral.* **58**, 50–59.

Vieillard, P. (1982). Modele de calcul des energies de formation des mineraux, bati sur la connaissance affinee des structures cristallines. *C.N.R.S. Mem.* **69**, 1–206.

Wagman, D.D., Evans, W.H., Parker, V.B., Halow, I., Bailey, S.M., and Schumm, R.H. (1981). Selected values of chemical thermodynamic properties. Nbs Tech. Note, 270–8, pp. 1–134.

Wood, B.J., Holland, T.J.B., Newton, R.C. and Kleppa, O.J. (1980). Thermochemistry of jadeite-diopside pyroxenes. *Geochim. Cosmochim. Acta* **44**, 1363–1371.

Zener, C. (1931). Interchange of translational, rotational and vibrational energy in molecular collisions. *Phys. Rev.* **37**, 556–569.

Chapter 8
Practical Problems in Calculating Thermodynamic Functions for Crystalline Substances from Empirical Force Fields

C.M. Gramaccioli and T. Pilati

Introduction

The correlation between the structure and thermodynamic properties of minerals has long been considered fundamental. Besides the first-law functions, it is possible to calculate entropy and free energy for a pure ideal crystal if the vibrational properties are taken into account. These calculations imply the precise evaluation of spectroscopic data (Raman, IR, phonon dispersion curves) and allow good estimates of elastic properties and atomic displacement parameters.

Whereas the improvement in computer design strongly favors development in this field, there are still several difficulties with respect to general application of these calculations. There is, however, evidence that the difficulties may be overcome, provided adequate simplification in computing techniques is considered. In this chapter, we show that close relationships can be observed between a Raphson–Newton process of energy minimization and the calculation of vibration frequencies; based on these techniques, it is possible to check the nature of a minimum and its physical significance. Thus, we can come much closer to a prediction of the entropy and free energy of a crystalline substance.

Crystallographic information (including optical data) has long been considered a collection of the "essential" physical properties of a certain mineral. This happens especially because of the well-known traditional connections between crystallography and mineralogy: For a great majority of minerals, there are accurate measurements of unit-cell parameters and X-ray powder diffraction data, together with the values of the refractive indices and the orientation of the indicatrix. For the most important species, these properties (including atomic coordinates, displacement parameters, and other information obtained from careful crystal-structure determination and refinement) have been investigated in a substantial number of specimens by different authors.

On the other hand, in general, the amount of crystallographic information available in the literature can hardly be compared with the relative scarcity of reliable measurements of thermodynamic properties. The latter are, however, continuously gaining in attention from mineralogists; for petrologists they are even more important for immediate use than the crystal-structure data. At the same time, spectroscopic information is also increasing in importance; for these reasons, in the last several years less unilateral knowledge of the physical-chemical properties of minerals has been developing.

Since all these data of a different nature for a certain mineral must be interconnected, the derivation of mutually consistent thermodynamic, spectroscopic, or elastic data from the crystal structure on a routine basis from a few parameters of general use would be a fundamental achievement; another very important (and closely related) problem is deriving the thermodynamically stable crystal structure under certain conditions, and possibly reproducing all the details, including thermal parameters, within the range of the experimental error. All this would emphasize the importance of crystallography in earth sciences, in a new role.

For some of these problems, or at least for some aspects of the situation, well-known solutions have already been available for a long time. For instance, there are satisfactory models (and computer programs) for deriving the stable crystal structure, starting from atomic charges and empirical atom–atom interaction parameters (see, e.g., Busing, 1981; Catlow and Mackrodt, 1982; Matsui and Matsumoto, 1982; Busing and Matsui, 1984; Matsui and Busing, 1984a; Parker et al., 1983; Price and Parker, 1984; Price et al., 1985; Burnham, 1990); most of these models and programs are based on an evaluation of the first derivatives of the crystal energy E with respect to atomic coordinates and unit-cell parameters. There are also good models for predicting the heat of formation for entire groups of compounds with surprising accuracy if their structure is known and similar results can be obtained for elastic constants (see, e.g., Vieillard, 1982; Catti, 1982, 1989; Matsui and Busing, 1984a, b; Ottonello, 1986; Au and Weidner, 1986). The Debye theory with adequate modifications is a "classic" example for deriving the specific heat or other temperature-dependent properties of many crystalline substances. In several cases, these calculations have also been successful in eliminating substantial experimental errors, or in providing a useful interpolation of experimental data (see, e.g., Kieffer, 1985).

Choice of an Adequate Lattice-Dynamical Model

However, in general, these procedures are not fully consistent with each other, and even when they are theoretically, it is often impossible to formulate a reasonable assumption about the value of corresponding parameters for different minerals; similarly, in general, it is difficult to be confident about the successful use of the same parameters in calculating different properties of a particular mineral. For instance, the Debye model implies at *least* one empirical parameter to be used for a certain substance only, and this parameter should be found only *a posteriori* on a "best-fit" basis to the experimental data unique to this substance. Similarly, wide differences in the atomic charge and VFF parameters (such as bond-stretching constants) or in repulsive short-range terms are encountered, depending not only on the particular mineral, but also the author, model, and/or particular properties that are the main thrust of the work.

In deriving consistent models, the same force field should be employed in different kinds of calculations, and the greater number of applications could provide a substantially wider basis for deducing the necessary empirical constants on best-fit grounds. For instance, the energy minimum criterion could be improved by verifying whether its "walls" are consistent with a real minimum of the energy function, and with the experimental values for the vibrational frequencies of the crystal. All this implies abandoning the Debye theory in favor of more complex models, mainly based on Born–von Karman's lattice-dynamical procedure (see, e.g., Born and Huang, 1954; Cochran, 1973; Willis and Pryor, 1975; Ghose, 1985). This procedure is based on the harmonic approximation and invariance of the equations of motion for appropriate translations within the crystal. The calculations can be brought to the following eigenvalue-eigenvector form:

$$\mathbf{D}(\mathbf{q})\mathbf{g} = \omega^2 \mathbf{g}, \tag{1}$$

where $\omega (= 2\pi\nu)$ are the angular frequencies, and $\mathbf{D}(\mathbf{q})$ is the dynamical matrix for a certain value of the wave vector $\mathbf{q}$. For $\mathbf{q} = \mathbf{0}$, most of the vibrational frequencies (ν's) correspond to IR- or Raman-active vibrational modes, and a check with experimental results is relatively easy. For nonzero values of $\mathbf{q}$, the experimental phonon dispersion curves can be used instead for comparison. However, very few experimental data of this kind have been obtained so far (e.g., by inelastic neutron scattering).

The elements of $\mathbf{D}(\mathbf{q})$ are given by

$$D_{ij}^{kk'}(\mathbf{q}) = (m_k m_{k'})^{-1/2} \Sigma_l \varphi_{ij}^{kk'}(0,l) \exp(2\pi i \mathbf{q} \cdot \Delta \mathbf{r}^l), \tag{2}$$

where

$$\varphi_{ij}^{kk'}(0,l) = \frac{\partial^2 E}{\partial x_{ik}^0 \partial x_{jk}^l}, \tag{3}$$

(see, e.g., Willis and Pryor, 1975, Eq. 3.10b). Here E is the potential energy, x_{ik}^0 is a coordinate of the kth atom in the unit cell; $x_{jk'}^l$ is a coordinate of the k'th atom in the crystal, related to $x_{jk'}^0$ by a lattice translation $\mathbf{r}^l$ ($l = 0$ when $\mathbf{r}^l = 0$), $\Delta \mathbf{r}^l$ is the distance between the two atoms involved, m_k and $m_{k'}$ are the masses of the atoms k and k', respectively. The summation Σ_l is extended (in principle) to all the translated units in the crystal.

By introducing mass-weighted coordinates

$$x_{ik}' = (m_k)^{1/2} x_{ik}, \tag{4}$$

Eq. (2) can be rewritten as

$$\mathbf{D}_{ij}^{kk'}(\mathbf{q}) = \Sigma_l \varphi_{ij}^{'kk'}(0,l) \exp(2\pi i \mathbf{q} \cdot \Delta \mathbf{r}^l), \tag{5}$$

where

$$\varphi_{ij}^{'kk'}(0,l) = \frac{\partial^2 E}{\partial x_{ik}^{'0} \partial x_{jk'}^{'l}}. \tag{6}$$

Since for a crystal, the contributions to Σ_l are nonzero only at definite points in the lattice, the elements of $\mathbf{D}(\mathbf{q})$ are the Fourier transforms of the $\varphi_{ij}^{'kk'}(0,l)/V$, where V is the volume of the crystal

$$\mathbf{D}_{ij}^{kk'}(\mathbf{q}) = \int_v \frac{1}{V} \varphi_{ij}^{'kk'}(0,l) \exp(2\pi i \mathbf{q} \cdot \Delta \mathbf{r}^l) \, dV. \tag{7}$$

Let us consider the nature of $\varphi_{ij}^{kk'}(0,l)$ or $\varphi_{ij}^{'kk'}(0,l)$: They are the second derivatives of the crystal energy with respect to the atomic coordinates or mass-weighted atomic coordinates, respectively. These second derivatives can be obtained like the corresponding first derivatives, starting from an adequate model, and the same empirical parameters could be used in both static and dynamical calculations. Because they can be derived from more general principles on an atom–atom basis, these parameters are no longer of one substance only and should be capable of transfer from one mineral to another. A notable result of this kind for minerals has been achieved by Price et al. (1987a, b) for the natural polymorphs of Mg_2SiO_4 (forsterite, wadsleyite, ringwoodite) and by Lam et al. (1990) for forsterite.

Thermodynamic Functions and Displacement Parameters

The use of consistent models for both the vibrational and static properties of crystals would imply definite advantages also in the range of the application and in the quality of the results. For instance, in agreement with the temperature-dependent character of thermodynamic functions, the vibrational contribution E_v to the total energy is the following:

$$E_v = E_0 + \Sigma_i g(v_i)\left(\frac{hv_i}{k}\right)\left[\exp\left(\frac{hv_i}{kT}\right) - 1\right]^{-1} \Delta v_i, \tag{8}$$

where v_i is the frequency of the ith mode; $g(v_i)$ is a normalized density of states function $[\Sigma_i g(v_i)\Delta v_i = n_f$, where n_f is the number of degrees of freedom considered in the calculations for the formula unit], and E_0 is the zero-point energy:

$$E_0 = R\Sigma_i g(v_i)\left(\frac{hv_i}{2k}\right)\Delta v_i, \tag{9}$$

where h and k are the Planck and Boltzmann constants, respectively. Therefore, the calculated value for energy becomes temperature-dependent, even in the harmonic approximation, and the zero-point term can also be evaluated. Contrary to widespread belief, this term is not quite negligible: For instance, according to our calculations, it amounts for forsterite to 14.1 kcal mol^{-1}, to be compared with a total vibrational energy of 18.2 kcal mol^{-1} at room temperature. Therefore, the packing energy minimum without any additional vibrational contribution does *not* correspond to the structure at 0 K, nor to any physically significant counterpart.

For entropy the vibrational contribution is the following:

$$S = \frac{E_{\text{vib}}}{T} - 3R\Sigma_i g(v_i) \ln\left[1 - \exp\left(\frac{hv_i}{kT}\right)\right]\Delta v_i. \tag{10}$$

This is also manifestly temperature-dependent, and in the absence of disorder in the atomic positions (as an exchange of different atoms or of spin alignment of similar atoms), it is the only one to be considered in practice for a pure compound. Similarly, for the specific heat c_v, we have

$$c_v = 3R\Sigma_i g(v_i)\left(\frac{hv_i}{kT}\right)^2 \exp\left(\frac{hv_i}{kT}\right)\left[\exp\left(\frac{hv_i}{kT}\right) - 1\right]^{-2}\Delta v_i. \tag{11}$$

As we have seen, the proper treatment of the vibrational contribution to thermodynamic functions also has the advantage of providing the necessary link with many branches of spectroscopy, especially, IR, Raman, or inelastic neutron scattering: Spectroscopic information of this kind is often invaluable in deriving adequate VFF constants (or the atomic charge) from best-fit to the experimental results. Feedback of crystallographic data is provided by the possibility of evaluating atomic displacement parameters (or **U**'s)

$$\mathbf{U}_p = (nm_p)^{-1}\Sigma_i E_i(2\pi v_i)^{-2}\mathbf{e}_{p,i} \cdot \mathbf{e}_{p,i}^{*T}, \tag{12}$$

where $\mathbf{e}_{p,i}$ is the mass-adjusted polarization vector for the atom p [which is part of an eigenvector **g** of the dynamical matrix **D(q)**], n the total number of unit cells in the crystal, and E_i the average energy of the ith mode

$$E_i = hv_i\left\{\tfrac{1}{2} + \left[\exp\left(\frac{hv_i}{kT}\right) - 1\right]^{-1}\right\}. \tag{13}$$

This evaluation of the **U**'s may be important not only for the "academic" exercise of reproducing the details of crystal structure and discussing their accuracy, but also for a general check of the behavior of the vibrational model *through the entire Brillouin zone*, and not at a certain point only, as for Raman- and infrared-active frequencies. Here, a further motive for using crystallographic information in connection with other properties is evident: For a detailed discussion of this point, see, e.g., Willis and Pryor (1975) or Gramaccioli (1987).

Since most minerals are not pure compounds, one might wonder whether this effort of calculating properties is indeed useful for wide application, e.g., the entropy of mixing might overshadow any vibrational contribution. However, one should consider that a deviation from ideal mixtures essentially depends on the difference in atom–atom interactions. Therefore, if consistent models can be obtained for pure crystals, the corresponding atom–atom interaction parameters might be used for developing adequate models for the nonideal behavior of solutions.

Practical Difficulties in Calculations

Why, in spite of all these possibilities, have only a modest number of lattice-dynamical calculations of this kind been carried out so far on minerals? Apart from tradition in mineralogy (see above) and chemistry, which has developed the tendency to confine these procedures to particular experts in solid-state physics, there are evident difficulties that must be overcome, at least if routine application on a wide basis is expected.

For instance, if the primitive unit cell of a certain mineral contains N atoms, vibrational frequency calculations in the harmonic approximation involve the diagonalization of a square matrix of order $3N$, and since in the most general case this matrix is complex, the order in practice becomes $6N$. A more worrying problem is the necessity of diagonalizing not only one matrix: It is essential to repeat the operation a considerable number of times by scanning the Brillouin zone appropriately. In the first work of this kind, the number of such matrices was on the order of several thousand.

For crystals involving bending, torsion, or many-body interactions extending outside a definite group (ion or molecule), it is not easy to codify an all-purpose routine in which the correct $\mathbf{q}\Delta\mathbf{r}^l$ values are obtained in any case. This is necessary when the calculations are performed not only at the center of the Brillouin zone ($\mathbf{q} = \mathbf{0}$), as for Raman- and infrared-active vibrational modes, but also for a general value of the wave vector $\mathbf{q}$, as it is required when the vibrational contributions to thermodynamic functions are also considered.

Other difficulties are connected with the apparent need to use complex models in lattice-dynamical calculations. In the Born–von Karman procedure, the simplest model is the so-called "short-range" model, which neglects Coulombic interactions as such, or the "rigid-ion" model, which considers Coulombic interactions but neglects ion polarizability. However, some authors such as Price (1987a, b) point out the inadequacy of the rigid-ion model, especially when it is necessary to account for the dielectric properties of a crystal correctly, and recommend instead the use of the so-called "shell" model (see, e.g., Cochran, 1971).

Other models addressing ion polarizability (e.g., the so-called "polarizable-ion" model) have been used for quartz by Iishi (1976, 1978a) or Iishi et al. (1983); applications to other minerals have also been considered (see, e.g., Iishi, 1978b for corundum). The harmonic approximation can also be expected to be inadequate for large amplitudes of motion, such as in phase-transition mechanisms or at high temperature, and the introduction of anharmonic models was consequently suggested [Choudhury et al. (1989)]. All these procedures are substantially more difficult to deal with, not only because of their complexity, but also because they require the introduction of a number of additional empirical constants, many of which are hardly significant. However, although the rigid-ion model does not interpret the dielectric properties of the crystal, it can nevertheless give good to excellent results in interpreting the vibrational frequencies of important minerals (see, e.g., Elcombe, 1967; Iishi, 1978b, c; Iishi et al., 1979). As an example, a comparison of the experimental Raman-active vibration frequencies for forsterite with the results of various calculations is given in Table 1: In spite of its greater complexity, the shell model (of Price et al.) does not yield decidedly better agreement with experimental data than the rigid-ion model

Table 1. Examples of vibrational frequencies (cm^{-1}) for forsterite at the Γ-point (for the complete set, see Pilati et al., 1990c). In the first column, the range of experimental measurements obtained by different authors is reported; the following columns, in sequence, report the calculated values of Iishi (1978b), Pilati et al. (1990c), Price et al. (1987a), and Rao et al. (1988).

	obs	Iishi	Pilati	Price	Rao
A_g	960–966	958	959	943	
	854–856	888	896	851	
Γ_1	822–826	837	841	807	
	606–609	606	659	630	
	541–545	535	566	580	
	420–424	414	504	427	
	334–340	358	374	360	380
	325–329	287	344	344	342
	304–305	269	295	312	293
	221–227	228	227	221	252
	181–183	157	193	184	178
B_{1g}	917–922	894	911	960	
(B_{3g})	588–595	580	596	606	
	407–412	411	492	417	
Γ_3	371–376	367	387	390	382
	312–318	304	333	327	328
	272	277	308	250	256
	226	234	257	133	174
B_{2g}	972–976	961	961	963	
(B_{1g})	863–866	891	898	864	
	835–839	846	845	821	
Γ_4	626–632	629	667	667	
	577–583	583	592	617	
	428–434	412	503	454	
	418	396	403	428	517
	314–318	316	349	337	367
	260–265	276	303	325	321
	215–224	230	247	269	264
	149–192	180	210	231	223

Table 2. Observed and calculated c_p and S (J/mol·K) for forsterite at various temperatures (K). The experimental data are from Robie et al. (1982), the calculated values were obtained from the rigid-ion and shell model, respectively, by Pilati et al. (1990c) and Price et al. (1987, in part interpolated).

	c_p			S		
T	obs	Pilati	Price	obs	Pilati	Price
20	0.27	0.31	0.22	0.08	0.02	0.08
77	18.64	19.21	18.83	6.72	6.50	6.78
298	118.5	118.4	115.2	94.0	94.0	92.07
453	143.3	145.6	141.4	148.9	148.3	145.7
578	154.1	157.4	153.0	185.5	185.0	182.1
723	162.9	165.6	160.9	221.2	220.2	217.4
883	170.5	171.5	166.0	254.6	252.9	250.1
1023	176.0	175.1	169.5	280.4	277.5	277.4
1173	180.7	178.1		304.5	300.8	
1293	183.9	180.2		322.4	317.6	

used by the other authors. Also the values of thermodynamic functions afforded by the rigid-ion model are quite satisfactory; an example for forsterite is given in Table 2.

In building up dynamical matrices to be diagonalized, Coulombic interaction between charged atoms is another difficult problem to be accounted for adequately, even with the simplest possible models like the rigid-ion one. This happens because of the very slow convergence of atom–atom interaction energy on increasing the maximum interaction distance; for this reason, summations are generally carried out on the reciprocal lattice, following a well-known procedure devised by Ewald (1921) in connection with the evaluation of Madelung constants. This procedure was extended to the evaluation of Fourier-transformed second derivatives for dynamical matrices in a correct way for the first time by Kellermann (1940). Although the papers just cited are "classic" and well established, the application to the general case involves considerable complexity, both in understanding the principle and in applying it to practice.

Another basic difficulty is the present virtual lack of empirical atom–atom energy fuctions that can be applied in general. For static energy calculations, the best-fit values for the atomic (ionic) radius and charge are often sufficient to obtain a reasonable value of packing energy and to have the actual crystal structure reproduced satisfactorily; the situation is, however, much more delicate when the same functions are tested on vibrational grounds.

Energy Minimization and Lattice Dynamics

In practice, the theoretical crystal structure corresponding to the empirical force field is obtained from a process of packing energy minimization, which is often based on first derivatives only with respect to the atomic positions within the unit cell and to the unit-cell parameters. As we have seen [see Eqs. (2, 3, 5–7)], vibrational calculations (in the harmonic approximation) are instead essentially based on *second* derivatives of energy with respect to atomic positions. The evaluation of the whole set of second derivatives surely is a much longer procedure than evaluating first derivatives; however, this brings definite advantages, even if calculations are performed only for a static model. Especially if the potential is obtained from the best fit of nonvibrational data, there is, in fact, no reason why the second derivatives for a certain minimum should be compatible with the actual vibration frequencies. The situation is not rarely still worse, the second derivatives' matrix being nonpositive definite and leading to imaginary frequencies or, in other words, to the instability of the calculated structure; a similar argument can be extended to any dynamical matrix, in general. In the practice of energy minimization, a process using first derivatives only, such as, for instance, the steepest descent, might be definitely simpler to consider and use than a method using also second derivatives, like the Raphson–Newton procedure. However, if mass-weighted atomic coordinates are used, the eigenvalues and eigenvectors of the second derivatives' matrix $\mathbf{D}(0)$ at convergence should correspond to the vibrational frequencies of the crystal for $\mathbf{q} = \mathbf{0}$, i.e., the results could be immediately compared with other easily accessible data, such as Raman- and infrared-spectra. A possible scheme could be the following:

$$\mathbf{D}(0)\Delta\mathbf{x}' = -\mathbf{F}. \tag{14}$$

The above equation represents a step in the Raphson–Newton iteration process, where $\mathbf{D}(0)$ is a matrix whose elements are the second derivatives of the energy E with respect to the mass-weighted coordinates, $\mathbf{F}$ is a first derivatives' "column" matrix, and $\Delta x_i'$ are the mass-weighted coordinate shifts. If all the atoms in the unit cell are considered independent, it is easy to notice that $\mathbf{D}(0)$ corresponds to the dynamical matrix for $\mathbf{q} = \mathbf{0}$ [see Eqs. (2–6)].

Therefore, after convergence has been reached, it is possible to check the vibration frequencies by diagonalizing $\mathbf{D}(0)$. In particular, if some of them are imaginary, we are dealing with a situation with no physical meaning, since it represents not even an energy minimum.

Sometimes, it might be useful to refer to the independent atoms only (i.e., to the so-called "asymmetric unit" in the crystal). In this case, the order of the matrices can be considerably reduced, with evident advantage. Here, the diagonalization of $\mathbf{D}(0)$ leads to the "all symmetric" vibrational modes for $\mathbf{q} = \mathbf{0}$: These modes correspond to shifts where all the symmetry operations of the space group are respected by the atomic positions at any instant. Although these modes are a fraction of the total number, a useful check is always possible and quite immediate.

In view of these advantages, the difficulties in finding a potential satisfactory in any respect might well be worth appropriate consideration, to be eventually overcome on a routine basis. This could be achieved only by combining a precise choice of the method of calculation together with the sequential order of the substances to be examined.

Rigid-Body Models

In all these procedures, the size of the matrices could be substantially reduced if some parts of the structure are considered "rigid bodies." This can be done most easily for molecular crystals (see, e.g., Gramaccioli, 1987), but even for ionic crystals such as olivine (forsterite) good results can be achieved (see, e.g., Ghose et al., 1987; Rao et al., 1988, and the corresponding column in Table 1). Figure 1 reports some of the lower phonon dispersion curves obtained by the authors using this approximation (and the rigid-ion model): The agreement with experimental measurements is striking.

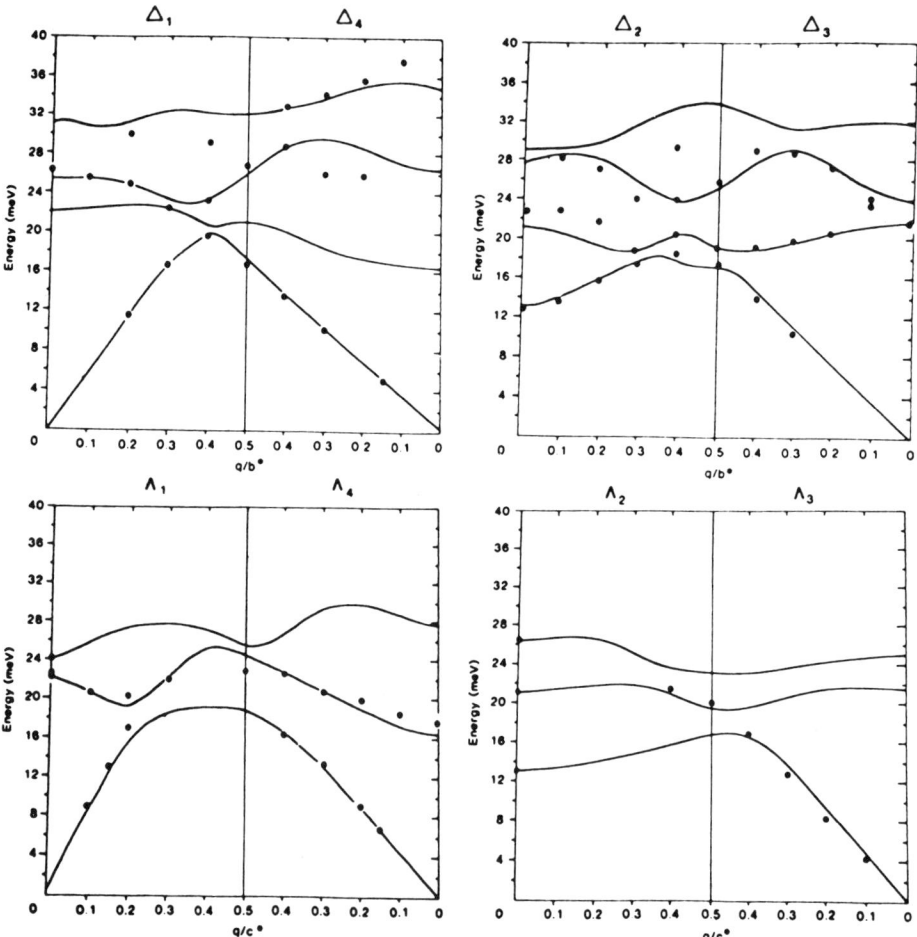

Fig. 1. Comparison of experimental and theoretical phonon dispersion curves for forsterite (after Ghose et al., 1987, Pergamon Press plc.); continuous lines are from calculations, full circles are experimental values. The calculated curves are *not* fitted to experimental data, but were *predictions* before the experiments were conducted.

Practical Problems in Calculating Thermodynamic Functions

The rigid-body model can be extended without difficulty to include some of the internal modes corresponding to the lowest vibrational frequencies in the isolated group, this has been explained in detail by Gramaccioli and Filippini (1983, 1985). According to these authors, the dynamical matrices $\mathbf{D}(\mathbf{q})$ as in Eq. (1) can be partitioned into blocks (or generalized "force constant" matrices) $\mathbf{G}_{st}(\mathbf{q})$, each of them corresponding to interactions between atoms of the group s with the group t, or their symmetric equivalents by translation. For each one of these blocks, it will be

$$\mathbf{G}'_{st}(\mathbf{q}) = \mathbf{M}_s^T \mathbf{G}_{st}(\mathbf{q}) \mathbf{M}_t. \tag{15}$$

Here, $\mathbf{M}_s$ and $\mathbf{M}_t$ are matrices whose columns are the eigenvectors $\mathbf{m}_i$ of the dynamical matrices $\mathbf{D}_s$ or $\mathbf{D}_t$ of the corresponding isolated molecule or group. Similarly to $\mathbf{D}(0)$, the elements of $\mathbf{D}_s$ (or $\mathbf{D}_t$) are of the kind

$$D_{ij}^{kk'} = (m_k m_{k'})^{1/2} \frac{\partial^2 E}{\partial x_{ik} \partial x_{jk'}} = \frac{\partial^2 E}{\partial x'_{ik} \partial x'_{jk}}, \tag{16}$$

where x'_{ik} and x'_{jk} are mass-weighted coordinates of the atoms k and k'. For a nonlinear isolated molecule or group, there are six eigenvalues of $\mathbf{D}_s$ equal to zero, corresponding to rigid translations or rotations of the group (or to their linear combinations).

For a rigid-body model, $\mathbf{M}_s$ consists of six columns only (five for a linear group), and the operation described in Eq. (15) reduces the order of $\mathbf{D}(\mathbf{q})$ to $6Z$, where Z is the number of these groups in the primitive unit cell. In view of the degeneracy corresponding to the zero eigenvalue, a diagonalization of $\mathbf{D}_s$ or similar matrices by any routine would not necessarily afford eigenvectors corresponding to a pure translational or rotational motion, or if this happens, these translations and rotations are not necessarily relative to the reference axes. For the sake of a better interpretation of the motion, it is advisable to replace these vectors by the following:

$\mathbf{m}_1 =$	$\mathbf{m}_2 =$	$\mathbf{m}_3 =$	$\mathbf{m}_4 =$	$\mathbf{m}_5 =$	$\mathbf{m}_6 =$
α_1	0	0	0	$x_1\beta_{21}$	$y_1\beta_{31}$
0	α_1	0	$-z_1\beta_{11}$	0	$x_1\beta_{31}$
0	0	α_1	$y_1\beta_{11}$	$-x_1\beta_{21}$	0
α_2	0	0	0	$z_2\beta_{22}$	$-y_2\beta_{32}$
0	α_2	0	$-z_2\beta_{12}$	0	$x_2\beta_{32}$
0	0	α_2	$y_2\beta_{12}$	$-x_2\beta_{22}$	0
$\vdots$	$\vdots$	$\vdots$	$\vdots$	$\vdots$	$\vdots$

(17)

where $\alpha_k = (m_k/m)^{1/2}$, $\beta_{ik} = (m_k/I_i)^{1/2}$ (m being the mass and I_i the ith principal moment of inertia of the group). This normal coordinate basis has the advantage that internal movements of the group corresponding to the lowest frequencies (which define the paths of easiest deformation) can be easily included, just by adding further columns to the $\mathbf{M}$'s.

By using this procedure, the reduction of the order of the matrix can be quite considerable for complex organic molecular crystals, but also for minerals the advantage is evident. For instance, for α-sulfur, the order of the dynamical matrix is 192 (96 for $\mathbf{q} = \mathbf{0}$), and it can be reduced to 80(40) by taking into account the two lowest-frequency internal modes (both twofold degenerate), practically without affecting the results.

The use of rigid or partially rigid models does not necessarily imply neglect of the contribution of the higher internal frequencies; these contributions (to the values of thermodynamic functions or the U's) can be simply added to the total, by assuming them to be constant throughout the entire Brillouin zone, with no mixing with the lattice modes (Gramaccioli and Filippini, 1983). This approximation is reasonable on both theoretical and experimental grounds.

An example of the success of this procedure in interpreting the experimental data for α-sulphur is given in Tables 3 (thermodynamic functions) and 4 (atomic displacement parameters). The tensor $\mathbf{W} = \langle \mathbf{q}_i \mathbf{q}_j \rangle$ indicates the importance and coupling of the external and lowest-frequency internal modes and is reported as an example in Table 5.

Table 3. Values of some thermodynamic functions for crystals of α-sulphur at room temperature (298 K) calculated by Gramaccioli and Filippini (1984) using an extended rigid-body model. Experimental data from Guthrie et al. (1954) and Eastman and McGavock (1937).

	calc	obs	
Entropy (S)	60.9	61.0(4)	cal mol^{-1}K^{-1}
Heat of sublimation	23.94 (24.36)^a	24.35(5)	kcal mol^{-1}
C_v	41.2	40.8(4)	cal mol^{-1}K^{-1}
C_p	43.3	43.2(2)	cal mol^{-1}K^{-1}

^{a} The calculated value for the heat of sublimation without considering the zero-point energy is reported within parentheses; the worse agreement with the experimental result of the corrected value derives from not having considered this effect in deriving the van der Waals potential (Rinaldi and Pawley, 1975).

Table 4. Atomic displacement parameters $\mathbf{B}$ ($\times 10^4$) for α-sulphur calculated by Gramaccioli and Filippini (1984); the corresponding experimental results are taken from Pawley and Rinaldi (1972) and are shown immediately below the calculated values.

	B_{11}	B_{22}	B_{33}	B_{12}	B_{13}	B_{23}
S1	114	53	12	13	5	2
	119	56	11	6	4	7
S2	107	51	16	−16	2	2
	108	49	17	−6	6	−8
S3	83	67	14	3	−4	7
	85	57	16	14	−3	2
S4	67	84	11	0	5	−2
	74	87	11	−12	6	0

Table 5. The tensor **W** for α-sulfur (atomic mass units $\times$ Å^2; referred to the principal axes of inertia of the molecule, after Gramaccioli and Filippini, 1983). For $q_7 - q_8$ and $q_9 - q_{10}$, which are the internal mode frequencies considered here, the calculated frequency for the free molecule is reported in the first column.

	$q_1(B)$	$q_2(A)$	$q_3(A)$	$q_4(B)$	$q_5(A)$	$q_6(A)$	$q_7(A)$	$q_8(B)$	$q_9(A)$	$q_{10}(B)$
q_1	6.420	0.000	0.000	−0.193	0.000	0.000	0.000	−0.119	0.000	0.035
q_2		8.637	−0.743	0.000	1.207	−1.349	−0.445	0.000	0.079	0.000
q_3			5.486	0.000	−0.980	0.075	0.128	0.000	−0.044	0.000
q_4				3.447	0.000	0.000	0.000	0.045	0.000	−0.054
q_5					3.713	−1.026	−0.023	0.000	−0.016	0.000
q_6						6.662	0.457	0.000	0.018	0.000
q_7 $(E_2; 67$ cm$^{-1})$							1.084	0.000	0.011	0.000
q_8								0.984	0.000	−0.017
q_9 $(E_2; 156$ cm$^{-1})$									0.280	0.000
q_{10}										0.285

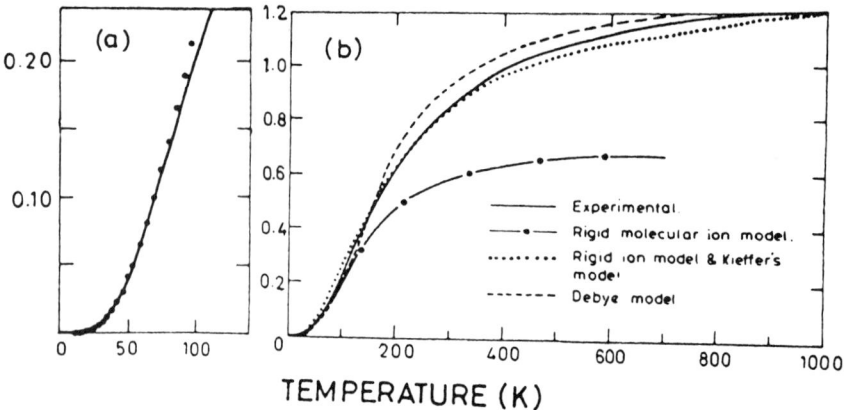

Fig. 2. Comparison of experimental specific heat data with calculated results (after Rao et al., 1988). (a) Low-temperature region showing good agreement of the continuous line given by the rigid-body (molecular ion) model with the experimental data shown by the filled circles. No fitting is attempted. (b) Comparison of experimental data (continuous line) with results of various theoretical models as indicated in the figure. The discrepancy between the results of the rigid-body model and those from the experiment is due to omission of the internal modes' contribution.

Likewise for forsterite, the order of the complete dynamical matrix is 168 (84), and it can be reduced to 48 (24), if the SiO_4 tetrahedron is considered rigid and the two Mg atoms independent points. The values of the specific heat at various temperatures calculated by Rao et al. (1988) are reported in Fig. 2: The disagreement at higher temperatures can probably be substantially eliminated if the contribution of the internal modes as specified above is introduced.

A similar procedure can be used (with evident advantage) in the Raphson–Newton minimization. For instance, if a single independent group is present in the crystal, for a single step we have [corresponding to Eq. (14)]:

$$\mathbf{M}^T \mathbf{D}(0) \mathbf{M} \Delta \mathbf{q} = -\mathbf{M}^T \mathbf{F}. \tag{18}$$

Here, **M** is the eigenvector matrix (in this case, unique), and $\Delta \mathbf{q}$ is the column matrix corresponding to the shifts in normal coordinates. These shifts can be easily referred to mass-weighted (and Cartesian) coordinates, since

$$\mathbf{M} \Delta \mathbf{q} = \Delta \mathbf{x}'. \tag{19}$$

Here, $\Delta \mathbf{x}'$ corresponds to shifts in mass-weighted coordinates, as is evident from the above equations.

Further Useful Steps in Simplification

Another important simplification involves the reduction of the number of sampled points in the Brillouin zone. For instance, 8000 symmetry-independent points were considered by Reid and Smith (1970) in one of the best theoretical

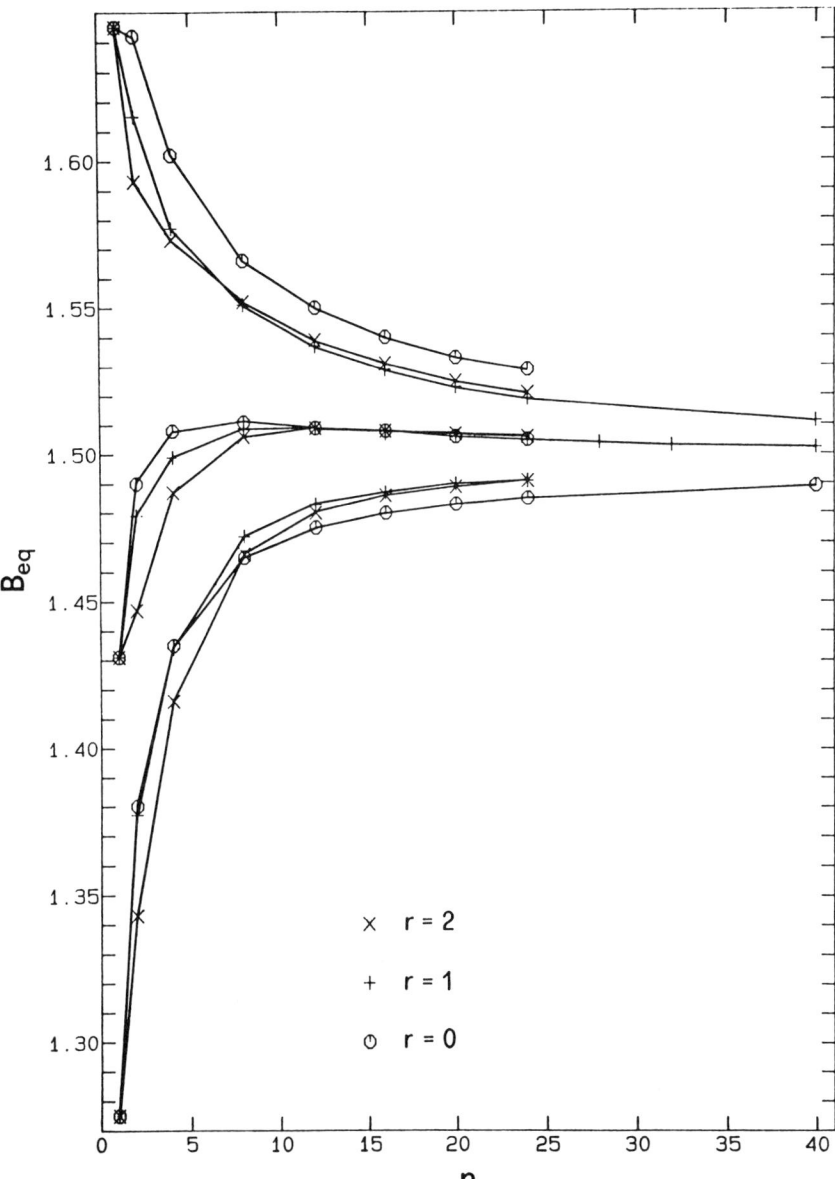

Fig. 3. Change in the estimated B_{eq} values (Å^2) for the Cl atom in NaCl as a function of the Brillouin-zone sampling (from Pilati et al., 1990a). Each curve joins the corresponding points belonging to the same sequence, but differing in the number n of sampling intervals along each reciprocal axis.

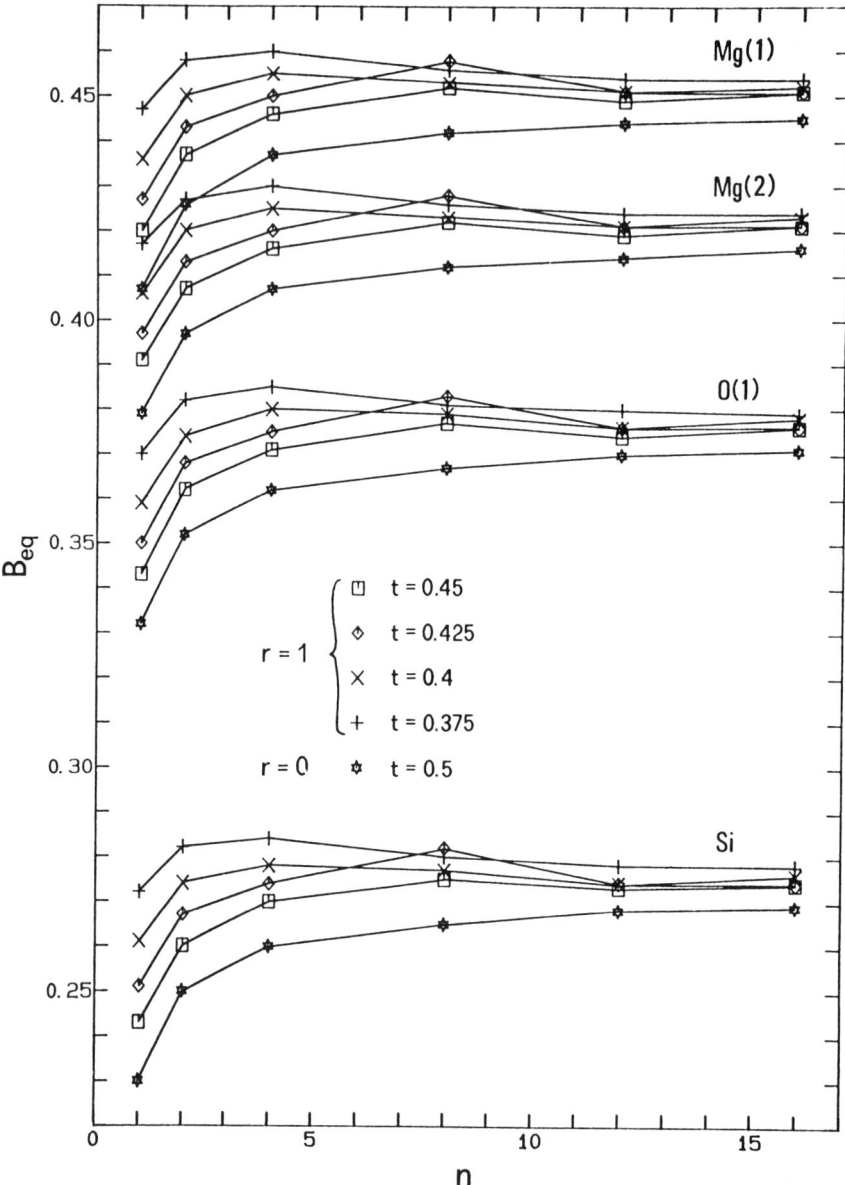

Fig. 4. Change of the B_{eq} values (Å^2) for various atoms in the forsterite structure at 298 K as a function of the Brillouin-zone sampling (from Pilati et al., 1990a). The kind of sequence is determined by the parameters r and t.

estimations of the atomic displacement parameters (**B**'s) for alkali halides; for thermodynamic functions, in general, a reasonable estimate can be obtained even with a considerably smaller number of points [on the order of 20–30, or even less (see, e.g., Price, 1987b, or Pilati et al., 1990a)]. However, if a suitable progression is chosen, even 100 to 200 points can be sufficient for determining all properties within the corresponding experimental accuracy (Filippini et al., 1976; Kroon and Vos, 1978, 1979; Pilati et al., 1990a). Figures 3 and 4 show convergence as a plot of the equivalent Debije–Waller factor B_{eq} against the number of grid spacing with several kinds of sequence for NaCl and forsterite.

In our computer programs, the above-mentioned simplifications have been considered; moreover, satisfactory coding for the atom–atom interactions has been developed, so that the correct contribution to dynamical matrices is obtained in any case even for nonzero values of the wave vector **q** and in the presence of interactions involving more than two atoms, such as bond-angle bending and torsion. For Coulombic interactions, a new routine has been written that can be easily understood on crystallographic terms; this routine performs summations in the reciprocal lattice, following a modification of Bertaut's method (Bertaut, 1952, 1978a, b, 1983, 1985, 1986; Pilati et al., 1990b). Besides its simplicity in application, this routine has the advantage of considering the contribution of the macroscopic field quite naturally, leading to the so-called TO–LO splitting for infrared-active modes.

An Example: Forsterite

Since forsterite has so far been the best example for such calculations, the atomic displacement parameters as calculated by Pilati et al. (1990c) are here reported in Table 6, together with the corresponding experimental values, obtained by different authors from accurate X-ray diffraction measurements and subsequent crystal-structure refinement. Similarly, a drawing of the thermal ellipsoids by the *ORTEP*II program (Johnson, 1976) is reported in Fig. 5. These results show remarkable agreement, with the exception of Hazen's (1976), in which, however, considerable systematic errors are present. An interesting feature is the considerable importance of the zero-point contribution (see Table 7), which corresponds to about one-half of the total mean-square displacement at room temperature.

All this confirms the validity of the rigid-ion model, at least for most purposes. The reason why the rigid-ion model works, in spite of its approximation, can be ascribed to the essential good agreement of the lowest calculated frequencies with the experimental data (see Ghose et al., 1987); for the highest values (where experimental phonon dispersion curves are not available), there might be considerable differences. This is similar to what happens for alkali halides, and especially here the shell model should prove its advantage (see, e.g., Cochran, 1973, Woods et al., 1960, 1963, and Fig. 6). However, since in practice only the lowest energy levels are occupied, at least for reasonable values of temperature, this explains the good agreement for thermodynamic functions and atomic displacement parameters.

Table 6. Lattice-dynamical evaluation of anisotropic atomic displacement parameters ($\times 10^5$) for forsterite at room temperature, compared with the corresponding results from crystal structure refinement. The cell parameters in our reference are chosen so that $a > b > c$. The temperature factor is in the form

$$T = \exp[-2\pi^2(U_{11}h^2a^{*2} + \cdots + 2U_{23}klb^*c^*)].$$

	U_{11}	U_{12}	U_{13}	U_{22}	U_{23}	U_{33}	$U_{eq}(\text{Å}^2)$
Pilati et al. (1990c), lattice dynamical							
Mg(1)	707	−88	−44	566	−52	544	606
Mg(2)	538	0	39	596	0	570	568
Si	400	0	9	383	0	292	358
O(1)	581	0	1	568	0	319	489
O(2)	421	0	−7	554	0	454	476
O(3)	553	78	34	452	6	446	484
Langen (1987), model G (high-order reflections)							
Mg(1)	702	−110	−10	469	−59	450	540
Mg(2)	466	0	15	547	0	590	534
Si	415	0	14	405	0	286	369
O(1)	699	0	36	576	0	315	530
O(2)	421	0	5	607	0	526	518
O(3)	677	163	31	523	−26	515	572
Bocchio et al. (1986) (crystal PB9, high-order reflections)							
Mg(1)	578	−118	0	575	−69	539	564
Mg(2)	339	0	17	635	0	692	555
Si	292	0	15	500	0	381	391
O(1)	573	0	−42	644	0	400	539
O(2)	329	0	−10	702	0	589	540
O(3)	530	149	30	630	−32	599	586
Hazen (1976) (data at 23°C and 1 atm)							
Mg(1)	540	−93	20	180	−40	280	333
Mg(2)	320	0	0	290	0	250	289
Si	230	0	12	150	0	90	157
O(1)	480	0	50	420	0	100	333
O(2)	370	0	−20	400	0	280	350
O(3)	490	120	20	290	−40	250	343
Fujino et al. (1981)							
Mg(1)	710	−111	−15	512	−55	477	566
Mg(2)	490	0	22	596	0	592	560
Si	437	0	2	422	0	300	385
O(1)	620	0	10	560	0	340	508
O(2)	430	0	10	600	0	500	510
O(3)	630	148	15	510	−27	510	551

Table 6. (Continued.)

	U_{11}	U_{12}	U_{13}	U_{22}	U_{23}	U_{33}	$U_{eq}(\text{Å}^2)$
Birle et al. (1968)							
Mg(1)							418
Mg(2)							456
Si							253
O(1)							443
O(2)							532
O(3)							519

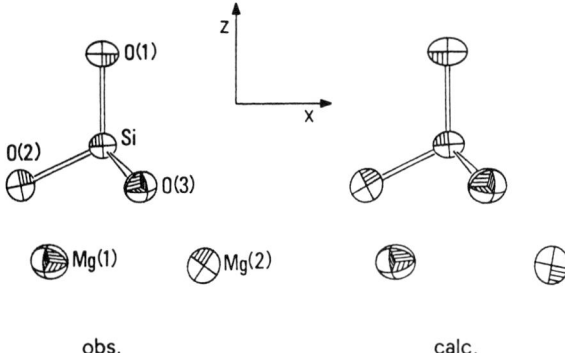

Fig. 5. An *ORTEP*II drawing (projection on the mirror plane {010}) of the calculated and observed thermal ellipsoids at 99.9% probability for the asymmetric unit of forsterite at room temperature (from Pilati et al., 1990c).

Table 7. Zero-point contribution to the atomic displacement parameters ($\times 10^5$) for forsterite, evaluated by Pilati et al. (1990c). The units, reference system, and conventions are the same as in Table 6.

	U_{11}	U_{12}	U_{13}	U_{22}	U_{23}	U_{33}	U_{eq}
Mg(1)	297	−25	−14	260	−16	263	273
Mg(2)	253	0	12	274	0	272	266
Si	171	0	3	166	0	143	160
O(1)	299	0	0	298	0	183	260
O(2)	212	0	−12	296	0	260	256
O(3)	284	41	17	235	−6	256	258

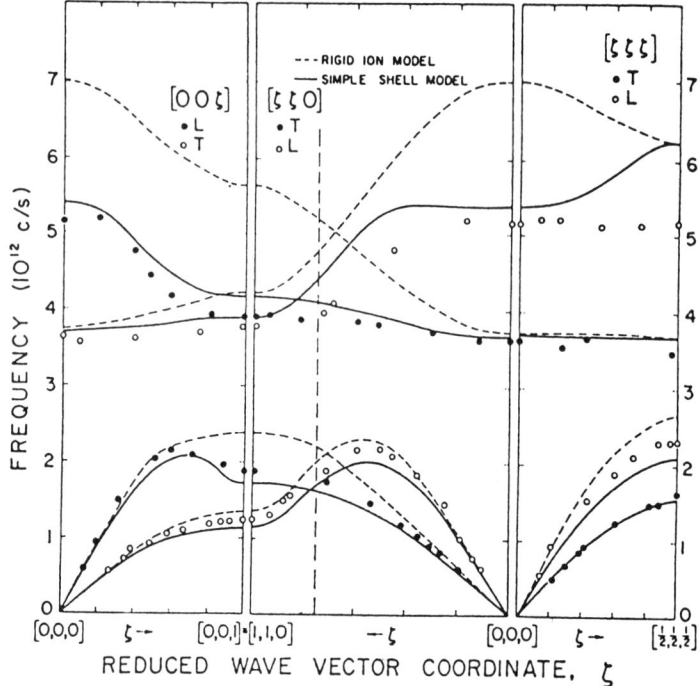

Fig. 6. Phonon dispersion curves for sodium iodide around 100 K (after Woods et al., 1963). The measured points are compared with calculations based on the rigid-ion (dashed) or simple shell model (continuous lines). The broken vertical line indicates the [110] zone boundary. The essential agreement of the two models for the lower branches is evident.

Forsterite can also be taken as an example of how an inconsistent model can be obtained if only energy minimization on a first derivatives' basis is considered. For instance, if the parameter set by Miyamoto et al. (1982) (countersigned as "W" in Price and Parker, 1984) is used, apparently satisfactory agreement with the observed unit-cell parameters and atomic coordinates in the crystal structure can be obtained. However, the second derivatives clearly indicate that the point is not a real minimum. In fact, if we use the coordinates corresponding to the equilibrium point and the same force field, the evaluation of vibration frequencies gives seven imaginary values (from $67i$ to $533i$ cm^{-1}); see Table 8. Therefore, neither the force field nor the proposed equilibrium structure are physically significant.

For instance, if energy minimization is carried out with respect to the mass-weighted coordinates of the asymmetric unit in forsterite, the eigenvalues of the second derivatives' matrix correspond to the so-called Ag frequencies. And using the force field mentioned above, we see that one of these is imaginary ($265i$ cm^{-1}) and the highest one (1874 cm^{-1}) is about twice its experimental value (966 cm^{-1}

Table 8. Frequencies (cm^{-1}) at the Γ point ($\mathbf{q} = \mathbf{0}$), corresponding to the packing energy "minimum" of Miyamoto et al. (1982) and evaluated using the same potential.

A_u	394i	196i	166	207	396	534	603	712	918	1678			
A_g	265i	238	287	426	449	613	1001	1136	1437	1817	1874	1797	1862
B_{1u}	195	310	331	429	493	578	683	784	922	1069	1641		
B_{2u}	336i	133	409	449	476	597	716	891	1646				
B_{3u}	104	263	347	395	585	636	676	737	910	1105	1477	1808	1878
B_{1g}	489i	67i	292	563	595	1206	1691						
B_{2g}	133	272	378	451	524	623	1144	1214	1457	1828	1908		
B_{3g}	533i	265	380	499	618	1231	1654						

according to Iishi, 1978c; see Table 1). Therefore, the lack of physical validity of this solution is evident even when a smaller matrix is used.

References

Au, A.Y. and Weidner, D.J. (1986). Theoretical modelling of the elastic properties of forsterite: A polyhedral approach. *Phys. Chem. Miner.* **13**, 360–370.

Bertaut, F. (1952). L'Energie electrostatique de reseaux ioniques. *J. Phys.* **13**, 499–505.

Bertaut, F. (1978a). The equivalent charge concept and its application to the electrostatic energy of charges and multipoles. *J. Phys.* **39**, 1331–1348.

Bertaut, F. (1978b). Electrostatic potentials, fields and field gradients. *J. Phys. Chem. Solids* **39**, 97–102.

Bertaut, F. (1983). Energie dipolaire d'une structure modulee. *C. R. Acad. Sci. Ser. II* **296**, 1123–1127.

Bertaut, F. (1985). Analyse de representation de la phase incommensurable de K_2SeO_4, cas d'un groupe d'espace non symmorphe. *C.R. Acad. Sci. Ser. II* **300**, 589–594.

Bertaut, F. (1986). Champs, energies coulombienne, dipolaire et de polarisation dans une structure incommensurable. *C.R. Acad. Sci. Ser. II* **302**, 1137–1142.

Birle, J.D., Gibbs, G.V., Moore, P.B., and Smith, J.V. (1968). Crystal structures of natural olivines. *Amer. Mineral.* **53**, 807–824.

Bocchio, R., Brajkovic, A., and Pilati, T. (1986). Crystal chemistry of the olivines in the peridotites from the Ivrea–Verbano zone (Western Italian Alps). *Neues Jahrb. Miner. Monatsh.* **7**, 313–324.

Born, M. and Huang, K. (1954). *Dynamical Theory of Crystal Lattices*. Oxford, Clarendon Press.

Burnham, C.W. (1990). The ionic model: Perceptions and realities in mineralogy. *Amer. Mineral.* **75**, 443–463.

Busing, W.R. (1981). WMIN, a computer program to model molecules and crystals in terms of potential energy functions. ORNL-5747, U.S. Technical Inform. Sev.

Busing, W.R. and Matsui, M. (1984). The application of external forces to computational models of crystals. *Acta Cryst.* **A40**, 532–538.

Catlow, C.R.A. and Mackrodt, W.C. (1982). Computer simulation of solids, in *Lecture Notes in Physics*, Vol. 166, Springer-Verlag, Berlin.

Catti, M. (1982). Atomic charges in Mg_2SiO_4 (forsterite), fitted to thermoelastic and structural properties. *J. Phys. Chem. Solids* **43**, 1111–1118.

Catti, M. (1989). Modelling of structural and elastic changes of forsterite (Mg_2SiO_4) under stress. *Phys. Chem. Miner.* **16**, 582–590.

Choudhury, N., Chaplot, S.L., and Rao, K.R. (1989). Equation of state and melting point studies of forsterite. *Phys. Chem. Miner.* **16**, 599–605.

Cochran, W. (1971). Lattice dynamics of ionic and covalent crystals. *C.R.C. Crit. Rev. Solid State Sci.* **2**, 1–83.

Cochran, W. (1973). *The Dynamics of Atoms in Crystals*. London, Arnold.

Eastman, E.D. and McGavock, W.C. (1937). The heat capacity and entropy of rhombic and monoclinic Sulfur. *J. Amer. Chem. Soc.* **59**, 145–151.

Elcombe, M. (1967). Some aspects of the lattice dynamics of quartz. *Proc. Phys. Soc.* **91**, 947–958.

Ewald, P.P. (1921). Die Berechnung optischer und elektrostatischer Gitterpotentiale. *Ann. Phys. (Leipzig)* **64**, 253–287.

Filippini, G., Gramaccioli, C.M., Simonetta, M., and Suffritti, G. B. (1976). Lattice-dynamical applications to crystallographic problems: Consideration of the Brillouin zone sampling. *Acta Cryst.* **A32**, 259–264.

Fujino, K., Sasaki, S., Takeuchi, Y., and Sadanaga, R. (1981). X-ray determination of electron distributions in forsterite, fayalite and tephroite. *Acta Cryst.* **B37**, 513–518.

Ghose, S. (1985). Macroscopic to microscopic, in *Lattice Dynamics, Phase Transitions and Soft Modes, Reviews in Mineralogy*, S.W. Kieffer and A. Navrotsky, eds., Mineralogy Society of America, Washington, D.C.; Vol. 14, Chap. 4.

Ghose, S., Hastings, J.M., Corliss, L.M., Rao, K.R., Chaplot, S.L., and Choudhury, L. (1987). Study of phonon dispersion relations in forsterite, Mg_2SiO_4 by inelastic neutron scattering. *Solid State Commun.* **63**, 1045–1050.

Gramaccioli, C.M. (1987). Spectroscopy of molecular crystals and crystallographic implication. *Int. Rev. Phys. Chem.* **6** (4), 337–349.

Gramaccioli, C.M. and Filippini, G. (1983). Lattice-dynamical evaluation of temperature factors in non-rigid molecular crystals: A first application to aromatic hydrocarbons. *Acta Cryst.* **A39**, 784–791.

Gramaccioli, C.M. and Filippini, G. (1984). Lattice-dynamical calculations for orthorhombic sulfur: A non-rigid molecular model. *Chem. Phys. Lett.* **108**, 585–588.

Gramaccioli, C.M. and Filippini, G. (1985). Thermal motion for non-rigid molecules in crystals: Symmetry of the generalized mean-square displacement tensor **W**. *Acta Cryst.* **A41**, 356–361.

Guthrie, G.B., Jr., Scott, D.W., and Waddington, G. (1954). Thermodynamic functions and heat of formation of S_8 (gas). *J. Amer. Chem. Soc.* **76**, 1488–1493.

Hazen, R.N. (1976). Effects of temperature and pressure on the crystal structure of forsterite. *Amer. Mineral.* **61**, 1280–1293.

Iishi, K. (1976). The analysis of the phonon spectrum of α-quartz based on a polarizable-ion model. *Z. Krist.* **144**, 289–303.

Iishi, K. (1978a). Lattice-dynamical study of the β-quartz phase transition. *Amer. Mineral.* **63**, 1190–1197.

Iishi, K. (1978b). Lattice Dynamics of Corundum. *Phys. Chem. Miner.* **3**, 1–10.

Iishi, K. (1978c). Lattice dynamics of forsterite. *Amer. Mineral.* **63**, 1198–1208.

Iishi, K., Salje, E., and Werneke, C. (1979). Phonon spectra and rigid-ion model calculations on andalusite. *Phys. Chem. Miner.* **4**, 173–186.

Iishi, K., Miura, M., Shiro, Y., and Murata, H. (1983). Lattice dynamics of α-quartz including the effect of the width of the atomic electron distribution. *Phys. Chem. Miner.* **9**, 61–66.

Johnson, C.K. (1976). *ORTEP II: A FORTRAN thermal-ellipsoid plot program for crystal structure illustrations*. Rept. ORNL-5138. Oak Ridge Nat. Lab. Tenn.

Kellermann, E.W. (1940). Theory of the vibrations of the sodium chloride lattice. *Phil. Trans. R. Soc. Lond.* **238**, 513–548.

Kieffer, S.W. (1985). Macroscopic to microscopic, in *Heat Capacity and Entropy: Systematic Relations to Lattice Vibrations, Reviews in Mineralogy*, 14, S.W. Kieffer and A. Navrotsky, eds., Mineralogy Society of America, Washington, D.C.; Vol. 14, Chap. 3.

Kroon, P.A. and Vos, A. (1978). Convergence of Brillouin zone summations. *Acta Cryst.* **A34**, 823–824.

Kroon, P.A. and Vos, A. (1979). Thermal diffuse scattering for molecular crystals: Error in X-ray diffraction intensities and atomic parameters. *Acta Cryst.* **A35**, 675–684.

Lam, P.K., Rici, Y., Lee, M.W., and Sharma, S.K. (1990). Structural distorsions and vibrational modes in Mg_2SiO_4. *Amer. Mineral.* **75**, 109–119.

Langen, R. (1987). Ph.D. Thesis, Die Abhängigkeit der Kationenverteilung in einem Mg–Fe-Olivin (Som Carlos, Arizona vom Sauerstoffpartialdruck. Rheinisches Friedrich-Wilhelm Universität, Bonn.

Matsui, M. and Busing, W.R. (1984a). Computational modeling of the structure and elastic constants of the olivine and spinel forms of Mg_2SiO_4. *Phys. Chem. Miner.* **11**, 55–59.

Matsui, M. and Busing, W.R. (1984b). Calculation of the elastic constants and high-pressure properties of diopside, $CaMgSi_2O_6$. *Amer. Mineral.* **69**, 1090–1095.

Matsui, M. and Matsumoto, T. (1982). An interatomic potential-function model for Mg, Ca and CaMg olivines. *Acta Cryst.* **A38**, 513–515.

Miyamoto, M., Takeda, H., Fujino, K., and Takeuchi, Y. (1982). The ionic compressibilities and radii estimated for some transition metals in olivine structure. *Miner. J.* **11**, 172–179.

Ottonello, G. (1986). Energetics of multiple oxides with spinel structure. *Phys. Chem. Miner.* **13**, 79–90.

Parker, S.C., Catlow, C.R.A., and Cormack, A.N. (1983). Prediction of mineral structure by energy minimisation techniques. *J. Chem. Soc., Chem. Commun.* **529**, 936–938.

Pawley, G.S. and Rinaldi, R.P. (1972). Constrained refinement of orthorhombic sulphur. *Acta Cryst.* **B28**, 3605–3609.

Pilati, T., Bianchi, R., and Gramaccioli, C.M. (1990a). Evaluation of atomic displacement parameters by lattice-dynamical calculations: Efficiency in Brillouin-zone sampling. *Acta Cryst.* **A46**, 485–489.

Pilati, T., Bianchi, R., and Gramaccioli, C.M. (1990b). Evaluation of Coulombic lattice sums for vibrational calculations in crystals: An extension of Bertaut's method. *Acta Cryst.* **A46**, 309–315.

Pilati, T., Bianchi, R., and Gramaccioli, C.M. (1990c). Lattice-dynamical estimation of atomic thermal parameters in silicates: Forsterite α-Mg_2SiO_4. *Acta Cryst.* **B46**, 301–311.

Price, G.D. and Parker, S.C. (1984). Computer simulations of the structural and physical properties of the olivine and spinel polymorphs of Mg_2SiO_4. *Phys. Chem. Miner.* **10**, 209–216.

Price, G.D., Parker, S.C., and Leslie, M. (1987a). The lattice dynamics of forsterite. *Miner. Magazine* **51**, 157–170.

Price, G.D., Parker, S.C., and Leslie, M. (1987b). The lattice dynamics and thermodynamics of the Mg_2SiO_4 polymorphs. *Phys. Chem. Miner.* **15**, 181–190.

Price, G.D., Parker, S.C., and Yeomans, J. (1985). The energetics of polytypic structures: A computer simulation of magnesium silicate spinelloids. *Acta Cryst.* **B41**, 231–239.

Rao, K.R., Chaplot, S.L., Choudhury, L., Ghose, S., Hastings. J.M. and Corliss L.M. (1988). Lattice dynamics and inelastic neutron scattering from forsterite, Mg_2SiO_4: Phonon dispersion relation, density of states and specific heat. *Phys. Chem. Miner.* **16**, 83–97.

Reid, J.S. and Smith, T. (1970). Improved Debye–Waller factors for some alkali halides. *J. Phys. Chem. Solids* **31**, 2689–2697.

Rinaldi, R. and Pawley, G.S. (1975). An investigation of the intermolecular modes in orthorhombic sulphur. *J. Phys. C* **8**, 599–616.

Robie, R.A., Hemingway, B.S., and Takei, H. (1982). Heat capacities and entropies of Mg_2SiO_4, Mn_2SiO_4, and Co_2SiO_4 between 5 and 380 K. *Amer. Miner.* **67**, 470–482.

Vieillard, P. (1982). Modele de Calcul des Energies de Formation des Mineraux, Bati sur la Connoissance Raffinée des Structures Cristallines. Memoire 69, Université Louis Pasteur de Strasbourg, Institut de Geologie.

Willis, B.T.M. and Pryor, A.W. (1975). *Thermal Vibrations in Crystallography.* Cambridge University Press, Cambridge, U.K.

Woods, A.D.B., Cochran, W., and Brockhouse, B.N. (1960). Lattice dynamics of alkali halide crystals. *Phys. Rev.* **119**, 980–999.

Woods, A.D.B., Brockhouse, B.N., and Cowley, R.A. (1963). Lattice dynamics of alkali halide crystals. II. Experimental studies of KBr and NaI. *Phys. Rev.* **131**, 1025–1029.

Chapter 9
Predictions of the Entropies of Molecules and Condensed Matter

M. Blander and C.R. Stover*

Introduction

In this paper, we discuss a statistical mechanical theory for calculating the standard nonelectronic entropies (S_T^0) and free energy functions [$(G_T^0 - H_{298}^0)/T$] of substances at high temperatures. These quantities are important and are often the only unknown data necessary for determining free energies of compounds, which is necessary for the calculation of chemical and phase equilibria. This lack of data is particularly significant at high temperatures that are important in the genesis of magmas and metamorphic rocks.

Accurate enthalpies of formation of compounds are relatively easy to measure calorimetrically. In order to calculate the free energies of compounds from enthalpies of formation at different temperatures, one needs a knowledge of the entropies as a function of temperature. On the other hand, with a knowledge of the free energy functions of a compound as a function of temperature, one needs only a single measurement of the enthalpy of formation of that compound to determine the free energies over a range of temperatures. Thus, since there are very many substances for which such enthalpies of formation are the only known thermodynamic data, a major expansion of standard tables of free energies of formation is possible if one can predict entropies or free energy functions for these substances.

In prior papers (Frurip and Blander, 1980, Frurip et al., 1982a, b; Blander and Stover, 1985), we have shown that the high-temperature (700 K) entropies and

---

*Summer 1983 Student Research Participant from Swarthmore College, Swarthmore, PA 19081.
The submitted manuscript has been authored by a contractor of the U.S. Government under contract No. W-31-109-ENG-38. Accordingly, the U.S. Government retains a nonexclusive, royalty-free license to publish or reproduce the published form of this contribution, or allow others to do so, for U.S. Government purposes.

free energy functions of all the nonhydrogenic vapor molecules in the JANAF tables (Chase et al., 1986; Barin and Knacke, 1973; Barin et al., 1977) and of molten alkali halides are in good agreement with the predictions of this theory. This means that one can predict these quantities with only a knowledge of a size parameter and of the masses of the atoms. For molecules, it means that one does not need to know the vibrational frequencies, which are unknown and/or difficult to measure for many high-temperature molecules. In this paper, we will review the theory and detail the proof of a modification that increases its accuracy (Blander and Stover, 1985; Stover, 1983). We will then explore the potential utility of the theory for solids by comparison with data on the entropies of alkali halides and alkaline earth oxides and sulfides. These preliminary calculations suggest that our approach could be used to make predictions of entropies and free energy functions of solids with applications to simple solids of importance in geology being possible at present. Ultimately, it should be possible to extend this theory to the more complex solids found in many geologic systems. The development and extension of the theory would enable one to greatly expand the available data on free energies of formation of compounds at high temperatures and improve the ability to calculate chemical and phase equilibria in geologic systems.

Statistical Mechanical Theory

The standard entropy of a molecular gas containing one or two different elements A and/or B, S_T^0, is given by the expression

$$\frac{S_T^0}{R} = \left\{ 1 + \frac{3(n_A + n_B)}{2} + \ln\left[\frac{ZQ_e}{N} \prod_{i=A,B} \left(\frac{2\pi m_i kT}{h^2}\right)^{3n_i/2} \right] + \frac{T\partial \ln Q_e}{\partial T} + \frac{Td \ln Z}{dT} \right\}, \tag{1}$$

where Z is the configurational integral

$$Z = \frac{1}{n_A! n_B!} \int \cdots \int_V e^{-\beta U} d\tau_1 \cdots d\tau_{n_A} d\tau_1 \cdots d\tau_{n_B}, \tag{2}$$

in which $\beta = 1/kT$, U is the total potential between the atoms of the molecule $A_{n_A} B_{n_B}$, and $d\tau_i$ are volume elements in real space that contain the center of an atom. The integral (2) is integrated over all possible configurations of the molecule in the volume V. In addition, Q_e is the electronic partition function, m_i the atomic mass, h Planck's constant, k Boltzmann's constant, T temperature, and N Avogadro's number. The molecule is thus considered to be a collection of the atoms that interact with the total potential U, which is a function of the atom coordinates. The total potential is assumed to be the sum of the pair potentials. In the theory, one first performs a dimensional analysis of the configurational integral for an ionic molecule $A_{n_A} B_{n_B}$, where A is a cation and B an anion. The total potential is then given by

$$U = \sum_i^{n_A} \sum_j^{n_B} U_{ij} + \sum_{i<i'}^{n_A} \sum^{n_A} U_{ii'} + \sum_{j<j'}^{n_B} \sum^{n_B} U_{jj'}. \quad (3)$$

When one considers two such ionic molecules, one with an interionic size parameter d and a second (test salt) with a size parameter d_o, then the relative values of the total potential are given by

$$U_d(r_{ij}) = \frac{d}{d_o} U_{d_o}\left(\frac{d_o}{d} r_{ij}\right) = g U_{d_o}\left(\frac{r_{ij}}{g}\right),$$

where g is a scaling parameter d/d_o. From this, one can show that the configurational integrals for the two molecules are related through the expression

$$Z(T, V, d) = g^{3(1-n_A-n_B)} Z_o(T/g, V), \quad (4)$$

where Z is the integral for any salt and Z_o the integral for the test salt. In the original paper (Frurip and Blander, 1980), it was shown that at constant V and T, the nonelectronic entropies for a molecule with the size parameter d are given by the expression

$$\frac{S(T, V)}{R} = \sigma_o(T, V) - \ln\left[\frac{n_A! n_B!}{(n_A + n_B)!}\right] + \tfrac{3}{2}\ln(m_A^{n_A} m_B^{n_B}) + 3(n_A + n_B - 1)\ln d, \quad (5)$$

where $\sigma_o(T, V)$ is a universal constant for all molecules with the given number of atoms $(n_A + n_B)$, and m_A and m_B are the atomic masses of the A and B atoms, respectively. A very similar expression has been shown to be valid for the free energy function $(G^0 - H_{298}^0)/T$. A correction to the coefficient of the $\ln d$ term was deduced (Blander and Stover, 1985; Stover, 1983) that improved the accuracy of the predictions for vapor molecules from the theoretical expression (5). Using the scaling property of Z with g for fixed d_o as given above and writing out $(\partial Z/\partial T)_g$ and $(\partial Z/\partial g)_T$ in terms of Zd_o, we can easily show that $(\partial Z/\partial g)T = 1/g[3(n_A + n_B - 1)Z + T(\partial Z/\partial T)_g]$ at constant V. By using an expression for S in terms of Z, it follows that

$$\left[\frac{\partial(S/R)}{\partial \ln g}\right]_{V,T} = \left[3(n_A + n_B - 1) + \frac{\partial}{\partial T}\left(T^2 \frac{\partial \ln Z}{\partial T}\right)\right],$$

where the expression on the right-hand side is the coefficient of the $\ln d$ term in the equation for S/R and $\partial/\partial T[T^2(\partial \ln Z/\partial T)]$ is a correction term to this coefficient in Eq. (5). Defining $\partial/\partial T[T^2(\partial \ln Z/\partial T)]$ as δ, we can show that $\delta = \langle(\beta U)^2\rangle - \langle\beta U\rangle^2$ and is a positive number equal to $[(C_V/R) - 3(n_A + n_B)/2]$, where C_V is the constant volume heat capacity. Because of the significant improvement in the accuracy of the predictions of relative values of entropies (and free energy functions), it is recommended that this correction term be included

Predictions of the Entropies

in calculations for vapor molecules by using the expression

$$\frac{S(T,V)}{R} = \sigma_o(T,V) - \ln\frac{n_A!n_B!}{(n_A+n_B)!} + \tfrac{3}{2}\ln(m_A^{n_A}m_B^{n_B})$$
$$+ \left[\frac{C_V(T)}{R} + \frac{3(n_A+n_B)}{2} - 3\right]\ln d. \tag{6}$$

Equation (6) has been derived for ionic molecules. It was shown that the difference between such ionic molecules and realistic molecules with the same number of atoms is equal to the differences between the constant volume heat capacities for these two types of molecules (Frurip and Blander, 1980). At high temperatures, these differences are very small for nonhydrogenic vapor molecules, indicating that Eq. (6) is valid for all such molecules at high temperatures.

Comparisons with Data

Successful comparisons of Eqs. (5) or (6) with data on all the molecules, clusters, and molecular ions in the JANAF tables and on all 20 molten alkali halides (Blander and Stover, 1985) have been made using Eqs. (5) (Frurip and Blander, 1980; Frurip et al., 1982a, b) and (6). (Blander and Stover, 1985) In this approach, the condensed phase is equivalent to a very large molecule with the order of Avogadro's number of atoms. For example, plots of the entropies vs. the last two terms in Eq. (6) were made. (Blander and Stover, 1985) Calculated values of the intercepts and standard deviations are given in Table 1 for AB_4, AB_5, and AB_6, as well as for the molten alkali halides. Part of the deviations from Eq. (6) are, of course, due to the inaccuracies in the known data. In any case, the data are consistent with the predictions of this simple statistical mechanical equation

Table 1. Values and standard deviations for σ_o.[a]

Molecule[b]	Intercept		SD	
	1000 K	5000 K	1000 K	5000 K
AB_4	17.347	38.139	0.605	0.596
AB_5	15.549	41.122	0.814	0.807
AB_6	12.046	42.379	1.159	1.174
	2.548[c]	—	0.768[c]	—
AB(liq.)	−0.338[d]	—	0.402[d]	—

[a] Taken from Blander and Stover (1985).
[b] AB_4 = $TiBr_4$, CCl_4, CF_4, $TiCl_4$, $PbCl_4$, SiF_4, SiI_4, SF_4, $ZrCl_4$, ZrI_4, $ZrBr_4$, ZrF_4, $PbBr_4$, PbI_4, PbF_4, $SiCl_4$; AB_5 = ClF_5, $NbCl_5$, PCl_5, $TaCl_5$, IF_5, PF_5, $NbBr_5$; AB_6 = $MoCl_6$, WCl_6, MoF_6, SF_6, WF_6, WBr_6; AB(liq.) = all 20 alkali halides.
[c] d = Interatomic spacing of the vapor molecules in Å.
[d] $d^3 = V$ in cm^3, where V is the molar volume at room temperature.

with relatively small standard deviations. In some prior cases, it was shown that free energy functions were also in accord with a similar equation. In cases in which dimensionless entropies or free energy functions deviated by more than 2 from these predictions, it was shown that the data were incorrect. Most significant for geologists is the fact that the entropies of molten alkali halides are consistent with Eq. (6). This presents the possibility that data for condensed phases might be consistent with Eq. (6). Equation (6) is valid for gaseous molecules that have different types of bonding and structures, with σ_o being a function only of the total number of atoms $(n_A + n_B)$ in the molecule. As will be seen below, this universality does not appear to hold for crystalline solids. However, our preliminary examination of some simple solids indicates that the theory could prove to be useful when σ_o is restricted to a subclass of solids of a given stoichiometry.

We performed calculations on simple 1 : 1 mostly cubic solids to illustrate the potential of Eq. (6). We do not have good values of C_V/R for such a calculation in which the coefficient of the $\ln d$ term in Eq. (6) is $(C_V/R + 3)$. The average values of C_p/R at 1200 K for the alkali halides and for alkaline earth oxides and sulfides are about 8.39 and 6.69, respectively. Since $C_p > C_V$, we will assume an effective value of 8 for $(C_V/R + 3)$ in our calculations on 1:1 solids. Since the correlations we are making are not very sensitive to this quantity, a more precise value is not warranted at present. In Fig. 1 we exhibit a plot of the dimensionless

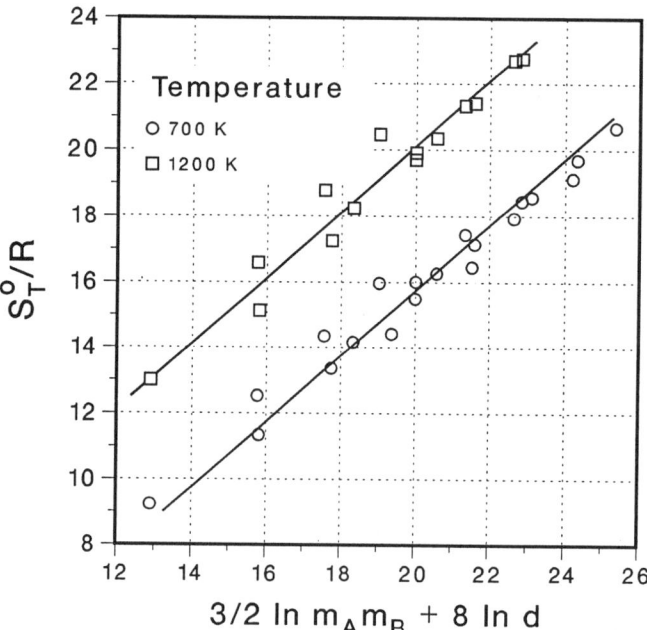

Fig. 1. Comparisons of measured entropies of solid alkali halides at 700 and 1200 K with the predictions of the theory. The theory is represented by the solid lines that have a unit slope. All 20 alkali halides are represented at 700 K, and, at 1200 K, data were not available for six alkali halides (RbF, RbCl, RbBr, RbI, CsBr, and CsI).

entropies of alkali halides at 700 and 1200 K vs. $3/2 \ln (m_A m_B) + 8 \ln d$, where m_A and m_B are the atomic weights of the cation and anion, respectively, and d is an interionic distance between the cation and the anion that is taken to be proportional to the cube root of the molar volume at room temperature. The average values of the first two terms on the RHS of Eq. (6) are -4.266 ± 0.592 at 700 K and 0.108 ± 0.622 at 1200 K. The points for LiCl, LiBr, and LiI in these two plots lie about 1–1.4 dimensionless entropy units above the predicted values plotted from these intercepts [i.e., the lines exhibited in Fig. 1 with a slope of 1 and intercepts given by the first two terms in Eq. (6)]. Because of high anion/cation radius ratios for these three salts, the cations will have considerably more freedom of motion in their lattice sites, which might be expected to increase the entropies of these salts. If these three salts are excluded from the averaging, the values of the intercepts are -4.449 ± 0.388 at 700 K and -0.181 ± 0.237 at 1200 K. We thus see that, except for LiCl, LiBr, and LiI, the alkali halides are consistent with the predictions of Eq. (6), with a relatively small standard deviation and a maximum deviation of about 0.5 dimensionless units for NaF. With all the lithium halides included, the standard deviations are larger and the maximum deviation is about 1.4 dimensionless units for LiI.

Analogous series of compounds are formed by alkaline earth oxides (BeO, MgO, CaO, SrO, BaO) and their corresponding sulfides. In Fig. 2 are plotted values of S^0/R Barin and Knacke (1973); Barin et al. (1977); M.W. Chase et al. (1986); L.B. Pankratz (1982) for these two series vs. $3/2 \ln m_A m_E + 8/3 \ln V$, where V, the molar volume near room temperature, is taken to be proportional to d^3. The average values of the intercepts are for the oxides -6.464 ± 0.735 at 700 K and -3.045 ± 0.768 at 1200 K and for the sulfides -6.690 ± 0.782 at 700 K and -3.128 ± 0.827 at 1200 K. The closeness of the values of these intercepts for the oxides to those for the sulfides at the same temperature suggests that these two populations can be combined to give common intercepts of -6.577 ± 0.725 at 700 K and of -3.086 ± 0.754 at 1200 K that were used to plot the lines in Fig. 2. It should be noted that the points for BeS in Fig. 2 deviate most from the predictions of theory. BeS has a very high anion/cation radius ratio (analogous to LiCl, LiBr, and LiI), which can lead to higher entropies than predicted in a manner that is similar to the lithium halides. If BeS is omitted from the calculation, the average values of the intercepts at 700 and 1200K are -6.713 ± 0.621 and -3.243 ± 0.754, respectively.

The difference between the intercepts for the alkali halides and those for these oxides and sulfides is 2.1 at 700 K and 3.2 at 1200 K. The reasons for these differences are not clear. A clue to part of the difference can be deduced from the relation mentioned earlier, which to first-order terms equates deviations from the theoretical predictions to differences in the constant volume heat capacities. For example, at 1200 K, the differences between the average value of C_p/R for the 14 alkali halides that are included in the plot of Fig. 2 and the average value of C_p/R for the alkaline earth oxides and sulfides is 1.70. This is substantial but not as large as the difference in the intercepts. Of course, the differences in C_p are not exactly equal to the differences in C_V but are not likely to be very different. In any case, the predicted correlations appear to be valid for each of these two

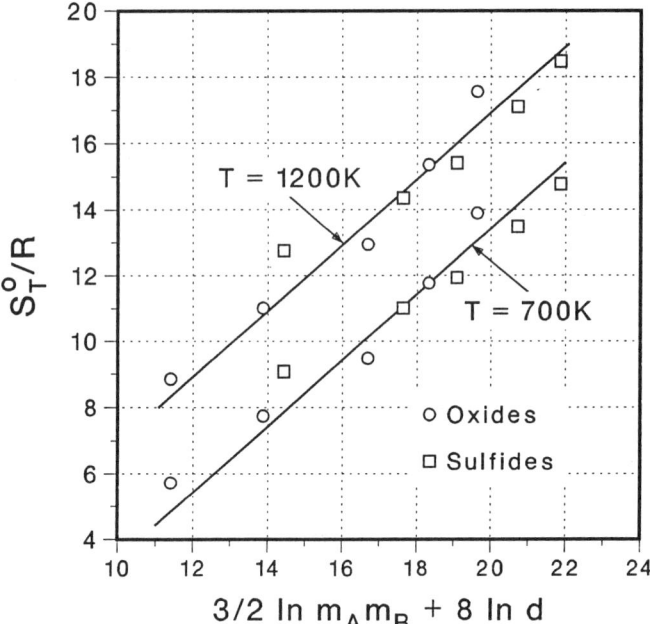

Fig. 2. Comparisons of measured entropies of alkaline earth oxides and sulfides at 700 and 1200 K with the predictions of the theory. The theory is represented by the solid lines that have a unit slope.

separate classes of materials. Further theoretical studies are necessary to understand this effect more completely. No such differences in the intercepts nor of the heat capacities at fairly high temperatures were encountered in our studies of gas molecules. (Frurip and Blander, 1980; Frurip et al., 1982a, b; Blander and Stover, 1985).

Discussion

A statistical mechanical theory that leads to a simple expression for the relative values of nonelectronic entropies (and free energy functions) of substances seems to apply to vapor molecules, simple liquids (alkali halides), and solids (alkali halides and alkaline earth oxides and sulfides). For vapor molecules, the theory is universal and applies to all non-hydrogenic molecules with the same total number of atoms, regardless of differences in structure and/or bonding. For such molecules, the theory greatly simplifies the calculation of entropies and free energy functions. This should permit one to greatly expand tables of thermodynamic data on standard free energies of formation of vapor compounds since there are many compounds for which the enthalpies of formation are known (mostly at 298 K) and entropies and free energy functions are unknown. From this theory, one can deduce that the universality is, at least in part, related to the fact that the high-temperature heat capacities of vapor molecules with any given total number of atoms differ little from each other.

For condensed matter, this universality is not likely to hold since the high-temperature heat capacities of, e.g., 1 : 1 and 1 : 2 halides (e.g., KBr and $CaBr_2$) are significantly larger than those of the corresponding 1 : 1 and 1 : 2 oxides and sulfides (e.g., CaO and TiO_2). As a consequence, this theory can only be applied to a class of materials such as, e.g., all halides or all oxides and sulfides of a given stoichiometry. Further study is required to define the range of the classes of solids (and liquids if enough data are available) to which such a theory applies. For example, our preliminary examination of 1 : 2 solid oxides (e.g., TiO_2, SiO_2, ThO_2, etc.) indicated that the theory applies. In addition, further theoretical studies of the differences in values of σ_o for different classes of solids of the same stoichiometry are needed. Differences in structure will have to be taken into account. For example, most of the compounds plotted in Figs. 1 and 2 are cubic, some being body-centered, some face-centered. Neither beryllium oxide nor beryllium sulfide are cubic. The aim is to clearly demonstrate that the theory is a useful tool for making predictions. For geologists, this theory has the potential to significantly expand tables of needed thermodynamic data on free energies of formation of compounds and to improve capabilities for calculations of chemical and phase equilibria. This potential can be realized by extending the theory to predict entropies and free energy functions for simple and complex solid compounds.

Acknowledgment

This work was performed under the auspices of the U.S. Department of Energy, Division of Materials Sciences, under Contract No. W-31-109-ENG-38.

References

Barin, I. and Knacke, O. (1973). *Thermochemical Properties of Inorganic Substances.* Springer-Verlag, New York.

Barin, I., Knacke, O., and Kubaschewski, O. (1977). Supplement, *Thermochemical Properties of Inorganic Substances.* Springer-Verlag, New York.

Blander, M. and Stover, C. (1985). *High Temp. Sci.* **19**, 231–241.

Chase, M.W. et al. (1986). *JANAF Thermochemical Tables*, 3rd ed., Parts I and II. American Chemical Society and American Institute Physics Washington, D.C.

Frurip, D.J. and Blander, M. (1980). *J. Chem. Phys.* **73**, 509.

Frurip, D.J., Blander, M., and Chatillon, C. (1982a). Statistical mechanical predictions of entropies and free energy functions for small clusters of atoms, in *Metal Bonding and Interactions in High Temperature Systems with Emphasis on Alkali Metals*, J.L. Gole and W.C. Stwalley, eds., ACS Symposium Series, Washington, D.C., Vol. 179.

Frurip, D.J., Chatillon, C., and Blander, M. (1982b). *J. Phys. Chem.* **86**, 647.

Pankratz, L.B. (1982). *Properties of elements and oxides*. U.S. Bureau of Mines Bull. 672, U.S. Govt. Printing Office, Washington, D.C.

Stover, C.R. (1983). *Entropies for MX_2—type molecules and a classical statistical mechanical dimensional model*. Undergraduate Res. Prog. Rep. to Div. Educational Prog., Argonne Nat. Lab. Argonne, IL.

Chapter 10
Systematics of Bonding Properties and Vibrational Entropy in Compounds

G. Grimvall and A. Fernández Guillermet

Introduction

The phase stability of a compound is determined by its Gibbs energy, $G(T,P) = H - TS$. The enthalpy (or enthalpy of formation) is often known from direct measurements or from estimates. The entropy S has been much less studied, and that will be the main theme of this paper. For details, the reader is referred to our recent work, which has dealt with, e.g., $3d$-transition metal carbides, nitrides, and oxides (Fernández Guillermet and Grimvall, 1989a, 1990, 1992); transition metal diborides (Fernández Guillermet and Grimvall, 1990b); and alkali halides and hydrides (Grimvall and Rosén, 1983; Häglund and Grimvall, 1990), as well as pure transition metals (Fernández Guillermet and Grimvall, 1989b, c; Grimvall et al., 1987). Here we shall add some material on geologically interesting solids.

The outline of the chapter is as follows. In the next section we introduce the concept of an entropy Debye temperature $\theta_S(T)$ with aluminum silicates chosen as an illustrative example. We also compare, for corundum, the temperature dependence of θ_S with that of a conventional Debye temperature derived from the temperature-dependent elastic constants. In the section that follows we show how the atomic masses separate from the interatomic forces in the quantity θ_S. This allows a comparison of chemically similar compounds, where the difference in atomic masses is the main reason for the difference in entropy. We also introduce an effective force constant k_S that is derived from the entropy data. We then devote the next section to a study of regularities in k_S and consider two groups of compounds that have the same crystal structure (NaCl) but very different chemical bonding, alkali halides and some transition metal carbides. We end with a brief account of the role of correlations for bonding properties in the modeling and prediction of phase diagrams.

Definition of an Entropy Debye Temperature θ_S

We shall only be concerned with the temperature-dependent part of the entropy and leave out the entropy due to, e.g., atomic disorder. Then, for nonmetallic and nonmagnetic systems, we can restrict ourselves to the entropy of atomic vibrations, S_{vib}. The textbook result for a single harmonic oscillator of frequency ω (energy $\hbar\omega$) is

$$S_{osc}(\omega, T) = k_B \left\{ \left(\frac{\hbar\omega}{2k_B T}\right) \coth\left(\frac{\hbar\omega}{2k_B T}\right) - \ln\left[2 \sinh\left(\frac{\hbar\omega}{2k_B T}\right)\right] \right\}. \quad (1)$$

In a solid with a frequency distribution (phonon density of states) $F(\omega)$, we get

$$S_{vib}(T) = \int_0^{\omega_{max}} S_{osc}(\omega, T) F(\omega) \, d\omega. \quad (2)$$

If $F(\omega)$ has the form of a Debye model,

$$F(\omega) = \frac{9\omega^2}{\omega_D^3}, \quad (3)$$

we get the Debye-model entropy S_D, here expressed per atom. The high-temperature expansion of S_D is

$$S_D(T) = 3k_B \left[\frac{4}{3} + \ln\left(\frac{T}{\theta_D}\right) + \left(\frac{1}{40}\right)\left(\frac{\theta_D}{T}\right)^2 - \left(\frac{1}{2240}\right)\left(\frac{\theta_D}{T}\right)^4 + \cdots \right], \quad (4)$$

where θ_D is the Debye temperature, i.e., $\hbar\omega_D = k_B \theta_D$.

It is often convenient to represent thermodynamic information through a Debye temperature. Suppose that we made the Debye-model expression for the entropy equal to the vibrational entropy of an actual system. For each temperature, that equation has a solution θ_D that we shall call an "entropy Debye temperature" $\theta_S(T)$. If we instead equate the Debye-model expression for the heat capacity with the actual heat capacity, we obtain a "heat capacity Debye temperature" $\theta_C(T)$. The two Debye temperatures are equal only if the real system has the Debye frequency distribution, Eq. (3), except for the limit $T \to 0$ and accidental coincidence at other temperatures. The entropy Debye temperature is often preferable to the heat capacity Debye temperature, because θ_S is always mathematically well defined, whereas the equation for θ_C has no real solution if anharmonic effects make the vibrational heat capacity larger than the maximum for harmonic vibrations, i.e., the Dulong–Petit limit $3k_B$ per atom.

As an illustration, we consider aluminum silicate, Al_2SiO_5. The JANAF thermochemical tables (1985) list entropies for three crystalline phases: andalusite, kyanite, and sillimanite. The result for $\theta_S(T)$ is given in Fig. 1. The dashed line in Fig. 1 gives $\theta_C(T)$ for andalusite and shows its failure to account for the heat capacity at high temperatures.

As another illustration, consider the high-temperature limit of Eq. (4) and

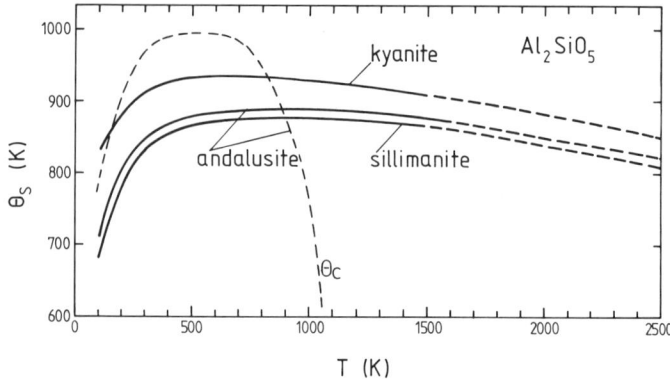

Fig. 1. The entropy Debye temperature $\theta_S(T)$ for three phases of aluminum silicate, Al_2SiO_5: andalusite, kyanite, and sillimanite. Experimental (solid line) and estimated (dashed line) data from the JANAF thermochemical tables (1985). We also show the heat capacity Debye temperature $\theta_C(T)$ for andalusite; dashed curve.

compare θ_S of two crystalline phases of the same compound. Their vibrational entropy difference ΔS is a constant at high T. When their θ_S differ by only a small amount $\Delta\theta_S$, we have (with ΔS being an average per atom)

$$\frac{\Delta\theta_S}{\theta_S} = -\frac{\Delta S}{3k_B}. \qquad (5)$$

The entropy difference between the α-phase and the β-phase of cryolite (Na_3AlF_6) at 1000 K is given (JANAF, 1985) as $\Delta S' = (509.179 - 497.868)$ J/(K·mol). Then, $\Delta\theta_S/\theta_S = -\Delta S'/(30R) = -0.045$.

There are two fundamentally different reasons for the temperature dependence of $\theta_S(T)$: one related to low temperatures and one to high temperatures. If the actual system does not have a Debye spectrum, the solution $\theta_S(T)$ will vary with the temperature when $T < \theta_S$. For high temperatures and strictly harmonic vibrations, $\theta_S(T)$ asymptotically approaches a value that represents the logarithmically averaged phonon frequency. This is seen if (2) is expanded at high temperatures and then compared with (4). For a real solid, the solution $\theta_S(T)$ derived from experiments almost always decreases with T when $T > \theta_S$. That is caused by anharmonic effects in the lattice vibrations, mainly indirectly through the thermal expansion.

The general shape of $\theta_S(T)$ in Fig. 1 is characteristic of most solids. In particular, one does not expect any humps or irregularities for $T > \theta_S$. Therefore, plots of $\theta_S(T)$ can be used as a check on experimental data, see a discussion of irregularities in the heat capacity of Ca (Grimvall and Rosén, 1982). Often, $\theta_S(T)$ has a minimum at low temperatures ($T/\theta_S \sim 0.1$) (Rosén and Grimvall, 1983). However, Be and Mg indicate that $\theta_S(T)$ may decrease continuously with increasing T from 0 K (Rosén and Grimvall, 1983). The precise behavior depends

Systematics of Bonding Properties and Vibrational Entropy

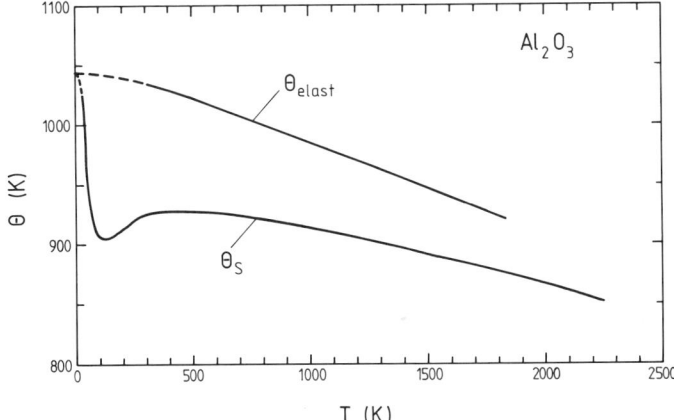

Fig. 2. Two Debye temperatures, $\theta_{\text{elast}}(T)$ and $\theta_S(T)$, for corundum (α-Al_2O_3). $\theta_{\text{elast}}(T)$ is from Goto et. al. (1989) and based on their measured elastic constants. $\theta_S(T)$ is evaluated by us from the entropy data of Ditmars and Douglas (1971) below 100 K and data given in the JANAF thermochemical tables (1985) at higher temperatures.

on how the Debye model can reproduce the properties of the real phonon spectrum. See also Fig. 2.

The maximum value of $\theta_S(T)$ in Fig. 1 gives a good account of the logarithmically averaged phonon frequency $\omega_{\log}$ through the relation

$$k_B \theta_S = \hbar \omega_{\log} \exp(\tfrac{1}{3}). \tag{6}$$

The initial decrease of θ_S with T at high temperatures reflects how $\omega_{\log}$ decreases. At very high temperatures, the anharmonic effects are complicated, but $\theta_S(T)$ is still a very useful parameter in the modeling of the high-temperature thermodynamic properties of solids.

For a quick calculation of $\theta_S(T)$ from a given table of entropy data, one may use Eq. (4) at high temperatures and an interpolation in the tables of the Debye model entropy at intermediate and low temperatures ($T < \theta_S/2$). See, e.g., tables in the *American Institute of Physics Handbook* (1972).

Another chapter in these proceedings deals with the Komada–Westrum modeling of realistic phonon spectra. In that approach, the thermal properties are described by a characteristic temperature $\theta_{KW}(T)$ that normally varies very little with T. However, their $\theta_{KW}(T)$ would also decrease with T if the model is used to represent the high-temperature vibrational entropy, and it would fail to represent a heat capacity per atom that is larger than $3k_B$.

The ordinary heat capacity Debye temperature referring to low temperatures, where the Debye T^3 law holds, can also be calculated from the elastic constants c_{ij}. Thus, we define the Debye temperature θ_{elast}. Since the elastic constants vary with temperature, for the same reason that $\theta_S(T)$ varies with T when $T > \theta_S$, we get a temperature-dependent $\theta_{\text{elast}}(T)$. Figure 2 compares $\theta_{\text{elast}}(T)$ from measurements of $c_{ij}(T)$ (Goto et al., 1989) with $\theta_S(T)$ from entropy data (JANAF, 1985; Ditmars and Douglas, 1971) for corundum (α-Al$_2$O$_3$). Because θ_S and θ_{elast} give different weights of the phonon spectrum, we do not expect their temperature dependence to be identical, but they have the same value in the limit $T \to 0$.

Separation of Atomic Masses and Interatomic Forces in the Entropy Debye Temperature

We noted above that a single parameter θ_S suffices to describe the high-temperature vibrational entropy, and that this parameter measures the logarithmically averaged phonon frequencies. A result from the theory of lattice vibrations (Grimvall and Rosén, 1983) says that in this average, the atomic masses are separated from the interatomic forces. That fact is important in a study of chemically similar compounds.

Let the temperature be so high that θ_S represents the logarithmic average. Mathematically, one must have $T \gg \theta_S$, but in practice, $T > \theta_S/2$ may suffice. We can formally write the separation of masses and forces as

$$k_B \theta_S = \hbar \left(\frac{k_S}{M_{\text{eff}}} \right)^{1/2}. \tag{7}$$

The effective mass M_{eff} is the logarithmically averaged mass in the compound. For a crystal $A_a B_b C_c$, one has

$$(a + b + c) \ln M_{\text{eff}} = a \ln M_a + b \ln M_b + c \ln M_c. \tag{8}$$

As an example, for sodium chloride $M_{\text{eff}} = (M_{\text{Na}} M_{\text{Cl}})^{1/2}$. The effective force constant k_S is a very complicated average over all interatomic forces in the solid. Here we only use the fact that k_S measures the bonding strength. One should note that k_S refers to an average over all bonds in a solid and not only to those that, e.g., are important for the hardness.

It is natural to assume that the interatomic forces are similar in chemically related compounds. If k_S is identical in two compounds, their θ_S (at high T) would differ only by the effect introduced through the effective mass. As an example, consider ZrTe$_5$ and HfTe$_5$ that have the same crystal structure. The heat capacities have been measured and at room temperature they are not far from the Dulong-Petit value 18R per formula unit (R is the gas constant). Therefore, the standard entropies $^0S(298.15 \text{ K})$ should give θ_S values that represent the logarithmically averaged phonon frequencies. Shaviv et al. (1989) obtained $^0S(\text{HfTe}_5) - {}^0S(\text{ZrTe}_5) = 32.99R - 31.96R = 1.03R$. The effective mass ratio is

$$\frac{M_{\text{eff}}(\text{HfTe}_5)}{M_{\text{eff}}(\text{ZrTe}_5)} = \left(\frac{M_{\text{Hf}}}{M_{\text{Zr}}} \right)^{1/6} \approx 1.118.$$

Hence, by Eqs. (4) and (7), the entropy difference per mol of formula units at high temperatures and from the mass effect alone would be $18R \ln(1.118)^{1/2} \approx 1.004R$, i.e., in a remarkable agreement with the measured entropy difference, $1.03R$. ZrTe$_5$ and HfTe$_5$ are examples of what is termed low-dimensional solids. Their crystal structure is chainlike and in some respects they can be viewed as having dimensions between 1 and 2. Therefore, the low-temperature heat capacity deviates from the usual T^3 law. At room temperature, however, one is in the classical regime of lattice vibrations for these compounds, and the separation of masses from force constants in θ_S holds in spite of the special crystal structure.

We remark here that Latimer's (1951) rule for the estimation of the vibrational entropy of compounds is closely related to the separation of interatomic forces and atomic masses. Briefly, this rule has had some success because it correctly accounts for the mass effect in vibrational entropy (Grimvall and Rosén, 1983; Grimvall 1983).

Regularities in the Variation of the Interatomic Forces

Anderson and Nafe (1965) noted, for the bulk modulus B, that $\log(B)$ varies almost linearly with $\log(\Omega)$ for many chemically related solids. Here, Ω is the volume per atom in the compound. We take a similar approach and investigate trends in the effective force constants k_S for compounds that have the same crystal structure, i.e., alkali halides and some transition metal compounds. Figures 3 and 4 show k_S vs. Ω in logarithmic plots. There is a clear correlation between k_S and Ω for the alkali halides, but it is not as good as for the bulk modulus, and lithium halides and alkali hydrides fall on trend lines different from those in Fig. 3 (Häglund and Grimvall, 1991). The force constants k_S of the NaCl-structure carbides, nitrides, and oxides formed with titanium and vanadium, respectively, follow two simple but different trends when plotted vs. Ω (Fig. 4).

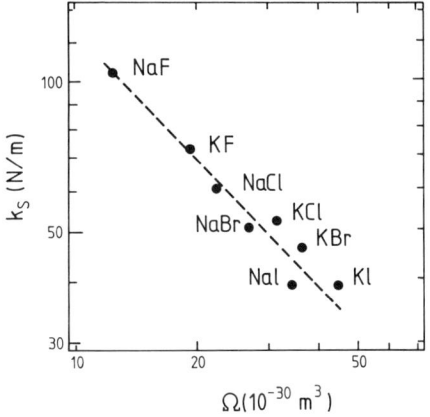

Fig. 3. A logarithmic plot of the average interatomic force constant k_S, calculated from entropy data, vs. the volume per atom Ω for some alkali halides. Data from Häglund and Grimvall (1991).

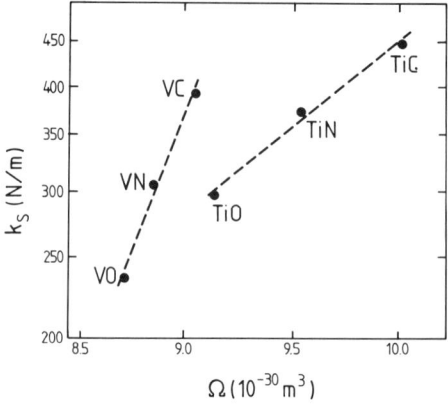

Fig. 4. A logarithmic plot of the average interatomic force constant k_S, calculated from entropy data, vs. the average atomic volume Ω. Data from Fernández Guillermet and Grimvall (1989a).

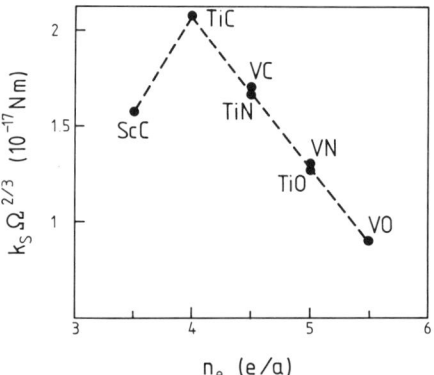

Fig. 5. The quantity $E_S \equiv k_S\Omega^{2/3}$ vs. the average number of valence electrons per atom n_e. Data from Fernández Guillermet and Grimvall (1989a).

We have noted in previous work that the quantity E_S with the dimension of energy

$$E_S \equiv k_S\Omega^{2/3} \qquad (9)$$

may show more regularity than k_S alone. Let the effective interatomic forces in a solid be described by a single central potential $V(r) = V_0\varphi(r/a)$, where the length parameter a but not the strength V_0 varies between solids. Then these solids would have atomic volumes Ω that scale as a^3 but E_S would be the same (Grimvall and Rosén, 1983). That model gives a crude description of the alkali halides (see Fig. 3, where the slope of the straight line corresponds to $k_S\Omega^{0.83}$ being a constant), but fails for the transition metal compounds considered in Fig. 4. This difference is not surprising since the two groups of solids have very different chemical bonding. However, when E_S from Eq. (9) is plotted vs. the average number of valence electrons per atom in the compound n_e, very regular behavior emerges for the transition metal compounds (Fig. 5). A discussion of the carbides has been given elsewhere (Fernández Guillermet and Grimvall, 1989a), and we only sketch the arguments. The density of states $N(E)$ for the valence electrons in the compound has a pronounced minimum that separates bonding and antibonding electron states. For TiC the Fermi level E_F falls near the minimum, and the bonding states are fully used. In ScC, which has one valence electron less per formula unit, the bonding states are not fully occupied. Similarly, for VC some antibonding states are occupied. Therefore, ScC and VC have lower k_S and E_S than TiC. (The point for ScC would fall far outside the range shown in Fig. 4, at $k_S = 283$ N/m and $\Omega = 13.144 \times 10^{-30}$ m^3/atom.) In a model with a common so-called rigid electron band for the carbides, nitrides, and oxides, the filling of the band is determined only by n_e, which is a reason why these compounds fall on a common curve. However, this approximation becomes less good for the (metastable) compounds with n_e larger than in Fig. 5 (Fernández Guillermet and Grimvall, 1989a; Häglund et al., 1991).

The strength of the interatomic forces should be reflected in the hardness of solids, and we end this section with a comment on such relations. Most quantities giving hardness have the dimension of force per area. We may then form $k_S\Omega^{-1/3}$, which has been shown (Grimvall and Thiessen, 1986) to correlate with the microhardness of several transition metal carbides and nitrides. The relation between $k_S\Omega^{-1/3}$ and the Mohs hardness is only qualitative. As an example, compare data for some alkali halides and some transition metal refractory compounds, all with the NaCl structure. For NaI, LiBr, NaBr, KBr, LiCl, NaCl, KCl, and NaF, our $k_S\Omega^{-1/3}$ lies between 120 and 452 GPa (Grimvall and Rosén, 1983; Häglund and Grimvall, 1991) and the Mohs hardness ranges from 1.75 to 2.75 (Szymański and Szymański, 1989). Similarly, for TiN, ZrN, TiC, VC, ZrC, NbC, and TaC, our $k_S\Omega^{-1/3}$ lies in the range 1510 to 2044 GPa (Fernández Guillermet and Grimvall, 1989a; Grimvall and Thiessen, 1986), whereas the Mohs hardness ranges from 8 to close to 10 (Szymański and Szymański, 1989). There is no strong correlation between $k_S\Omega^{-1/3}$ and the Mohs hardness within these two groups of compounds.

Role of Correlations for Bonding Properties in Thermodynamic Calculations of Phase Diagrams

The phase diagram of an alloy system can be calculated if one knows the molar Gibbs energy

$$G = H - TS \tag{10}$$

of its various competing phases as a function of temperature, pressure, and composition. In Eq. (10), H is the enthalpy and S the entropy. For a stoichiometric compound A_aB_b at constant atmospheric pressure P_0, one can write the Gibbs energy $^0G_{A_aB_b}$ as

$$^0G_{A_aB_b}(T, P_0) = {^0G_{A_aB_b}}(T_0, P_0) - \int_{T_0}^{T} {^0S_{A_aB_b}}(T', P_0)\, dT', \tag{11}$$

where T_0 is a reference temperature, usually, $T_0 = 298.15$ K. Often, one refers the $^0G_{A_aB_b}(T,P_0)$ values to the weighted sum of the enthalpies $^0H_i(T_0,P_0)$ of the elements in a certain reference state, e.g., their stable modifications at T_0, P_0. Then Eq. (11) becomes

$$^0G_{A_aB_b}(T,P_0) - a^0H_A(T_0,P_0) - b^0H_B(T_0,P_0)$$
$$= \Delta^0H_{A_aB_b} - T_0\,^0S_{A_aB_b}(T_0,P_0) - \int_{T_0}^{T} {}^0S_{A_aB_b}(T',P_0)\,dT', \qquad (12)$$

where

$$\Delta^0H \equiv {}^0H_{A_aB_b}(T_0,P_0) - a^0H_A(T_0,P_0) - b^0H_B(T_0,P_0). \qquad (13)$$

Equation (12) summarizes the relations between the thermodynamic quantities involved in an account of the stability of a stoichiometric compound at P_0, i.e., the enthalpy of formation $\Delta^0H_{A_aB_b}$ at room temperature T_0 [Eq. (13)] and the total entropy at and above T_0. For stable compounds these quantities can, in principle, be taken from direct measurements. In practice, the experimental information is often scarce, conflicting, or lacking. As a consequence, there is a strong interest in the development of methods for predicting the quantities in Eq. (12) and for judging available estimates.

Miedema's formula (de Boer et al., 1989) has been very successful in accounting for the enthalpy of formation of various groups of compounds, but it does not consider the entropy part of G, which is necessary in the treatment of the phase-equilibrium at high temperatures. Other empirical methods for estimating S, such as Latimer's (1951) rule or the well-known Lindemann melting formula, are not accurate enough for phase diagram calculations of practical interest.

We shall now refer to our recent work on transition metal compounds to exemplify how bonding properties can be related to phase diagram calculations. Reasonably accurate estimates of Δ^0H and ^0S for various groups of $3d$-transition metal compounds with C, N, and O (Fernández Guillermet and Grimvall, 1989a) and $3d$-, $4d$-, and $5d$-transition metals with B (Fernández Guillermet and Grimvall, 1991) can be obtained in a consistent way by interpolations and extrapolations in plots of a quantity related to the strength of bonding [Δ^0H or E_S, see Eq. (9)] vs. the average number of valence electrons per atom in the compound. That has allowed thermodynamic calculations of phase diagrams also when they contain phases whose stabilities are not known from experiments, e.g., a "CrC" phase with the (cF8) NaCl structure (Fernández Guillermet and Grimvall, 1990) and the "cementite" phase Cr_3C (oP16) (Fernández Guillermet, 1991) of the Cr-C system. Moreover, by combining predicted thermodynamic quantities with some information from experiments, it has been possible to construct, by calculation, the phase diagrams of transition metal systems that are not known from direct measurements, i.e., the Ni-N (Fernández Guillermet and Frisk, 1991), Co-N (Fernández Guillermet and Jonsson, 1992), and Mn-V-C (Fernández Guillermet and Huang, 1991) systems.

Acknowledgments

This work was supported in part by the Swedish Board for Technical Development and by the Swedish Natural Science Research Council.

References

American Institute of Physics Handbook (1972). D.E. Gray, ed., McGraw-Hill, New York, 4–114.

Anderson, O.L. and Nafe, J.E. (1965). The bulk modulus-volume relationship for oxide compounds and related geophysical problems. *J. Geophys. Res.* **70**, 3951–3963.

de Boer, F.R., Boom, R., Mattens, W.C.M., Miedema, A.R., and Niessen, A.K. (1989). *Cohesion in Metals*. North Holland, Amsterdam.

Ditmars, D.A. and Douglas, T.B. (1971). Measurement of the relative enthalpy of pure α-Al_2O_3 (NBS heat capacity and enthalpy standard reference material no. 720) from 273 K to 1173 K. *J. Res. NBS.* **75A**, 401–420.

Fernández Guillermet, A. (1991). Predictive approach to thermodynamic properties of metastable Cr_3C carbide *Int. J. Thermophys.* **12**, 919–936.

Fernández Guillermet, A. and Frisk, K. (1991). Thermodynamic properties of Ni nitrides and phase stability in the Ni-N system. *Int. J. Thermophys.* **12**, 417–431.

Fernández Guillermet, A. and Grimvall, G. (1989a). Cohesive properties and vibrational entropy of $3d$ transition-metal compounds: MX (NaCl) compounds (X = C, N, O, S), complex carbides, and nitrides. *Phys. Rev.* **B40**, 10582–10593.

Fernández Guillermet A. and Grimvall, G. (1989b). Homology of interatomic forces and Debye temperatures in transition metals. *Phys. Rev.* **B40**, 1521–1527.

Fernández Guillermet, A. and Grimvall, G. (1989c). Thermodynamic properties of technetium. *J. Less-Common Metals* **147**, 195–211.

Fernández Guillermet, A. and Grimvall, G. (1990). Correlations for bonding properties and vibrational entropy in $3d$-transition metal compounds, with application to the CALPHAD treatment of a metastable Cr-C phase. *Z. Metallk.* **81**, 521–524.

Fernández Guillermet, A. and Grimvall, G. (1991). Bonding properties and vibrational entropy of transition metal MeB_2 (AlB_2) diborides. *J. Less-Common Metals* **169**, 257–281.

Fernández Guillermet, A. and Grimvall, G. (1992). Cohesive properties and vibrational entropy of $3d$-transition metal carbides. J. Phys. Chem. Solids **53**, 105–125

Fernández Guillermet, A. and Huang, W. (1991). Thermodynamic analysis of stable and metastable carbides in the Mn-V-C system and predicted phase diagram. *Int. J. Thermophys.* **12**, 1077–1102.

Fernández Guillermet, A. and Jonsson, S. (1992). Predictive approach to thermodynamical properties of Co nitrides and phase stability in the Co-N system. *Z. Metallk.* **83**, 21–31.

Goto, T., Anderson, O.L., Ohno, I., and Yamamoto, S. (1989). Elastic constants of corundum up to 1825 K. *J. Geophys. Res.* **94**(B6), 7588–7602.

Grimvall, G. (1983). Standard entropies of compounds: Theorretical aspects of Latimer's rule. *Int. J. Thermophys.* **4**, 363–367.

Grimvall, G. and Rosén, J. (1982). Heat capacity of fcc calcium. *Int. J. Thermophys.* **3**, 251–257.

Grimvall, G. and Rosén, J. (1983). Vibrational entropy of polyatomic solids: Metal carbides, metal borides, and alkali halides. *Int. J. Thermophys.* **4**, 139–147.

Grimvall, G. and Thiessen, M. (1986). The strength of interatomic forces, in *2nd International Conference on Science Hard Materials, Institute Physics Conference Series*, Vol. 75, Adam Hilger, Bristol, pp. 61–67.

Grimvall, G., Thiessen, M., and Fernández Guillermet, A. (1987). Thermodynamic properties of tungsten. *Phys. Rev.* **B36**, 7816–7826.

Häglund, J. and Grimvall, G. (1991). Lattice vibrations and bonding in alkali hydrides and alkali halides. Unpublished.

Häglund, J., Grimvall, G., Jarlborg, T., and Fernández Guillermet, A. (1991). Band structure and cohesive properties of NaCl-structure transition-metal carbides and nitrides. *Phys. Rev.* **B43**, 14400–14408.

JANAF Thermochemical Tables, 3rd ed. (1985). M.W. Chase, C.A. Davies, J.R. Downey, Jr., D.J. Frurip, R.A. McDonald, and A.N. Syverud, eds., *J. Phys. Chem. Ref. Data* 14, Supplement 1.

Latimer, W.M. (1951). Methods of estimating the entropies of solid compounds. *J. Am. Chem. Soc.* **73**, 1480–1481.

Rosén, J. and Grimvall, G. (1983). Anharmonic lattice vibrations in simple metals. *Phys. Rev.* **B27**, 7199–7208.

Shaviv, R., Westrum, Jr., E.F., Fjellvåg, H., and Kjekshus, A. (1989). $ZrTe_5$ and $HfTe_5$: The heat capacity and derived thermophysical properties from 6 to 344 K. *J. Solid State Chem.* **81**, 103–111.

Szymański, A. and Szymański, J.M. (1989). *Hardness Estimation of Minerals, Rocks and Ceramic Materials*. Materials Science Monographs, No. 49, Elsevier, Amsterdam.

Chapter 11
Phonon Density of States and Thermodynamic Properties of Minerals

Subrata Ghose, Narayani Choudhury, S.L. Chaplot, and K.R. Rao

Introduction

An important objective of the earth sciences is to develop the capability of predicting the thermodynamic properties of minerals and their phase relations under various pressure-temperature conditions in the earth, moon and the terrestrial planets. Experimentally, thermodynamic properties such as specific heat can be measured by adiabatic and differential scanning calorimetry. However, there are cases, in which the determination of the low temperature specific heat by adiabatic calorimetry is not feasible due to the paucity of materials. Such is the case for the high-pressure and high-temperature magnesium silicate phases: Mg_2SiO_4 (β- and γ-spinel) and $MgSiO_3$ (ilmenite, perovskite and garnet) considered to be stable in the earth's mantle. These phases are synthesized in cubic-anvil and split-sphere apparati in 10 to 20 mg quantities, that are barely adequate for the specific heat measurement by differential scanning calorimetry (DSC), usually in the range 300 to 900 K. Hence, the ability to correctly predict the low temperature specific heat of these phases would be extremely useful. Such a theoretical treatment should at the same time provide an understanding of the thermodynamic properties of minerals at the atomistic level. This was the goal aimed at the Mineralogical Society of America Short Course organized by S.W. Kieffer and A. Navrotsky on "Macroscopic to microscopic: Atomic environments to mineral thermodynamics" held at Washington College, Chestertown, Maryland in spring, 1985. At least for minerals of medium structural complexity, this goal is in sight. It is now possible to theoretically explore the entire spectrum of thermal vibrations of crystals by lattice dynamical methods. The vibrational average, known as the phonon density of states, $g(\omega)$, is the basis for the calculation of thermodynamic properties such as internal energy, free energy, specific heat, and entropy. The accuracy of the theoretical calculation of $g(\omega)$ can be tested against the generalized density of states $G(E)$, which can be measured by inelastic neutron scattering on large powder samples. Such complete theoretical and experimental lattice dynamical studies of the phonon density of states and specific heat have been carried out on the olivine end members, forsterite, Mg_2SiO_4 (Rao et al., 1987, 1988), and fayalite, Fe_2SiO_4 (Price et al., 1991). In this chapter, we discuss various theoretical and experimental methods of determination of the phonon density of states with an emphasis on lattice dynamics and inelastic neutron scattering. The results obtained on other minerals to date are also reviewed.

Phonon Density of States and Thermodynamic Properties of Crystals

The thermodynamic properties of a crystal are based on the averages of energies associated with $3rN$ vibrations corresponding to the number of degrees of freedom of r atomic constituents in the N unit cells of the crystal. This average is given by the phonon density of states $g(\omega)$, defined such that there are $g(\omega)\,d\omega$ modes in the frequency range between ω and $\omega + d\omega$, where ω is the vibrational frequency of a particular mode.

$$g(\omega) = A \int \sum_j \delta[(\omega - \omega_j)(\mathbf{q})]\,d\mathbf{q}, \qquad (1)$$

where $\omega_j(\mathbf{q})$ gives the dispersion relation, i.e., the frequency dependence with wave vector $\mathbf{q}$ for the jth phonon branch in the Brillouin zone. A is a normaliza-

Table 1. Thermodynamic properties of a crystal.

Property	Symbol	Expression
Partition function	Z	$e^{-u(v)/k_BT} \prod^i e^{-\hbar\omega_i/k_BT}/(1 - e^{\hbar\omega_i/k_BT})$
Free energy	$F = -k_B T \ln Z$ $= U(V) + F_{vib}$	$U(V) + \int \left\{\frac{1}{2}\hbar\omega + k_B T \ln\{1 - e^{-\hbar\omega/k_BT}\}\right\} g(\omega)\,d\omega$
Internal energy	$E = F - (dF/dT)$ $= U(V) + E_{vib}$	$U(V) + \int \left\{\frac{1}{[e^{(\hbar\omega/k_BT)} - 1]} + \frac{1}{2}\right\} \hbar\omega\, g(\omega)\,d\omega$
Specific heat	$C_v = (dE/dT)$	$k_B \int \left(\frac{\hbar\omega}{k_BT}\right)^2 \frac{e^{(\hbar\omega/k_BT)}}{\{e^{(\hbar\omega/k_BT)} - 1\}^2} g(\omega)\,d\omega$
Entropy	$S = (dF/dT)$	$k_B \int \left[-\ln\{1 - e^{(-\hbar\omega/k_BT)}\} + \frac{(\hbar\omega_i/k_BT)}{\{e^{(\hbar\omega_i/k_BT)} - 1\}}\right] g(\omega)\,d\omega$

tion constant such that

$$\int g(\omega)\,d\omega = 1. \quad (2)$$

Once $g(\omega)$ is known, the free energy, internal energy, specific heat and entropy can be calculated (Table 1). The specific heat at constant volume, $C_v(T)$, e.g., is given by:

$$C_v(T) = k_B \int \left(\frac{\hbar\omega}{k_B T}\right)^2 \frac{\exp(\hbar\omega/k_B T)}{[\exp(\hbar\omega/k_B T) - 1]^2} g(\omega)\,d\omega, \quad (3)$$

where, $\hbar$ is the Planch's constant, $h/2\pi$; k_B the Boltzman constant, and T the temperature.

Einstein Model

The simplest approximation of the phonon density of states is given by the Einstein model, in which all the atoms are assumed to oscillate independently of each other with the same mean frequency, ω_E, the Einstein frequency. In this model, each atom is connected by an elastic spring to its equilibrium position, but there are no interactions with the neighboring atoms. The mean energy of a linear harmonic oscillator of frequency ω in thermal equilibrium at temperature T is given by the Planck's formula if we assume no zero-point energy

$$\frac{\hbar\omega}{\exp(\hbar\omega/k_B T) - 1}. \quad (4)$$

Hence, the internal energy E of the crystal of r atoms in the primitive unit cell and N unit cells at temperature T is given by

$$E = \frac{3rN\hbar\omega}{\exp(\hbar\omega/k_B T) - 1}. \quad (5)$$

The density of states is given by

$$g_E(\omega) = 3rN\delta(\omega - \omega_E). \quad (6)$$

In terms of the Einstein temperature θ_E, the Einstein specific heat

$$C_E = 3rNk_B \left(\frac{\theta_E}{T}\right)^2 \frac{e^{\theta_E/T}}{(e^{\theta_E/T} - 1)^2}, \quad (7)$$

where $\theta_E = \hbar\omega_E/k_B$. At high temperatures, the specific heat is correctly predicted, where it approaches the Dulong-Petit limit, $C_v \to 3rNk_B$, but it is underestimated at low temperatures, giving an exponential decrease instead of T^3 dependence as experimentally observed. This deficiency is due to the fact that the assumed single Einstein frequency lies somewhere in between the optic modes and the acoustic modes, and at low temperatures mostly the low frequency acoustic modes and not the high frequency optic modes are excited.

Debye Model

To alleviate this problem, Debye (1912) proposed an alternative model, in which the crystal was assumed to be an elastic continuum due to the coupling of the atoms. In this model, only the three acoustic branches were considered, each of which has the same linear dispersion

$$\omega = vq, \qquad (8)$$

where v is the mean sound velocity and q the wave vector in the Brillouin zone. If we sum over the three acoustic branches, the density of states $g_D(\omega)$ is given by

$$g_D(\omega) = \frac{3V}{2\pi^2} \frac{\omega^2}{v^3}, \qquad (9)$$

where V is the volume of the crystal. In a crystal with r atoms in the primitive cell and N unit cells, the total number of frequencies is $3rN$, and hence

$$\int g_D(\omega)\,d\omega = 3rN \qquad (10)$$

where the intergration is over all allowed frequencies. Since this integral would diverge to infinity if it is extended to include infinitely high frequencies, Debye assumed that it should be used only to a maximum or cut-off frequency ω_D known as the Debye frequency. Hence, in the Debye model the allowed vibrational modes exist in the frequency range 0 to ω_D with no modes of higher frequency. Therefore for $\omega > \omega_D$, $g_D(\omega) = 0$, and for $\omega < \omega_D$, $g_D(\omega) = 9rN\omega^2/\omega_D^3$. In terms of the Debye temperature θ_D, defined as

$$\theta_D = \frac{\hbar \omega_D}{k_B}, \qquad (11)$$

the Debye specific heat $C_D(T)$ is given by

$$C_D(T) = 9rNk_B \left(\frac{T}{\theta_D}\right)^3 \int_0^{\theta_D/T} \frac{x^4 e^x}{(e^x - 1)^2}\,dx, \qquad (12)$$

where $x = \hbar\omega/k_B T$.

At temperatures much higher than the Debye temperature, $C_D = 3rNk_B$. At low temperatures ($T \ll \theta_D$), the Debye specific heat

$$C_D(T) = \frac{12\pi^4}{5} rNk_B \left(\frac{T}{\theta_D}\right)^3, \qquad (13)$$

which gives the experimentally observed T^3 dependence. For simple materials such as metals and alkali halides, the agreement between theory and experiment is surprisingly good, which explains the wide acceptance of Debye theory. The difficulty, however, lies in the fact that for structurally complex materials such as silicates, the Debye density of states does not accurately represent the true density of states. Another difficulty is with the Debye temperature θ_D that must be used in the above equation, such that $C_D(T/\theta_D) = C_v(T)$, is not a constant but is a function of temperature. The T^3 law that depends on a constant θ_D is only accurate over a small temperature range. θ_D is still a useful parameter characteristic of a given material and reflects the strength of the interatomic forces. For example, for diamond with strong covalent forces, θ_D is close to 2000 K, whereas for the soft potassium metal it is 100 K.

Hybrid and Kieffer Models

To circumvent the deficiencies of both the Einstein and Debye models, a hybrid model can be developed (Brüesch, 1982) in which the density of states $g_{DE}(\omega)$ is given by

$$g_{DE}(\omega) = \tilde{g}_D(\omega) + \tilde{g}_E(\omega). \tag{14}$$

In this model the acoustic modes are described by a Debye model, whereas the optic modes are described by an Einstein model. The density of states corresponding to the three models and the general case are illustrated in Fig. 1. In the hybrid model,

$$\tilde{g}_E(\omega) = (3r - 3)N\delta(\omega - \tilde{\omega}_E), \tag{15}$$

where $\tilde{\omega}_E$ is the mean optical frequency. The Debye density of states $\tilde{g}_D(\omega)$ is the same as before, but the normalizing condition is now

$$\int_0^{\tilde{\omega}_D} \tilde{g}_D(\omega)\, d\omega = 3rN. \tag{16}$$

The resulting specific heat $C_{DE}(T)$ gives the classical result $3rNk_B$ at high temperatures. At low temperatures the Einstein term can be neglected, and

$$C_{DE}(T) = \frac{12\pi^4}{5} Nk_B \left(\frac{T}{\tilde{\theta}_D}\right)^3, \tag{17}$$

where $\tilde{\theta}_D^3 = \theta_D^3/r$, again showing the T^3 dependence of the specific heat.

Kieffer (1985) adopted a modified version of this hybrid model to predict the specific heats of a large number of structurally complex minerals, including rock forming silicates such as olivines, pyroxenes, amphiboles, micas, and feldspars. In her model, the phonon density of states is constructed from the acoustic modes derived from the experimentally measured elastic constants and the optic modes from the experimental measurements of infra-red (IR) and Raman spectra. The acoustic modes are assumed to have a simple sine-wave dispersion from the center toward the edge of the Brillouin zone. The $3N-3$ optic modes span a broad range of frequencies, estimated from far-IR, near-IR, and Raman spectra. The high-frequency optic modes are assumed to be dispersionless, whereas the lowest-frequency optic mode is assumed to vary inversely with a characteristic mass ratio across the Brillouin zone. The optic modes are distributed in an optic continuum between the lower and upper cut-off frequencies ω_1 and ω_2, specified

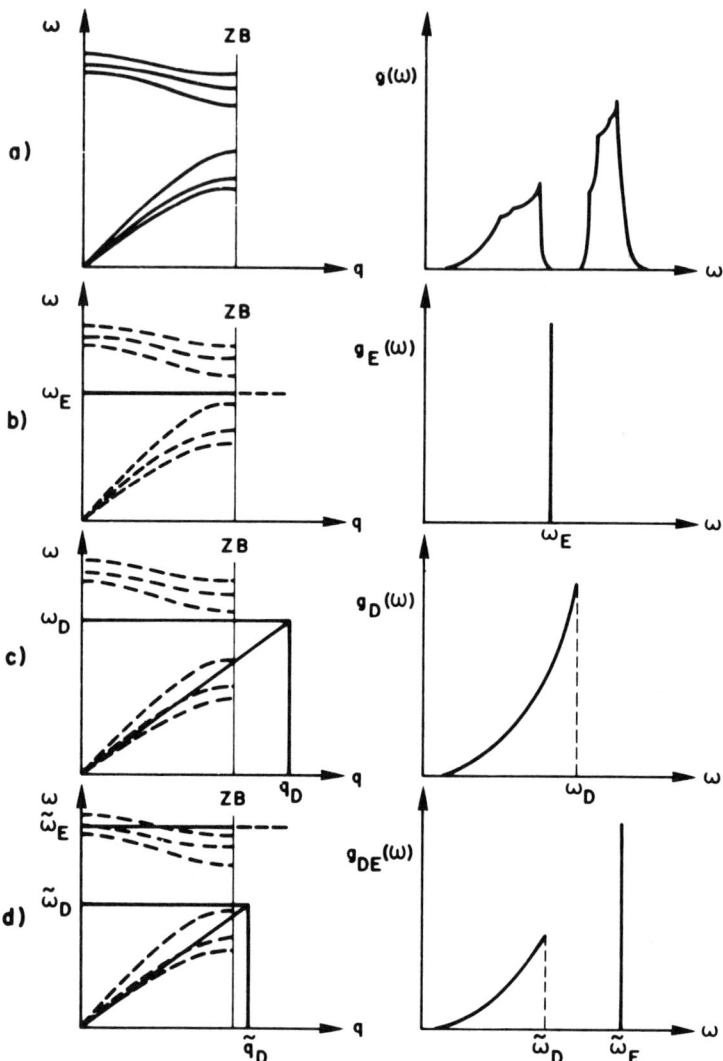

Fig. 1. Phonon dispersion and phonon density of states for a crystal with two atoms in the unit cell (a) qualitative general behavior, (b) Einstein approximation, (c) Debye approximation, and (d) hybrid Einstein-Debye model (after Brüesch, 1982).

from spectral data. Isolated high-frequency optic modes (e.g., Si–O and O–H stretching modes) are represented as separate Einstein oscillators with frequencies ω_{E1}, ω_{E2}, ... etc., or by a second optic continuum. A phonon density of states so constructed for forsterite is shown in Fig. 2. This model has been successfully used to estimate the specific heats of a large number of rock-forming silicates (Kieffer, 1985) and a number of the high-pressure magnesium silicate

Fig. 2. Phonon density of states of forsterite according to Kieffer (1980) model.

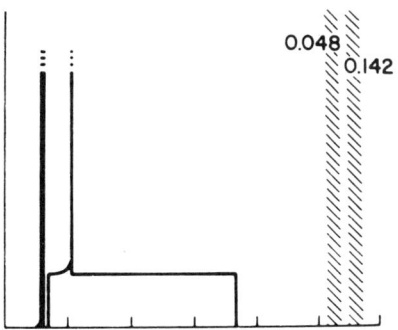

phases considered to be stable in the earth's mantle (Akaogi et al., 1984; Ashida et al., 1988). In spite of its success, this model is empirical in nature and requires extensive prior knowledge of elastic constants and frequencies of optic phonons from IR and Raman data. In case inelastic neutron scattering results are not available, the neglect of low-frequency optic phonons that are IR- and Raman-inactive may result in an underestimation of the low-temperature specific heat. On the other hand, at high temperatures where the Einstein oscillator model would be valid, this model might give satisfactory results.

Lattice Dynamics

A correct formulation of the phonon density of states requires a knowledge of the frequencies of all phonon branches and their dispersion in the Brillouin zone. This is the subject of lattice dynamics, the foundations of which were laid in a paper by Born and von Kármán (1912). Although lattice dynamics provides a more fundamental approach to the determination of the thermodynamic properties of crystals, the early success of Debye theory hampered progress in this field, until it was possible to experimentally measure the phonon dispersion in a crystal by inelastic neutron scattering. This was accomplished in the early 1960s by using thermal neutrons from a nuclear reactor and a triple axis spectrometer developed by B.N. Brockhouse at Chalk River in Canada (Brockhouse, 1962). Since this development, lattice dynamics in conjunction with inelastic neutron scattering has been established as a very powerful tool for studying the dynamical properties of crystals. The measurement of the phonon density of states by inelastic neutron scattering is now possible by high-energy neutrons (200 meV) available from a spallation source such as the Intense Pulsed Neutron Source (IPNS) at the Argonne National Laboratory. The high energies of pulsed neutrons are needed to measure the high-frequency internal modes such as Si–O stretching (~ 1000 cm^{-1} or ~ 124 meV) in silicates. On the other hand, thermal neutrons (25 meV) from a nuclear reactor are better suited for the measurement of the acoustic and low frequency optic modes in a single crystal. The energies of these low-frequency modes are particularly important for the estimation of low-temperature specific heat. Furthermore, the zone-center optic modes, which are IR- and Raman-inactive, can also be measured by inelastic neutron scattering. These measurements provide a valuable test for the accuracy of the interatomic potential used for phonon dispersion and density of states calculations by lattice dynamical methods.

Lattice Dynamics in the Harmonic Approximation

In the harmonic approximation of lattice dynamics, only small oscillations around the equilibrium atomic positions are considered. It is also assumed that the electronic wave functions change adiabatically during nuclear motion. Any collective motion can be described as a superposition of normal vibrational modes. In a solid, the number of normal modes is equal to the number of vibrational degrees of freedom, $3rN$. The frequencies of $3rN$ normal modes can be determined by diagonalizing the dynamical matrix (see below). For crystals, the presence of translational symmetry requires that the displacements in different unit cells must be equivalent. Hence, it is useful to represent the normal modes as propagating vibrational waves (phonons) with wave vector $\mathbf{q}$ determined by the size and symmetry of the unit cell. The problem reduces to that of determining the motions of r particles in the unit cell. For each value of the wave vector $\mathbf{q}$, the (circular) frequencies $\omega_j(\mathbf{q})$ are determined by diagonalizing the $3r \times 3r$ dynamical matrix. The relation

$$\omega = \omega_j(\mathbf{q}) \qquad (j = 1, 2, 3, \ldots r) \tag{18}$$

between the frequency and the wave vector is known as the dispersion relation. The index j, which distinguishes the various frequencies corresponding to the same propagation vector, characterizes the various branches of the dispersion relation. The pattern of motion of the nuclei in the cell is determined by the polarization vectors e_d^j ($d = 1, 2, \ldots, r$). A certain branch of the dispersion relation is called longitudinal or transverse if the polarization vectors are parallel or perpendicular to the wave vector $\mathbf{q}$ respectively. Because of translational symmetry, the atomic displacements and frequencies, $\omega_j(\mathbf{q})$ are periodic functions of $\mathbf{q}$ with the period of the reciprocal lattice. Hence, all physically distinct allowed values of $\mathbf{q}$ are restricted to the primitive cell of the reciprocal lattice, known as the Brillouin zone. The first Brillouin zone of an orthorhombic lattice is shown in Fig. 3.

The energy $E_j(\mathbf{q})$ and momentum $\mathbf{p}$ of the phonon are related to the frequency and wave vector of the corresponding vibrational wave by the relations

$$E_j(\mathbf{q}) = \hbar\omega_j(\mathbf{q}) \tag{19}$$

and

$$\mathbf{p} = \hbar\mathbf{q}. \tag{20}$$

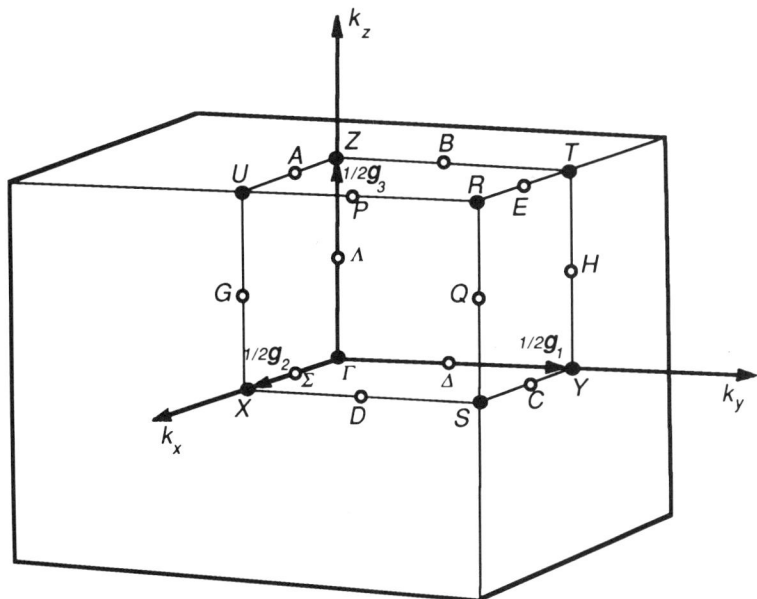

Fig. 3. Brillouin zone for an orthorhombic lattice. The principal symmetry directions Σ, Δ, and Λ are shown.

The harmonic approximation assumes completely free motion of noninteracting phonons. However, anharmonic interactions are always present and elastic and inelastic collisions between the phonons occur, which provide a mechanism for the thermal equilibrium of the phonon gas. The lifetime of a phonon is thus finite. Since any number of identical phonons can be excited at any given time and this number is determined by the equilibrium condition, the phonon gas obeys Bose-Einstein statistics. Hence, the average number $n_j(\mathbf{q})$ of phonons in a given state at thermal equilibrium is given by

$$n_j(\mathbf{q}) = \frac{1}{\exp(E_j(\mathbf{q})/k_B T) - 1}. \tag{21}$$

At high temperatures $(E_j(\mathbf{q}) \ll k_B T)$, the number of phonons in a given state is proportional to the temperature and inversely proportional to their energy.

A comprehensive account of the formal theory of lattice dynamics is given by Born and Huang (1954). We present here a summary of the mathematical formalism for a perfect crystal in harmonic approximation. The potential energy of a crystal ϕ can be expanded in a power series of the displacements of the ions $u(l,k)$ from their equilibrium positions $r(lk)$, where l denotes the lth unit cell $(l = 1, \ldots, N)$, and k is the kth type of ion $(k = 1, \ldots, r)$ within the unit cell. The potential is given by

$$\phi = \phi_0 + \frac{1}{2} \sum_{ll'} \sum_{kk'} \sum_{\alpha\beta} \phi_{\alpha\beta}\begin{pmatrix} ll' \\ kk' \end{pmatrix} u_\alpha\begin{pmatrix} l \\ k \end{pmatrix} u_\beta\begin{pmatrix} l' \\ k' \end{pmatrix} + \ldots, \tag{22}$$

where the suffices α, β, γ denote the Cartesian coordinates. The second derivative $\phi_{\alpha\beta}\begin{pmatrix} ll' \\ kk' \end{pmatrix}$ is the negative of the force constant and gives the force in the α-direction on the ion (lk) when the ion $(l'k')$ is displaced by a (small) unit displacement in the β-direction. For a perfect crystal with uniform translational symmetry, no residual stresses remain. Hence,

$$\sum_{l'k'} \phi_{\alpha\beta}\begin{pmatrix} ll' \\ kk' \end{pmatrix} = 0. \tag{23}$$

The equation of motion of the (lk) th atom is then

$$M_k \ddot{u}_\alpha \begin{pmatrix} l \\ k \end{pmatrix} = -\sum_{l'k'\beta} \phi_{\alpha\beta}\begin{pmatrix} ll' \\ kk' \end{pmatrix} u_\beta \begin{pmatrix} l' \\ k' \end{pmatrix}. \tag{24}$$

This equation is solved in a perfect crystal by finding a solution in terms of plane waves, such that

$$u_\alpha \begin{pmatrix} l \\ k \end{pmatrix} = \frac{1}{\sqrt{M_k}} u_\alpha(k|\mathbf{q}) \exp i\left(\mathbf{q}\cdot\mathbf{r}\begin{pmatrix} l \\ k \end{pmatrix} - \omega(\mathbf{q})t\right), \tag{25}$$

where,

$$\omega(\mathbf{q})^2 u_\alpha(k|\mathbf{q}) = \sum_{k'\beta} D_{\alpha\beta}\begin{pmatrix} \mathbf{q} \\ kk' \end{pmatrix} u_\beta(k'|\mathbf{q}). \tag{26}$$

Here, the Fourier-transformed dynamical matrix $D_{\alpha\beta}\begin{pmatrix} \mathbf{q} \\ kk' \end{pmatrix}$ is given by

$$D_{\alpha\beta}\begin{pmatrix} \mathbf{q} \\ kk' \end{pmatrix} = \frac{1}{\sqrt{M_k M_{k'}}} \sum_{l'} \phi_{\alpha\beta}\begin{pmatrix} ll' \\ kk' \end{pmatrix} \exp i\mathbf{q}\cdot\left[\mathbf{r}\begin{pmatrix} l' \\ k' \end{pmatrix} - \mathbf{r}\begin{pmatrix} l \\ k \end{pmatrix}\right]. \tag{27}$$

We then have $3rN$ coupled equations of motion. The standard technique that is used in solving these equations involves making a transformation to a set of normal coordinates. The squares of the frequencies are the eigenvalues of the dynamical matrix. The displacements of the atoms in one of these normal modes labeled by $(\mathbf{q}_j)$ correspond to a wavelike displacement of atoms and is given by

$$u_\alpha \begin{pmatrix} l \\ k \end{pmatrix} = \frac{1}{\sqrt{M_k}} e_\alpha(k|\mathbf{q}_j) \exp i\left[\mathbf{q}\cdot\mathbf{r}\begin{pmatrix} l \\ k \end{pmatrix}\right] P\begin{pmatrix} \mathbf{q} \\ j \end{pmatrix}, \tag{28}$$

where $P\begin{pmatrix} \mathbf{q} \\ j \end{pmatrix}$ are the normal coordinates and $e(k|\mathbf{q}_j)$ is the eigenvector of the normal mode $(\mathbf{q}_j)$, where j runs from 1 to $3N$ and is used to distinguish between the $3N$ normal modes at $\mathbf{q}$; $\omega_j(\mathbf{q})/2\pi$ is the frequency of the normal mode. The frequencies of the normal modes (eigenvalues) and their eigenvectors are determined by diagonalizing the dynamical matrix through a solution of the secular equation

$$\det\left|\omega_j(\mathbf{q})^2 \delta_{kk'}\delta_{\alpha\beta} - D_{\alpha\beta}\begin{pmatrix} \mathbf{q} \\ kk' \end{pmatrix}\right| = 0. \tag{29}$$

The eigenvectors are normalized such that

$$\sum_{k\alpha} e_\alpha^*(k|\mathbf{q}_j) e_\alpha(k|\mathbf{q}_j)/M_k = \delta_{jj'}. \tag{30}$$

Because the dynamical matrix is Hermitian (i.e., $\mathbf{D}^* = \mathbf{D}^T$, where $\mathbf{D}^*$ and $\mathbf{D}^T$ indicate the complex conjugate and transpose of $\mathbf{D}$), the eigen-frequencies are real and the eigenvectors can be chosen as orthonormal.

Lattice Dynamics in the Quasi-harmonic Approximation

In the quasi-harmonic formulation, the internal energy E can be expressed in terms of the static lattice energy, $U(V)$ and the sum of vibrational energy, $E_{vib}(\omega_i)$ of phonons of frequency ω_i:

$$E = U(V) + \sum E_{vib}(\omega_i) \tag{31}$$

and

$$E_{vib}(\omega_i) = \left(n_i + \frac{1}{2}\right)\hbar\omega_i,$$

where i represents phonons in state $\mathbf{q}_j$. Here, n_j is the thermal occupancy number of phonons given by eqn. (21). The total vibrational energy of a crystal is given by a summation over all the phonon modes in the Brillouin zone.

The thermodynamic properties of a crystal, namely, the free energy, internal energy, specific heat, and entropy, are obtained from the partition function Z, defined as

$$Z = \mathrm{Tr}\left[\exp\left(-\frac{\hat{H}}{k_B T}\right)\right]. \tag{32}$$

The thermodynamic properties of the crystal obtained from the partition function can be expressed as averages over the phonon density of states $g(\omega)$ and are given in Table 1.

The external pressure P acting on the crystal is related to the volume derivative of the free energy F

$$P = -\frac{\partial F}{\partial V} = -\left[\frac{\partial U(V)}{\partial V} + \frac{\partial F_{vib}}{\partial V}\right],$$

$$= -\frac{\partial U(V)}{\partial V} + \frac{1}{V_0}\sum_j \Gamma_j E_{vib}(\omega_j). \tag{33}$$

This equation is known as the Mie-Grüneisen equation of state. The Grüneisen parameter Γ_i for the ith phonon is given by

$$\Gamma_i = -\frac{\partial \ln \omega_i}{\partial \ln V} \cong -\frac{\Delta \ln \omega_i}{\Delta \ln V}. \tag{34}$$

The bulk modulus

$$B = -V\frac{dP}{dV}. \tag{35}$$

The negative volume derivative of the static lattice energy, $U(V)$ can be expressed in terms of the bulk modulus B at zero pressure (Brüesch, 1982)

$$-\frac{\partial U(V)}{\partial V} = -B\frac{(V - V_0)}{V_0} \tag{36}$$

where V_0 is the crystal volume at 0 K.

The coefficient of volume thermal expansion α_v of the crystal, expressed as the fractional change of volume with the temperature at constant pressure, is given by

$$\alpha_v(T) = \frac{1}{V}\left(\frac{\partial V}{\partial T}\right)_P = \frac{1}{BV_0}\sum_j \Gamma_i C_{v_i}(T), \tag{37}$$

where $C_{v_i}(T)$ is the contribution from phonons in the state i to the specific heat at constant volume. In the Grüneisen approximation, all phonon modes in the Brillouin zone are assumed to be characterized by an average Grüneisen constant $\bar{\Gamma}$. In this approximation

$$\alpha_v(T) = \frac{\bar{\Gamma} C_v(T)}{BV_0}, \tag{38}$$

where $C_v(T)$ is the specific heat at constant volume and temperature T. In the Grüneisen approximation, the thermal expansion has the same temperature dependence as the specific heat. Physically, this means that both quantities are affected in the same way by anharmonicity.

The anharmonic corrections to the heat capacity, i.e., the differences in the heat capacities at constant pressure $C_p(T)$ and at constant volume $C_v(T)$ are given by

$$C_p(T) - C_v(T) = [\alpha_v(T)]^2 BVT. \tag{39}$$

These corrections are needed to compare the calorimetrically measured $C_p(T)$ with $C_v(T)$ calculated from the phonon density of states derived from lattice dynamics in the harmonic approximation.

Inelastic Neutron Scattering: Measurement of Phonon Dispersion and Phonon Density of States

The scattering of a monochromatic beam of neutrons by a crystal is described by the scattering cross-section σ, which consists of an incoherent and a coherent part. The coherent one phonon scattering of neutrons from a single crystal is the most powerful technique for the detailed investigation of lattice dynamics of single crystals (Dorner, 1982; Sköld and Price, 1986). The coherent scattering cross-section of the neutron dependent on energy E and scattering angle Ω due to the creation ($+$) and annihilation ($-$) of a single phonon of frequency $\omega_j(\mathbf{q})$ is given by

$$\frac{d^2\sigma_{\text{coh}}}{d\Omega\,dE}(\mathbf{Q},\omega)$$

$$= \frac{k_1}{k_0}\frac{(2\pi)^3}{2V}\hbar\frac{[n_j(\mathbf{q}) + \tfrac{1}{2} \pm \tfrac{1}{2}]}{\omega_j(\mathbf{q})}\sum_\tau |F_j(Q)|^2\delta[E_1 - E_0 \mp \hbar\omega_j(\mathbf{q})]\delta(\mathbf{Q} \pm \mathbf{q} - \tau). \quad (40)$$

Here, the scattering vector, $\mathbf{Q} = \mathbf{k}_1 - \mathbf{k}_0$, where $\mathbf{k}_1$ and $\mathbf{k}_0$ are scattered and incident neutron wave vectors and $[n_j(\mathbf{q}) + \tfrac{1}{2} \pm \tfrac{1}{2}]$ is the population factor, the upper and lower $-$ and $+$ signs indicate phonon annihilation and phonon creation respectively, and τ is a reciprocal lattice vector.

The delta functions indicate the conservation of energy and momentum during the scattering process

$$\mathbf{Q} = \pm\mathbf{q} + \tau, \quad (41)$$

$$E_1 - E_0 = \pm\hbar\omega_j(\mathbf{q}). \quad (42)$$

These two conditions enable the phonon dispersion relation to be measured in a single crystal.

The inelastic structure factor $F_j(\mathbf{Q})$ is given by

$$F_j(\mathbf{Q}) = \sum_k \bar{b}_k \frac{\mathbf{Q}\cdot\mathbf{e}_j(q)}{\sqrt{M_k}} e^{-W_k(Q)} e^{i\mathbf{Q}\cdot\mathbf{r}_k}. \quad (43)$$

The sum extends over all atoms of the unit cell. Here M_k is the mass of the kth atom, $\bar{b}_k$ = neutron scattering length averaged over various isotopes, $\mathbf{r}_k$ the position vector, and e^{-W_k} the Debye-Waller factor for the kth atom. The dynamical structure factor can be calculated from polarization vectors based on a lattice dynamical model. Good agreement between the observed and calculated dynamical structure factors along with the phonon dispersion and phonon density of states provides a stringent test for the accuracy of the interatomic potential used in the lattice dynamical calculation.

In the technique known as coherent inelastic neutron spectroscopy, scattered neutron groups are observed whenever the energy and momentum conservation laws are simultaneously satisfied at certain ω and $\mathbf{q}$. The wavelength or velocity of the neutrons can be measured in two ways: (1) Bragg scattering from a single crystal monochromator or analyzer using the Bragg equation, $\lambda = 2d_{hkl}\sin\theta$, and (2) time of flight method, where the transit time of neutrons over a finite flight path is electronically measured, from which the neutron velocity can be determined. For phonon measurements, a single crystal triple axis spectrometer based on the Bragg law is most widely used. The most commonly used method is the constant $\mathbf{Q}$ method, where the momentum transfer is constant, as opposed to the constant E method, where the energy transfer is constant. For further details, see Ghose (1988) and the references therein.

The generalized phonon density of states $G(E)(E = \hbar\omega)$, as determined by inelastic neutron scattering from a large powder sample, is different from $g(\omega)$, in that $G(E)$ is a sum of the partial components of the density of states due to the various species of atoms, weighted by their scattering cross sections. $G(E)$ also needs correction for multiphonon scattering. Hence, a direct determination of $g(\omega)$ from $G(E)$ is not possible. However, when $g(\omega)$ is available from a lattice dynamical calculation, $G(E)$ can always be calculated and compared with the experimentally measured values.

An incoherent approximation to the coherent scattering (Placzek and Van Hove, 1955) is usually used to derive the scattering from a powder sample

$$\frac{d^2\sigma_{coh}}{d\Omega\,dE} = \frac{1}{2}\sum_k e^{-2W_k(Q)}\frac{1}{M_k}(b_k^{coh})^2 \cdot \sum_{\tau q j}\frac{1}{3}Q^2|e_j^k|^2\frac{\hbar}{\omega_j(\mathbf{q})}$$
$$\left[n_j(\mathbf{q}) + \frac{1}{2} \pm \frac{1}{2}\right]\delta[\mathbf{Q} - \tau \pm \mathbf{q}]\delta[\omega \mp \omega_j(\mathbf{q})], \quad (44)$$

where the sum over $\mathbf{q}$ is over all $\mathbf{q}$ satisfying the relation $\mathbf{Q} = \tau \pm \mathbf{q}$, in which τ is the reciprocal lattice vector. The scattering then provides the one phonon density of states weighted by the scattering length and the population factor. Any inelastic neutron scattering spectrum contains a contribution from multiphonon scattering that has to be estimated and subtracted from the experimentally observed spectrum $G(E)$ to obtain the one-phonon density of states. Usually, multiphonon scattering contributes a continuous spectrum and effectively increases the background. The coherent neutron scattering measures the scattering function in terms of Q and E, which in the conventional harmonic phonon expansion can be written as:

$$S(Q, E) = S^{(0)} + S^{(1)} + S^{(n)}, \quad (45)$$

where $S^{(0)}$, $S^{(1)}$ and $S^{(n)}$ represent elastic, one-phonon, and multiphonon scattering, respectively. The multiphonon contribution to the total scattering is estimated in the incoherent approximation using Sjolander's formalism (Rao et al., 1988), in which the total scattering is treated as a sum of the partial components of the density of states from the various species of atoms, weighted by the squares of their scattering lengths.

The time-of-flight method using a pulsed spallation neutron source is the most commonly used technique, which is capable of the necessary energy resolution and covers the energy range needed for both external and internal vibrations in solids. The low-resolution medium-energy chopper spectrometer (LRMECS) at the Argonne National Laboratory is such an instrument using the Intense Pulsed Neutron Source (IPNS) (Fig. 4). The chopper is phased to the source and gives pulses of monochromatic neutrons at the sample. Scattered neutrons are time-analyzed over a heavily shielded flight path of 2.5 m, whereas the incident flight path is 7.5 m. The neutron chopper has a 5-μs burst time to give around 6% resolution of the incident energy. The range of detector angles from -10 to $120°$ gives a wide range of simultaneous $\mathbf{Q}$ vectors from 1 to 20 Å^{-1}

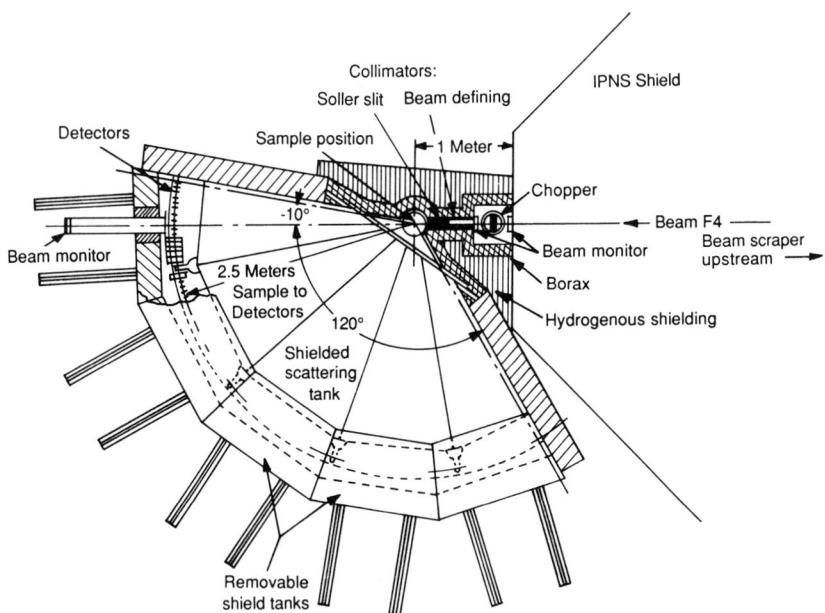

Fig. 4. The low-resolution-medium-energy chopper spectrometer (LRMECS) at the Intense Pulsed Neutron Source (IPNS) at Argonne National Laboratory. (Courtesy Price, 1988.)

at 100 meV energy transfers. This instrument has been used to measure the generalized density of states from forsterite and fayalite powder samples at several temperatures.

Lattice Dynamics of Forsterite and Fayalite: Phonon Density of States and Specific Heat

Olivine, $(Mg,Fe)_2SiO_4$, is a major component of the earth's upper mantle. Because of its geophysical significance, its two end-members, forsterite (Mg_2SiO_4) and fayalite (Fe_2SiO_4), have been extensively studied by theoretical and experimental lattice dynamical methods. Theoretical lattice dynamical calculations of zero-center phonon frequencies in forsterite and their comparison with IR and Raman measurements were carried out by Iishi (1978), Price et al. (1987a), and Rao et al. (1988); phonon dispersion along certain symmetry directions, phonon density of states, and specific heat in forsterite by Rao et al. (1987, 1988) and Price et al. (1987b) and in fayalite by Ghose et al. (1991) and Price et al. (1991). Inelastic neutron scattering measurements of phonon dispersion and phonon density of states in forsterite were reported by Ghose et al. (1987) and Rao et al. (1987, 1988) and in fayalite by Ghose et al. (1991) and Price et al. (1991).

Both forsterite and fayalite are orthorhombic, space group *Pnma* (a reorientation of the usual *Pbnm* setting) with four formula units, i.e., 28 atoms in the unit cell. The unit cell dimensions of forsterite are $a = 10.190$, $b = 5.987$, and $c = 4.753$ Å, and of fayalite: $a = 10.471$, $b = 6.086$ and $c = 4.818$ Å. They are isostructural with two types of [MO_6] octahedra forming serrated edge-sharing chains running parallel to the *b*-axis, cross-linked by isolated [SiO_4] tetrahedra sharing edges with adjacent octahedra (Fig. 5). Since the unit cell contains 28 atoms, there are $3 \times 28 = 84$ independent phonon branches! To simplify the task, a rigid-molecular-ion model was adopted by Rao et al. (1988) for the calculation of the external modes, in which the "molecule", i.e., the [SiO_4] group, was assumed to be rigid with only three translational and three rotational degrees of freedom, whereas the two M atoms (M1 and M2) each have three degrees of translational freedom, thereby reducing the number of independent phonon branches to be calculated to 48. The potential energy of the crystal was assumed to contain two-body potentials containing a Coulombic and a repulsive term

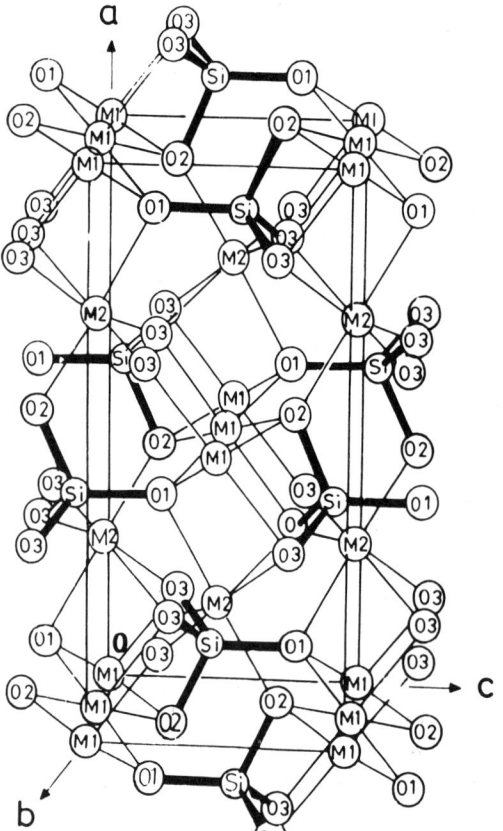

Fig. 5. Crystal structure of forsterite, Mg_2SiO_4 (Matsui and Busing, 1984).

$$V(r) = \frac{e^2}{4\pi\varepsilon_0} \frac{Z(Kk)Z(K'k')}{r} + a\exp\left[-\frac{br}{R(Kk) + R(K'k')}\right], \quad (46)$$

where Z_k and R_k are the "effective charge" and "effective radius" of the kth ion, and K is the rigid unit, r the interatomic separation, e the electron charge and ε_0 is the vacuum permittivity. Constants a and b were set equal to 1822 eV and 12.364, respectively. Z_k values were adjusted to maintain charge neutrality. The effective charges and radii were treated as adjustable parameters, which were optimized with respect to the known crystal structure and the elastic constants (Rao et al., 1987, 1988; Ghose et al., 1987, 1991). The optimized values used for the phonon dispersion calculations of forsterite are given below

	Mg1	Mg2	Si	O1	O2	O3
Charge	1.60	1.80	1.0	−1.2	−1.0	−1.10
Radius (Å)	1.68	1.73	1.00	1.55	1.45	1.50

Similar values were used for the preliminary calculations on fayalite. The phonon dispersion relations for the external modes of forsterite calculated along the three principal symmetry directions are shown in Fig. 6 (Rao et al., 1988). One-phonon dynamical structure factors were also calculated using the rigid-molecular-ion model, which were used as guides for assigning the observed phonon peaks during the inelastic neutron-scattering experiments.

Phonon dispersion relations in single crystals of forsterite (3.4 × 1 × 0.6 cm) and fayalite (cylinder of 6 × 0.6 cm length) (kindly provided by Prof. H. Takei, University of Tokyo) were measured by inelastic neutron scattering at the Brookhaven National Laboratory, using a triple-axis spectrometer and the High Flux Beam Reactor (HFBR) (Ghose et al., 1988, 1991; Rao et al., 1988). Over 150 phonon measurements were made for forsterite and 70 for fayalite along the three principal directions in the xy-, yz-, and xz-planes. Some of the selected

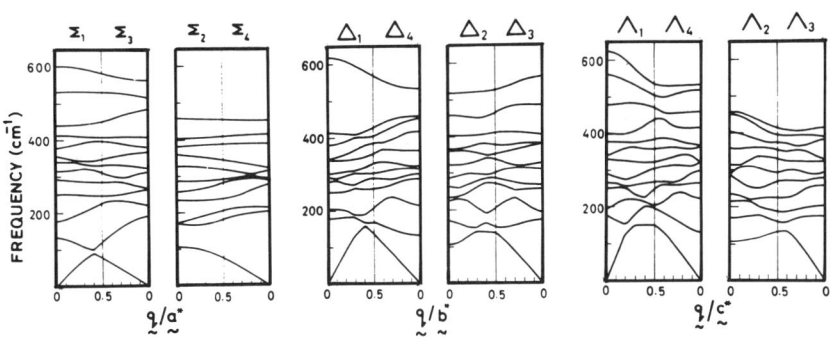

Fig. 6. The calculated phonon dispersion relation of the external modes in forsterite (Rao et al., 1988).

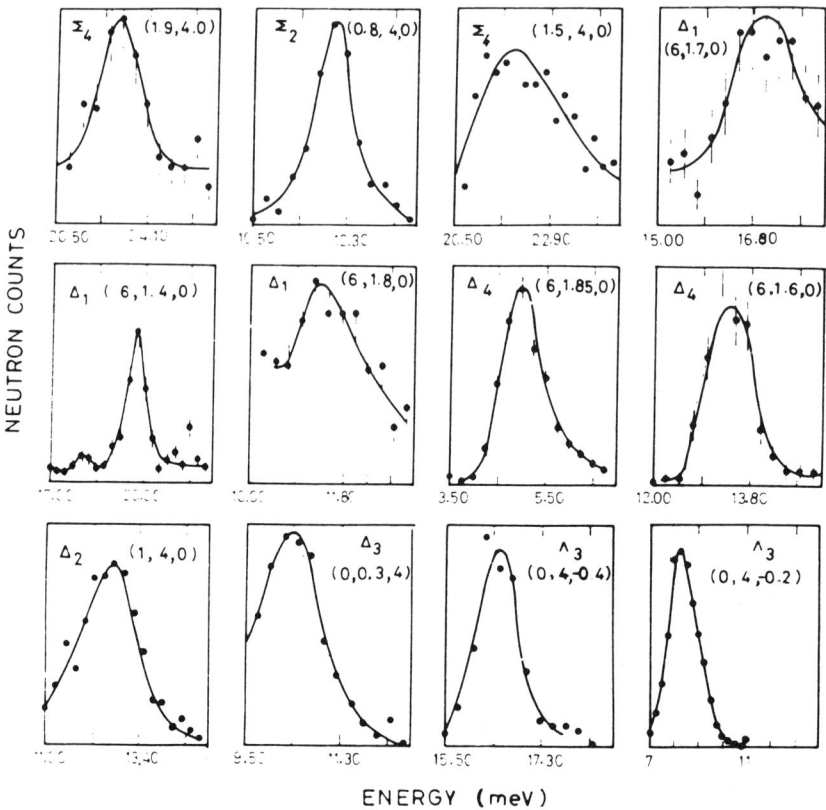

Fig. 7. Selected acoustic and optic phonon measurements in a single crystal of forsterite. The numbers in parentheses indicate the coordinates of the wave vector involved in the neutron scattering process in units of reciprocal lattice vectors (Ghose et al., 1987 Pergamon Press, plc.).

measurements of acoustic and optic phonons in forsterite are shown in Fig. 7. The calculated (full lines) and measured (filled circles) phonon branches along the three principal symmetry directions Σ, Δ and Λ in forsterite are shown in Fig. 8. The frequencies of the zone-center optic phonons in forsterite measured by inelastic neutron scattering are compared with those measured by Raman and IR spectroscopy in Table 2, which shows very good agreement. Note that the optic phonon with a frequency of 104 cm^{-1} measured by inelastic neutron scattering is both Raman- and IR-inactive, which makes an important contribution to the low temperature specific heat.

The rigid-ion model was used for the calculation of the one-phonon density of states $g(\omega)$, which includes both external and internal vibrations. For the covalent tetrahedral [SiO$_4$] group, the Si–O stretching potential is of the form

$$v(r_{\text{Si-O}}) = CD \exp\left\{ -\frac{n}{2C} \frac{(r_{\text{Si-O}} - r_0)^2}{r_{\text{Si-O}}} \right\}, \tag{47}$$

Fig. 8. Comparison of experimental and theoretical phonon dispersion curves: theory (full lines); experiment (filled circles). The theoretical curves are predictions, and not fits to experimental data (Rao et al., 1988).

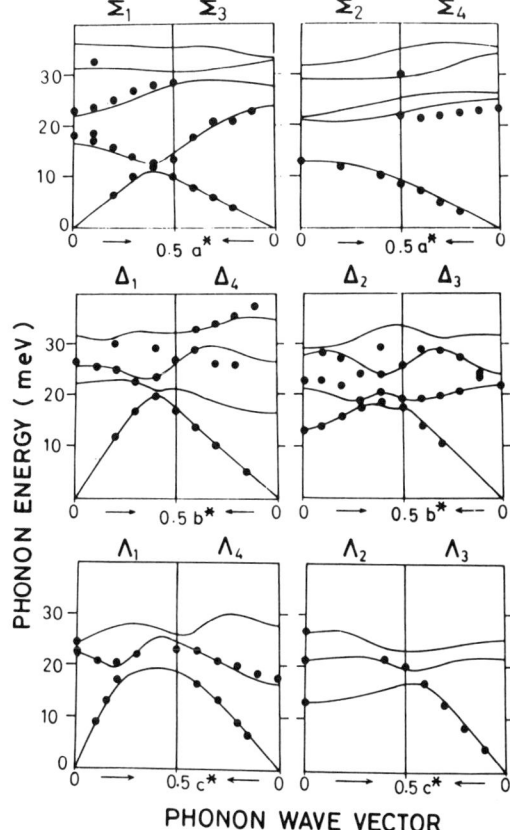

Table 2. Frequencies of zone-center phonons (cm^{-1}) measured by inelastic neutron scattering compared with Raman and IR measurements under ambient conditions.

INS	Raman	IR
104	—	—
144	142	144
184	183	—
192	192	—
200	—	201
258	260	—
315	318	313
325	324	323

Note: For sources of Raman and IR data, see Iishi (1978).

with parameters C, D, n, and r_O. To simulate the O–Si–O bond bending potential, the O–O interaction within the same silicate group is of the form

$$v(r_{O-O}) = \frac{e^2}{4\pi\varepsilon_0} \frac{Z^2(O)}{r_{O-O}} + sa\exp\left[\frac{-br_{O-O}}{2R(O)}\right] - \frac{w}{r_{O-O}^6} \quad (48)$$

with parameters s and w. The parameters $Z(O)$, $R(O)$, a, and b were the same as in the rigid-molecular-ion model. The other parameters were optimized to the following values: $C = 1$, $D = 3.4$ eV, $n = 10.5$ Å^{-1}, $r_O = 1.61$ Å, $s = 55$, and $w = 1250$ eV Å^6. The computer program DISPR (Chaplot, 1978) was used in the numerical computations. The total one-phonon density of states, $g(\omega)$ of forsterite derived from the rigid-ion model by sampling 27 wave vectors in the irreducible Brillouin zone is shown in Fig. 9. The partial density of states in the rigid-molecular-ion model from external modes only due to the translations of the two Mg atoms and the translation and rotation of the silicate group are shown in Fig. 10a, b.

Figures 11 and 12 show the comparison of the observed and calculated neutron spectra, including the one phonon and multiphonon contributions and intensity broadening due to instrumental resolution at 300 K for forsterite and at 300 and 17 K in fayalite respectively. The inelastic neutron-scattering measurement of $G(E)$ was made on synthetic powder samples (~ 50 g) in both cases by the time-of-flight method using the low-resolution medium-energy chopper spectrometer and the Intense Pulsed Neutron Source at the Argonne National Laboratory. The agreement between the theoretical and experimental curves at 300 K for both forsterite and fayalite is very satisfactory. The peaks in the 20 to 80 meV region in the neutron spectra of fayalite are shifted to lower energies compared to those in forsterite. However, the neutron spectrum of fayalite at 17 K in the 20 to 40 meV region shows a marked deviation from that at 300 K and the calculated spectrum at 17 K based only on lattice contribution. Since fayalite is paramagnetic at 300 K that undergoes a transition to the antiferromagnetic state at 65 K (Fuess et al., 1988), this deviation is most likely due to the spin-wave (magnon) contribution. The magnetic transition causes a λ-type anomaly in the specific heat curve of fayalite at 65 K (Robie et al., 1982b). In both sets of

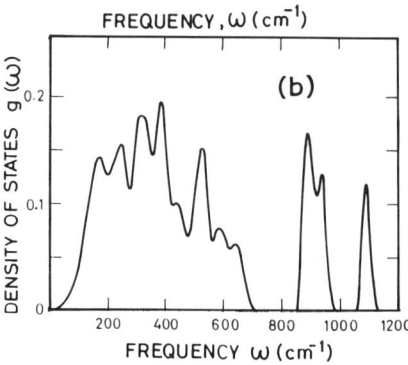

Fig. 9. Total one phonon density of states calculated from the rigid-ion model of forsterite. Note the energy gap in the 700 to 850 cm^{-1} region (Rao et al., 1988).

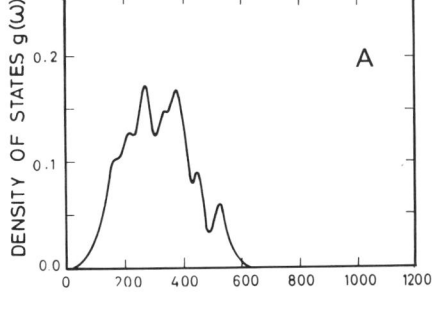
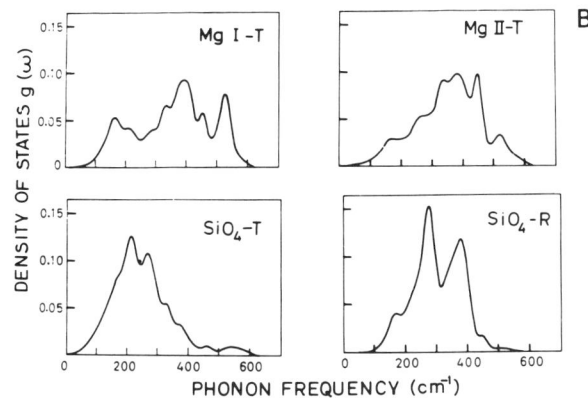

Fig. 10. (a) total and (b) partial density of states, $g(\omega)$ from the external modes (translation of Mg atoms, and translation and rotation of [SiO$_4$] group) in forsterite (Rao et al., 1988).

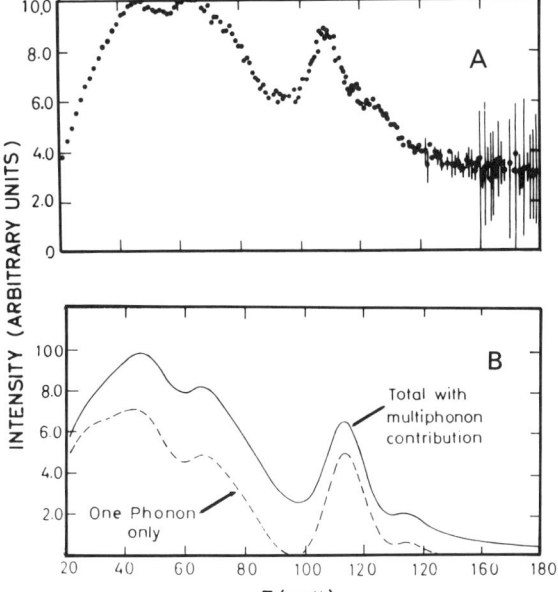

Fig. 11. Generalized density of states, $G(E)$ of forsterite from a powder sample (a) inelastic neutron scattering spectrum and (b) theoretical spectrum from the rigid-ion model, including multiphonon contribution and resolution broadening (Rao et al., 1988).

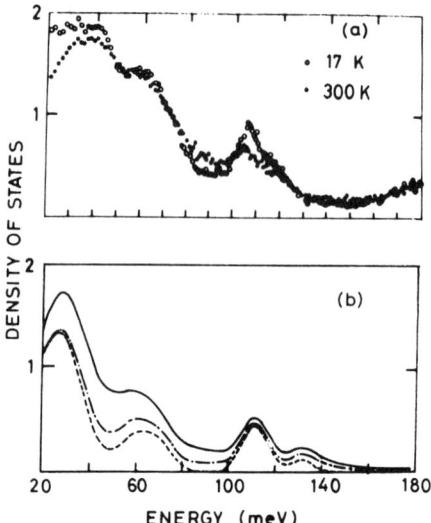

Fig. 12. Comparison of the inelastic neutron scattering spectra of fayalite with the calculated spectra based on the rigid-ion model. (a) Experimental spectra at 300 and 17 K, including multiphonon contributions. (b) Calculated phonon spectra with one-phonon and multi-phonon contributions at 300 K (full line) and at 17 K (dash-dot line). The dashed line gives the one phonon contribution to the phonon spectra (Price et al., 1991).

spectra, the peaks below 80 meV are mainly from translation of the Mg or Fe atoms and external vibrations of the silicate groups, whereas above 80 meV the peaks are due to the internal vibrations of the silicate group. The tail above 140 meV is mainly due to the multiphonon contributions in both systems. The significant aspect about the one-phonon density of states $g(\omega)$ in both cases is the band gap between 700 cm^{-1} (87 meV) and about 850 cm^{-1} (105 meV), principally separating the external from internal vibrations. This band gap has significant implications for the thermodynamic properties of olivines.

Once $g(\omega)$ is known, the lattice specific heat at constant volume C_v can be calculated. Results of these calculations for forsterite using the rigid-molecular-ion, rigid-ion, Kieffer, and Debye models are shown along with the calorimetrically measured values at constant pressure C_p by Robie et al. (1982a) in Fig. 13. The specific heat computed from the rigid-molecular-ion model shows excellent agreement (within 1%) with experimental values in the low-temperature region (0–100 K). The discrepancy between the two increases with increasing temperature because this model neglects the high-frequency internal modes, which are excited only at high temperatures. Since all degrees of freedom are not considered, the Dulong-Petit limit at high temperature is never reached. In the rigid-ion model all degrees of freedom are taken into account and the agreement is much more satisfactory. Beyond 400 K, the difference between the calculated C_v and the experimental C_p is due to the neglect of anharmonic corrections. Choudhury et al. (1989) estimated the bulk modulus B and the thermal expansion α_v from quasiharmonic lattice dynamical calculations and used these values to estimate the anharmonic corrections according to the equation (39). At 300 K, the correction is on the order of 1% increasing to about 5% at 1200 K. With these corrections, which are important at high temperatures, the agreement between computed and experimental values of $C_p(T)$ is within 1% at high temperatures up to 1200 K. The estimated weighted average of mode Grüneisen

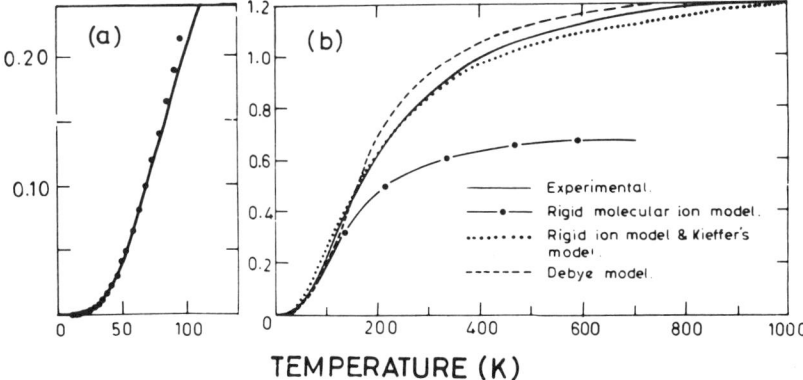

Fig. 13. Comparison of experimental specific heat, C_p of forsterite (Robie et al., 1982a) with theoretical specific heat, C_v (a) Theoretical results (0–100 K) based on rigid-molecular-ion model (full line) and experimental data (filled circle). (b) Comparison of experimental data (full line) with theoretical models. The discrepancy between the results of the rigid-molecular-ion model and experimental results above 100 K is due to the neglect of internal vibrations that are excited at higher temperatures (Rao et al., 1988).

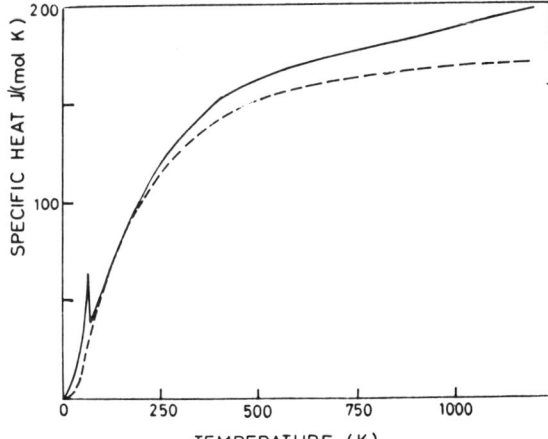

Fig. 14. Comparison of calculated C_v of fayalite in the rigid-ion model (dashed line) with the experimental specific heat, C_p (Robie et al., 1982b) (full line) (Price et al., 1991).

parameters, $\bar{\Gamma}$ of all modes in the entire frequency range is 1.14. The specific heat calculated from the Debye model shows considerable deviation from the experimental values. This is not surprising in view of the fact that the Debye density of states is a poor approximation of the true density of states of forsterite. Furthermore, considerable variation of the Debye temperature with temperature (Kieffer, 1985) indicates that the Debye model is inadequate for forsterite. On the other hand, the Kieffer model predicts the specific heat quite well, although agreement in the low-temperature region (0–100 K) is not as good as that predicted by the rigid-molecular-ion model.

The computed specific heat C_v of fayalite is compared with the experimental values (Robie et al., 1982b) in Fig. 14. The measured specific heat shows a λ-peak at 65 K corresponding to the Néel point. Below 70 K, the specific heat is dominated by the spin-wave (magnon) excitations associated with ordering of the Fe^{2+} spins (magnetic moments) from the paramagnetic to antiferromagnetic state. Below 65 K, the Fe^{2+} spins on M2 are ordered antiparallel along **b**, whereas the Fe^{2+} spin on M1 is canted away from **b** and is still fluctuating. The computed C_v from lattice contributions alone shows a large discrepancy with the experimental values in the 0 to 70 K range because it does not include the contribution to specific heat from spin-wave excitation. Between 70 and 260 K, agreement is very good, above which the anharmonic corrections $(C_p - C_v)$ become significant and the computed C_v deviates again from the experimental C_p. These corrections can be easily computed from the experimentally determined bulk modulus B (Sumino, 1979) and thermal expansion α_v (Smyth, 1975) of fayalite.

Price et al. (1987a,b) also used an ionic model for their lattice dynamical calculations on forsterite. In their model, the static lattice energy U is given by the sum of Coulombic and short-range interactions of the form

$$U = \tfrac{1}{2}\sum_{ij} e^2 q_i q_j r_{ij}^{-1} + A_{ij}\exp\left(-\frac{r_{ij}}{B_{ij}}\right) - C_{ij}r_{ij}^{-6}, \tag{49}$$

where e is the electron charge, q_i and q_j are ionic charges of ions i and j, A_{ij}, B_{ij}, and C_{ij} are parameters of the short-range potential, and r_{ij} is the ionic separation. Included was Si–O–Si bond bending term of the type

$$U = \tfrac{1}{2}\sum_{ijk} k_{ijk}(\theta_{ijk} - \theta_0)^2, \tag{50}$$

where k_{ij} is a spring constant, θ_{ijk} the O–Si–O bond angle, and θ_0 the tetrahedral angle. The polarizability of the oxygen atom was described by a shell model. The phonon dispersion relations along the Γ-X-R-Γ direction were calculated using this potential. The one-phonon density of states constructed by sampling 8 and 27 grid points in the Brillouin zone (Price et al., 1987b) is very similar to that constructed by Rao et al. (1987, 1988). The calculated specific heat shows good agreement with the calorimetric measurements between 80 to 300 K, but the discrepancy in the 10 to 80 K range is quite large (10–20%).

Price et al. (1987) have also calculated the specific heats of β and γ (spinel) phases of Mg_2SiO_4 using the same potential, as for forsterite. The predicted bulk moduli are up to 35% too stiff and the thermal expansion coefficients too low, indicating that the interatomic potential needs further improvement. Nevertheless, the calculated specific heats show fair agreement with those measured by differential scanning calorimetry in the 300 to 1000 K range by Watanabe (1982). Because of the lack of experimental data in the low temperature region (0–300 K), it is not possible to evaluate the accuracy of the predicted low temperature specific heat.

Phonon Density of States and Specific Heat of Orthorhombic MgSiO₃ Perovskite

(Mg,Fe)SiO$_3$ perovskite is considered to be the most abundant mineral in the earth's lower mantle below the 670 km discontinuity (Knittle and Jeanloz, 1987). Because of its geophysical importance, MgSiO$_3$ perovskite has been subjected to considerable theoretical and experimental investigations. The material synthesized above 27 GPa and about 2000 K is orthorhombic under ambient conditions with the space group *Pnma*. The unit cell dimensions are: $a = 4.9313$, $b = 6.9083$, and $c = 4.7787$ Å (Horiuchi et al., 1987). There are four formula units, i.e., 20 atoms in the unit cell. The crystal structure is a distorted version of the cubic ABO$_3$-type perovskite, in which the [SiO$_6$] octahedra are alternately rotated with respect to the cubic structure (Fig. 15). The Mg^{2+} ion occurs in a cavity with 12 coordination.

Lattice dynamical calculations on the cubic and orthorhombic phases of MgSiO$_3$ perovskite have been carried out by Wolf and Jeanloz (1985), Wolf and Bukowinski (1987), Cohen (1987), Hemley et al. (1987, 1989), Bukowinski and Wolf (1988), and Choudhury et al. (1988). The calculated phonon dispersion relations of the cubic phase indicate imaginary phonon frequencies associated with coupled librations of the [SiO$_6$] octahedra (Wolf and Jeanloz, 1985; Bukowinski and Wolf, 1988; Cohen, 1987; Hemley et al., 1987). Hence, the cubic MgSiO$_3$ phase is unstable under ambient conditions. The phonon spectrum of the orthorhombic phase shows no soft (unstable) modes. From lattice dynamical calculations, Bukowinski and Wolf (1988) have shown that at 950 K, softening of modes associated with the librations of the [SiO$_6$] octahedra (M$_2$ and neigh-

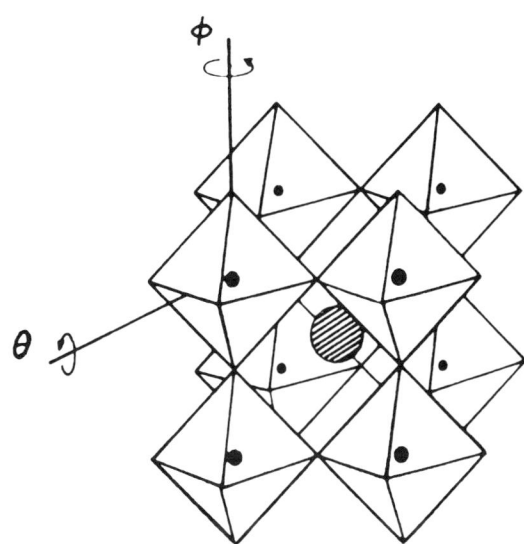

Fig. 15. Crystal structure of orthorhombic MgSiO$_3$ perovskite.

boring modes) leads to a transition to the tetragonal structure and at about 1500 K to the cubic structure due to the unstable R_{25} mode. MgSiO$_3$ perovskite synthesized at 26 GPa and 1873 K shows twin domains with {110} and {112} reflection planes, which may be associated with the cubic to tetragonal and tetragonal to orthorhombic phase transitions, respectively, on cooling (Wang et al., 1990). The equation of state of orthorhombic perovskite has been calculated by Wolf and Bukowinski (1985), Hemley et al. (1987), and Cohen (1987). The experimentally observed equation of state (volume dependence on pressure at ambient temperature) (Knittle and Jeanloz, 1987) has been fitted very well by the potential-induced breathing (PIB) model (Cohen, 1987). This model also has been used successfully to calculate the zone-center optic phonons in the orthorhombic perovskite (Hemley et al., 1989). In this Gordon-Kim-type model, the electron density of the crystal is approximated by a superposition of component ion electron densities, which are calculated by using the local density approximation with self-interaction corrections. The electrostatic field of the crystal is simulated by a Watson sphere in each atomic calculation. The radius of the Watson sphere is chosen such that it gives an electrostatic potential within the sphere equal to the Madelung potential at the crystallographic site.

Choudhury et al. (1988) have calculated the phonon dispersion relations along the three principal symmetry directions (Σ, Δ, Λ) of orthorhombic MgSiO$_3$ perovskite using the rigid-ion model in the quasiharmonic approximation. The interatomic potential, consisting of a Coulomb interaction and a Born-Mayer-type repulsive interaction, is similar to that used by Rao et al. (1988) for forsterite. The final parameters $Z(k)$ and $R(k)$ used are 2.2, 2.3, and -1.5 for charges and 1.66, 1.20, and 1.70 Å for the radii of Mg, Si, and O respectively; these have been revised from our initial report (Choudhury et al., 1988) as they yielded elastic constants that were systematically lower than the experimental values (Yeganeh-Haeri et al., 1989). The phonon dispersion relations are shown in Fig. 16 and the phonon density of states in Fig. 17. The phonon density of states has peaks around 400, 600, and 900 cm^{-1} and no band gap, as in forsterite. The band gap in the phonon density of states of forsterite is essentially due to the large

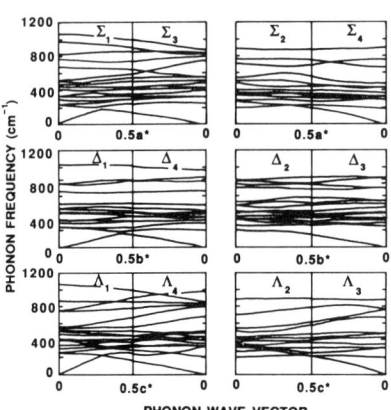

Fig. 16. MgSiO$_3$ perovskite (orthorhombic), phonon dispersion relation along the three principal symmetry directions: **a*** (Σ), **b*** (Δ), and **c*** (Λ) directions. There are 60 phonon branches in each direction.

Fig. 17. The phonon density of states of orthorhombic MgSiO$_3$ pervoskite calculated from the rigid-ion model.

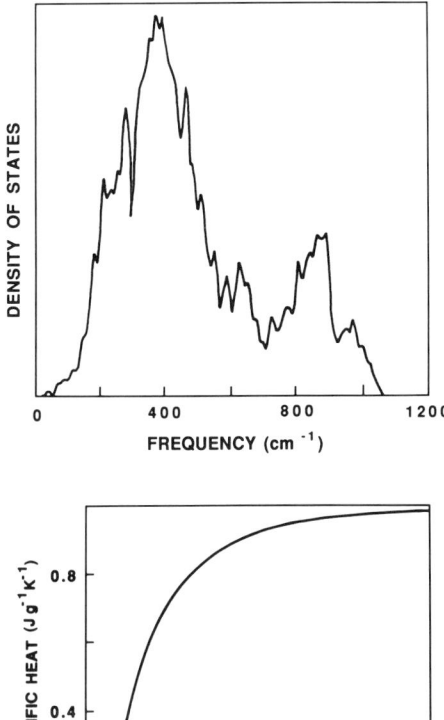

Fig. 18. The calculated specific heat of orthorombic MgSiO$_3$ perovskite as a function of temperature.

difference between the vibrational frequency ranges corresponding to external (intermolecular) and internal (intramolecular) vibrations of the silicate "molecule." In contrast, the octahedral [SiO$_6$] group cannot be considered a rigid molecular unit, which explains the absence of any band gap in the phonon density of states of MgSiO$_3$ perovskite. Hence, there is a fundamental difference in the nature of the thermodynamic properties of forsterite and MgSiO$_3$ perovskite. The computed specific heat as a function of temperature is shown in Fig. 18.

Phonon Density of States and Specific Heat of Andalusite, Al$_2$SiO$_5$

The only other mineral whose phonon density of states is calculated by lattice dynamical methods is andalusite (Winkler and Buehrer, 1990). Andalusite is orthorhombic, space group *Pnnm*, with unit-cell dimensions $a = 7.798$,

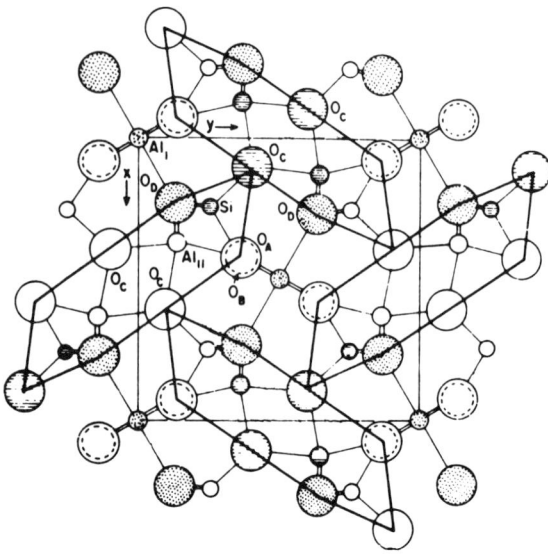

Fig. 19. Crystal structure of andalusite, Al_2SiO_5.

$b = 7.903$, and $c = 5.557$ Å. There are four formula units, i.e., 32 atoms in the unit cell. The crystal structure consists of chains of edge-sharing $[AlO_6]$ octahedra running parallel to the c-axis, cross-linked by dimers of edge-sharing $[AlO_5]$ trigonal bipyramids and tetrahedral $[SiO_4]$ groups (Fig. 19). A transferable set of interatomic potential valid for all three Al_2SiO_5 polymorphs (andalusite, sillimanite and kyanite) was used for the lattice dynamical calculations, which was very similar in nature to the one used for the Mg_2SiO_4 polymorphs by Price et al. (1987b). Because of the variable coordination of half of the Al atoms (four, five, and six in sillimanite, andalusite, and kyanite, respectively), the potential used for andalusite is not expected to be very accurate. The calculated phonon frequencies at the zone center show moderate agreement with experimental IR and Raman measurements (see also Iishi et al., 1979). The phonon dispersion relation along c^* was calculated and some of the low-frequency phonon branches were measured by inelastic neutron scattering. Although the qualitative agreement is good and the calculated acoustic branches match the experimental measurements well, the optic branches show large discrepancies between the calculated and measured ones. The calculated phonon density of states is shown in Fig. 20, which also illustrates the phonon density of states based on IR and Raman measurements by Kieffer (1985) and Salje and Werneke (1982). Apparently, there are no band gaps in the phonon density of states. Perhaps this is due to the fact that the internal vibration frequencies of the $[AlO_5]$ group span the band gap between the ranges of external vibration frequencies (translations of Al atoms, and translation and rotation of the $[SiO_4]$ group) and the internal vibration frequencies of the $[SiO_4]$ group. If this argu-

Fig. 20. The phonon density of states of andalusite according to (a) Kieffer (1980, (b) Salje and Werneke (1982), and (c) lattice dynamical study by Winkler and Buehrer (1990).

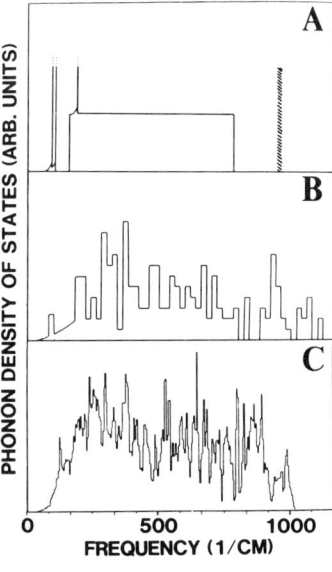

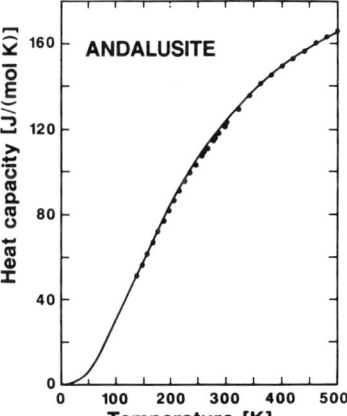

Fig. 21. The calculated specific heat of andalusite (full line) with experimental DSC measurements of Salje and Werneke (1982) (Winkler and Buehrer, 1990).

ment is correct, small band gaps are expected to occur in the phonon density of states of sillimanite and kyanite. The calculated specific heat of andalusite (Fig. 21) shows good agreement with the experimental DSC measurements by Salje and Werneke (1982) in the 150 to 500 K range. However, a comparison of the calculated low-temperature specific heat (5–300 K) with the experimental measurements by adiabatic calorimetry (Robie et al., 1984) is necessary to evaluate the accuracy of the potential model used. It will also be very interesting to calculate the phonon density of states of sillimanite to see if the predicted band gap exists and, if so, to explore its effect on thermodynamic properties. Winkler et al. (1991) have also calculated the specific heat of sillimanite by lattice dynamical methods, but no details are given. The calculated values are in good agreement with DSC measurements by Salje and Werneke (1982), but the agreement in the low-temperature region (0–80 K) is poor. The reason for this discrepancy again is most likely due to the approximate nature of the interatomic potential used rather than the poor sampling of the Brillouin zone.

Conclusions

We have shown that lattice dynamics in the quasi-harmonic approximation is a very good technique for the calculation of the phonon density of states of minerals, which determines the thermodynamic properties, such as internal energy, free energy, specific heat, and entropy. The calculated one-phonon density of states, $g(\omega)$ can be compared with the generalized phonon density of states measured by inelastic neutron scattering on a large powder sample. The mode Grüneisen parameter, bulk modulus, and thermal expansion can also be determined by quasi-harmonic lattice dynamics. This technique has been successfully used to predict the phonon density of states and lattice specific heats of forsterite, fayalite, andalusite, and the $MgSiO_3$ perovskite. The success of the lattice dynamical calculations depends directly on the accuracy of the interatomic potential used. Due to the formidable nature of the computing involved, even for a structure such as forsterite with 28 atoms in the unit cell, fairly simple sets of optimized potentials have been used so far for such calculations. With the advent of faster computers and fast algorithms, in the near future it would be possible to derive more accurate potentials from *ab initio* quantum mechanical calculations on large enough clusters (~ 100 atoms). In the meantime, further theoretical and experimental lattice dynamical studies including inelastic neutron-scattering measurements of phonon dispersion relation and phonon density of states (the latter with higher resolution) should be made on more minerals of geological and geophysical interest, with a view toward a fundamental understanding of their thermodynamic properties and stability relations.

References

Akaogi, M., Ross, N.L., McMillan, P., and Navrotsky, A. (1984). The Mg_2SiO_4 polymorphs (olivine, modified spinel and spinel)—thermodynamic properties from oxide melt solution calorimetry, phase relations, and models of lattice vibrations. *Amer. Mineral.* **69**, 499–512.

Ashida, T., Kume, S., Ito, E., and Navrotsky, A. (1988). $MgSiO_3$ ilmenite: Heat capacity, thermal expansivity, and enthalpy of transformation. *Phys. Chem. Miner.*, **16**, 239–245.

Born, M. and Huang, K. (1954). *The Dynamical Theory of Crystal Lattices*. Clarendon Press, Oxford.

Born, M. and von Kármán, Th. (1912). Über Schwingungen in Raumgittern. *Physik. Zeit.* **13**, 297–309.

Brockhouse, B.N. (1962). Interatomic forces from neutron scattering. *J. Phys. Soc. Jap.*, Suppl. 17, **BII**, 363–367.

Brüesch, P. (1982). *Phonons: Theory and Experiment I*. Springer-Verlag, Berlin.

Bukowinski, M.S.T. and Wolf, G.H. (1988). Equation of state and possible critical phase transitions in $MgSiO_3$ perovskite at lower mantle conditions, in *Structural and Magnetic Phase Transitions in Minerals*, S. Ghose, J.M.D. Coey, and E. Salje, eds., Springer-Verlag, New York, pp. 91–112.

Chaplot, S.L. (1978). A computer program for external modes in complex molecular-ionic crystals. Re. 972, Bhabha Atomic Research Centre, Bombay, India.

Choudhury, N., Chaplot, S.L., and Rao, K.R. (1989). Equation of state and melting point studies of forsterite. *Phys. Chem. Miner.* **16**, 599–605.

Choudhury, N., Chaplot, S.L., Rao, K.R. and Ghose, S. (1988). Lattice dynamics of $MgSiO_3$ perovskite. *Pramāna-J. Phys.* **30**, 423–428.

Cohen, R.E. (1987). Elasticity and equation of state of $MgSiO_3$ perovskite. *Geophys. Res. Lett.*, **14**, 1053–1056.

Debye, P. (1912). Zur Theorie der spezifischen Wärmen. *Ann. der Physik* **39**, 789.

Dorner, B. (1982). *Coherent Inelastic Neutron Scattering in Lattice Dynamics*. Springer-Verlag, Berlin.

Fuess, H., Ballet, O., and Lottermoser, W. (1988). Magnetic phase transitions in olivines Mg_2SiO_4 (M = Mn, Fe, Co, Fe_xMn_{1-x}), in *Structural and Magnetic Phase Transitions in Minerals*, S. Ghose, J.M.D. Coey, and E. Salje, eds., Springer-Verlag, New York, pp. 185–207.

Ghose, S. Hastings, J.M., Corliss, L.M., Rao, K.R., Chaplot, S.L., and Choudhury, N. (1987). Study of phonon dispersion relations in forsterite, Mg_2SiO_4 by inelastic neutron scattering. *Solid State Comm.* **63**, 1045–1050.

Ghose, S. (1988). Inelastic neutron scattering, in *Spectroscopic Methods in Mineralogy and Geology*, F.C. Hawthorne, ed., *Reviews in Mineralogy*, vol. 18, 161–192. Mineralogical Society America, Washington, D.C.

Ghose, S., Hastings, J.M., Choudhury, N., Chaplot, S.L., and Rao, K.R. (1991) Phonon dispersion relation in fayalite, Fe_2SiO_4. *Physica B* **174**, 83–86.

Hemley, R.J., Cohen, R.E., Yeganeh-Haeri, A., Mao, H.K., Weidner, D.J., and Ito, E. (1989). Raman spectroscopy and lattice dynamics of $MgSiO_3$ perovskite at high pressure, in *Perovskite: A Structure of Great Interest to Geophysics and Materials Science*, pp. 35–44. American Geophysical Union, Washington, D.C.

Hemley, R.J., Jackson, M.D., and Gordon, R.G. (1987). Theoretical study of the structure, lattice dynamics and equation of state of perovskite-type $MgSiO_3$ and $CaTiO_3$. *Phys. Chem. Miner.* **14**, 2–12.

Horiuchi, H., Ito, E., and Weidner, D.J. (1987). Perovskite-type $MgSiO_3$: Single crystal x-ray diffraction study. *Amer. Mineral* **72**, 357–360.

Iishi, K. (1978). Lattice dynamics of forsterite. *Amer. Mineral.* **63**, 1198–1208.

Iishi, K., Salje, E., and Werneke, C. (1979). Phonon spectra and rigid-ion model calculations on andalusite. *Phys. Chem. Miner.* **4**, 173–188.

Kieffer, S.W. (1980). Thermodynamics and lattice vibrations of minerals 4: Application to chain and sheet silicates and orthosilicates. *Rev. Geophys. Space Phys.* **18**, 862–886.

Kieffer, S.W. (1985). Heat capacity and entropy: Systematic relations to lattice vibrations, in *Microscopic to Macroscopic: Atomic Environments to Mineral Thermodynamics*, S.W. Kieffer and A. Navrotsky, eds., *Reviews in Mineralogy*, vol. 14, pp. 65–126, Mineralogical Society of America, Washington, D.C.

Knittle, E. and Jeanloz, R. (1987). Synthesis and equation of state of $(Mg,Fe)SiO_3$ perovskite to over 100 GPa. *Science* **235**, 669–670.

Placzek, G. and Van Hove, L. (1955). Interference effects in total neutron scattering cross-section of crystals. *Nuovo Cimento* **1**, 233–244.

Price, D.L., Ghose, S., Choudhury, N., Chaplot, S.L., and Rao, K.R. (1991). Phonon density of states in fayalite, Fe_2SiO_4. *Physica B174*, 87–90.

Price, G.D., Parker, S.C., and Leslie, M. (1987a). The lattice dynamics of forsterite. *Mineral. Mag.* **51**, 157–170.

Price, G.D., Parker, S.C., and Leslie, M. (1987b) The lattice dynamics and thermodynamics of the Mg_2SiO_4 polymorphs. *Phys. Chem. Miner* **15**, 181–190.

Rao, K.R., Chaplot, S.L., Choudhury, N., Ghose, S., and Price, D.L. (1987). Phonon density of states and specific heat of forsterite, Mg_2SiO_4. *Science* **236**, 64–65.

Rao, K.R., Chaplot, S.L., Choudhury, N., Ghose, S., Hastings, J.M., Corliss, L.M., and Price, D.L. (1988). Lattice dynamics and inelastic neutron scattering from forsterite, Mg_2SiO_4; phonon dispersion relations, phonon density of states and specific heat. *Phys. Chem. Miner.* **16**, 83–97.

Robie, R.A., Finch, C.B., and Hemingway, B.S. (1982a). Heat capacity and entropy of fayalite (Fe_2SiO_4) between 5.1 and 383 K: Comparison of calorimetric and equilibrium values for the QFM buffer. *Amer. Mineral.* **67**, 463–469.

Robie, R.A., Hemingway, B.S., and Takei, H. (1982b). Heat capacities and entropies of Mg_2SiO_4, Mn_2SiO_4 and Co_2SiO_4 between 5 and 380 K. *Amer. Mineral.* **67**, 470–482.

Robie, R.A. and Hemingway, B.S. (1984). Entropies of kyanite, andalusite and sillimanite: additional constraints on pressure and temperature of the Al_2SiO_5 triple point. *Amer. Mineral.* **69**, 298–306.

Salje, E. and Werneke, C. (1982). The phase equilibrium between sillimanite and andalusite as determined from lattice vibrations. *Contrib. Mineral. Petrol.* **79**, 56–76.

Sköld, K. and Price, D.L., eds. (1986). *Neutron Scattering, Part I. Methods of Experimental Physics*, vol. 15, Academic Press, New York.

Smyth, J.R. (1975). High temperature crystal chemistry of fayalite. *Amer. Mineral.* **60**, 1092–1097.

Sumino, Y. (1979). The elastic constants of Mn_2SiO_4, Fe_2SiO_4 and CO_2SiO_4, and the elastic properties of olivine group minerals at high temperature. *J. Phys. Earth* **27**, 209–238.

Wang, Y., Guyot, F., Yeganeh-Haeri, A., and Lieberman, R.C. (1990). Twinning in $MgSiO_3$ perovskite. *Science* **248**, 468–471.

Watanabe, H. (1982). Thermochemical properties of synthetic high pressure compounds relevant to the earth's mantle, in *High Pressure Research in Geophysics*, S. Akimoto and M.H. Manghnani, eds., Center for Academic Publications, Tokyo, pp. 441–464.

Winkler, B. and Buehrer, W. (1990). Lattice dynamics of andalusite: Prediction and experiment. *Phys. Chem. Miner.* **17**, 453–461.

Winkler, B., Dove, M.T., and Leslie, M. (1991). Static lattice energy minimization and lattice dynamics calculations on aluminosilicate minerals. *Amer. Mineral.* **76**, 313–331.

Wolf, G. and Bukowinski, M. (1985). *Ab initio* structural and thermoelastic properties of orthorhombic $MgSiO_3$ perovskite. *Geophys. Res. Lett.* **12**, 809–812.

Wolf, F.H. and Jeanloz, R. (1985). Lattice dynamics and structural distortions of $CaSiO_3$ and $MgSiO_3$ perovskites. *Geophys. Res. Lett.* **12**, 413–416.

Yeganeh-Haeri, A., Weidner, D.J., and Ito, E. (1989). Single-crystal elastic moduli of magnesium metasilicate perovskite, in *Perovskite: A Structure of Great Interest to Geophysics and Materials Science*, A. Navrotsky and D.J. Weidner, eds., pp. 13–25. American Geophysical Union, Washington, D.C., pp. 13–25.

Chapter 12
Thermal Expansion Studies of $(Mg,Fe)_2SiO_4$-Spinels Using Synchrotron Radiation

L.C. Ming, M.H. Manghnani, Y.H. Kim,* S. Usha-Devi,[†] J.-A. Xu, and E. Ito

Introduction

On the basis of high-pressure-temperature studies of the systems $(Mg,Fe)_2SiO_4$, $(Mg,Fe)SiO_3$, and $(Mg,Fe)_3Al_2Si_3O_{12}$, it is now well recognized that seismic discontinuity at 400-km depth is due to the olivine-spinel transformation in $(Mg,Fe)_2SiO_4$ and that the $(Mg,Fe)_2SiO_4$-spinel is one of the most abundant mineral phases in the transition zone (e.g., see Ringwood, 1975; Weidner, 1985; Weidner and Ito, 1987; Bina and Wood, 1987; Katsura and Ito, 1989). An accurate knowledge of the density (ρ) as a function of pressure and temperature of this phase is therefore of paramount importance in realistically modeling the Earth's transition zone in terms of mineralogical assemblages. The density-pressure relationship for $(Mg,Fe)_2SiO_4$-spinels and their structural analogs has been studied quite extensively using ultrasonic interferometry (e.g., see Mizutani et al., 1970; Chung, 1971; Wang and Simmons, 1972; Liebermann, 1975), Brillouin scattering (Weidner et al., 1984), and X-ray diffraction (Mizukami et al., 1975; Mao et al., 1969; Wilburn and Bassett, 1976; Sato, 1977). However, data for the density-temperature relationship for $(Mg,Fe)_2SiO_4$-spinels are not so well established, mainly due to one of three problems: 1) Data for the Mg_2SiO_4-spinel are limited to a single experiment (Suzuki et al., 1979); 2) the thermal expansion measurements have been made in a narrow range of temperature, as in the case of the Fe_2SiO_4-spinel where data have been obtained only between 8 and 398°C (Mao et al., 1969); and 3) anomalously high thermal expansion values have been obtained in the case of single-crystal measurement of the Fe_2SiO_4-spinel at temperatures above $\sim 400°C$ (Takeuchi et al., 1984; Yamanaka, 1986).

The first two problems are related to the small amounts of the high-pressure phases that are generally available for the powder X-ray diffraction technique employed. In addition, the high-pressure phase, which is metastable at ambient conditions, converts back to the low-pressure phase on heating at 0.1 MPa. Therefore, unless the data are acquired rapidly so as to outpace the conversion rate at some high temperature, the thermal expansion measurements of such a metastable high-pressure phase above the transition temperature are not possible. The use of extremely intense X-rays from synchrotron radiation makes it feasible to conduct high-temperature X-ray diffraction measurements on a tiny sample (e.g., 10^{-6}–10^{-7} gm) in a relatively short time (2 to 10 min) (e.g., see Furnish and Bassett, 1983; Manghnani et al., 1987; Ming et al., 1983, 1987; Mao et al., 1991); we can largely circumvent the first two problems.

Concerning the anomalously high thermal expansion reported for the Fe_2SiO_4-spinel by Takeuchi et al. (1984) and Yamanaka (1986), Yagi et al. (1987) carried out thermal expansion measurements on the Fe_2SiO_4-spinel at temperatures of up to 1200°C at 5 GPa and did not observe the anomalous behavior. They suggested that the anomalously high thermal expansion is due to the metastability of the Fe_2SiO_4-spinel, causing a partial retrogressive transition within the crystal and thus changing the apparent volume of the crystal. The question then arises that if this is true for the Fe_2SiO_4-spinel, should it not also be true for other $(Mg,Fe)_2SiO_4$-spinels? Since no anomalous high thermal expansion has been observed in the Mg_2SiO_4-spinel at temperatures of up to 700°C at 1 atm (Suzuki et al., 1979), the suggestion provided by Yagi et al. (1987) needs to be verified for the Mg_2SiO_4-spinel at higher temperatures.

In a recent study of the spinel-olivine back transformation in $(Mg,Fe)_2SiO_4$, we have found that on heating in a vacuum, all the $(Mg,Fe)_2SiO_4$-spinels convert back to olivine (Ming et al., 1991). However, when these spinels are heated in air, the results are remarkably different. The spinels with $X_{Fe} \geq 0.2$ are oxidized into hematite $\pm$ magnetite $\pm$ enstatite + amorphous silica, depending on the original composition and temperature (Ming et al., 1991). It is important to know if and to what extent the oxidation of a spinel does affect its thermal expansivity. In order to address these issues, we have carried out thermal expansion measurements on $(Mg,Fe)_2SiO_4$-spinels in air as well as in a vacuum.

The main objectives of this paper therefore are 1) to present thermal expansion data for four $(Mg,Fe)_2SiO_4$-spinels, 2) to investigate the systematics between the thermal expansion coefficient (α) and the composition (i.e., Fe/Mg ratio) of $(Mg,Fe)_2SiO_4$-spinels, and 3) to investigate the effect of oxidation of iron-rich spinels on thermal expansion properties.

Experimental Methods

Samples

Polycrystalline spinel samples of composition Mg_2SiO_4, $(Mg_{0.8},Fe_{0.2})_2SiO_4$, $(Mg_{0.6},Fe_{0.4})_2SiO_4$, and $(Mg_{0.4},Fe_{0.6})_2SiO_4$ were synthesized at pressures between 10 and 16 GPa and at 1200°C at the Institute for Study of the Earth's Interior, Okayama University, Japan. All the samples were examined by X-ray diffraction and found to be of a pure spinel phase.

Instrumentation

The spinel sample was heated either in air or a vacuum (10^{-4}–10^{-5} torr) in an improved diamond-anvil cell (DAC) equipped with a micro-resistive heater (Ming et al., 1987) and X-rayed under in-situ high-temperature conditions using energy-dispersive X-ray diffraction techniques. All the measurements were carried out at Stanford Synchrotron Radiation Laboratory (SSRL). The experimental setup at SSRL and the temperature calibration procedures have been described previously (Ming et al., 1983, 1987).

The sample temperature was monitored within $\pm 3°C$ by a precalibrated Pt–Pt 10% Rh thermocouple placed adjacent to the sample (Ming et al., 1983, 1987). All the X-ray diffraction data were collected by an intrinsic Ge detector with $2\theta \approx 13°$. Runs conducted in a vacuum were carried out to 900°C and those in air were limited to less than 650°C in order to prevent the oxidation of the tungsten carbide and the graphitization of diamond anvils. With the storage ring operating at 3.0 GeV and 30 to 60 mA, each spectrum was collected for 5 min in the live-time mode.

Data Reduction

A typical energy-dispersive X-ray diffraction spectrum (Fig. 1) indicates that the spinel phase is characterized by five to seven well-defined diffraction peaks—(111), (220), (311), (400), (331), (511), (440)—and that with increasing temperature, all the diffraction peaks shift toward the lower energy. The shift of the peak position with temperature is expected from the Bragg equation

$$E_{(hkl)} d_{(hkl)} = \frac{6.199}{\sin \theta} = \text{constant}, \tag{1}$$

where $E_{(hkl)}$ and $d_{(hkl)}$ are the photon energy (in keV) of the diffraction peak (hkl) and the corresponding interplanar distance (in Å), respectively; 2θ is the angle between the direct beam and the detector. At a fixed 2θ angle, the constant in Eq. (1) can be determined by using a standard sample (e.g., Au) at ambient conditions. As the temperature increases, d-spacings increase and the corresponding photon energies decrease. The relative change in the d-spacing at a given high temperature with respect to that at room temperature (d_0) can be represented by

$$\left(\frac{d}{d_0}\right)_{(hkl)} = \left(\frac{E_0}{E}\right)_{(hkl)}. \tag{2}$$

In the case of a cubic material, where $(d/d_0)_{(hkl)} = (a/a_0)$, the thermal expansion can be evaluated by using either Eq. (1) or (2). For noncubic materials, either Eq. (1) or (2) first provides the values for each $d_{(hkl)}$, which can, in turn, be fitted by a least-squares program to yield the lattice parameters and thus the thermal expansion. It is evident that the same result will be obtained if the same diffrac-

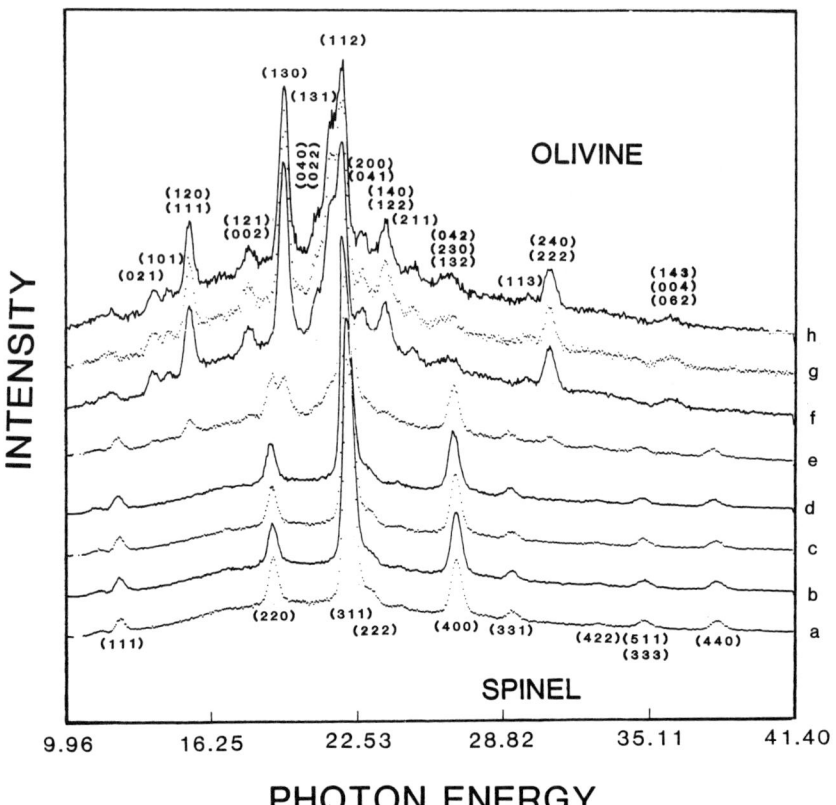

Fig. 1. Energy dispersive X-ray diffraction spectrum of a $(Mg_{0.4},Fe_{0.6})_2SiO_4$-spinel in a vacuum ($10^{-4}$–$10^{-5}$ torr) and at a temperature (in °C) of (a) 330, (b) 408, (c) 490, (d) 570, (e) 650; (f), (g), and (h) are at 740°C at 2-min intervals. The data also show the back transition of spinel → olivine at high temperatures (Ming and Manghnani, 1988).

tion lines are used in both equations. Based on four to five diffraction peaks observed at high temperatures, the values of a/a_0 and thus V/V_0 in this study have been determined to be ± 0.1 to 0.15% and 0.3 to 0.45%, respectively.

Results and Discussion

Results in a Vacuum

Thermal expansion measurements on Mg_2SiO_4-, $(Mg_{0.8},Fe_{0.2})_2SiO_4$-, and $(Mg_{0.4},Fe_{0.6})_2SiO_4$-spinels have been carried out to 900, 735, and 768°C, respectively. The results are listed in Table 1 and plotted in Figs. 2 to 4. These data can be expressed in terms of second-order polynomials. For Mg_2SiO_4,

$$\frac{V}{V_0} = 1 + (12.8 \pm 2.34)10^{-6}(T-21) + (17.6 \pm 3.34)10^{-9}(T-21)^2. \qquad (3)$$

For $(Mg_{0.8}, Fe_{0.2})_2SiO_4$,

$$\frac{V}{V_0} = 1 + (14.1 \pm 2.90)10^{-6}(T-21) + (22.7 \pm 4.90)10^{-9}(T-21)^2. \qquad (4)$$

And for $(Mg_{0.4}, Fe_{0.6})_2SiO_4$,

$$\frac{V}{V_0} = 1 + (18.3 \pm 2.10)10^{-6}(T-21) + (9.43 \pm 3.4)10^{-9}(T-21)^2. \qquad (5)$$

The room-temperature thermal expansion coefficients (α_0, 10^{-6} deg^{-1}) calculated from polynomial functions are 12.8, 14.1, and 18.3 for Mg_2SiO_4, $(Mg_{0.8}, Fe_{0.2})_2SiO_4$, and $(Mg_{0.4}, Fe_{0.6})_2SiO_4$, respectively. Because polynomial fit is only a mathematical expression for the experimental data without involving any solid-state theory, it does not have the capability to discern any subtle deviations from the normal behavior for a solid at high temperature and thus may not provide the true intrinsic physical properties of the material under investigation. Moreover, the least-squares fit does not permit us to extrapolate the data to very high temperatures.

Thermal expansion of a solid material not undergoing phase transformation has been successfully treated by Grüneisen theory. Using this theory and retaining the Taylor expansion of the thermal strain to the second order, Suzuki (1975)

Table 1. Thermal expansion of $(Mg,Fe)_2SiO_4$-spinels in a vacuum.

Run #	T (°C)	V/V_0	α (10^{-6} deg^{-1})
Mg_2SiO_4			
1339	21	1.000	15.1
1341	220	1.0045(12)	21.1
1342	286	1.0046(35)	22.1
1343	314	1.0058(15)	22.5
1344	396	1.0081(18)	23.3
1345	416	1.0069(12)	23.5
1346	450	1.0075(17)	23.8
1347	498	1.0101(22)	24.2
1348	584	1.0126(11)	24.8
1349	648	1.0137(52)	25.1
1350	673	1.0152(43)	25.2
1351	704	1.0182(12)[a]	
1352	761	1.0202(13)[a]	
1353	795	1.0226(22)[a]	
1354	810	1.0193(27)[a]	
1355	900	1.0246(66)[a]	

Table 1. (Continued.)

Run #	T (°C)	V/V_0	α (10^{-6} deg^{-1})
$(Mg_{0.8}Fe_{0.2})_2SiO_4$			
1246	21	1.0000	17.6
1247	154	1.0019(21)	22.3
1248	182	1.0018(18)	23.0
1249	229	1.0032(26)	23.8
1250	262	1.0065(46)	24.4
1251	285	1.0060(19)	24.7
1252	342	1.0065(12)	25.4
1254	374	1.0074(13)	25.8
1255	426	1.0118(25)	26.2
1256	478	1.0117(27)	26.7
1257	532	1.0113(43)	27.0
1258	564	1.0142(27)	27.3
1259	596	1.0155(20)	27.4
1260	614	1.0140(22)	27.6
1261	632	1.0185(21)	27.7
1262	650	1.0168(17)	27.8
1263	670	1.0208(20)[a]	
1264	682	1.0208(20)[a]	
1265	700	1.0224(09)[a]	
1266	735	1.0205(45)[a]	
1267	748	1.0241(17)[a]	
$(Mg_{0.4}Fe_{0.6})_2SiO_4$			
1302	21	1.0000	18.2
1303	128	1.0017(21)	21.7
1304	228	1.0040(13)	23.4
1305	292	1.0057(17)	24.2
1306	350	1.0060(18)	24.8
1307	407	1.0088(20)	25.2
1308	443	1.0117(14)	25.5
1309	471	1.0123(18)	25.7
1310	486	1.0106(32)	25.8
1311	503	1.0113(37)	25.8
1312	536	1.0124(36)	26.0
1313	571	1.0136(42)	26.2
1314	600	1.0138(15)	26.4
1315	644	1.0155(13)	26.6
1316	650	1.0149(22)	26.8
1317	685	1.0153(21)	26.8
1318	712	1.0155(23)	26.9
1319	724	1.0170(12)	26.9
1320	733	1.0169(15)	27.0
1321	753	1.0196(10)	27.0
1322	768	1.0196(05)	27.1

[a] Not included in the final refined analysis; see explanation in the text.

Thermal Expansion Studies of $(Mg,Fe)_2SiO_4$-Spinels

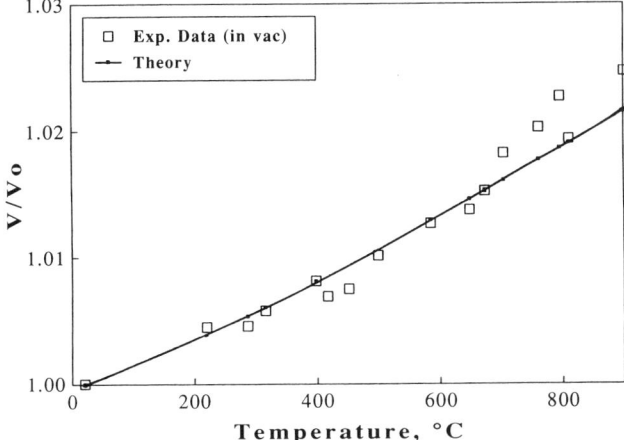

Fig. 2. Thermal expansion data of a Mg_2SiO_4-spinel (in a vacuum). Results obtained on the basis of Grüneisen theory (i.e., the solid line) indicate that data at a temperature above 650°C are anomalous and are not included for the final calculations of γ and α (refer to the text for discussion).

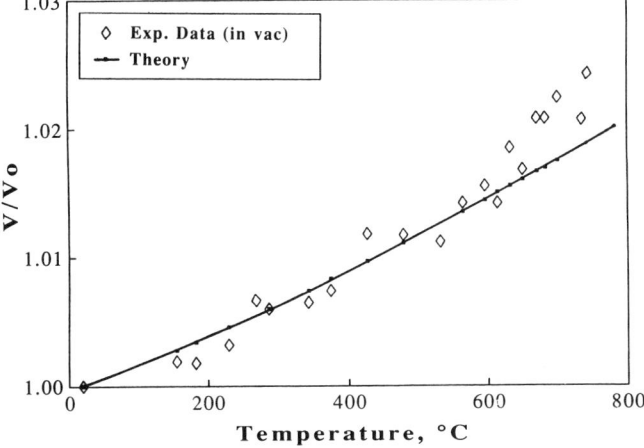

Fig. 3. Thermal expansion data of a $(Mg_{0.8},Fe_{0.2})_2SiO_4$-spinel (in a vacuum). Results obtained on the basis of Grüneisen theory (i.e., the solid line) indicate that data at temperatures above 600°C are anomalous and are not included for the final calculations of γ and α (refer to the text for discussion).

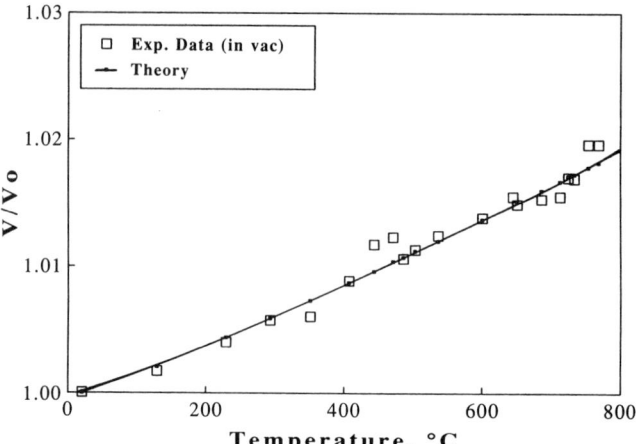

Fig. 4. Thermal expansion data of a $(Mg_{0.4},Fe_{0.6})_2 SiO_4$-spinel (in a vacuum). Results obtained on the basis of Grüneisen theory are shown in a solid line.

first derived a formula for thermal expansion

$$\frac{V(T)}{V_0} = \frac{\{2k + 1 - \sqrt{[1 - 4k\,E(T)/Q]}\}}{2a}, \quad (6)$$

where $E(T)$ is the thermal energy of lattice vibrations, $k = (dK_0/dP - 1)/2$, $Q = K_0 V_0/\gamma$, and a is the lattice parameter. Here, K_0, dK_0/dP, and γ are, respectively, the isothermal bulk modulus at 0 K, its first pressure derivative, and the Grüneisen's parameter.

The thermal energy $E(T)$ is generally approximated by the Debye function such that

$$E(\theta_D, T) = 9p\,R\,T\left(\frac{T}{\theta_D}\right)^3 \int x^3 \frac{dx}{(e^x - 1)}, \quad (7)$$

where p is the number of atoms in the molecular formula, R the gas constant, and θ_D the Debye temperature. Hill and Jackson (1990) suggested that the thermal expansion of a material at a given temperature T with respect to the reference temperature (room temperature T_R) can be written as

$$\frac{V(T)}{V_R} = \frac{\{2k + 1 - \sqrt{[1 - 4k\,E(\theta_D, T)/Q]}\}}{\{2k + 1 - \sqrt{[1 - 4k\,E(\theta_D, T_R)/Q]}\}}, \quad (8)$$

where $V(T)$ and V_R are the molar volumes at T and T_R. Because the values of $V(T)/V_R$ calculated from Eq. (8) are insensitive to the value of dK_0/dP, which is usually in the range of 4 to 5 for most of the silicates, we have assumed a value of 4 for dK_0/dP in this study. Using the experimental $V(T)/V_R$ data and T given in Table 1, we can obtain the best-fit values for Q and θ_D from Eq. (8), and the

Table 2. Values of various parameters used for calculating the Grüneisen parameter for (Mg,Fe)$_2$SiO$_4$-spinels.

	Mole fraction of Mg$_2$SiO$_4$			
	1.0	0.8	0.4	0
K_T (GPa)	182.8[a]	184.5[b]	188.9[b]	193[c]
dK_T/dP	4[d]	4[d]	4[d]	4[d]
V_0 (cm^3/mol)[e]	39.65	40.14	41.13	42.03

[a] Calculated from $K_s = 184$ GPa (Weidner et al., 1984) using $K_T = K_s/(1 + \alpha\gamma T)$, where $\alpha = 19 \times 10^{-6}$ K^{-1} (Suzuki et al., 1979) and $\gamma = 1.2$ (Watanabe, 1987).
[b] Interpolated values between two end-members.
[c] An average value of 197 GPa (Sato, 1977) and 189 GPa (Wilburn and Bassett, 1976).
[d] Assumed value.
[e] Based on the relationship $V_0 = 39.65 + 2.37$ X$_{Fe}$ (Jeanloz and Thompson, 1983).

Table 3. A comparison of the Debye temperature (θ_D) and the Grüneisen parameter (γ) of (Mg,Fe)$_2$SiO$_4$-spinels between the best-fit solutions obtained in this study and those calculated on the basis of the thermochemical data (Watanabe, 1987).

	This study		Watanabe (1987)	
	θ_D	γ	θ_D	γ
Mg$_2$SiO$_4$	1800	1.594	1019	1.20
(Mg$_{0.8}$,Fe$_{0.2}$)$_2$SiO$_4$	1900	1.971	944	1.20
(Mg$_{0.4}$,Fe$_{0.6}$)$_2$SiO$_4$	1075	1.275	840	1.23
Fe$_2$SiO$_4$	1200[a]	1.491[a]	781[a]	1.29[a]

[a] From Manghnani et al., 1991.

values of γ are calculated using the identity $\gamma = V_0 K_0/Q$. As VK = constant generally holds for most of the ionic isostructural solids at room temperature (Anderson, 1972), we use room-temperature data instead of the zero K data for both the molar volume and the bulk modulus to calculate γ. Data used in these calculations are summarized in Table 2; the best-fit calculated values of γ and θ_D are listed in Table 3.

Comparison of Results Between This and Previous Studies

On the basis of the heat capacity measurement (C_p) for the temperature range between 350 and 700 K for four (Mg,Fe)$_2$SiO$_4$-spinels, Watanabe (1987) calculated their thermal gamma (γ_{th}) and the Debye temperature (θ_D) using the pub-

Table 4. A comparison of the values of the Grüneisen parameter of $(Mg,Fe)_2SiO_4$-spinels reported by Watanabe (1987) and this study.

Ref.	1.0	0.8	0.4	0.0
Watanabe (1987)	1.20	1.20	1.23	1.29
This study[a]	1.13	1.18	1.17	1.26[c]
This study	1.05[b]	1.15[b]	1.17	1.26[c]

[a] All the V/V_0 data are used in the calculation.
[b] Some V/V_0 data at high temperature (marked ^{a} in Table 1) are excluded in the final calculation.
[c] From Manghnani et al., 1991.

lished values of γ, V_0, and K_s. His results are also given in Table 4 for comparison. As can be seen, the best values of γ an θ_D obtained in this study are too high. The most probable reason is that the present experimental data are not accurate enough to permit a two-parameter fit. This situation can be greatly improved if Watanabe's θ_D is assumed and we allow only the γ_{th} to be evaluated using Eq. (8). Results given in Table 4 show fairly good agreement with those of Watanabe (1987). Both data sets indicate that the value of γ increases slightly with an increasing Fe content, in spite of the fact that the values obtained in this study are systematically lower that those of Watanabe's. The slightly higher values of Watanabe (1987) are probably a result of the higher α_0 used for both $MgSiO_4$- and Fe_2SiO_4-spinels in his calculation.

The thermal expansion of Mg_2SiO_4 (in polycrystalline form) has been studied in air to 700°C using an angular-dispersive X-ray diffraction technique (Suzuki et al., 1979). They performed two different fitting procedures for evaluating their results. Their unconstrained solution I yields 1.43 and 1198 K for γ and θ_D, respectively. By assuming the value of θ_D to be 849 K based on ultrasonic measurements (Liebermann, 1975), they obtained in their solution II a value of 1.27 for γ. By adjusting the values of V_0 and K_0 to those used in this study, we obtain values of γ as 1.27 and 1.13 for their solutions I and II, respectively. Our thermal expansion data yield a value of $\gamma = 1.06$ at $\theta_D = 849$ K, which is in fairly good agreement with the $\gamma = 1.13$ obtained by Suzuki et al. in their solution II. Table 5 summarizes the values of θ_D and γ obtained by different investigators using various experimental methods and/or assumptions.

Once the γ and θ_D are known, the value of $V(T)/V_R$ as a function of temperature can be calculated using Eq. (8). Plotting the calculated $V(T)/V_R$ for Mg_2SiO_4, $(Mg_{0.8},Fe_{0.2})_2SiO_4$, and $(Mg_{0.4},Fe_{0.6})_2SiO_4$ in Figs. 2, 3, and 4, respectively, we found some anomalous high values of $V(T)/V_R$ for Mg_2SiO_4 and $(Mg_{0.8},Fe_{0.2})_2SiO_4$, but not $(Mg_{0.4},Fe_{0.6})_2SiO_4$. A similar study of the Fe_2SiO_4-spinel also did not show such an anomaly at high temperatures (Manghnani et al., 1991). This phenomenon may be related to some unknown structural change to the Mg-rich spinels at high temperatures, as indicated in the case of $MgSiO_3$-perovskite (Wang et al., 1991). It may also be attributed to the $\gamma \rightarrow \beta$

Table 5. Values of θ and γ obtained from various methods for the Mg_2SiO_4-spinel.

$\theta(K)$	γ	Method	Ref.
1019[a]	1.2[b]	DSC	Watanabe (1987)
1198[c]	1.43[c]	XRD	Suzuki et al. (1979)
1198[d]	1.27[d]	XRD	Suzuki et al. (1979)
849[e]	1.27[e]	XRD	Suzuki et al. (1979)
849[f]	1.13[f]	XRD	Suzuki et al. (1979)
1800[g]	1.59[g]	XRD	This study
1019[i]	1.13[h]	XRD	This study
849[j]	1.06[i]	XRD	This study

DSC = Differential scanning calorimetry, where C_p was measured between 350 and 700 K (Watanabe, 1987).

XRD = X-ray diffraction, where V/V_0 was measured.

[a] Calculated by fitting the measured C_p data to Debye's theoretical expression for heat capacity.

[b] Calculated using the thermodynamic relationship, together with the experimental data of C_p (Watanabe, 1987), V_0 (Sasaki et al., 1982), α_0 (Suzuki, 1979), and K_s (Weidner et al., 1984).

[c] Solution I of Suzuki et al. (1979).

[d] Same as Solution I except new values of $K_0 = 1.828$ Mbar (Weidner et al., 1984) and $V_0 = 39.65$ cm^3/mol (Jeanloz and Thompson, 1983) are used in the calculation of γ.

[e] Solution II of Suzuki et al. (1979).

[f] Same as Solution II except new values of $K_0 = 1.828$ Mbar (Weidner et al., 1984) and $V_0 = 39.65$ cm^3/mol (Jeanloz and Thompson, 1983) are used in the calculation of γ.

[g] The best-fit values using the V/V_0 data obtained in this study and eq. (8) with given values of $K_0 = 1.828$ Mbar (Weidner et al., 1984), $dK/dP = 5$ (assumed), and $V_0 = 39.65$ cm^3/mol (Jeanloz and Thompson, 1983).

[h] Calculated using the V/V_0 data obtained in this study and Eq. (8) with given values of $K_0 = 1.828$ Mbar (Weidner et al., 1984), $dK/dP = 4$ (assumed), $V_0 = 39.65$ cm^3/mol (Jeanloz and Thompson, 1983), and $\theta = 1019$ K (Watanabe, 1987).

[i] Calculated using the V/V_0 data obtained in this study and Eq. (8) with given values of $K_0 = 1.828$ Mbar (Weidner et al., 1984), $dK/dP = 4$ (assumed), $V_0 = 39.65$ cm^3/mol (Jeanloz and Thompson, 1983), and $\theta = 849$ K (Liebermann, 1975).

[j] Assumed value.

transition prior to finally reverting to the olivine, as can be expected from the well-established $P - X$ phase diagram of $(Mg,Fe)_2SiO_4$-spinels (e.g., Akimoto, 1987; Katsura and Ito, 1989). The reason that we fail to distinguish the β phase from the γ phase in our X-ray diffraction spectrum is probably because the five or seven observed diffraction peaks are the major ones common to both phases (Ito et al., 1974; Moore and Smith, 1970). Although the β phase has more lines, they are not observed in this study. This may be due, in part, to the weak intensity of those additional peaks of the β phase and due, in part, to the low resolution power of the energy-dispersive X-ray diffraction method employed. Further studies with an angular-dispersive X-ray diffraction method employing a wide-angle position-sensitive detector would be valuable in elucidating the cause of such behavior. In order to deduce accurately the value of γ and hence α for the spinel phase, we repeated the calculation described earlier without including the anomalous data points at high temperatures. The final recalculated values are given at the last row of Table 4. The relationship between γ and the composition of $(Mg,Fe)_2SiO_4$-spinels can be expressed linearly as $\gamma = 1.09 + 0.16\, X_{Fe}$ with a correlation coefficient of 0.946.

Using the refined values of γ and θ_D, we calculated the thermal expansion coefficient (α) as a function of temperature for Mg_2SiO_4, $(Mg_{0.8},Fe_{0.2})_2SiO_4$, and $(Mg_{0.4},Fe_{0.6})_2SiO_4$. Results are listed in the last column of Table 1 and are also plotted together with those of Fe_2SiO_4 (Manghnani et al., 1991) in Fig. 5. This result shows that the values of α for $(Mg,Fe)_2SiO_4$-spinels increase slightly with increasing Fe content at room temperature, but that such a relationship becomes much less obvious at some high temperatures. Values of α_0 of $(Mg,Fe)_2SiO_4$-spinels obtained in this study and those from other studies are plotted in Fig. 6. In general, there is good agreement among the various studies. However, our results for Mg_2SiO_4 and Fe_2SiO_4 are slightly lower than previous experimental results (Suzuki et al., 1979; Mao et al., 1969), and the theoretical value obtained for Mg_2SiO_4 is slightly lower than experimental values. On the basis of results obtained from synchrotron studies and treated with Grüneisen theory (i.e., this study; Manghnani et al., 1991), the compositional dependence of

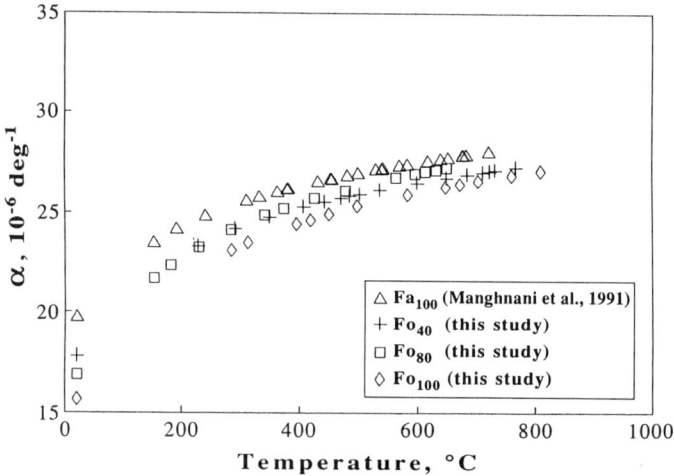

Fig. 5. Thermal expansion (α) as a function of temperature for Mg_2SiO_4, $(Mg_{0.8},Fe_{0.2})_2SiO_4$, and $(Mg_{0.4},Fe_{0.6})_2SiO_4$ of this study and Fe_2SiO_4 of Manghnani et al. (1991).

Thermal Expansion Studies of $(Mg,Fe)_2SiO_4$-Spinels

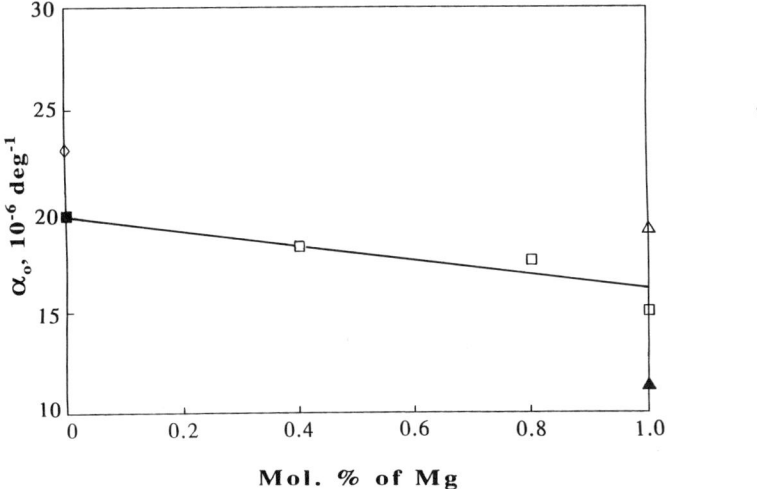

Fig. 6. Room-temperature thermal expansion coefficient (α_0) as a function of the Mg content in $(Mg,Fe)_2SiO_4$-spinels, where □ are from this study, ◇ from Mao et al. (1969), △ from Suzuki et al. (1979), ▲ from Reynard and Price (1990), and ■ from Manghnani et al. (1991). The linear fit to those obtained with synchroton radiation give α_0 $(10^{-6}, \deg^{-1}) = 15.74 + 4.14\, X_{Fe}$.

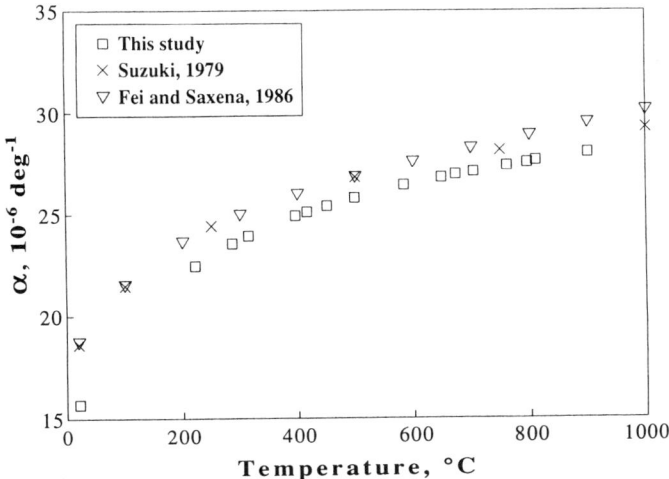

Fig. 7. A comparison of the α of Mg_2SiO_4-spinels in this study and those previously reported.

α_0 in $(Mg,Fe)_2SiO_4$-spinels can be approximated with a linear relationship such that $\alpha_0(10^{-6}, deg^{-1}) = 15.74 + 4.14\, X_{Fe}$ with a correlation coefficient of 0.935.

Figure 7 shows good agreement between the thermal expansion coefficient (α) values for Mg_2SiO_4 as a function of temperature obtained from this study and those obtained by Suzuki et al. (1979). Fei and Saxena (1986) have expressed Suzuki's thermal expansion data using a polynomial function such that $\alpha = 367 \times 10^{-5} + 5.298 \pm 10^{-9}\, T - 0.57/T^2$, which served as a convenient form for other thermodynamic calculations (e.g., Akaogi et al., 1989).

Observations in Air

The results in air of three ion-bearing spinels with compositions $(Mg_{0.8}, Fe_{0.2})_2SiO_4$, $(Mg_{0.6}, Fe_{0.4})_2SiO_4$, and $(Mg_{0.4}, Fe_{0.6})_2SiO_4$ are given in Table 6 and plotted in Fig. 8. One of the most striking features in Fig. 8 is that in the cases of $(Mg_{0.6}, Fe_{0.4})_2SiO_4$ and $(Mg_{0.4}, Fe_{0.6})_2SiO_4$, the thermal expan-

Table 6. Thermal expansion of $(Mg,Fe)_2SiO_4$-spinels in air.

Run #	T (°C)	V/V_0
$(Mg_{0.8}, Fe_{0.2})_2SiO_4$		
1174	21	1.0000
1175	92	1.0030(15)
1177	115	1.0042(30)
1177	160	1.0030(15)
1178	186	1.0030(06)
1179	200	1.0051(15)
1180	234	1.0045(09)
1181	260	1.0063(09)
1182	296	1.0063(15)
1183	330	1.0063(09)
1184	376	1.0078(12)
1185	408	1.0099(21)
1186	460	1.0093(18)
1187	492	1.0105(21)
1188	530	1.0105(12)
1189	586	1.0144(21)
1190	625	1.0177(18)
$(Mg_{0.6}, Fe_{0.4})_2SiO_4$		
1076	21	1.0000
1077	108	1.0024(12)
1078	176	1.0018(18)
1079	222	1.0045(09)
1080	295	1.0033(09)

Table 6. (Continued.)

Run #	T (°C)	V/V_0
$(Mg_{0.6}, Fe_{0.4})_2 SiO_4$		
1081	372	0.9998(13)
1082	440	1.0060(09)
1083	470	1.0057(12)
1084	540	1.0227(18)
1085	602	1.0289(21)
$(Mg_{0.4}, Fe_{0.6})_2 SiO_4$		
1130	21	1.0000
1131	110	1.0030(18)
1132	148	1.0018(12)
1133	170	1.0033(15)
1134	196	1.0036(09)
1135	228	1.0036(09)
1136	262	1.0042(09)
1137	298	1.0045(09)
1138	350	1.0045(15)
1139	388	0.9949(15)
1140	430	0.9970(18)
1141	472	1.0036(30)
1142	509	1.0081(30)
1143	545	1.0110(30)
1144	582	1.0183(06)
1145	594	1.0216(30)

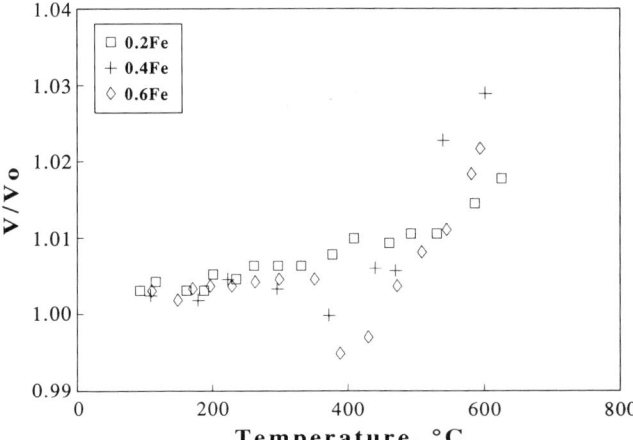

Fig. 8. Thermal expansion of $(Mg,Fe)_2 SiO_4$-spinels indicating some anomalous behavior at high temperatures for Fe-rich spinels when heated in air.

sion decreases at temperatures between 300 and 400°C and then increases very rapidly at temperatures above 400°C.

Figures 9 and 10 show the comparison of the thermal expansion data obtained in air and a vacuum for $(Mg_{0.8},Fe_{0.2})_2SiO_4$ and $(Mg_{0.4},Fe_{0.6})_2SiO_4$, respectively. It is found that in the case of a Fe-poor spinel such as $(Mg_{0.8},Fe_{0.2})_2SiO_4$, the thermal expansion data obtained in air are similar to those obtained in a vacuum; however, in the case of Fe-rich spinels such as $(Mg_{0.6},Fe_{0.4})_2SiO_4$ and $(Mg_{0.4},Fe_{0.6})_2SiO_4$, the results obtained in air are very much different from those obtained in a vacuum. Similar anomalous behavior has also been recently observed in a Fe_2SiO_4-spinel (Manghnani et al., 1991).

Anomalous thermal expansion has also been observed in other materials such as $BaTiO_3$ (Shirane and Takeda, 1950), $PbTiO_3$ (Shirane and Hoshino, 1951), and ZrO_2 (Adams et al., 1985) and has been correlated to phase transition at high temperature. This is because the high-temperature phase has a smaller volume than that of the low-temperature phase. In view of the fact that the molar volume of the olivine (α) phase is ~8 to 10% larger than that of the spinel (γ) phase (Ringwood and Major, 1970), it is difficult to reconcile our anomalous thermal expansion data with the $\gamma \rightarrow \alpha$ transition for the Fe-rich spinels.

In a recent study on the back transition of $(Fe,Mg)_2SiO_4$-spinels at high temperatures, we found that the Fe-rich spinels are oxidized when heated in air (Ming et al., 1991). Figure 11 shows the energy spectrum of $(Mg_{0.4},Fe_{0.6})_2SiO_4$ at various temperatures, indicating that the formation of the Fe_2O_3 (hematite) phase starts between 298 and 388°C and its amount increases with increasing temperature. Because all the diffraction peaks of the spinel phase are present up

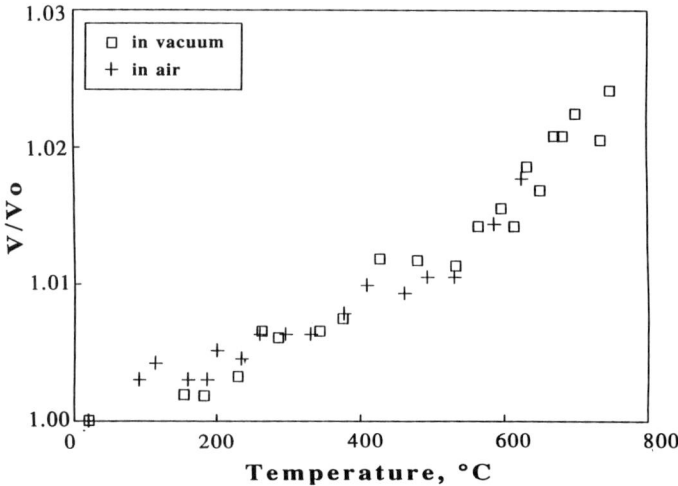

Fig. 9. The thermal expansion data of a $(Mg_{0.8},Fe_{0.2})_2SiO_4$-spinel obtained in air appear to be the same as those obtained in a vacuum.

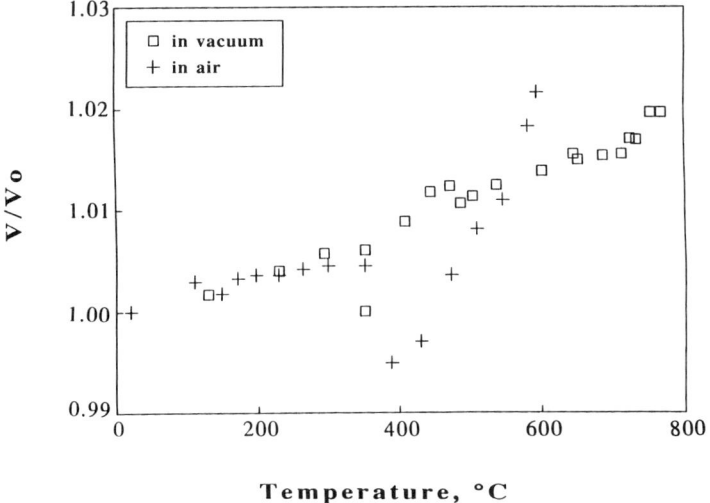

Fig. 10. The thermal expansion data of a $(Mg_{0.4},Fe_{0.6})_2SiO_4$-spinel obtained in air appear to be quite different from those obtained in a vacuum.

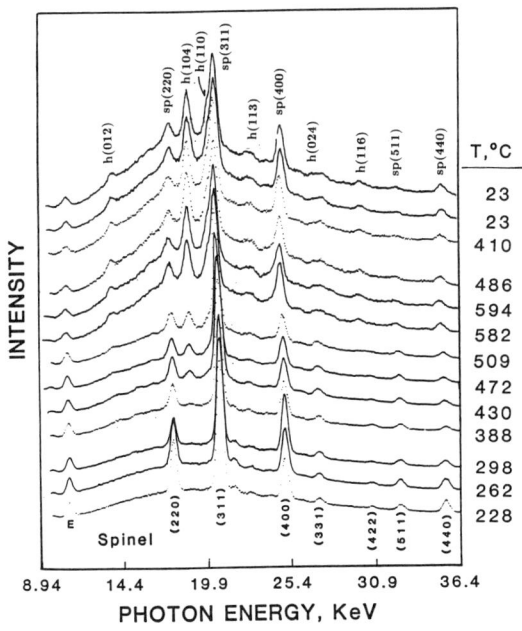

Fig. 11. Energy-dispersive spectrum as a function of temperature for the $(Mg_{0.4},Fe_{0.6})_2SiO_4$-spinel. The appearance of the hematite phase indicates that the oxidation of the spinel occurs at temperatures between 298 and 388°C (Ming and Manghnani, 1989).

to 596°C (see Fig. 11), it is a certainty that as the oxidation of the Fe-rich spinel phase proceeds, the spinel structure remains intact. Most likely, the Fe atoms on the surface of the spinel grains combine with oxygen to form Fe_2O_3, thus leaving vacant sites behind. The Fe atoms next to the vacant sites then hop into these sites and the oxidation process continues. It is clear that as this oxidation proceeds at high temperatures, the spinel phase becomes a defect structure with many vacant Fe sites inside the structure. Such a defect structure will result in a decrease of the zero-pressure density and at the same time in an increase of the thermal expansion coefficient. If the defect structure is formed fast enough at some temperature T_D, it is anticipated that, as the temperature increases, the thermal expansion will first increase slowly, then will have a sudden decrease drop at T_D, and finally above T_D, will increase rapidly. This represents the most probable scenario of the thermal expansivity of the Fe-rich spinels in air.

Acknowledgment

The authors thank the staff of the Stanford Synchrotron Radiation Laboratory for providing the facilities for this research. The authors would also like to thank S.K. Saxena, R. Jeanloz, and W.A. Bassett for reviewing this paper; X. Li and V. Askarpour for helpful discussions; L.-J. Wang for some data analysis, and J. Balogh for modifying the high P-T DAC used in our study. This research was supported by the Earth Science Division, National Science Foundation (NSF Grant EAR87-08482). School of Ocean and Earth Science and Technology contribution number No. 2683.

References

Adams, J., Nakamura, H.H., Ingel, R.P., and Rice, R.W. (1985). Thermal expansion behavior of single-crystal zirconia. *J. Amer. Ceram. Soc.* **68** (9), C228–C231.

Akaogi, M., Ito, E., and Navrotsky, A. (1989). Olivine-modified spinel-spinel transition in the system Mg_2SiO_4–Fe_2SiO_4 calorimetric measurements, thermodynamic calculation and geophysical application. *J. Geophys. Res.* **94**, 15671–15685.

Akimoto, S. (1987). High-pressure research in geophysics: Past, present, and future, in *High-Pressure Research in Mineral Physics (The Akimoto Volume)*, M.H. Manghnani, and Syono, Y. eds., Terra Scientific Publishing Co., and Tokyo/American Geophysical Union, Washington, D.C., pp. 1–16.

Anderson, O.L. (1972). Pattern in elastic constants of minerals important to geophysics, in *Nature of Solid Earth*, E.C. Robertson, ed., McGraw-Hill, New York, pp. 575–612.

Bina, G.R. and Wood, B.J. (1987). Olivine-spinel transitions: Experimental and thermodynamic constraints and implications for the nature of the 400-km seismic discontinuity. *J. Geophys. Res.* **92**, 4853–4866.

Chung, D.H. (1971). Elasticity and equation of state of olivine in Mg_2SiO_4–Fe_2SiO_4 system. *Geophys. J.R. Astron. Soc.* **25**, 511–538.

Fei, Y. and Saxena, S.K. (1986). A thermochemical data base for phase equilibria in the system Fe–Mg–Si–O at high pressure and temperature. *Phys. Chem. Miner.* **13**, 311–324.

Furnish, M.D. and Bassett, W.A. (1983). Investigation of the mechanism of the olivine-spinel transition in fayalite by synchrotron radiation. *J. Geophys. Res.* **88**(B12), 10333–10341.

Hill R.J. and Jackson, I. (1990). The thermal expansion of $ScAlO_3$—a silicate perovskite analogue. *Phys. Chem. Miner.* **17**, 89–96.

Ito, E., Matsui, Y., Suito, K., and Kawai, N. (1974). Synthesis of γ-Mg_2SiO_4. *Phys. Earth Planet. Inter.* **8**, 342–344.

Jeanloz, R. and Thompson, A.B. (1983). Phase transitions and mantle discontinuities. *Rev. Geophys.* **21**, 51–74.

Katsura, T. and Ito, E. (1989). The system Mg_2SiO_4–Fe_2SiO_4 at high pressure and temperatures: Precise determination of stabilities of olivine, modified spinel and spinel. *J. Geophys. Res.* **94**(10), 15663–15670.

Liebermann, R.C. (1975). Elasticity of olivine (α), beta (β) and spinel (γ) polymorphs of germanates and silicates. *Geophys. J.R. Astr. Soc.* **42**, 899–929.

Manghnanai, M.H., Ming, L.C., and Nakagiri, N. (1987). Investigation of α-Fe–ε-Fe phase transition and the equation of the ε-Fe phase by synchrotron radiation, in *High Pressure Research in Mineral Physics* (*the Akimoto Volume*), M.H. Manghnani and Y. Syono, eds., Terra Scientific Publishing Co., and Tokyo/American Geophysical Union, Washington, D.C. pp. 155–164.

Manghnani, M.H., Ming, L.C., and Wang, L.J. (1991). Back transition, thermal expansion and equation of state of Fe_2SiO_4 (spinel): A study using synchrotron radiation, in *Program with Abstract for the U.S.–Japan Seminar on High-Pressure Research in Mineral Physics: Application to Earth and Planetary Sciences*, Jan. 15–18, Ise, Mie-ken, Japan.

Mao, H.K., Takahashi, T., Basset, W.A., Weaver, J.S., and Akimoto, S. (1969). Effect of pressure and temperature on the molar volume of wustite and of three $(Mg,Fe)_2SiO_4$ spinel solid solutions. *J. Geophys. Res.* **74**, 1061–1069.

Mao, H.K., Hemley, R.J., Fei, Y., Shu, J.F., Chen, L.C., Jephcoat, A.P., and Wu, Y. (1991). Effect of pressure, temperature, and composition on lattice parameters and density of $(Mg,Fe)_2SiO_3$-perovskite to 30 GPa. *J. Geophys. Res.* **96**, 8069–8079.

Ming, L.C. and Manghani, M.H. (1989). High pressure and high temperature X-ray diffraction studies using the diamond anvil cell with synchrotron radiation, in *Synchrotron Radiation Applications in Mineralogy and Petrology*, A. Bartko-Kyriakidis, ed., Theophrastus Publications, Athens, Greece, pp. 33–48.

Ming, L.C., Manghnani, M.H., and Balogh, J. (1987). Resistive heating in the diamond anvil cell under vacuum conditions, in *High-Pressure Research in Mineral Physics* (*The Akimoto Volume*), M.H. Manghnani, and Y. Syono, eds., Terra Scientific Publishing Co., and Tokyo/American Geophysical Union, Washington, D.C., pp. 69–74.

Ming, L.C., Manghnani, M.H., Qadri, S.Q., Skelton, E.F., Jamieson, J.C., and Balogh, J. (1983). Gold as a rich reliable internal pressure calibrant at high temperatures. *J. Appl. Phys.* **54**(18), 4390–4397.

Ming, L.C., Manghnani, M.H., Usha-Devi, S., Ito, E., and Xie, H.-S. (1991). Back transformation and oxidation of $(Mg,Fe)_2SiO_4$-spinels at high temperatures. *Phys. Chem. Miner.* **18**, 171–179.

Mizukami, S., Ohtani, A., and Kawai, N. (1975). High-pressure X-ray diffraction studies on β and γ-Mg_2SiO_4. *Phys. Earth Planet. Inter.* **10**, 177–182.

Mizutani, H., Hamano, Y., Ida, Y., and Akimoto, S. (1970). Compressional-wave velocities of fayalite, Fe_2SiO_4 spinel and coesite. *J. Geophys. Res.* **75**, 2741–2747.

Moore, P.B. and Smith, J.V. (1970). Crystal structure of β Mg_2SiO_4: Crystal-chemical and geophysical implications. *Phys. Earth Planet. Inter.* **3**, 166–177.

Reynard, B. and Price, G.D. (1990). Thermal expansion of mantle minerals at high pressure—a theoretical study. *Geophys. Res. Lett.* **17**(6), 689–692.

Ringwood, A.E. (1975). *Composition and Petrology of the Earth's Mantle*. McGraw-Hill, New York.

Ringwood, A.E. and Major, A. (1970). The system Mg_2SiO_4–Fe_2SiO_4 at high pressure and temperatures. *Phys. Earth Planet. Inter.* **3**, 89–108.

Sasaki, S., Prewitt, C.T., Sato, Y., and Ito, E. (1982). Single-crystal X-ray study of γ-Mg_2SiO_4. *J. Geophys. Res.* **87**, 7829–7832.

Sato, Y. (1977). Equation of state of mantle minerals determined through high-pressure X-ray study, in *High Pressure Research: Application to Geophysics*, M.H. Manghnani, and S. Akimoto, eds., Academic Press, Inc., New York, N.Y. pp. 307–323.

Shirane, G. and Hoshino, S. (1951). On the phase transition in lead titanite. *J. Phys. Soc. Jap.* **6**, 256–270.

Shirane, G. and Takeda, A. (1950). Volume change at three transitions in $BaTiO_3$ ceramics. *J. Phys. Soc. Jap.* **26**, 128–129.

Suzuki, I. (1975). Thermal expansion of periclase and olivine, and their anharmonic properties. *J. Phys. Earth* **23**, 145–159.

Suzuki, I., Ohtani, E., and Kumazawa, M., (1979). Thermal expansion of γ-Mg_2SiO_4. *J. Phys. Earth* **27**, 53–61.

Takeuchi, Y., Yamanaka, T., Naga, N., and Hirano, M. (1984). High temperature crystallography of olivines and spinels, in *Materials Science of the Earth's Interior*, I. Sunagawa, ed., TERRAPUB, Tokyo, pp. 191–231.

Wang, H. and Simmons, G. (1972). Elasticity of some mantle crystal structures 1, pleonaste and hercynite spinel. *J. Geophys. Res.* **77**, 4379–4392.

Wang, Y., Weidner, D., Liebermann, R., Liu, X., Ko, J., Vaughan, M., Zhao, Y., Yeganeh-Haeri, A., and Pacalo, R.E.G. (1991). Phase transition and thermal expansion of $MgSiO_3$ perovskite. *Science* **251**, 410–431.

Watanabe, H. (1987). Physical-chemical properties of olivine and spinel solid solutions in the system Mg_2SiO_4–Fe_2SiO_4, in *High-Pressure Research in Mineral Physics* (*The Akimoto Volume*), M.H. Manghnani, and Y. Syono, eds., Terra Scientific Publishing Co., Tokyo/American Geophysical Union, Washington, D.C., pp. 275–278.

Weidner, D.J. (1985). A mineral physics test of a pyrolite mantle. *Geophys. Res. Lett.* **12**, 417–420.

Weidner, D.J. and Ito, E. (1987). Mineral physics constraints on uniform mantle composition, in *High-Pressure Research in Mineral Physics* (*The Akimoto Volume*), M.H. Manghnani, and Y. Syono, eds., Terra Scientific Publishing Co., Tokyo/American Geophysical Union, Washington, D.C., pp. 439–446.

Weidner, D.J., Sawamoto, H., Sasaki, S., and Kumazawa, M. (1984). Single-crystal elastic properties of the spinel phase of Mg_2SiO_4. *J. Geophys. Res.* **89**(B9), 7852–7860.

Wilburn, D.R. and Bassett W.A. (1976). Isothermal compression of spinel (Fe_2SiO_4) up to 75 kbar under hydrostatic condition. *High Temp.-High Press.* **8**, 343–348.

Yagi, T., Akaogi, M., Shimomura, O., Suzuki, T., and Akimoto, S. (1987). In situ observation of the olivine-spinel phase transformation in Fe_2SiO_4 using synchrotron radiation. *J. Geophys. Res.* **92**(B7), 6207–6213.

Yamanaka T. (1986). Crystal structures of Ni_2SiO_4 and Fe_2SiO_4 as a function of temperature and heating duration. *Phys. Chem. Miner.* **13**, 227–232.

Appendix
Tables of Molar Volumes and Fugacities of Fluids on C–O–H System (H_2O, CO_2, CH_4, CO, O_2, H_2) up to 1 Mbar and 4000 K

The values of calculated molar volume and fugacity have been calculated. Molar volume was calculated as a solution of Eq. (9) in Chap. 3. The equation has been solved with a Newton–Raphson technique.

The value of fugacity under the specified temperature and pressure was calculated with the numerical integration of the VdP term as follows:

$$RT \ln f(P, T) = RT \ln f(5 \text{ Kbar}, T) + \int_{5 \text{ kbar}}^{P} V dP.$$

The coefficients for the calculation of $RT \ln f(5 \text{ Kbar}, T)$ with expression (10) are listed in Table 5. The computer program in Fortran for the calculation of fluid properties (ΔG, fugacity, molar volume) is available on request.

Table 1. Molar volume of water (cm^3) under T and P.

T (K)	P (GPa)					
	0.5	0.6	0.7	0.8	0.9	1.0
673.15	19.378	18.672	18.105	17.632	17.229	16.887
773.15	20.683	19.765	19.052	18.472	17.985	17.567
873.15	22.182	20.998	20.106	19.397	18.812	18.317
973.15	23.859	22.361	21.262	20.404	19.707	19.124
1073.15	25.669	23.832	22.504	21.484	20.664	19.985
1173.15	27.548	25.373	23.811	22.620	21.671	20.891
1273.15	29.431	26.943	25.156	23.795	22.716	21.832
1500.00	33.508	30.444	28.212	26.503	25.145	24.034
2000.00	41.042	37.216	34.344	32.097	30.281	28.780
2500.00	46.956	42.696	39.445	36.865	34.758	32.996
3000.00	51.809	47.253	43.740	40.928	38.615	36.668
3500.00	55.938	51.160	47.448	44.460	41.988	39.899
4000.00	59.541	54.587	50.718	47.587	44.987	42.783

Table 1. (Continued.)

T (K)	P (GPa)					
	2.0	3.0	4.0	5.0	6.0	7.0
673.15	14.746	13.622	12.872	12.314	11.875	11.514
773.15	15.131	13.898	13.089	12.495	12.030	11.650
873.15	15.537	14.185	13.315	12.682	12.190	11.791
973.15	15.964	14.485	13.548	12.875	12.355	11.935
1073.15	16.412	14.769	13.790	13.074	12.525	12.083
1173.15	16.878	15.119	14.040	13.279	12.699	12.235
1273.15	17.362	15.453	14.297	13.489	12.878	12.391
1500.00	18.515	16.245	14.907	13.987	13.300	12.758
2000.00	21.196	18.121	16.355	15.170	14.302	13.628
2500.00	23.859	20.064	17.885	16.433	15.377	14.564
3000.00	26.364	21.974	19.428	17.727	16.491	15.542
3500.00	28.677	23.796	20.936	19.013	17.613	16.536
4000.00	30.806	25.514	22.384	20.267	18.719	17.526
	8.0	9.0	10.0	20.0	30.0	40.0
673.15	11.209	10.945	10.715	9.299	8.550	8.052
773.15	11.331	11.056	10.815	9.353	8.587	8.080
873.15	11.456	11.169	10.919	9.409	8.625	8.108
973.15	11.585	11.285	11.025	9.466	8.664	8.137
1073.15	11.716	11.404	11.133	9.523	8.703	8.167
1173.15	11.851	11.525	11.243	9.582	8.742	8.196
1273.15	11.990	11.650	11.356	9.642	8.783	8.226
1500.00	12.315	11.942	11.622	9.781	8.876	8.296
2000.00	13.084	12.632	12.248	10.106	9.094	8.458
2500.00	13.914	13.377	12.923	10.455	9.327	8.630
3000.00	14.785	14.162	13.637	10.827	9.574	8.813
3500.00	15.676	14.970	14.376	11.219	9.836	9.006
4000.00	16.572	15.787	15.127	11.628	10.112	9.210
	50.0	60.0	70.0	80.0	90.0	100.0
673.15	7.684	7.395	7.158	6.959	6.788	6.638
773.15	7.706	7.413	7.174	6.972	6.799	6.648
873.15	7.729	7.432	7.189	6.985	6.810	6.657
973.15	7.752	7.450	7.205	6.999	6.822	6.667
1073.15	7.775	7.469	7.220	7.012	6.833	6.677
1173.15	7.798	7.488	7.236	7.026	6.845	6.687
1273.15	7.822	7.508	7.253	7.039	6.857	6.698
1500.00	7.877	7.552	7.290	7.017	6.884	6.721
2000.00	8.004	7.656	7.376	7.144	6.946	6.775
2500.00	8.139	7.765	7.467	7.221	7.012	6.832
3000.00	8.282	7.881	7.563	7.302	7.082	6.892
3500.00	8.433	8.003	7.665	7.388	7.155	6.956
4000.00	8.592	8.133	7.772	7.478	7.232	7.022

Table 2. Molar volume of carbon dioxide (cm^3) under T and P.

T (K)	P (GPa)					
	0.5	0.6	0.7	0.8	0.9	1.0
373.15	34.101	32.589	31.417	30.463	29.664	28.978
473.15	35.828	34.005	32.623	31.519	30.605	29.829
573.15	37.732	35.546	33.925	32.651	31.610	30.733
673.15	39.794	37.206	35.321	33.858	32.676	31.691
773.15	41.983	38.969	36.799	35.135	33.801	32.698
873.15	44.257	40.811	38.347	36.471	34.978	33.751
973.15	46.574	42.706	39.948	37.857	36.200	34.845
1073.15	48.892	44.629	41.584	39.279	37.458	35.972
1173.15	51.182	46.555	43.238	40.725	38.742	37.126
1273.15	53.422	48.465	44.895	42.184	40.042	38.298
1500.00	58.261	52.675	48.600	45.482	43.009	40.991
2000.00	67.674	61.104	56.208	52.400	49.341	46.822
2500.00	75.640	68.393	62.924	58.623	55.136	52.243
3000.00	82.542	74.777	68.867	64.187	60.370	57.184
3500.00	88.653	80.463	74.193	69.203	65.115	61.690
4000.00	94.155	85.603	79.026	73.772	69.453	65.825
	2.0	3.0	4.0	5.0	6.0	7.0
373.15	24.988	22.972	21.651	20.681	19.921	19.301
473.15	25.440	23.288	21.895	20.881	20.092	19.449
573.15	25.911	23.614	22.147	21.087	20.266	19.600
673.15	26.401	23.951	22.406	21.298	20.445	19.755
773.15	26.909	24.298	22.673	21.515	20.627	19.913
873.15	27.436	24.657	22.946	21.737	20.814	20.075
973.15	27.980	25.025	23.227	21.964	21.005	20.240
1073.15	28.541	25.404	23.514	22.197	21.201	20.408
1183.15	29.118	25.792	23.809	22.434	21.400	20.580
1273.15	29.709	26.190	24.110	22.677	21.604	20.755
1500.00	31.095	27.125	24.817	23.247	22.080	21.165
2000.00	34.295	29.315	26.480	24.585	23.200	22.127
2500.00	37.547	31.613	28.248	26.019	24.403	23.162
3000.00	40.723	33.940	30.075	27.517	25.669	24.256
3500.00	43.761	36.238	31.917	29.049	26.977	25.393
4000.00	46.641	38.473	33.742	30.588	28.304	26.558

Table 2. (Continued.)

T (K)	P (GPa)					
	8.0	9.0	10.0	20.0	30.0	40.0
373.15	18.779	18.330	17.937	15.549	14.296	13.466
473.15	18.910	18.448	18.044	15.602	14.330	13.490
573.15	19.044	18.567	18.152	15.657	14.365	13.514
673.15	19.180	18.689	18.263	15.712	14.400	13.539
773.15	19.320	18.814	18.375	15.768	14.435	13.563
873.15	19.462	18.941	18.490	15.824	14.471	13.588
973.15	19.607	19.071	18.607	15.882	14.507	13.614
1073.15	19.755	19.203	18.726	15.940	14.544	13.639
1173.15	19.906	19.337	18.847	15.999	14.581	13.665
1273.15	20.060	19.474	18.970	16.059	14.619	13.692
1500.00	20.419	19.794	19.258	16.199	14.707	13.752
2000.00	21.261	20.542	19.931	16.523	14.909	13.892
2500.00	22.168	21.348	20.655	16.869	15.125	14.041
3000.00	23.129	22.204	21.425	17.238	15.354	14.198
3500.00	24.134	23.101	22.235	17.629	15.596	14.365
4000.00	25.168	24.030	23.077	18.042	15.853	14.542
	50.0	60.0	70.0	80.0	90.0	100.0
373.15	12.855	12.375	11.982	11.652	11.368	11.120
473.15	12.872	12.388	11.992	11.660	11.374	11.124
573.15	12.890	12.401	12.002	11.667	11.380	11.128
673.15	12.908	12.415	12.013	11.675	11.385	11.132
773.15	12.926	12.428	12.023	11.683	11.391	11.137
873.15	12.944	12.442	12.033	11.691	11.397	11.141
973.15	12.963	12.456	12.044	11.699	11.403	11.146
1073.15	12.981	12.470	12.055	11.707	11.409	11.150
1173.15	13.000	12.484	12.065	11.715	11.415	11.154
1273.15	13.019	12.498	12.076	11.723	11.422	11.159
1500.00	13.064	12.532	12.101	11.742	11.436	11.169
2000.00	13.166	12.608	12.159	11.786	11.468	11.193
2500.00	13.273	12.688	12.220	11.832	11.503	11.218
3000.00	13.388	12.773	12.284	11.880	11.539	11.245
3500.00	13.508	12.863	12.351	11.931	11.577	11.273
4000.00	13.636	12.958	12.423	11.985	11.618	11.302

Tables of Molar Volumes and Fugacities of Fluids on C–O–H System 339

Table 3. Molar volume of methane (cm^3) under T and P.

T (K)	P (GPa)					
	0.5	0.6	0.7	0.8	0.9	1.0
373.15	35.050	33.376	32.085	31.040	30.166	29.418
473.15	36.617	34.652	33.169	31.988	31.012	30.183
573.15	38.325	36.026	34.327	32.994	31.903	30.986
673.15	40.171	37.497	35.557	34.056	32.841	31.828
773.15	42.143	39.058	36.857	35.173	33.825	32.708
873.15	44.223	40.702	38.221	36.343	34.851	33.625
973.15	46.386	42.415	39.642	37.559	35.917	34.576
1073.15	48.604	44.183	41.111	38.818	37.020	35.558
1173.15	50.850	45.990	42.618	40.111	38.153	36.568
1273.15	53.101	47.820	44.154	41.433	39.313	37.603
1500.00	58.129	51.986	47.690	44.497	42.015	40.019
2000.00	68.430	60.820	55.379	51.283	48.078	45.492
2500.00	77.546	68.867	62.562	57.757	53.961	50.879
3000.00	85.680	76.149	69.153	63.776	59.499	56.005
3500.00	93.054	82.796	75.216	69.355	64.670	60.826
4000.00	99.832	88.928	80.832	74.548	69.505	65.355
	2.0	3.0	4.0	5.0	6.0	7.0
373.15	25.090	22.914	21.493	20.451	19.636	18.972
473.15	25.500	23.204	21.719	20.638	19.797	19.113
573.15	25.923	23.500	21.950	20.829	19.960	19.255
673.15	26.359	23.804	22.187	21.024	20.126	19.401
773.15	26.809	24.115	22.428	21.222	20.295	19.548
873.15	27.271	24.434	22.674	21.423	20.466	19.698
973.15	27.745	24.759	22.924	21.629	20.641	19.850
1073.15	28.231	25.092	23.180	21.838	20.818	20.004
1173.15	28.730	25.431	23.440	22.050	20.998	20.160
1273.15	29.239	25.777	23.705	22.265	21.180	20.319
1500.00	30.432	26.584	24.321	22.767	21.605	20.687
2000.00	33.208	28.465	25.754	23.930	22.585	21.537
2500.00	36.105	30.451	27.271	25.161	23.623	22.435
3000.00	39.031	32.498	28.849	26.446	24.709	23.375
3500.00	41.923	34.567	30.462	27.769	25.831	24.350
4000.00	44.742	36.629	32.090	29.115	26.978	25.350

Table 3. (Continued.)

T (K)	P (GPa)					
	8.0	9.0	10.0	20.0	30.0	40.0
373.15	18.414	17.934	17.515	14.977	13.655	12.784
473.15	18.539	18.048	17.619	15.033	13.693	12.812
573.15	18.666	18.162	17.723	15.089	13.731	12.840
673.15	18.795	18.279	17.829	15.145	13.769	12.869
773.15	18.926	18.397	17.937	15.202	13.807	12.897
873.15	19.059	18.516	18.046	15.260	13.846	12.926
973.15	19.194	18.638	18.156	15.318	13.886	12.956
1073.15	19.331	18.761	18.268	15.377	13.925	19.985
1173.15	19.470	18.885	18.381	15.436	13.965	13.015
1273.15	19.610	19.011	18.496	15.496	14.005	13.045
1500.00	19.936	19.304	18.761	15.635	14.098	13.113
2000.00	20.686	19.976	19.370	15.950	14.309	13.270
2500.00	21.479	20.686	20.013	16.280	14.529	13.432
3000.00	22.309	21.428	20.685	16.625	14.757	13.600
3500.00	23.170	22.200	21.384	16.983	14.995	13.775
4000.00	24.056	22.996	22.106	17.355	15.241	13.956
	50.0	60.0	70.0	80.0	90.0	100.0
373.15	12.144	11.643	11.235	10.892	10598	10.342
473.15	12.166	11.661	11.250	10.905	10.609	10.351
573.15	12.188	11.679	11.265	10.918	10.620	10.361
673.15	12.210	11.697	11.280	10.931	10.631	10.370
773.15	12.233	11.716	11.296	10.944	10.642	10.380
873.15	12.256	11.734	11.311	10.957	10.653	10.389
973.15	12.279	11.753	11.326	10.970	10.665	10.399
1073.15	12.302	11.772	11.342	10.983	10.676	10.409
1173.15	12.325	11.791	11.358	10.996	10.687	10.418
1273.15	12.349	11.810	11.374	11.010	10.699	10.428
1500.00	12.403	11.854	11.410	11.040	10.725	10.451
2000.00	12.525	11.953	11.492	11.109	10.784	10.502
2500.00	12.652	12.055	11.577	11.181	10.845	10.554
3000.00	12.783	12.162	11.665	11.255	10.908	10.608
3500.00	12.919	12.272	11.756	11.332	10.973	10.665
4000.00	13.061	12.385	11.850	11.411	11.041	10.722

Table 4. Molar volume of carbon monoxide (cm^3) under T and P.

T (K)	P (GPa)					
	0.5	0.6	0.7	0.8	0.9	1.0
373.15	33.032	31.364	30.069	29.018	28.139	27.387
473.15	34.632	32.682	31.194	30.003	29.017	28.181
573.15	36.365	34.097	32.395	31.049	29.945	29.017
673.15	38.226	35.607	33.669	32.153	30.923	29.894
773.15	40.200	37.204	35.012	33.315	31.947	30.812
873.15	42.271	38.380	36.419	34.529	33.016	31.768
973.15	44.418	40.621	37.883	35.791	34.127	32.760
1073.15	46.620	42.415	39.394	37.096	35.275	33.785
1173.15	48.857	44.250	40.944	38.436	36.455	34.840
1273.15	51.113	46.112	42.524	39.806	37.663	35.920
1500.00	56.232	50.384	46.176	42.989	40.480	38.446
2000.00	67.187	59.701	54.266	50.125	46.856	44.203
2500.00	77.497	68.602	62.101	57.125	53.180	49.968
3000.00	87.209	77.037	69.580	63.852	59.300	55.585
3500.00	96.430	85.065	76.717	70.296	65.184	61.006
4000.00	105.255	92.749	83.558	76.481	70.843	66.231
	2.0	3.0	4.0	5.0	6.0	7.0
373.15	23.075	20.946	19.573	18.575	17.800	17.172
473.15	23.496	21.240	19.801	18.754	17.961	17.313
573.15	23.930	21.542	20.035	18.956	18.125	17.456
673.15	24.380	21.852	20.275	19.152	18.292	17.601
773.15	24.843	22.170	20.519	19.352	18.462	17.749
873.15	25.320	22.495	20.769	19.556	18.634	17.900
973.15	25.811	22.829	21.024	19.763	18.810	18.052
1073.15	26.315	23.169	21.283	19.974	18.989	18.207
1173.15	26.832	23.517	21.548	20.189	19.170	18.365
1273.15	27.361	23.873	21.818	20.408	19.354	18.524
1500.00	28.603	24.704	22.447	20.917	19.783	18.895
2000.00	31.502	26.646	23.914	22.099	20.776	19.752
2500.00	34.542	28.705	25.472	23.355	21.829	20.660
3000.00	37.635	30.836	27.098	24.669	22.932	21.611
3500.00	40.720	33.001	28.766	26.025	24.075	22.599
4000.00	43.761	35.172	30.456	27.409	25.246	23.614

Table 4. (Continued.)

T (K)	P (GPa)					
	8.0	9.0	10.0	20.0	30.0	40.0
373.15	16.646	16.196	15.804	13.454	12.244	11.452
473.15	16.772	16.309	15.907	13.510	12.283	11.481
573.15	16.899	16.424	16.012	13.566	12.321	11.511
673.15	17.028	16.541	16.118	13.623	12.360	11.540
773.15	17.159	16.659	16.226	13.680	12.400	11.570
873.15	17.293	16.779	16.335	13.738	12.439	11.600
973.15	17.428	16.900	16.445	13.797	12.479	11.630
1073.15	17.565	17.023	16.557	13.856	12.520	11.661
1173.15	17.704	17.148	16.670	13.916	12.560	11.692
1273.15	17.845	17.275	16.785	13.976	12.601	11.723
1500.00	18.172	17.568	17.051	14.115	12.696	11.794
2000.00	18.927	18.243	17.662	14.431	12.910	11.954
2500.00	19.726	18.956	18.306	14.762	13.132	12.121
3000.00	20.562	19.703	18.981	15.106	13.363	12.294
3500.00	21.432	20.480	19.683	14.464	13.602	12.472
4000.00	22.328	21.281	20.408	15.834	13.849	12.656
	50.0	60.0	70.0	80.0	90.0	100.0
373.15	10.873	10.421	10.053	9.745	9.481	9.251
473.15	10.896	10.440	10.070	9.759	9.493	9.262
573.15	10.920	10.460	10.086	9.774	9.506	9.273
673.15	10.943	10.479	10.103	9.788	9.519	9.284
773.15	10.967	10.499	10.120	9.803	9.531	9.295
873.15	10.991	10.519	10.137	9.817	9.544	9.307
973.15	11.015	10.539	10.154	9.832	9.557	9.318
1073.15	11.040	10.559	10.171	9.847	9.570	9.330
1173.15	11.064	10.580	10.188	9.862	9.583	9.341
1273.15	11.089	10.600	10.206	9.877	9.596	9.353
1500.00	11.146	10.647	10.245	9.911	9.627	9.380
2000.00	11.274	10.753	10.335	9.989	9.694	9.440
2500.00	11.407	10.863	10.428	10.068	9.764	9.501
3000.00	11.544	10.975	10.523	10.151	9.836	9.564
3500.00	11.685	11.092	10.621	10.235	9.909	9.629
4000.00	11.831	11.212	10.722	10.322	9.985	9.696

Table 5. Molar volume of oxygen (cm^3) under T and P.

	\multicolumn{6}{c}{P (GPa)}
T (K)	0.5	0.6	0.7	0.8	0.9	1.0
373.15	28.506	26.673	25.362	24.353	23.538	22.858
473.15	30.164	27.985	26.457	25.299	24.375	23.611
573.15	31.895	29.356	27.601	26.286	25.246	24.394
673.15	33.653	30.764	28.781	27.305	26.146	25.202
773.15	35.399	32.186	29.980	28.345	27.067	26.029
873.15	37.100	33.599	31.186	29.397	28.001	26.871
973.15	38.737	34.987	32.384	30.449	28.940	27.720
1073.15	40.297	36.336	33.564	31.495	29.879	28.572
1173.15	41.777	37.640	34.717	32.527	30.811	29.421
1273.15	43.179	38.893	35.840	33.538	31.730	30.264
1500.00	46.084	41.544	38.253	35.742	33.754	32.132
2000.00	51.389	46.537	42.917	40.097	37.825	35.949
2500.00	55.604	50.607	46.810	43.805	41.356	39.312
3000.00	59.076	54.013	50.113	46.992	44.425	42.267
3500.00	62.013	56.928	52.967	49.771	47.123	44.883
4000.00	64.551	59.468	55.474	52.228	49.522	47.221
	2.0	3.0	4.0	5.0	6.0	7.0
373.15	19.164	17.413	16.293	15.482	14.853	14.342
473.15	19.564	17.696	16.516	15.667	15.012	14.482
573.15	19.974	17.986	16.743	15.856	15.175	14.625
673.15	20.396	18.282	16.976	16.049	15.340	14.771
773.15	20.828	18.585	17.212	16.245	15.508	14.918
873.15	21.268	18.893	17.453	16.444	15.678	15.068
973.15	21.716	19.207	17.698	16.646	15.852	15.220
1073.15	22.170	19.525	17.946	16.852	16.027	15.374
1173.15	22.630	19.848	18.198	17.060	16.206	15.530
1273.15	23.094	20.175	18.454	17.271	16.386	15.688
1500.00	24.154	20.927	19.043	17.758	16.803	16.053
2000.00	26.483	22.615	20.376	18.865	17.752	16.887
2500.00	28.725	24.298	21.728	19.998	18.731	17.748
3000.00	30.833	25.933	23.068	21.136	19.720	18.625
3500.00	32.796	27.501	24.376	22.260	20.708	19.506
4000.00	34.618	28.991	25.640	23.359	21.681	20.381

Table 5. (Continued.)

T (K)	P (GPa)					
	8.0	9.0	10.0	20.0	30.0	40.0
373.15	13.914	13.548	13.229	11.307	10.311	9.656
473.15	14.040	13.662	13.334	11.365	10.351	9.687
573.15	14.168	13.778	13.440	11.423	10.392	9.718
673.15	14.298	13.896	13.547	11.483	10.433	9.749
773.15	14.430	14.015	13.656	11.542	10.475	9.781
873.15	14.563	14.136	13.767	11.603	10.517	9.813
973.15	14.699	14.259	13.879	11.664	10.559	9.845
1073.15	14.837	13.383	13.992	11.726	10.602	9.878
1173.15	14.976	14.509	14.107	11.789	10.645	9.911
1273.15	15.117	14.636	14.224	11.852	10.689	9.944
1500.00	15.442	14.930	14.492	11.998	10.789	10.020
2000.00	16.186	15.602	15.106	12.330	11.018	10.194
2500.00	16.957	16.301	15.744	12.677	11.256	10.375
3000.00	17.745	17.016	16.400	13.037	11.504	10.563
3500.00	18.540	17.742	17.068	13.409	11.761	10.757
4000.00	19.335	18.470	17.740	13.790	12.025	10.958
	50.0	60.0	70.0	80.0	90.0	100.0
375.15	9.176	8.800	8.494	8.237	8.017	7.824
473.15	9.200	8.821	8.512	8.252	8.030	7.836
573.15	9.225	8.841	8.529	8.268	8.043	7.848
673.15	9.251	8.862	8.547	8.283	8.057	7.860
773.15	9.276	8.884	8.565	8.299	8.070	7.872
873.15	9.302	8.905	8.583	8.314	8.084	7.884
973.15	9.328	8.926	8.601	8.330	8.098	7.896
1073.15	9.354	8.948	8.620	8.346	8.112	7.908
1173.15	9.380	8.970	8.638	8.362	8.126	7.921
1273.15	9.407	8.992	8.657	8.378	8.140	7.933
1500.00	9.468	9.043	8.700	8.415	8.712	7.962
2000.00	9.607	9.158	8.798	8.499	8.246	8.027
2500.00	9.752	9.278	8.899	8.587	8.322	8.094
3000.00	9.902	9.402	9.004	8.677	8.401	8.164
3500.00	10.058	9.531	9.113	8.771	8.483	8.236
4000.00	10.219	9.664	9.226	8.869	8.568	8.311

Table 6. Molar volume of hydrogen (cm^3) under T and P.

T (K)	P (GPa)					
	0.5	0.6	0.7	0.8	0.9	1.0
373.15	20.245	18.549	17.325	16.387	15.637	15.017
473.15	21.848	19.826	18.388	17.299	16.436	15.730
573.15	23.539	21.175	19.510	18.259	17.276	16.478
673.15	25.292	22.582	20.681	19.262	18.153	17.257
773.15	27.083	24.030	21.891	20.299	19.061	18.064
873.15	28.893	25.505	23.129	21.364	19.994	18.895
973.15	30.708	26.995	24.388	22.450	20.948	19.744
1073.15	32.517	28.491	25.657	23.550	21.916	20.609
1173.15	34.313	29.987	26.993	24.658	22.895	21.485
1273.15	36.094	31.477	28.209	25.772	23.881	22.368
1500.00	40.067	34.821	31.089	28.297	26.126	24.388
2000.00	48.489	41.966	37.294	33.778	31.033	28.828
2500.00	56.524	48.811	43.267	39.083	35.807	33.170
3000.00	64.269	55.410	49.033	44.214	40.436	37.390
3500.00	71.794	61.818	54.632	49.197	44.934	41.496
4000.00	79.151	68.075	60.095	54.058	49.322	45.501
	2.0	3.0	4.0	5.0	6.0	7.0
373.15	11.797	10.375	9.502	8.888	8.422	8.050
473.15	12.153	10.619	9.691	9.044	8.555	8.167
573.15	12.520	10.870	9.884	9.202	8.690	8.285
673.15	12.899	11.127	10.082	9.364	8.828	8.405
773.15	13.289	11.391	10.283	9.529	8.968	8.527
873.15	13.690	11.660	10.489	9.696	9.110	8.651
973.15	14.100	11.935	10.699	9.867	9.254	8.777
1073.15	14.518	12.215	10.912	10.040	9.401	8.905
1173.15	14.944	12.500	11.128	10.216	9.550	9.034
1273.15	15.376	12.790	11.348	10.394	9.701	9.165
1500.00	16.378	13.461	11.857	10.807	10.049	9.468
2000.00	18.648	14.994	13.023	11.752	10.846	10.160
2500.00	20.946	16.568	14.226	12.730	11.673	10.877
3000.00	23.230	18.153	15.447	13.727	12.519	11.613
3500.00	25.480	19.732	16.672	14.733	13.374	12.359
4000.00	27.690	21.293	17.891	15.738	14.232	13.109

Table 6. (Continued.)

T (K)	P (GPa)					
	8.0	9.0	10.0	20.0	30.0	40.0
373.15	7.743	7.484	7.260	5.955	5.309	4.895
473.15	7.847	7.578	7.346	6.003	5.343	4.922
573.15	7.953	7.673	7.433	6.052	5.378	4.949
673.15	8.060	7.770	7.521	6.101	5.413	4.976
773.15	8.169	7.868	7.611	6.151	5.448	5.003
873.15	8.279	7.967	7.701	6.201	5.483	5.031
973.15	8.391	8.086	7.793	6.251	5.519	5.059
1073.15	8.504	8.170	7.886	6.302	5.555	5.087
1173.15	8.619	8.274	7.980	6.353	5.591	5.116
1273.15	8.735	8.378	8.076	6.405	5.628	5.144
1500.00	9.003	8.619	8.295	6.525	5.712	5.209
2000.00	9.615	9.169	8.795	6.795	5.901	5.357
2500.00	10.251	9.740	9.314	7.074	6.096	5.508
3000.00	10.903	10.327	9.848	7.362	6.297	5.664
3500.00	11.565	10.923	10.391	7.656	6.502	5.823
4000.00	12.233	11.526	10.941	7.955	6.712	5.986
	50.0	60.0	70.0	80.0	90.0	100.0
373.15	4.596	4.366	4.180	4.026	3.894	3.780
473.15	4.618	4.385	4.197	4.401	3.908	3.793
573.15	4.641	4.404	4.214	4.056	3.921	3.805
673.15	4.663	4.423	4.231	4.071	3.935	3.817
773.15	4.686	4.443	4.248	4.086	3.949	3.830
873.15	4.709	4.462	4.265	4.101	3.962	3.842
973.15	4.732	4.482	4.282	4.116	3.976	3.855
1073.15	4.755	4.502	4.299	4.132	3.990	3.868
1173.15	4.779	4.522	4.317	4.147	4.004	3.880
1273.15	4.802	4.542	4.334	4.163	4.018	3.893
1500.00	4.856	4.588	4.374	4.198	4.050	3.922
2000.00	4.977	4.691	4.464	4.278	4.122	3.988
2500.00	5.102	4.797	4.557	4.360	4.196	4.055
3000.00	5.229	4.905	4.651	4.444	4.271	4.231
3500.00	5.360	5.016	4.748	4.530	4.348	4.194
4000.00	5.493	5.130	4.847	4.618	4.427	4.265

Table 7. Fugacity of water (kJ/mol) under T and P.

T (K)	P (GPa)					
	0.5	0.6	0.7	0.8	0.9	1.0
673.15	40.09	41.99	43.83	45.62	47.36	49.06
773.15	49.29	51.31	53.25	55.12	56.94	58.72
873.15	58.15	60.31	62.36	64.34	66.25	68.10
973.15	66.76	69.07	71.25	73.33	75.33	77.28
1073.15	75.17	77.64	79.95	82.15	84.25	86.28
1173.15	83.40	86.04	88.49	90.81	93.03	95.15
1273.15	91.49	94.30	96.90	99.35	101.67	103.90
1500.00	109.42	112.61	115.54	118.27	120.85	123.31
2000.00	147.51	151.42	154.99	158.30	161.42	164.37
2500.00	184.34	188.82	192.92	196.73	200.31	203.69
3000.00	200.38	225.32	229.87	234.10	238.07	241.83
3500.00	255.87	261.21	266.14	270.73	275.05	279.14
4000.00	290.96	296.65	301.91	306.82	311.45	315.83
	2.0	3.0	4.0	5.0	6.0	7.0
673.15	64.73	78.87	92.09	104.68	116.76	128.45
773.15	73.89	89.35	102.82	115.60	127.86	139.69
873.15	84.81	99.61	113.34	126.32	138.75	150.73
973.15	94.56	109.71	123.70	136.90	149.50	161.64
1073.15	104.17	119.70	133.96	147.37	160.16	172.46
1173.15	113.67	129.59	144.13	157.77	170.75	183.21
1273.15	123.08	139.39	154.22	168.10	181.27	193.89
1500.00	144.05	161.31	176.84	191.26	204.89	217.91
2000.00	188.63	208.12	225.29	241.02	255.74	269.69
2500.00	231.28	253.03	271.92	289.04	304.92	391.87
3000.00	272.43	296.36	316.97	335.49	352.57	368.57
3500.00	312.45	338.43	360.69	380.61	398.89	415.94
4000.00	351.60	379.49	403.32	424.59	444.05	462.14

Table 7. (Continued.)

T (K)	P (GPa)					
	8.0	9.0	10.0	20.0	30.0	40.0
673.15	139.81	150.88	161.71	260.82	349.76	432.63
773.15	151.18	162.37	173.30	273.14	362.53	445.72
873.15	162.35	173.66	184.70	285.29	375.14	458.65
973.15	173.40	184.83	195.98	297.33	387.64	471.49
1073.15	184.35	195.91	207.17	309.30	400.09	484.28
1173.15	195.24	206.23	218.31	321.23	412.49	497.02
1273.15	206.08	217.89	229.39	333.11	424.87	509.75
1500.00	230.44	242.56	254.34	350.95	452.84	538.53
2000.00	283.04	295.89	308.32	418.37	513.90	601.46
2500.00	334.10	347.74	360.88	475.70	574.07	663.63
3000.00	383.72	398.19	412.08	531.97	633.36	725.04
3500.00	432.03	447.34	462.01	587.21	691.79	785.72
4000.00	479.18	495.34	510.79	641.47	749.39	845.68
	50.0	60.0	70.0	80.0	90.0	100.0
673.15	511.23	586.57	659.30	729.86	798.57	865.69
773.15	524.57	600.11	673.01	743.71	812.55	879.76
873.15	537.75	613.50	686.57	757.41	826.37	893.69
973.15	550.85	626.80	700.84	771.03	840.11	907.54
1073.15	563.89	640.05	713.46	784.60	853.80	921.34
1173.15	576.91	653.28	726.86	798.14	867.47	935.12
1273.15	589.90	666.49	740.25	811.68	881.13	948.89
1500.00	619.29	696.38	770.55	842.32	912.07	980.08
2000.00	638.65	761.88	836.99	909.56	979.98	1048.57
2500.00	747.34	826.79	902.89	976.30	1047.43	1116.63
3000.00	810.37	891.10	968.26	1042.55	1114.44	1184.29
3500.00	872.76	954.85	1033.13	1108.35	1181.03	1251.56
4000.00	934.53	1018.05	1097.50	1173.71	1247.23	1318.48

Tables of Molar Volumes and Fugacities of Fluids on C–O–H System   349

Table 8. Fugacity of carbon dioxide (kJ/mol) under T and P.

T (K)	P (GPa)					
	0.5	0.6	0.7	0.8	0.9	1.0
373.15	33.98	37.31	40.51	43.60	46.60	49.54
473.15	42.66	46.15	49.48	52.68	55.79	58.81
573.15	50.96	54.62	58.09	61.42	64.63	67.75
673.15	59.01	62.85	66.47	69.93	73.25	76.47
773.15	66.87	70.91	74.69	78.28	81.73	85.05
873.15	74.58	78.83	82.78	86.52	90.09	93.52
973.15	82.19	86.64	90.77	94.66	98.36	101.91
1073.51	89.71	94.37	98.68	102.71	106.55	110.22
1173.15	97.16	102.03	106.51	110.70	114.67	118.46
1273.15	104.54	109.62	114.28	118.63	122.74	126.65
1500.00	121.11	126.64	131.70	136.40	140.82	145.01
2000.00	157.03	163.45	169.30	174.72	179.81	184.61
2500.00	192.40	199.58	206.14	212.20	217.89	223.25
3000.00	227.43	235.28	242.45	249.09	255.31	261.19
3500.00	262.23	270.66	278.38	285.54	292.35	298.59
4000.00	296.85	305.82	314.04	321.67	328.82	335.58
	2.0	3.0	4.0	5.0	6.0	7.0
373.15	76.22	100.12	122.39	143.54	163.82	183.43
473.15	86.10	110.37	132.92	154.29	174.76	194.52
573.15	95.68	120.34	143.17	164.77	185.43	205.35
673.15	105.07	130.13	153.26	175.09	195.94	216.03
773.15	114.35	139.83	163.26	185.33	206.38	226.64
873.15	123.55	149.46	173.20	195.51	216.77	237.20
973.15	132.68	159.04	183.10	205.67	227.13	247.74
1073.15	141.77	168.59	192.98	215.80	237.48	258.27
1173.15	150.82	178.10	202.83	225.92	247.81	268.79
1273.15	159.82	187.58	212.66	236.01	258.13	279.30
1500.00	180.08	208.97	234.86	258.85	281.49	303.09
2000.00	233.94	255.47	283.26	308.74	332.60	355.24
2500.00	266.75	300.99	330.79	357.86	383.03	406.79
3000.00	308.61	345.57	377.43	406.14	432.69	457.63
3500.00	349.68	389.28	423.19	453.58	481.55	507.70
4000.00	390.10	432.22	468.15	500.22	529.61	557.01

Table 8. (Continued.)

T (K)	P (GPa)					
	8.0	9.0	10.0	20.0	30.0	40.0
373.15	202.46	221.01	239.14	404.93	553.65	692.22
473.15	213.69	232.37	250.61	417.14	566.29	705.15
573.15	224.66	243.46	261.82	429.11	578.69	717.84
673.15	235.49	254.42	272.89	440.96	590.97	730.41
773.15	246.25	265.31	283.90	452.74	603.20	742.94
873.15	256.96	276.16	294.87	464.51	615.41	755.45
973.15	267.66	286.99	305.82	467.27	627.63	767.97
1073.15	278.34	297.81	316.77	488.04	639.87	780.51
1173.15	289.02	308.63	327.72	499.83	652.12	793.08
1273.15	299.69	319.45	338.67	551.62	664.39	805.66
1500.00	323.87	323.97	363.49	538.42	692.29	834.29
2000.00	376.92	397.81	418.04	597.57	753.99	897.68
2500.00	429.44	451.18	472.18	656.64	815.79	961.26
3000.00	481.30	503.96	525.76	715.48	877.53	1024.90
3500.00	542.44	556.04	578.70	773.99	939.11	1088.49
4000.00	582.85	607.43	630.97	832.09	1000.46	1151.98
	50.0	60.0	70.0	80.0	90.0	100.0
373.15	823.69	949.75	1071.47	1189.60	1304.67	1417.08
473.15	836.82	963.03	1084.87	1203.08	1318.22	1430.68
573.15	849.71	976.08	1098.03	1216.34	1331.54	1444.05
673.15	862.50	989.02	1111.09	1229.48	1344.75	1457.31
773.15	875.24	1001.91	1124.10	1242.59	1357.92	1470.54
873.15	887.96	1014.80	1137.11	1255.69	1371.09	1483.76
973.15	900.70	1027.70	1150.13	1268.80	1384.28	1496.99
1073.15	913.47	1040.62	1163.18	1281.94	1397.48	1510.25
1173.15	926.25	1053.57	1176.25	1295.11	1410.72	1523.54
1273.15	939.06	1066.55	1189.36	1308.31	1423.99	1536.87
1500.00	968.21	1096.09	1219.18	1338.35	1454.20	1567.20
2000.00	1032.79	1161.55	1285.30	1404.97	1521.20	1634.48
2500.00	1097.64	1227.33	1351.79	1471.99	1588.62	1702.19
3000.00	1162.62	1293.30	1418.50	1539.26	1656.31	1770.19
3500.00	1227.64	1359.36	1485.34	1606.69	1724.19	1838.40
4000.00	1292.63	1425.46	1552.27	1674.24	1792.21	1906.76

Tables of Molar Volumes and Fugacities of Fluids on C–O–H System 351

Table 9. Fugacity of methane (kJ/mol) under T and P.

T (K)	P (GPa)					
	0.5	0.6	0.7	0.8	0.9	1.0
373.15	37.84	41.26	44.53	47.68	50.74	53.72
473.15	45.85	49.41	52.80	56.06	59.02	62.26
573.15	53.62	57.33	60.85	64.21	67.45	70.60
673.15	61.22	65.10	68.74	72.22	75.56	78.80
773.15	68.70	72.75	76.54	80.14	83.59	86.91
873.15	76.09	80.32	84.26	87.98	91.54	94.96
973.15	83.40	87.83	91.92	95.78	99.45	102.97
1073.15	90.66	95.29	99.54	103.53	107.32	110.95
1173.15	97.87	102.70	107.12	111.25	115.16	118.89
1273.15	105.04	110.07	114.66	118.94	122.97	126.81
1500.00	121.20	126.68	131.66	136.26	140.58	144.68
2000.00	156.40	162.84	168.64	173.96	178.92	183.60
2500.00	191.26	198.55	205.11	211.12	216.70	221.93
3000.00	225.89	233.96	241.21	247.84	254.00	259.77
3500.00	260.38	269.14	277.02	284.24	290.93	297.20
4000.00	294.75	304.16	312.63	320.38	327.58	334.31
	2.0	3.0	4.0	5.0	6.0	7.0
373.15	80.65	104.56	126.72	147.67	167.70	187.00
473.15	89.74	113.99	136.41	157.57	177.77	197.21
573.15	98.64	123.25	145.92	167.99	187.67	207.26
673.15	107.43	132.40	155.34	176.92	197.48	217.23
773.15	116.16	141.50	164.71	186.51	207.25	227.16
873.15	124.85	150.56	174.06	196.08	217.00	237.07
973.15	133.51	159.62	183.40	205.64	226.76	246.99
1073.15	142.16	168.66	192.74	215.21	236.52	256.91
1173.15	150.79	177.71	202.07	224.78	246.29	266.85
1273.15	159.42	186.75	211.42	234.36	256.07	276.80
1500.00	178.94	207.25	232.62	256.12	278.28	299.41
2000.00	211.68	252.25	279.26	304.05	327.28	349.32
2500.00	263.91	296.86	325.60	351.75	376.11	399.11
3000.00	305.57	340.96	371.49	399.06	424.60	448.61
3500.00	346.69	384.52	416.87	445.90	472.65	497.71
4000.00	387.34	427.56	461.74	492.25	520.24	546.37

Table 9. (Continued.)

T (K)	P (GPa)					
	8.0	9.0	10.0	20.0	30.0	40.0
373.15	205.68	223.85	241.57	402.28	544.90	676.84
473.15	216.03	234.32	252.14	413.59	556.67	688.94
573.15	226.22	244.62	262.56	424.76	568.30	700.89
673.15	236.42	254.85	272.90	435.86	579.86	712.78
773.15	246.39	265.04	283.20	446.94	591.40	724.66
873.15	256.44	275.22	293.50	458.01	602.95	736.54
973.15	266.50	285.41	303.80	469.10	614.51	748.44
1073.15	276.57	295.61	314.12	480.22	626.11	760.37
1173.15	286.66	305.82	324.45	491.35	637.73	772.34
1273.15	296.75	316.06	334.80	502.52	649.39	784.35
1500.00	319.71	339.32	358.35	527.95	675.94	811.69
2000.00	370.42	390.74	410.41	584.31	734.87	872.44
2500.00	421.06	442.13	462.47	640.90	794.13	933.58
3000.00	471.43	493.29	514.34	697.50	853.53	994.93
3500.00	521.45	544.12	565.90	753.98	912.91	1056.36
4000.00	571.05	594.56	617.10	810.28	972.22	1117.76
	50.0	60.0	70.0	80.0	90.0	100.0
373.15	801.34	920.18	1034.50	1145.09	1252.51	1357.19
473.15	813.68	931.72	1047.21	1157.94	1265.48	1370.25
573.15	825.88	945.12	1059.78	1170.65	1278.30	1383.18
673.15	838.03	957.47	1072.29	1183.30	1291.07	1396.05
773.15	850.16	969.80	1084.79	1195.94	1303.83	1408.91
873.15	862.29	982.15	1097.31	1208.59	1316.61	1421.79
973.15	874.46	994.52	1109.84	1221.28	1329.41	1434.70
1073.15	886.65	1006.92	1122.42	1233.99	1342.25	1447.64
1173.15	898.88	1019.36	1135.03	1246.75	1355.13	1460.63
1273.15	911.15	1031.84	1147.69	1259.55	1368.05	1473.66
1500.00	939.11	1060.28	1176.53	1288.73	1397.51	1503.36
2000.00	1001.23	1123.50	1240.65	1353.60	1463.03	1595.42
2500.00	1063.81	1187.22	1305.30	1419.04	1529.13	1636.09
3000.00	1126.65	1251.24	1370.29	1484.83	1595.60	1703.15
3500.00	1189.61	1315.43	1435.48	1550.85	1662.33	1770.49
4000.00	1252.61	1379.70	1500.79	1617.02	1729.23	1838.01

Table 10. Fugacity (kJ/mol) of carbon monoxide under T and P.

T (K)	P (GPa)					
	0.5	0.6	0.7	0.8	0.9	1.0
373.15	39.03	42.25	45.32	48.27	51.13	53.90
473.15	46.65	50.01	53.20	56.26	59.21	62.06
573.15	54.14	57.65	60.97	64.14	67.19	70.14
673.15	61.54	65.22	68.68	71.97	75.12	78.16
773.15	68.88	72.74	76.35	79.76	83.02	86.16
873.15	76.17	80.22	83.98	87.52	90.90	94.14
973.15	83.43	87.67	91.56	95.27	98.76	102.11
1073.15	90.66	95.10	99.18	103.00	106.62	110.07
1173.15	97.86	102.50	106.76	110.72	114.46	118.02
1273.15	105.05	109.89	114.32	118.43	122.30	125.97
1500.00	121.28	126.60	131.41	135.86	140.03	143.98
2000.00	156.86	163.19	168.87	174.08	178.92	183.47
2500.00	192.26	199.54	206.06	212.01	217.52	222.67
3000.00	227.55	235.73	243.05	249.71	255.86	261.60
3500.00	262.76	271.80	279.87	287.21	293.98	300.28
4000.00	297.91	307.78	316.57	324.56	331.92	338.76
	2.0	3.0	4.0	5.0	6.0	7.0
373.15	78.80	100.72	120.94	139.99	158.16	175.64
473.15	87.53	109.80	130.28	149.53	167.88	185.51
573.15	96.19	118.82	139.56	159.03	177.56	195.34
673.15	104.83	127.83	148.84	168.53	187.23	205.17
773.15	113.46	136.83	158.13	178.04	196.93	215.02
873.15	122.10	145.87	167.44	187.57	206.65	224.91
973.15	130.75	154.92	176.78	197.15	216.41	234.83
1073.15	139.41	163.99	186.16	206.75	226.21	244.79
1173.15	148.09	173.09	195.56	216.39	236.05	254.80
1273.15	156.78	182.21	204.98	226.06	245.91	264.84
1500.00	176.51	202.95	226.45	248.09	268.41	287.73
2000.00	220.02	248.82	273.99	296.94	318.35	338.59
2500.00	263.34	294.62	321.58	345.93	368.49	389.71
3000.00	306.37	340.20	369.02	394.83	418.58	440.83
3500.00	349.08	385.49	416.20	443.50	468.51	491.81
4000.00	391.49	430.45	463.07	491.90	518.18	542.57

Table 10. (Continued.)

T (K)	P (GPa)					
	8.0	9.0	10.0	20.0	30.0	40.0
373.15	192.54	208.96	224.95	369.58	497.58	615.83
473.15	202.54	219.08	235.18	380.56	509.01	627.59
573.15	212.51	229.16	245.38	391.50	520.42	639.34
673.15	222.47	239.25	255.58	402.47	531.85	651.11
773.15	232.47	249.37	265.81	413.47	543.32	662.92
873.12	242.49	259.52	276.07	424.52	554.85	674.79
973.15	252.56	269.72	286.39	435.62	566.43	686.72
1073.15	262.67	279.96	296.74	446.77	578.07	698.71
1173.15	272.82	290.24	307.14	457.98	589.77	710.76
1273.15	283.01	300.57	317.59	469.24	601.52	722.87
1500.00	306.25	324.12	341.42	494.96	628.38	750.54
2000.00	357.92	376.49	394.44	552.29	688.30	812.31
2500.00	409.88	429.21	447.83	610.21	748.91	874.85
3000.00	461.90	482.02	501.35	668.46	809.96	937.88
3500.00	513.81	534.75	554.82	726.84	871.25	1001.24
4000.00	565.52	587.31	608.14	785.26	932.67	1064.78
	50.0	60.0	70.0	80.0	90.0	100.0
373.15	727.32	833.70	936.01	1034.96	1131.05	1224.69
473.15	739.34	845.94	948.43	1047.53	1143.76	1237.51
573.15	751.35	858.16	960.83	1060.09	1156.45	1250.32
673.15	763.39	870.41	973.26	1072.67	1169.18	1263.17
773.15	775.47	882.71	985.74	1085.31	1181.95	1276.06
873.15	787.60	895.07	998.28	1098.01	1194.78	1289.01
973.15	799.80	907.49	1010.89	1110.07	1207.69	1302.04
1073.15	812.07	919.97	1023.56	1123.60	1220.66	1315.13
1173.15	824.39	932.52	1036.30	1136.50	1233.69	1328.29
1273.15	836.78	945.13	1049.10	1149.46	1246.79	1341.51
1500.00	865.08	973.95	1078.35	1179.08	1276.74	1371.74
2000.00	928.29	1038.32	1143.69	1245.26	1343.64	1439.28
2500.00	992.31	1103.54	1209.92	1312.35	1411.47	1507.77
3000.00	1056.88	1169.35	1276.77	1380.08	1479.97	1576.94
3500.00	1121.82	1235.57	1344.06	1448.28	1548.96	1646.62
4000.00	1187.00	1302.08	1411.66	1516.82	1618.31	1716.68

Table 11. Fugacity of oxygen (kJ/mol) under T and P.

T (K)	P (GPa)					
	0.5	0.6	0.7	0.8	0.9	1.0
373.15	35.77	38.52	41.12	43.60	46.00	48.31
473.15	43.47	46.37	49.09	51.68	54.16	56.56
573.15	51.02	54.08	56.92	59.61	62.19	64.67
673.15	58.47	61.68	64.65	67.45	70.12	72.69
773.15	65.83	69.20	72.30	75.22	77.98	80.64
873.15	73.14	76.66	79.90	82.92	85.79	88.53
973.15	80.40	84.08	87.44	90.58	93.54	96.37
1073.15	87.63	91.45	94.94	98.18	101.25	104.17
1173.15	94.83	98.78	102.39	105.75	108.91	111.92
1273.15	102.00	106.09	109.82	113.28	116.54	119.64
1500.00	118.19	122.56	126.54	130.23	133.70	137.00
2000.00	153.62	158.51	162.97	167.12	171.01	174.69
2500.00	188.83	194.13	199.00	203.52	207.77	211.80
3000.00	223.90	229.55	234.75	239.60	244.16	248.49
3500.00	258.88	264.81	270.30	275.43	280.27	284.87
4000.00	293.78	299.97	305.71	311.09	316.17	321.01
	2.0	3.0	4.0	5.0	6.0	7.0
373.15	69.02	87.23	104.05	119.92	135.08	149.66
473.15	77.80	96.34	113.41	129.48	144.81	159.55
573.15	86.46	105.35	122.67	138.95	154.45	169.34
673.15	95.05	114.29	131.87	148.36	164.04	179.09
773.15	103.58	123.18	141.03	157.74	173.60	188.80
873.15	112.07	132.03	150.16	167.08	183.12	198.49
973.15	120.51	140.85	159.25	176.39	192.63	208.15
1073.15	128.92	149.63	168.31	185.68	202.10	217.79
1173.15	137.29	158.38	177.34	194.94	211.56	227.41
1237.15	145.61	167.09	186.35	204.18	220.98	237.01
1500.00	164.36	186.73	206.64	225.01	242.27	258.68
2000.00	205.01	229.35	250.76	270.34	288.63	305.92
2500.00	244.85	271.12	294.04	314.85	334.19	352.41
3000.00	284.03	312.15	336.55	358.59	378.99	398.14
3500.00	322.66	352.54	378.36	401.62	423.07	433.15
4000.00	360.86	392.38	419.58	444.01	466.50	487.50

Table 11. (Continued.)

T (K)	P (GPa)					
	8.0	9.0	10.0	20.0	30.0	40.0
373.15	163.79	177.51	190.90	312.23	419.91	519.56
473.15	173.81	187.65	201.15	325.25	431.41	531.40
573.15	183.73	197.70	211.31	334.18	442.83	543.18
673.15	193.62	207.71	221.43	345.08	454.22	554.93
773.15	203.47	217.68	231.51	355.97	465.60	566.67
873.15	213.29	227.64	241.59	366.84	476.98	578.41
973.15	223.10	237.58	251.64	377.71	488.35	590.16
1073.15	232.89	247.49	261.68	388.57	499.72	601.90
1173.15	242.66	257.39	271.70	399.42	511.09	613.65
1273.15	252.40	267.27	281.70	410.27	522.46	625.39
1500.00	274.42	289.60	304.30	434.82	548.21	652.02
2000.00	322.45	338.33	353.68	488.63	604.77	710.57
2500.00	369.75	386.37	402.38	541.96	660.96	768.83
3000.00	416.31	433.68	450.38	594.75	716.72	826.74
3500.00	462.16	480.29	497.69	646.96	772.00	884.25
4000.00	507.34	526.23	544.33	698.59	826.79	941.34
	50.0	60.0	70.0	80.0	90.0	100.0
373.15	613.61	703.42	789.84	873.46	954.70	1033.89
473.15	625.73	715.77	802.38	886.16	967.55	1046.85
573.15	637.79	728.05	814.85	898.80	980.33	1059.76
673.15	649.82	740.31	827.31	911.42	993.09	1072.65
773.15	661.84	752.57	839.76	924.04	1005.86	1085.55
873.15	673.87	764.83	852.22	936.67	1018.63	1098.45
973.15	685.90	777.10	864.69	949.30	1031.42	1111.36
1073.15	697.94	789.37	877.16	961.95	1044.21	1124.29
1173.15	709.98	801.65	889.64	974.60	1057.01	1137.22
1273.15	722.02	813.93	902.13	987.26	1069.82	1150.17
1500.00	749.33	841.80	930.46	1015.99	1098.90	1179.55
2000.00	809.43	903.16	992.88	1079.32	1163.02	1244.35
2500.00	869.31	964.35	1055.17	1142.56	1227.07	1309.12
3000.00	928.89	1025.31	1117.27	1205.63	1290.98	1373.78
3500.00	988.14	1085.97	1179.12	1268.49	1354.72	1438.29
4000.00	1047.03	1146.33	1240.70	1331.12	1418.26	1502.62

Tables of Molar Volumes and Fugacities of Fluids on C–O–H System 357

Table 12. Fugacity of hydrogen (kJ/mol) under T and P.

T (K)	P (GPa)					
	0.5	0.6	0.7	0.8	0.9	1.0
373.15	34.04	35.98	37.77	39.45	41.05	42.58
473.15	41.10	43.18	45.09	46.87	48.55	50.16
573.15	48.15	50.38	52.41	54.30	56.07	57.76
673.15	55.02	57.59	59.75	61.74	63.61	65.38
773.15	62.25	64.79	67.08	69.19	71.15	73.01
873.15	69.29	71.99	74.42	76.64	78.71	80.65
973.15	76.32	79.20	81.76	84.10	86.26	88.30
1073.15	83.36	86.40	89.10	91.55	93.82	95.95
1173.15	90.39	93.60	96.43	99.01	101.38	103.60
1273.15	97.43	100.79	103.77	106.46	108.94	111.25
1500.00	113.38	117.11	120.39	123.36	126.07	128.60
2000.00	148.52	153.03	156.98	160.52	163.76	166.75
2500.00	183.66	188.90	193.49	197.60	201.34	204.78
3000.00	218.79	224.74	229.95	234.60	238.83	242.71
3500.00	253.91	260.56	266.36	271.54	276.24	280.56
4000.00	289.02	296.35	302.74	308.44	313.60	318.33
	2.0	3.0	4.0	5.0	6.0	7.0
373.15	55.70	66.72	76.63	85.81	94.45	102.68
473.15	63.77	75.08	85.20	94.55	103.34	111.70
573.15	71.88	83.49	93.83	103.36	112.29	120.77
673.15	80.03	91.95	102.51	112.22	121.30	129.91
773.15	88.20	100.44	111.24	121.12	130.36	139.10
873.15	96.41	108.97	120.00	130.07	139.46	148.33
973.15	104.63	117.53	128.80	139.05	148.60	157.61
1073.15	112.87	126.10	137.62	148.07	157.77	166.92
1173.15	121.12	134.70	146.46	157.10	166.97	176.25
1273.15	129.37	143.30	155.32	166.16	176.19	185.61
1500.00	148.11	162.86	175.45	186.75	197.16	206.91
2000.00	189.38	205.98	219.90	232.25	243.53	254.02
2500.00	230.52	249.00	264.30	277.73	289.91	301.16
3000.00	271.49	291.87	308.56	323.09	336.18	348.22
3500.00	312.32	334.57	352.64	368.28	382.29	395.14
4000.00	353.01	377.11	396.55	413.29	428.24	441.88

Table 12. (Continued.)

T (K)	P (GPa)					
	8.0	9.0	10.0	20.0	30.0	40.0
373.15	110.58	118.19	125.55	190.65	246.69	297.58
473.15	119.70	127.41	134.87	200.59	257.04	308.23
573.15	128.89	136.70	144.25	210.61	267.46	318.96
673.15	138.14	146.05	153.69	220.70	277.96	329.77
773.15	147.44	155.45	163.19	230.86	288.53	340.65
873.15	156.79	164.91	172.74	241.07	299.16	351.59
973.15	166.19	174.41	182.34	251.33	309.85	362.59
1073.15	175.61	183.95	191.97	261.64	320.58	373.64
1173.15	185.07	193.51	201.64	271.99	331.35	384.73
1273.15	194.56	203.11	211.33	282.37	342.17	395.87
1500.00	216.14	224.94	233.40	306.02	366.81	421.25
2000.00	263.89	273.28	282.26	358.48	421.52	477.62
2500.00	311.72	321.70	331.22	411.18	476.54	534.36
3000.00	359.47	370.07	380.15	463.94	531.69	591.27
3500.00	407.09	418.32	428.97	516.67	586.86	648.24
4000.00	454.54	466.41	477.63	569.31	641.99	705.21
	50.0	60.0	70.0	80.0	90.0	100.0
373.15	344.97	389.73	432.43	473.44	513.02	551.38
473.15	355.86	400.83	443.70	484.87	524.60	563.08
573.15	366.83	412.01	455.07	496.39	536.26	574.88
673.15	377.89	423.27	466.51	507.99	548.01	586.75
773.15	389.02	434.61	478.03	519.67	559.83	598.71
873.15	400.21	446.02	489.62	531.42	571.72	610.73
973.15	411.46	457.48	501.27	543.24	583.68	622.82
1073.15	422.77	469.00	512.97	555.10	595.69	634.97
1173.15	434.12	480.57	524.72	567.02	607.76	647.16
1273.15	445.51	492.18	536.52	578.98	619.86	659.40
1500.00	471.48	518.65	563.42	606.26	647.48	687.32
2000.00	529.19	577.47	623.21	666.89	708.87	749.41
2500.00	587.30	636.72	683.45	728.00	770.76	812.00
3000.00	645.62	696.22	743.96	789.40	832.96	874.91
3500.00	704.03	755.83	804.61	850.96	895.33	938.02
4000.00	762.46	815.49	865.32	912.61	957.80	1001.24

Index

A
α-Al$_2$O$_3$, 104, 105
ab initio calculation, 49, 312
α-β quartz transition, 101
acmite, 195
adiabatic calorimetry, 100, 104, 283, 311
adiabatic compressibility, 137, 138
alabandite, 188
albite, 108, 147
Al$_2$SiO$_5$, 273, 274
aluminosilicate glasses, 153
amorphous H$_2$O, 156
andalusite, 309, 312
    band gap, 310, 311
    phonon density of states, 309–311
    phonon dispersion relation, 310
    Raman measurement, 310
    specific heat, 311
    unit cell, 310
angular-dispersive X-ray diffraction, 324, 326
anharmonic contribution, 110, 119, 123
anharmonic correction, 294, 306
anharmonic model, 244
anharmonic oscillators, 106, 121
anharmonicity, 99, 119, 121, 123, 124, 139, 205
anomalous thermal expansion, 316, 330
    BaTiO$_3$, 330
    Fe$_2$SiO$_4$, 316
    ZrO$_2$, 330
anorthite, 109, 147

aqueous solutions, 1–3, 12, 41, 49
aqueous systems, 5, 6, 11, 12, 17, 49
Ar, 64, 84
Arrhenius laws, 149
α-sulfur, 250, 251
asymmetry parameter, 166
    KCl-RbCl, 166
    KI-RbI, 166
    NaBr-KBr, 166
    NaBr-NaI, 166
    NaCl-NaBr, 166
atomic mobility, 132, 136
atomization energy, 166
Avogadro's number, 177, 265

B
BaSi$_2$O$_5$, 142
beryllium-bearing minerals, 114
binary melts, 185
Birch-Marnaghan equation, 125, 134
Bohr's radius, 232
Boltzmann constant, 82, 112, 285
Boltzman's constant, 184, 243, 265
bonding energy, 25, 26, 29
Born-Mayer approach, 202
Born-Harbr cycle, 219
Born-Haber treatment, 234
Born-Haber-Fayans energy, 206
Born-Haber-Fayans thermochemical cycle, 200, 207, 209
Burn-Lande model, 163

Born-Oppenheimr approximation, 82
Born-Von Karman's lattice-dynamical procedure, 241, 244
Bose-Einstein statistics, 291
Bragg equation, 295, 317
Bragg law, 295
Bragg scattering, 295
Brillouin scattering, 315
Brillouin zone, 111, 113, 243, 244, 250, 252–254, 284, 286, 287, 289–291, 293, 302, 306, 311
bulk moduli, 99, 127
bulk modulus, 277, 294, 304, 312
  alkali halides, 277
  transition metal compounds, 277
bulk properties, 62
bulk viscosities, 38

C

Ca-Tschermak, 206
$CaAl_2SiO_6$, 153, 207, 208
$CaAl_2Si_2O_8$, 145, 149, 150, 153
$Ca_3Al_2Si_3O_{12}$, 145
calorimetric measurements, 113
$CaFeSi_2O_6$, 191, 206, 207, 213
$CaMgSi_2O_6$, 151, 194, 206, 207–213
$CaMgSi_2O_6$-$CaAl_2Si_2O_6$, 227, 228
$CaMgSi_2O_6$-$NaAl_2Si_2O_6$, 227, 228, 230
$CaMgSi_2O_6$ glass, 132, 133, 144, 145, 148
$CaMnSi_2O_6$, 207–209, 231
CaO, 120, 142
carbon dioxide CO_2, 62, 64, 81, 84–89, 91
carnegieite, 109, 110, 147
$CaSiO_3$, 110, 145
$CaSiO_3$ glass, 148
$CaTiAl_2O_6$, 207–209, 231
cell volume, 73, 196
CH_4, 64, 81, 84–89, 91
chemical potential, 7
  standard state, 203
Clausius-Clapeyron equation, 155
clinoenstatite, 114
clinozoisite, 114, 116–118
coesite, 109, 156, 157
CO, 81, 85–89, 91
collision diameter σ, 64, 66
compressibility, 84, 88, 136, 138, 139, 155, 189, 210
  isothermal, 7, 164, 205
  AgBr, 170
  AgCl, 170
  NaBr, 170
  NaCl, 170
  solid solution, 183
configurational disorder, 205
configurational enthalpy, 144
configurational entropy, 134, 144, 146–148, 150–152, 162
configurational heat capacity, 134, 135, 139, 142, 143, 146, 150, 151
configurational states, 137–139
convergence, 11
corundum, 118, 126, 272
Coulomb term, 13
Coulombic, 200, 219, 298, 306
covalency, 170
cristobalite, 110
critical point, 62
critical solution parameter, 184
critical solution temperatures, 185, 189
crystal field, 111
crystal field stabilization energy, 200, 216
CsCl-type structure, 164

D

Debiji-Waller factor, 255
Debye density of states, 286, 287
Debye frequency, 286
Debye model, 273, 286
Debye temperatures, 183
  alkali halides, 286
  corundum, 275
  definition, 286
  diamond, 287
  elastic constants, 275
  metals, 286
  soft potassium metal, 287
  specific heat, 286
  spinel, 322, 323
Debye theory, 240, 241
Debye-Waller factor, 295
decomposition, 162
defects, 111
deformation energy, 170
density of states, 112–116, 118, 119, 123, 124

Index 361

diamond anvils, 317
differential scanning, 98, 100, 283
diffusion, 86
diffusion coefficients, 38
diopside, 110, 132, 133, 234
dipole-dipole, 200
dipole-quadrupole, 200
disordering, 145–147
dispersion relation, 111, 113, 284, 290
dispersive terms, 200
displacement parameters, 242
drop calorimetry, 98, 99, 102, 104–106, 141, 146
Dulong-Petit limit, 109, 124, 139, 140, 273, 285

E

effective charge, 165, 299
Einstein frequency, 285
Einstein model, 285
Einstein oscillator, 111, 112
Einstein specific heat, 285
Einstein temperature, 285
elastic distortion energy, 169
electrical-conductivity, 165
electronic entropy contributions, 216
electronic wave functions, 290
energetic asymmetry parameters, 185
energy-dispersive X-ray diffraction, 317, 318, 326
  experimental procedure, 317
  temperature calibration, 317
enstatite, 114, 148
enthalpy, 7, 10, 19, 98, 99, 102, 104, 105, 140, 141, 143, 205, 219, 234
  binary mixture, 204
  formation, 264
  high P, T, 205
  interaction parameters, 167, 168
  melting, 186
  mixing, 163–165, 167–169, 170, 183, 184, 185, 188, 204, 209
  NaBr AgBr, 171, 172
  NaCl-AgCl, 171, 172
  of formation, 200
  polymorphous transition, 172
entropy, 98, 106, 108, 109, 111, 114, 115, 118, 119, 146, 153, 239
  absolute, 144
  alkaline earth oxides, 269, 270
  alkali halides, 265, 268, 269
  α-sulfur, 250
  $CaMgSi_2O_6$, 133
  chemical, 146, 148
  corundum, 276
  cryolite (Na_3AlF_6), 274
  debye temperature, 272–274
  diopside, 133
  effective force constant, 272
  expression, 284
  forsterite, 246
  high temperature, 264, 275, 276
  insulators, 182
  low temperature, 147
  melting, 145, 148
  mixing, 189, 243
  mixtures, 219
  nonconfigurational, 183
  of fusion, 144, 148
  residual, 144–148
  random solid solution, 182
  semiconductors, 182
  sulfides, 269, 270
  topological, 146, 148
  vibrational, 145, 147, 153, 183, 184, 243, 274, 277
  viscosity, 152
epidotes, 116–118
equation of state, 79, 200
Ewald method, 8, 66
EXAFS spectroscopy, 174, 175
EXAGS, 175
exponential-6 potential, 83

F

fayalite, 156, 312
  bulk modulus, 306
  inelastic neutron scattering, 299
  λ-type anomaly
  phonon dispersion, 299
  specific heat, 304, 305
  thermal expansion, 306
$Fe_2Si_2O_6$, 194, 210–213
feldspars, 194
Fermi level E_F, 279

ferrosilite,
  clino, 206, 210
Fersman's rule, 162
forsterite, 110, 119–126, 255, 312
  atomic displacement parameters, 255–25
  inelastic neutron scattering, 299, 300
  partial density of state, 303
  phonon dispersion, 299
  rigid-ion model, 255
  specific heat, 301, 305
  total one-phonon density of state, 302
  unit cell, 298
four-point charge rigit model, 12
Fourier transforms, 242
fractional ionicity, 165
free energy functions, 266, 268
  high temperature, 264
free ion polarizability, 201, 232
fugacities, 88, 91, 92

G
garnets, 114, 121, 152
GeO_2 polymorphs, 108, 110, 118, 119
generized density of states, 283
Gibbs energy, 61, 98, 279
Gibbs free energy, 150, 156
  effect of pressure, 227
  $Fe_2Si_2O_6$-$CaFeSi_2O_6$, 222, 223
  $Mg_2Si_2O_6$-$CaMgSi_2O_6$, 222, 223
  $Mg_2Si_2O_6$-$Fe_2Si_2O_6$, 222, 223
  mixing, 229, 231, 234
  mixture, 203, 219, 223
  molar, 203, 222, 224, 226, 227, 229, 230
  at $T = 1773$ K, $P = 20$ kbar, 219
Gibbs space, 194
glass transition, 136–140, 143–146, 151, 154, 155, 157
glaucophanc, 114, 115
Goldschmidt's polarity rule, 162, 187
graphitization, 317
Green-Kubo relation, 38
Grüncisen parameters, 123, 293
Grüncisen theory, 318, 321

H
H_2, 81, 85–89, 91
H_2O, 64, 81, 85–89, 91

hardness factor, 202
harmonic approximation, 111, 119, 121, 243, 244, 247, 290
heat capacities
  andalusite, 273
  α-sulfur, 250
  Ca, 274
  condensed matter, 271
  forsterite, 246, 252
  $HfTe_5$, 276
  high temperature, 110, 271, 273
  isobatic, 7, 19, 98
  isochoric, 99, 111, 139
  solid solution, 183
  spinels, 323
  volume, 5, 19
  $ZrTe_5$, 276
He, 64, 82
Hess additivity rule, 200
heterogeneous equilibria, 210
Hietala's theory, 164, 165
Huggis-Mayer formulation, 233, 234
Huggis-Mayer treatment, 202, 216
high pressure fusion, 155
Hugoniots, 87, 91
Hume-Roseri's rule, 162, 187
hybrid model, 287
hydrides, 272
hydrogen bonds, 2, 11, 12, 31, 33, 35, 42, 83
hydrothermal solutions, 3, 4

I
ice calorimeter, 104
ice VII, 73, 75, 76
ice VIII, 73,
ideal gas, 61, 83
impurities, 111
inelastic neutron scattering, 112, 113, 241, 243, 283, 284, 289, 294, 296, 299, 312
inelastic structure factor, 295
infrared spectroscopy, 114
interaction parameters, 167–170
  alkali halide, 168
  oxide, 168
  semiconductor, 168
internal energy, 10, 19, 205
intrinsic anharmonicity, 119, 123
ionicity degrees, 172, 188

Index

ionization energy, 200
IR spectroscopic data, 45, 48
isoenergetic lines, 226
isomorphous mixtures, 162
isothermal bulk modulus, 132

J
jadeite, 108, 109, 195

K
$KAlSi_3O_8$, 145, 149, 150
KCl, 42, 49
Kieffer model, 112, 113, 123, 287, 304, 305
kinetic energy, 5, 66
Kohler model, 224
Kohler-type combination, 223
Komada-Westrum modeling, 275
Kr, 64
$K_2TiSi_2O_7$, 142

L
λ-type anomaly, 302
λ-type Cp function, 206
Latimer's rule, 277, 280
lattice energy, 165
least squares fit, 104
least square method, 79, 196
Legendre transforms, 205
Lennard-Jones potential, 13, 63, 83
LiCl solutions, 3
LiI solution, 37, 38, 42, 43
lime, 119, 120
Lindemann melting formula, 280
Li_2O, 142
liquid argon, 80
liquid glass transition, 132
Li_2SiO_3, 142
London forces, 63
long range attractive forces, 61
long range Coulombic term, 66
long range interactions, 8
long range order, 135, 136

M
Madelung constant, 170, 246
magnetic contributions, 111

Maier-Kelley-type function, 204
Margules parameters, 166, 203
Markov chain, 6
Maxwell relationship, 183
mean excitation energy, 201
mean heat capacities, 140
mean polyhedral distances, 196, 198
Mee potential, 167
melting point, 100, 118, 144
melting relationships, SiO_2 polymorphs, 157
melting volume, 155
metastable phases, 104
mercury, 103
$Mg_2Al_4Si_5O_{18}$, 143
$Mg_3Al_2Si_3O_{12}$, 145
MgO, 120
$MgSiO_3$, 145
  garnet, 283
  ilmenite, 118, 283
  perovskite, 283
Mg_2SiO_4
  β-phase, 283, 306
  forsterite, 242, 243, 245
  γ-spinel, 283, 306
  polymorphs, 242
  ringwoodite, 242
  wadsleyite, 242
$Mg_2Si_2O_6$, 194, 200, 206, 207–213
Mie-Grüneisen equation of state, 293
Miedema's formula, 280
mineral-fluid equilibria, 3
mineral solubility, 50
miscibility, 162, 194
miscibility gaps, 184, 185, 189, 210, 226
mode Grüneisen parameter, 304, 305, 312
Mohs hardness, 279
molar volume, 7, 20, 61, 62, 86, 88, 89, 109, 124, 153, 154
molecular dynamics(MD), 1–9, 69, 80, 81, 84, 92
Monte Carlo(MC), 1–4, 6, 8, 18, 63, 80
Mott-Littleton's theory, 165, 174
multianvil apparatus, 101
multicomponent mixture, 231

N
N_2, 63, 64
$NaAlSiO_4$, 140, 145–147

N

$NaAlSi_2O_6$, 195, 207
$NaAlSi_3O_8$, 140, 145–147, 149–151
NaCl solution, 33, 35–37, 40, 41, 50
NaCl-type structure, 163, 164
$NaCrSi_2O_6$, 195, 207, 208, 231
$NaFeSi_2O_6$, 195, 207, 208, 231
Na_2O, 142
$Na_2Si_2O_5$, 138
$Na_2TiSi_2O_7$, 142
Ne, 64
Neél point, 306
nephcline, 108–110
Newton-Euler equations, 63
Newtonian equation of motion, 81
neutron diffraction, 65
nonpolar fluids, 11
nuclear magnetic resonance, 134, 146

O

O_2, 64, 81, 85–89, 91
octahedral coordination, 176
olivines, 156
Onuma's diagrams, 186
optic branches, 111
optical data, 239
organic molecular crystals, 249
orthozoisite, 114, 116, 118

P

P Z/n omphacites, 195
pair energy, 26, 27, 29
partition function, 284
Pauling number, 202
Pauling scale of electronnegativitics, 172
Pauling's strength of the bond, 199
periclase, 119, 120, 126
perovskite, 167, 312, 324
    band gap, 309
    equation of state, 308
    $MgSiO_3$, 307
    phonon density of states, 308, 309
    phonon dispersion relation, 307, 308
    rigid-ion model, 244–246, 255, 298
    specific heat, 309
    unit cell, 298
    zone-center optic phonons, 308

phase transitions, 136
Phillips spectroscopic definition of ionicity, 172
phonon density of states, 273, 283, 284, 288, 312
    Debye, 288
    Einstein, 288
    generized, 296
    inelastic neutron scattering, 289
    forsterite, 284, 297
    fayalite, 284, 297
    Kieffer model of forsterite, 289
phonon dispersion curves, 239, 241, 301
    forsterite, 248
    sodium iodide, 258
phonon dispersion relation, 295, 299, 312
phonons, 111
photon energy, 317
piston-cylinder apparatus, 101
Planck's constant, 72, 112, 201, 243, 265, 285
Planck's formula, 285
polarizable-ion model, 244
post-spinel, 156
pseudowollastonite, 109, 110
Pl-PlRh 10% thermocouples, 103
P-V-T properties, 79, 80
pyrope, 109, 147, 148
pyroxene, 152, 194
    C2/c structure, 194, 196–198
    Ca-Tschermak, 206
    clino, 195, 210
    crystal structure of C2/c, 195
    enthalpy, 206
    $Fe_2Si_2O_6$, 206
    lattice energy, 216
    molar volume, 209–211
    ortho, 195
    Pbca, 195, 226
    quadrilateral, 195, 224
    system $Mg_2Si_2O_6$-$CaMgSi_2O_6$-$CaFeSi_2O_6$-$Fe_2Si_2O_6$, 210, 216, 217, 220

Q

quantum effects, 9
quantum mechanical calculations, 232

Index 365

quartz, 108–110, 112, 113, 121, 156, 157
quasiharmonic, 121, 293, 312

R

radial distribution functions, 29, 30, 32, 35–37
Raman spectrometry, 92, 243
Raman spectroscopic data, 45, 48, 146, 239
Raman spectroscopy, 114
Raphson-Newton process, 239, 247
Redlich-Kister formulation, 229
Redlich-Kister parameters, 231
Redlich-Kwong equations, 62, 72, 76
Redlich-Kwong model, 71
relaxation parameters, 173, 175, 176
repulsive factor, 202
repulsive terms, 200
Retger's rule, 182, 183, 189
rigid-body model, 248–250, 252
rigid ion model, 244–246, 248, 300
rigid-molecular ion model, 304, 305
rutile, 118

S

sapphire, 100, 104
Schrödinger equation, 82
self-diffusion coefficients, 38–43
semiconductor alloys, 169, 175
separation distance, 64
shear viscosities, 28
shell model, 244–246
shock-wave data, 84
short range interactions, 8, 232, 306
short range model, 244
short range order, 136, 143, 204
short range repulsive, 61, 209
silicate glasses, 143, 146, 153
silicate melts, 5, 6, 50, 60, 137, 141, 144
silicate minerals, 60
sillimanite, 311
SiO_2, 153
site preference energy, 172, 188
solid-fluid transformation, 92
solid solution, 118, 162
    α-Mg_2SiO_4-α-Fe_2SiO_4, 168
    Al_2O_3-Fe_2O_3, 188
    alkali hilide, 163, 165, 168, 170
    atomization energy, 166
    binary ionic, 163
    bonding character, 170
    $Ca_3Al_2Si_3O_{12}$-$Mg_3Al_2Si_3O_{12}$, 187
    $CaCO_3$-$MgCO_3$, 187
    CaO-MgO, 187
    CaO-SrO, 188
    CsCl type, 178, 180
    composition dependence, 181
    γ-Mg_2SiO_4-γ-Fe_2SiO_4, 168
    energetic properties, 172
    FeS solubility in ZnS, 189
    garnet, 189
    high pressures, 189
    ilmenite, 189
    ionic, 165, 175
    KBr-Ki, 165
    (K, Rb)Br, 175
    lattice energy, 163
    local structure, 174
    metallic, 184
    $MgSiO_3$-$FeSiO_3$, 189
    mixing properties, 189
    MoO_3-WO_3, 188
    NaBr-KI, 165
    NaBr-NaI, 165
    NaCl structure, 178, 180
    NaI-KI, 165
    NeCl-KCl, 165
    NeCl-NaBr, 165
    oxides, 168
    perovskite, 189
    pyroxene, 189
    Rb(Br, I), 175
    size parameter, 8, 184, 187
    stability conditions, 163
    stability field, 187, 190
    TiO_2-SnO_2, 188
    vibrational entropy, 182
    x-ray diffraction study, 174
    ZnS-structure, 178, 180
    $ZnWO_4$-$ZnMoO_4$, 188
sound velocities, 138
specific heat, 243, 252
    at constant Pressure Cp, 204
    at constant volume Cv, 285, 304
    Debye model, 304, 305

specific heat (*cont.*)
  expression, 284
  low temperature, 283, 300, 306
sphalerite, 172, 188, 189
sphalerite geo- and cosmobarometer, 189
spin-wave contribution, 302
spin-wave excitation, 306
spinel, 110, 156
  defect structure, 332
  density-pressure relationship, 315
  density-temperature relationship, 315
  Fe_2SiO_4, 315, 316
  Grüneisen parameter, 322, 323
  isothermal bulk modulus, 322
  $(Mg, Fe)_2SiO_4$, 315, 316
  Mg_2SiO_4, 318
  silicate, 167
  thermal expansion, 315, 316, 321, 322, 326–331
static lattice energy, 200, 219, 234, 294
static potential model, 205
statistical error, 9, 19, 87
statistical mechanical theory, 265
Stirling's approximation, 203
stishovite, 118, 119, 121, 157
structural relaxation, 144
subregular Margules formulation, 227, 229, 231
supercritical fluids, 50, 85
surface effects, 8

T
Taylor's series, 164
tetrahedral coordination, 176, 177
thermal conductivity, 38
thermal expansion coefficients, 99, 126, 132, 135, 136, 138, 139, 153, 155, 294, 312
  anharmonic effects, 274, 304
  complex mixtures, 220
  isobaric, 205
  K_2O, 154
  Li_2O, 154
  low temperature, 318
  Mg_2SiO_4, 324
  $(Mg, Fe)_2SiO_4$, 316, 326–331
  minimization, 173
  molten plagioclase, 155
  Na_2O, 154
  silicate liquids, 153, 154
  solid solution, 183
  temperature dependence, 326
  vacuum, 330, 331
thermal expansion, 127, 153, 154, 200, 210
thermal expansivity, 7
third-law entropies, 99
3d-transition metal, 280
  cabides, 272
  nitrides, 272
  oxides, 272
TO-LO splitting, 255
transition metal diborides, 272
transport properties, 165
tridymite, 110

U
ultrasonic interferometry, 315
ureyite, 195

V
vacuum permittivity, 299
Van der Waals equation, 61, 62
Van Laar formulation, 230
Vegard's rule, 162, 163, 172, 173, 178–182
velocity autocorrection function, 38–45
vibrational internal energy, 111
vibrational modes, 111
virial equation, 62
virial theorem, 5
viscosity, 86, 132, 134, 137, 138, 143–145, 148, 150, 152
  $CaAl_2$-Si_2O_8, 149
  $CaMgSi_2O_6$, 149
  composition dependence, 151
  glass, 148
  $KAlSi_3O_8$, 149
  $K_2O.nSiO_2$, 149
  liquids ($CaSiO_3$-$MgSiO_3$), 151
  $Mg_3Al_2Si_3O_{12}$, 149
  molten pyroxenes, 151
  $NaAlSiO_4$, 149
  $NaAlSi_2O_6$, 149
  $NaAlSi_3O_8$, 149
  $Na_2O.nSiO_2$, 149
  silicate melts, 149
volatile fluids, 61, 62

Index

W
Wasastjerna-Hovi model, 164, 165
Wasastjerna's assumption, 173
water,
  ab-initio, 16, 21, 40, 72, 76
  BJH model, 17, 19, 23, 31, 33, 34, 43, 45, 48
  β-tridymite, 50
  central face model, 15
  compressibility, 3, 21, 24
  critical point, 4, 13, 15, 21, 23, 40
  density, 20–22, 67
  dipole moment, 12
  enthalpy, 22
  equation of state, 71, 72, 76
  freezing point, 21
  heat capacity Cp, 23
  heat capacity Cv, 23
  HGK equation of state, 18–20, 24
  in quartz, 50
  intermolecular potentials, 11, 15
  internal energy, 22
  liquid-vapor coexistence, 4, 7, 15, 21
  Lennard-Jones model, 65
  MC simulation, 9, 10, 15, 17, 20, 22, 23, 28, 31, 33
  MCY model, 17
  MD simulation, 9, 10, 13, 15, 17, 19, 22, 27, 31, 33, 34, 45, 48, 68, 70, 73
  melting point, 21
  molar volume, 22
  neutron diffraction, 31
  phase diagram, 4, 21
  P-V-T, 22, 63, 65, 69–72, 76
  quantum corrections, 9
  self-diffusion coefficients, 39
  shock wave data, 11, 70
  simple point charge model, 15
  specific volume, 71
  supercritical, 4, 8, 17, 25–29, 32, 39, 48, 49
  thermal expansivity, 24, 25
  TIP4P model, 14, 17, 20–23, 29, 65–73, 76
  X-ray diffraction, 31, 32
well depth, 66
wollastonite, 110, 147
wurtzite, 172, 189

X
xe, 64
x-ray measurement, 179

Z
zero-point conditions, 204
zero-point energy, 135, 242
zone-center optic phonons, 308
zone-center phonons, 301